ISBN 978-0-282-82146-3
PIBN 10866814

FB &c Ltd, Dalton House, 60 Windsor Avenue, London, SW19 2RR.
Company number 08720141. Registered in England and Wales.

For support please visit www.forgottenbooks.com

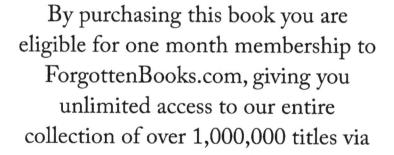

REPORT

OF

CHIEF ENGINEER J. W. KING,

UNITED STATES NAVY,

ON

EUROPEAN SHIPS OF WAR AND THEIR ARMAMENT, NAVAL ADMINISTRATION AND ECONOMY, MARINE CONSTRUCTIONS, TORPEDO-WARFARE, DOCK-YARDS, ETC., ETC.

SECOND EDITION,

REVISED, ENLARGED, AND ILLUSTRATED.

WASHINGTON:
GOVERNMENT PRINTING OFFICE.
1878.

IN THE SENATE OF THE UNITED STATES,
February 18, 1878.

Resolved by the Senate (the House of Representatives concurring therein,) That there be printed 1,000 copies of the second edition of the report of Chief Engineer King on European ships of war, for the use of the Navy Department.

LETTER OF TRANSMITTAL.

NAVY DEPARTMENT,
Washington, January 27, 1877.

SIR: In compliance with a resolution of the Senate, passed on the 26th instant, I have the honor to transmit herewith the report of Chief Engineer J. W. King, of the United States Navy, on European ships of war, &c.

Very respectfully,

GEO. M. ROBESON,
Secretary of the Navy.

Hon. THOMAS W. FERRY,
President pro tempore of the United States Senate.

CONTENTS.

PART I.

PART II.

PART III.

PART IV.

PART V.

PART VI.

PART VII.

PART VIII.

PART IX.

PART X.

PART XI.

PART XII.

PART XIII.

PART XIV.

PART XV.

PART XVI.

PART XVII.

PART XVIII.

PART XIX.

PART XX.

PART XXI.

PART XXII.

PART XXIII.

PART XXIV.

APPENDIX.

LIST OF ILLUSTRATIONS.

PART I.

LETTERS; PREFACE TO SECOND EDITION; INTRODUCTION; THE BRITISH NAVY; ADMIRALTY DESIGNS FOR HER BRITANNIC MAJESTY'S SHIPS OF WAR.

LETTERS.

FIRST LETTER.

WASHINGTON, D. C., *January* 16, 1877.

SIR: In obedience to your order, dated July 29, 1875, received at Bethlehem, N. H., August 6, directing me to proceed to Europe for the purpose of personally observing and reporting upon recent constructions and mechanical appliances for ships of war, with the view of utilizing the information to the advantage of the naval service, the accompanying is respectfully submitted.

I sailed from New York August 14, 1875, and returned to the United States July 30, 1876.

While in Europe I was steadily employed, and no time was unnecessarily lost in the discharge of the duties assigned, as may be seen from the information collected and contained in the report.

All the navies of Europe have been recently undergoing reconstruction, and there has never been a time, during peace, when such large expenditures for naval purposes were made as at present, and such radical changes effected in the construction of ships of war, in steam-machinery, in machinery for working guns, and for various other purposes on board ship, and in offensive torpedo-warfare. It is, therefore, expedient for the department to be correctly informed of the extent and character of improvements in European naval warfare and economy in order that it may take advantage of the results.

I am indebted for kind attentions to the dock-yard naval officers and proprietors of iron-ship yards and engine-factories visited in Great Britain and on the continent of Europe.

I have the honor to be, sir, respectfully, your obedient servant,

J. W. KING,
Chief Engineer, United States Navy,
Late Chief of Bureau of Steam-Engineering.

HON. GEO. M. ROBESON,
Secretary of the Navy.

SECOND LETTER.

WASHINGTON, D. C., *March* 6, 1878.

SIR: I have the honor to transmit herewith the second edition of my Report on European Ships of War, &c. The labor of preparing this work has been no easy task, it having employed my time wholly since the end of May last, except that portion which has been devoted to other official duties.

Respectfully, your obedient servant,

J. W. KING,
Chief Engineer, United States Navy.

HON. R. W. THOMPSON,
Secretary of the Navy.

PREFACE TO THE SECOND EDITION.

The flattering reception of the first edition of the *Report on European Ships of War, &c.*, evidenced by the fact that the 2,900 copies printed by order of the United States Senate were exhausted in a few months, and that the Navy Department has not recently been able to supply the book to officers and others requesting copies, has encouraged the writer to prepare the second edition, containing more matter and better illustrations.

This edition includes the revision of a considerable portion of the matter contained in the first, also the additions of the ships, dock-yards, and *personnel* of the navies of Turkey, Holland, Spain, Denmark, Sweden, Norway, and Portugal. Besides, there is additional matter relating to the British ships *Inflexible*, *Ajax*, and *Agamemnon*, *Téméraire*, *Alexandra*, and *Dreadnought;* cruisers of the rapid type, such as the *Iris*, of which a detailed description is given, as her construction, fittings, speed, &c., are interesting and worthy of note; composite vessels; the *personnel* of the British navy, and the cost of maintaining that branch of the British service; additional particulars of the French navy, its *personnel* and dock-yards; new ships of the German navy, both armored and unarmored; the establishment of Herr Krupp, and his great breech-loading gun, with observations on the manufacture of heavy guns; the new Italian ships with their ponderous armor and armaments; the new Brazilian sea-going armor-clad *Independencia;* and the recently-constructed ships for the Japanese Government; torpedo-warfare, and the latest improvements in the same; torpedo-boats, with descriptions of the weapons used, the manner of using them, and defense against such craft; and additional facts relating to compound engines, and inspection of boilers.

The merit of originality cannot be claimed for the contents of this work; they are in the main the result of personal observation during European travel in 1875–'76, also during two other tours of Europe, one in 1871 and the other in 1873, amounting in all to nearly two years of employment abroad on official duties. A considerable portion of the facts and figures have been obtained by much study of blue-books, parliamentary returns, the reports of committees and commissions, and by correspondence with naval architects, engineers, officers of navies, and other scientific men. I have also availed myself freely of the valuable and trustworthy information furnished by *Engineering* and *The Engineer*, published in London, well and deservedly known as the leading European journals devoted to engineering science and the mechanic arts. I am also indebted for information to the naval columns of the London *Times* and the *Naval and Military Gazette*, and for several items to the excellent periodical *L'Année Maritime.*

While such a compilation of facts as are here collected cannot be

2 K

esteemed a brilliant achievement, it may prove useful to many persons desiring to be informed of the navies of Europe and matters pertaining to them.

POSTSCRIPT.—After this report was fully prepared, a copy of an English edition of the former one was received, claiming to be a reprint of that one, "revised and corrected by an English naval architect." The revision consists almost solely of the substitution of pounds and shillings for dollars and cents, and the change in form of some words, the spelling of which the preface says is peculiarly American. The corrections (for which I am grateful) have reference almost entirely to the productions of Mr. E. J. Reed, C. B., M. P. I believe that every important correction, as well as every stricture, has been noted in its proper place in the present edition, and has been acknowledged, excepting where it had already been made; and my main endeavor has been to make these remarks as brief as possible, because the criticisms themselves form, in some cases, inconveniently long foot-notes. As my reviewer has preferred to conceal his well-known name, I have alluded to him under his title of "An English Naval Architect."

J. W. K.

INTRODUCTION.

For the study of naval construction and marine engineering, the most important field of observation is Great Britain. England is in the forefront as the leader and model to all European naval powers. In no other country can there be found so many scientific constructors and engineers. The Institution of Naval Architects reckons among its members many of the greatest masters of their art in the world, and the institutions of civil engineers, of mechanical engineers, and other scientific bodies, contain on their lists the names of men possessing the highest engineering talent in Europe. In addition to their magnificently-equipped public dock-yards, the patronage of the British Government has sufficed to keep in existence and to increase the supplemental resources which relieve and aid the national establishments in time of peace, and which, in time of war, would be to them of priceless value. It is owing to this patronage, and to foreign orders for ships of war, in no small measure, that on the Thames, the Mersey, the Clyde, and the Tyne are found unrivaled establishments fully equipped, with expanded and developed resources, requisite for modern war-ship construction. Besides the numerous ships designed and built yearly for the British flag, English ship-yards have produced, and are still producing, war-ships for other nations. Nearly every considerable naval power, except the United States and France, has employed English designers, English ship-builders, engineers, and gun-manufacturers. It was here that the *König Wilhelm*, *Kaiser*, *Deutschland*, and other ships for the German navy were built. Turkey obtained from the Clyde and the Thames a large proportion of her armored fleet, including all the most powerful vessels. Russia, Spain, Holland, Italy, Denmark, Greece, Portugal, Brazil, Chili, Peru, and Japan all come to England to have armored ships of war constructed. The Sheffield works not only supply armor-plates for these ships, but also plates and materials for war-vessels built in continental countries. The Elswick works and Whitworth manufacture guns solely for foreign orders. The armaments of many foreign ships, including the monster guns for the Italian service and the machinery for working them, also the formidable pieces for the last-built Brazilian ships, were made at these works. Besides this, all the nations above named are customers of the English ship-yards and engineering works, to supply vessels, machinery, and appliances for their mercantile marine.

In London may be found naval *attachés* of nearly every important nation, watching and studying with ceaseless vigilance the principles and science of naval architecture and engineering, especially the newer and later inventions, the experiments in artillery practice, and the progress made every year in the science of warfare, offensive and defensive.

In consequence of these facts, and for the additional reason that comparatively little of value, strictly novel, originating with continental naval architects or engineers was found on the continent of Europe, the larger portion of my time abroad was employed in Great Britain.

THE BRITISH NAVY.

In contemplating the power of England, the navy is always regarded as her bulwark. On her navy England depends for security at home and respect abroad. Everything concerning it excites eager interest, and it never fails to receive support, whatever party may be in power. No censure is ever passed upon the large expenditures for maintenance and additions to the fleets; but the criticisms of the press and the people are constantly directed to the administration of the admiralty, and the types of vessels constructed. If any condition proposed in a design be not realized in the completed ship, the fact is certain to be exposed by the press, and severely commented upon. This influence, together with the watchfulness over the progress made elsewhere, has not been without effect. In the House of Commons, at a recent session of Parliament, the first lord of the admiralty said, "It is our policy to keep pace with the inventions of the day, and ahead of all maritime powers." The most able constructive ability and engineering talent in the kingdom is employed in producing designs for new types of vessels, for machinery, and for appliances of offense and defense. It may be confidently asserted that never since the application of steam propulsion to ships of war has the British navy been relatively so strong as at the present time, and yet the complaints are that it is not more powerful.

The fleets of former beautiful wooden screw-ships, like their predecessors of the old sailing line-of-battle ship period, and the subsequent paddle-wheel steamers, are fast disappearing from the navy list for either fighting or cruising purposes. Numbers of wooden line-of-battle ships and frigates provided with auxiliary steam-power, but whose days were passed mainly under canvas; and others that never made a cruise—indeed, antiquated before completed—vessels in whose outlines the beauty of naval architecture may be said to have culminated, are in the same category. In fact, whole squadrons may be seen in the harbors of Portsmouth, Devonport, and other dock-yards, some bearing famous names, and "pierced for" from fifty to one hundred and one guns, but now as useless for purposes of modern warfare as the old paddle-wheel frigates or the fifty-nine sailing-vessels borne on the British navy list.

The effective force of the British navy may now be divided into ships for great naval battles, ships for coast defense, and unarmored cruising-vessels. There are so many different types that it is quite impossible to classify them according to any former standard. The present collective fleet as presented in the navy list consists of nearly four hundred vessels of all kinds. This includes those building, but does not include one hundred and thirty-four laid up or employed in permanent harbor service, and not ever likely to be sent to sea. The total tonnage of these four hundred vessels is about 900,000.

From published returns it appears that during the eight years from 1866 to 1874, ten and a half millions of pounds sterling were expended in the construction of new ships, six millions of which were for armored ships, and four and a half millions for unarmored vessels. During the same period, one and one-third million pounds were expended on the repairs of armored ships, and nearly four millions on the repairs of vessels of all other kinds, and it is now estimated that about a million pounds sterling is expended annually on new armored ships, and three-quarters of a million on all other new vessels.

ARMORED SHIPS.

It is to the production of the most powerful sea-going fighting-ships that the resources of the navy are first directed; ships sufficiently armored to resist projectiles of any ordinary kind, sufficiently armed to silence forts or to meet the enemy under any conditions proffered; sufficiently fast to choose the time and place to fight, and sufficiently buoyant to carry coal and stores into any ocean. Of this class, according to official statement in the House of Commons, there will be, when those now under construction shall have been completed, eighteen, placed in the order following, according to their power, the *Inflexible* ranking first.*

TURRET-SHIPS.

Inflexible, building at Portsmouth, to be completed in 1878.
Dreadnought, launched March 8, 1875, commissioned 1877.
Thunderer, launched March 25, 1872, first commission 1877.
Devastation, launched July 12, 1871, first commission 1873.
Agamemnon, building at Chatham, date of completion uncertain.
Ajax, building at Pembroke, date of completion uncertain.
Monarch, launched May 25, 1868, first commission 1870.

BROADSIDE-SHIPS.

Alexandra, launched April 7, 1875, first commission 1877.
Téméraire, launched 1876, first commission 1877.
Sultan, launched May 31, 1870, first commission 1872.
Hercules, launched February 10, 1868, first commission 1871.
Bellerophon, launched April 26, 1865, first commission 1867.
Swiftsure, launched June 15, 1870, first commission 1871.
Triumph, launched September 27, 1870, first commission 1872.
Audacious, launched February 27, 1869, first commission 1872.
Invincible, launched May 29, 1869, first commission 1873.
Iron Duke, launched March 1, 1870, first commission 1872.
Penelope, launched June 18, 1867, first commission 1870.

OCEAN-CRUISING SHIPS OF THE ARMOR-BELTED TYPE.

Shannon, built at Pembroke, not yet commissioned.
Nelson, building at Glasgow, to have been completed September, 1877.
Northampton, building at Glasgow, completed about September, 1877.

VESSELS FOR COAST DEFENSE.

These are for the most part turret-vessels, built on the breastwork system, and are named *Glatton*, *Hotspur* (a ram), *Rupert* (a ram), *Prince*

* When wooden vessels were first plated with armor, they were known as "iron-clads"; now that all hulls are built of iron and plated with heavy armor upon a wooden backing, the term "armored ships" is used, as being more proper than "iron-clads."

Albert, *Cyclops*, *Gorgon*, *Hecate*, *Hydra*, *Scorpion*, *Wivern*; also the broadside-gunboats *Viper* and *Vixen*. Besides these for home defense, there are the *Abyssinia* and *Magdala*, stationed at Bombay, and the *Cerberus*, in one of the Australian harbors. Any of these monitors are capable of going to sea, but they are unfit for cruising; all are low free-board vessels, each provided with a single revolving turret rising above the breastwork, the *Hotspur* and *Rupert* excepted. The former of these was built to be used solely as a powerful ram,* and the latter has a fixed turret in which the gun revolves on a platform. This vessel is also fitted especially for ramming.

SHIPS OF THE ORIGINAL ARMORED TYPE.

These vessels are built of iron, and are of the broadside variety. They have become antiquated, and are not now regarded as competent to meet in line of battle the armored ships of the present period. They consist of the *Agincourt*, *Northumberland*, *Achilles*, *Black Prince*, *Warrior*, *Hector*, *Valiant*, *Resistance*, and *Defence*.

WOODEN ARMORED SHIPS.

Most of these ships were under construction as line-of-battle ships or frigates at the time of the battle between the little *Monitor* and *Merrimac* in Hampton Roads; they were subsequently altered and converted into sea-going iron-clads. Many of them are decayed and relegated to harbor service, and it is not probable that any of them will be much longer continued as cruisers, or extensively repaired. They are as follows: *Prince Consort*, *Royal Oak*, *Caledonia* (which has an iron upper deck), *Research*, *Zealous* (which has an iron upper deck), *Lord Clyde* (which has had her machinery removed, and is fitted for a drill-ship), *Royal Alfred* (which has an iron deck), *Royal Sovereign* (a turret-ship with an iron upper deck), *Favorite*, *Enterprise* (with iron top-sides), *Lord Warden* (with iron inner skin), *Ocean*, *Pallas*, and *Repulse*. The *Pallas* was built for an iron-clad in 1865, and the *Repulse* in 1868; they were the last armored wood-built ships for the royal navy, and are still cruising, the former in the Mediterranean and the latter in the Pacific.

MASTLESS ARMORED SEA-GOING SHIPS.

The *Devastation*, the *Thunderer*, and the *Dreadnought* come under the above heading, and take rank first as the most powerful fighting-ships armed and now afloat in the world. The *Inflexible*, now building, designed as a still more powerful ship, is intended to be masted only during time of peace, as also the *Agamemnon* and *Ajax*, ships of the same type but of smaller dimensions.

Descriptions and particulars of these powerful ships will be given presently.

ADMIRALTY DESIGNS FOR SHIPS OF WAR.

But before proceeding to describe Her Britannic Majesty's ships, it will, perhaps, be interesting to examine the details of the system by which they are produced.

The designs of ships for the royal navy are prepared by a staff of

* The *Hotspur* was not intended solely as a ram, as she was designed and built to carry a 25-ton gun in a revolving turret, plated with 11-inch armor. When she was built this was the most powerful gun that was being made.—AN ENGLISH NAVAL ARCHITECT.

draughtsmen at Whitehall, under the direction of a council of construction. This council consists of the director of naval construction as the president, three chief constructors, and three assistants; the engineering department being represented by the engineer-in-chief and an engineer officer.

In preparing a new design, the initiative is taken by the sea lords of the admiralty, who consult with the controller, the director of naval construction, and the director of naval ordnance. It having been decided to add a vessel of a certain type to the navy, the director of construction is ordered to prepare the plans. This he does after first discussing the question in the council with the other members of that body. The draughtsmen are then set to work about the preliminary calculations and the salient features of the design, after which the controller and director of ordnance are again consulted. From time to time their lordships are referred to, and throughout the whole period of the preparation of the plans the latter are continually being modified, so as to comply with the decisions arrived at during the discussions of the officials interested. When prepared, the design represents the collective opinions of these officials, or, at all events, it is supposed to be the nearest possible approach thereto, as absolute unanimity can scarcely be expected upon every question.

The director of construction is, frequently, the originator of the type, and in every case, after all important conditions have been settled, he is responsible for the realization, in the completed ship, of the design decided upon.

In professional skill the members of the council of construction have high standing. Every one of them has served his apprenticeship in a royal dock-yard, was sent to the Royal School of Naval Architecture and Engineering after a competitive examination, and has won his way to his present position through the possession of superior ability and attainments.

It would seem that the course of procedure here set forth as adopted at the admiralty, in the preparation of designs for ships of war, would meet with public favor, but professional traditions and prejudices are difficult to overcome. In a pamphlet, attributed to the Duke of Somerset, which was published a few years ago, it was stated that "the mind of man does not go back to the time when the management of the navy by the admiralty was not a subject of dissatisfaction," and this is proved by the continuous succession of parliamentary inquiries, commissions, and committees on the subject. No part of the admiralty administration has been so constantly and so seriously questioned as its management of the designing, building, arming, and equipping of ships, including the materials required and the maintenance of the dock-yards.

Whether the unfavorable criticisms, to which the officials have been subjected, have originated from anything that needs reform, or whether it is owing to the eager interest of the English people in the welfare of the navy, is a question not to be considered here.

PART II.

THE INFLEXIBLE; THE AJAX AND AGAMEMNON.

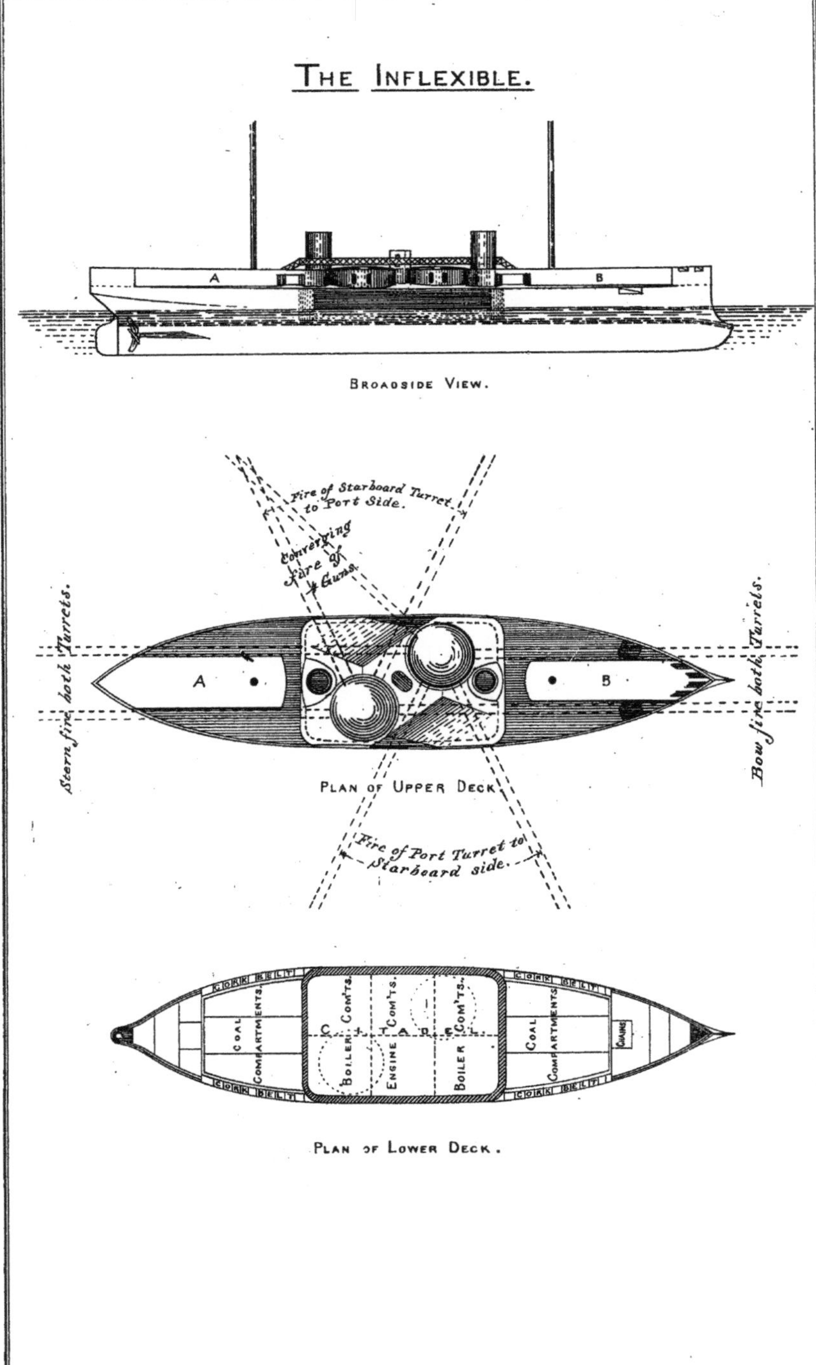

BROADSIDE VIEW.

PLAN OF UPPER DECK.

PLAN OF LOWER DECK.

THE INFLEXIBLE.

The modern man of-war is much more than an armored steamer; she is a great engine of destruction, clad in heavy armor, provided with huge guns which are operated by machinery, driven by powerful engines, and fitted with machinery for purposes of all kinds. Year by year the thickness of armor and weight of naval artillery go on increasing together. Mechanical appliances have more and more replaced manual labor, and at the same time the forms of ships have been adapted to the work they have to do and to the conditions under which they must act. No war-vessel yet designed has departed so widely from pre-existing types, and in none has so enormous a stride been made, in offensive and defensive power, as in the one about to be described.

The *Inflexible*, which was commenced at Portsmouth dock-yard in February, 1874, and launched April, 1876, is a twin screw, double-turret ship, with a central armed citadel. She was designed by Mr. Barnaby, the director of naval construction at the admiralty, and at a meeting of the Institution of Naval Architects in London he describes the vessel in the following language:

> Imagine a floating castle 110 feet long and 75 feet wide, rising 10 feet out of water, and having above that again two round turrets, planted diagonally at its opposite corners. Imagine this castle and its turrets to be heavily plated with armor, and that each turret has two guns of about eighty tons each. Conceive these guns to be capable of firing, all four together, at an enemy ahead, astern, or on either beam, and in pairs toward every point of the compass. Attached to this rectangular armored castle, but completely submerged, every part being 6 to 7 feet under water, there is a hull of ordinary form with a powerful ram bow, with twin screws and a submerged rudder and helm. This compound structure is the fighting part of the ship. Seaworthiness, speed, and shapeliness would be wanting in such a structure if it had no addition to it; there is therefore an unarmored structure lying above the submerged ship and connected with it, both before and aft the armored castle, and as this structure rises 20 feet out of water from stem to stern without depriving the guns of that command of the horizon already described, and as it moreover renders a flying deck unnecessary, it gets over the objections which have been raised against the low free-board and other features in the *Devastation, Thunderer*, and *Dreadnought*. These structures furnish also most luxurious accommodations for officers and seamen. The step in advance has therefore been from 14 inches of armor to 24 inches; from 35-ton guns to 80 tons; from two guns ahead to four guns ahead; and from a height of 10 feet for working the anchors to 20 feet. And this is done without an increase in cost, and with a reduction of nearly 3 feet in draught of water. My belief is that in the *Inflexible* we have reached the extreme limit in thickness of armor for sea-going vessels.

The length of the vessel between perpendiculars is 320 feet, and she has the extraordinary breadth of 75 feet at the water-line; depth of hold, 23 feet 3½ inches; free-board, 10 feet; mean draught of water, 24 feet 5 inches (23 feet 5 inches forward and 25 feet 5 inches aft); area of midship section, 1,658 square feet; and displacement when all the weights are on board, 11,407 tons, being the largest of any man-of-war hitherto constructed. She is, as before described, a rectangular armored castle; the whole of the other parts of the vessel which are unprotected by armor have been given their great dimensions for the simple purpose of floating and moving this invulnerable citadel and the turrets by which it is surmounted. Her immense bulk, unprecedented armament, powerful machinery, and the provision for ramming, and for resisting the impact of rams as well as of shot and shell, have made it necessary that strength and solidity should enter into every part of the structure.

HULL AND APPENDAGES.

While the cellular compartments of the double bottom have a little less depth than in the *Devastation* class, they are built up of heavier angle-irons and plating, and steel has been very largely employed for the purpose of securing great strength with comparative lightness of material. The hull is composed of flat and vertical keels, transverse and longitudinal frames, inner and outer bottom-plating. The vertical keel is formed of steel plates $\frac{3}{4}$ inch thick by 40 inches deep, and the flat keel-plates are of iron in two thicknesses of $\frac{13}{16}$ inch and $\frac{3}{4}$ inch, the two being connected by angle-irons 5 by 5 inches by $\frac{3}{4}$ inch. On the upper edge of the vertical keel the angle-irons by which it is fastened to the inner bottom-plates are 3 inches by $3\frac{1}{2}$ by $\frac{1}{2}$ inch. The framework of the vessel below the armor is composed of longitudinal and transverse frames. The former, eight in number, are formed of steel plates $\frac{7}{16}$ inch in thickness, the shelf-plate being of iron $\frac{1}{2}$ inch thick. These frames extend as far forward and aft as is deemed practicable. Within the double bottom, which extends through 212 feet of the ship's length, the transverse frames are solid, and are made water-tight at intervals of 20 feet. There are also intermediate bracket-frames placed 4 feet apart. Throughout the double bottom the transverse frames, which are likewise 4 feet apart, are of the thickness of $\frac{3}{8}$ inch, but are considerably lightened by having holes cut through them, the upper parts at the same time being much narrowed. Additional intermediate frames are worked in the engine-room in order to secure greater strength. The angle-irons forming the frames vary from $5\frac{1}{2}$ inches by 3 inches by $\frac{1}{2}$ inch to 3 inches by $3\frac{1}{2}$ by $\frac{1}{2}$ inch. The outer skin plating of the bottom varies from $\frac{15}{16}$ inch in the garboard strakes to $\frac{5}{8}$ inch, with the exception of the ends, where the thickness is increased to $\frac{7}{8}$ inch, and behind the anchors, where the plating is doubled. The plating of the inner bottom, which extends through the length of the double bottom, and which, like the outer bottom, is made perfectly water-tight, is of the uniform thickness of $\frac{3}{8}$ inch, except under the engines, where it is $\frac{7}{16}$ inch. As is usual in iron vessels, the stern of the *Inflexible* consists of a solid iron forging, scarfed at its lower end to the keel-plates. The stern-post and after pieces of keel, which are formed of the best angle-iron, were also made in a single forging. The rudder is a solid iron frame filled in with wood and covered with iron plates. In consequence of its immense weight—some 9 tons—it is made to work upon double pintles in combination with the ordinary pintles and braces. It is moved by a tiller 4 feet 6 inches below the water. Indeed, the whole of the steam steering-gear will be placed below the water-line and armored deck, so that it will be impossible for the rudder-head to be injured by shot or shell during an engagement. To receive the propeller-shafts two iron tubes are constructed, one under each quarter. The fore parts of these tubes, where they leave the run of the ship, are supported by the framework of the hull, which is bossed out in a suitable form for the purpose, the after parts being supported by struts from the ship's bottom. There are four decks—the lower, middle, upper, and superstructure decks—the last being a middle-line erection placed forward and aft above the upper deck for working the ship, carrying and lowering the boats, &c. Outside the citadel the lower-deck beams are covered with iron 3 inches

THE INFLEXIBLE.

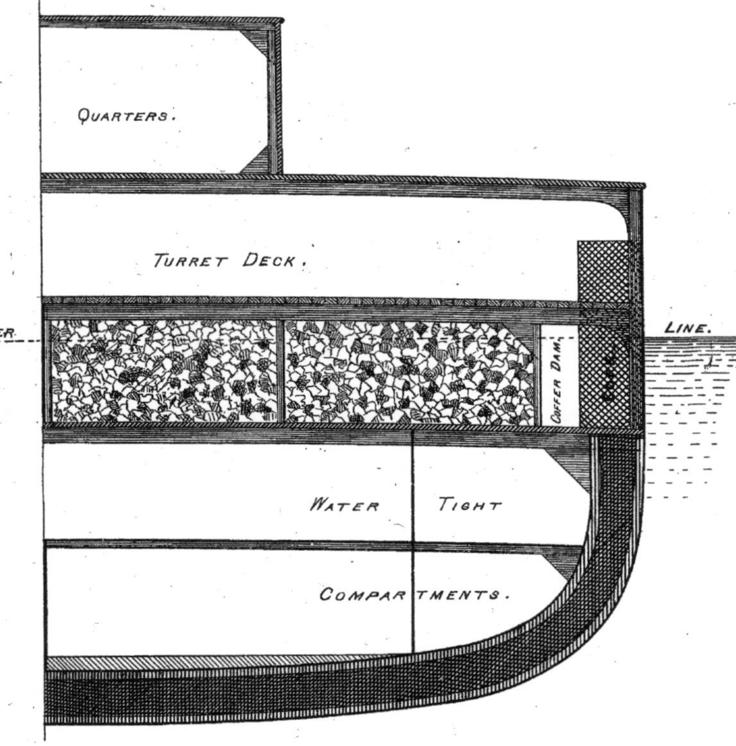

SECTION ABAFT CITADEL.

thick. This deck is depressed at the fore end so as to meet that part of the bow which is intended for ramming, thus conferring upon it greatly increased strength and resistance when engaged in butting an enemy's ship. It may be here stated that the ram of the *Inflexible* is of the spur kind, and though it is fixed at the present time, it will eventually be made to unship during ordinary cruises. The middle-deck flat consists of ¼-inch plating covered with 3-inch deal planks; while the upper-deck beams in the vicinity of the citadel are covered with 3-inch plating, and in other places with ½-inch plating. The beams, pillars, and bulkheads for supporting the various decks and platforms, and forming the different compartments and rooms, are arranged and fitted so as to give the greatest possible strength to the sides of the vessel. The largest beams are on the main deck. They are 14 inches deep, while those on the upper deck are 10 inches, and those on the lower deck are 12 inches deep. Every beam is either supported by wrought-iron tube-pillars or is trussed where pillars cannot be erected, the strongest being under the turrets. The two superstructures themselves in no wise add to the power of the ship, either for attack or defense. Their purpose in the economy of the ship is to afford accommodation for the officers and crew; and as the structures are erected on the upper deck, this will be of the best kind, with abundance of air and natural light. Their dimensions are: fore superstructure, extreme length, 104 feet 4 inches; breadth, 21 feet 4 inches; after superstructure, extreme length, 105 feet 4 inches; breadth, 30 feet. The frames are formed of angle-iron, 7 inches by 3 inches, placed 4 feet apart, and between them are intermediate frames made of angle-iron 4 inches by 3 inches. The ends are covered with ⅜-inch plates, and the whole surface with 3-inch deals. The cabin-walls are all coated with Welch's wood-faced cement, as a protection against the results of atmospheric condensation. The officers and men together will number 350. As a protection against the casualties of war and the sea, the hull is divided by means of the transverse and longitudinal bulkheads into no fewer than 135 water-tight compartments, and arrangements will be made for quickly removing therefrom any water that may collect within them through collision or other cause. Powerful steam-pumps, among which may be mentioned two of Friedmann's patent ejectors, capable of discharging 300 tons of water each per hour, will be fitted. All the bulkheads are provided with water-tight doors of an improved pattern, sluice-valves, manholes, and water-tight scuttles. Water-tight doors can also be fitted, when necessary, to the bulkheads passing through the coal-bunkers. Each of the water-tight compartments has been tested by hydraulic pressure. Great attention has been bestowed upon the question of ventilation, which in ships of the *Devastation* class, and indeed in all monitors of low free-board, has been a source of considerable discomfort and embarrassment. In the *Inflexible* the fresh air will be drawn into the midship part of the vessel through a series of downcast shafts, by means of eight powerful fans, worked by four of Messrs. Brotherhood & Hardingham's patent three-cylinder engines. The air is then conducted into main pipes, which run around the sides of the hull to the extremities, and from these, subsidiary or branch pipes discharge the air in ample quantities to every part of the ship.

DEFENSE.

The annexed drawings will give a good idea of the design of the ship.

Over the shot-proof deck, at a level a little above the water-line, comes the middle deck, and, as may be seen from the plates, the entire space between the two decks is divided into compartments arranged partly to carry coals and partly stores packed in water-tight tanks, forming further subdivisions of the space. Next to the sides of the ship the compartments are about 4 feet wide, and are filled with cork, and inside this again are compartments 2 feet wide, filled with layers of canvas and oakum, which, by experiment, are found partially to close holes made by shot passing through, and to check the passage of water. The cork and canvas compartments are carried above the main deck 4 feet and 2 feet respectively, and 30 feet forward of the citadel and 37 feet aft of it. Thus, if a shot hit the unarmored ends of the vessel at right angles to the water-line, it would travel through, first, 4 feet of cork, then 2 feet of canvas and oakum, then such coal and stores as were unconsumed, and finally pass through oakum and cork to the sea, on the opposite side from which it entered. The cork is, of course, intended as a life-belt to the ship, to give her additional buoyancy when the unprotected ends are riddled and filled with water.

The protected portion of the ship is confined to the citadel or battery, within whose walls are inclosed the engines and boilers, the turrets, the hydraulic loading-gear, the magazines, and in fact all the vital parts of the vessel. It measures 110 feet in length, 75 feet in breadth, and is armored to the depth of 6 feet 5 inches below the water-line, and 9 feet 7 inches above it. The sides of the citadel consist of an outer thickness of 12-inch armor-plating, strengthened by vertical angle-iron guides 11 inches wide and 3 feet apart, the space between them being filled in with teak backing. Behind these girders, in the wake of the water-line, is another thickness of 12-inch armor, backed by horizontal girders 6 inches wide, and supported by a second thickness of teak backing. Inside this are two thicknesses of 1-inch plating, to which the horizontal girders are secured; the whole of the armor backing and plating being supported by and bolted to transverse frames 2 feet apart, and composed of plates and angle-irons. It will thus be seen that the total thickness of armor at the water-line strake is not less than 24 inches. The armor-belt, however, is not of uniform strength throughout, but varies in accordance with the importance of the protection required and the exposure to attack. Consequently, while the armor at the water-level is 24 inches in two thicknesses of 12 inches each, above the water-line it is 20 inches in two thicknesses of 12 inches and 8 inches, and below the water-line it is reduced to 16 inches in two thicknesses of 12 inches and 4 inches. The teak backing with which it is supported also varies inversely as the thickness of the armor, being respectively 17 inches, 21 inches, and 25 inches in thickness, and forming with the armor, with which it is associated, a uniform wall 41 inches thick. The depth of armor below the load water-line is 6 feet 5 inches, but as the vessel will be sunk a foot on going into action by letting water into its double bottom, the sides will thus have armor protection to the depth of 7 feet 5 inches below the fighting-line. The outside armor is fastened by bolts 4 inches in diameter, secured with nuts and elastic washers on the inside. The shelf-plate on which the armor rests is formed of ½-inch steel plates, with angle-iron on the outer edge 5 inches by 3½ inches by $\frac{9}{10}$ inch. The armor on the fore bulkhead of the citadel is exactly the same in every respect as that on the sides, but the armor of the rear bulkhead is somewhat thinner, being of the respective gradations of 22, 18, and 14 inches, and forming with the teak backing, which is 16, 20, and 24 inches, a uniform thickness of 38 inches. It may also be useful to mention that before and abaft the citadel the frames

THE INFLEXIBLE.

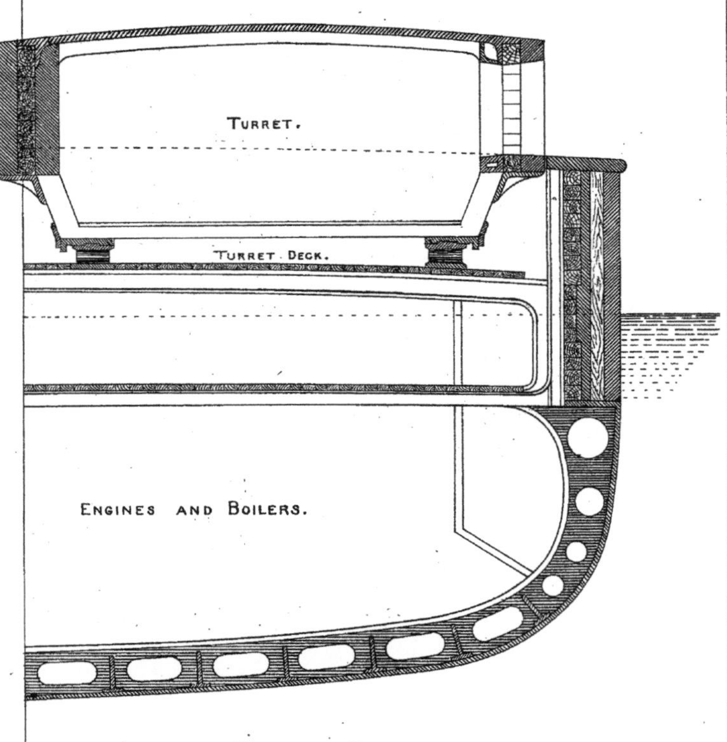

SECTION THROUGH CITADEL.

are formed of 7-inch and 4-inch angle-irons, covered with $\frac{9}{16}$-inch plates. The total weight of the armor, exclusive of deck, is 2,250 tons, and the total weight of armor, inclusive of deck, is 3,155 tons.

TURRETS.

But the most singular feature in the design of the ship is the situation of the turrets. In the *Devastation* and *Thunderer*, and in fact all monitors afloat, the turrets are placed on the middle line, an arrangement which, though advantageous in some respects, possesses this signal disadvantage, that in double-turreted monitors only one-half of the guns can be brought to bear on the enemy either right ahead or directly astern. In the *Inflexible*, however, the turrets rise up on either side of the ship *en échelon* within the walls of the citadel, the forward turret being on the port side and the after turret on the starboard side, while the superstructures are built up along a fore-and-aft line of the deck. By these means the whole of the four guns can be discharged simultaneously at a ship right ahead or right astern, or on either beam, or in pairs toward any point of the compass. Besides these important advantages, the guns of each turret can be projected clear of the ship's side—in the case of the one turret to port, and in the case of the other turret to starboard. They can then be depressed enough not only to strike a vessel at close quarters, below the line of her armor, but even to fire down upon her deck, should the enemy be ranged alongside. The walls of the turrets, which last have an internal diameter of 28 feet and an external diameter of about 33 feet 10 inches, are formed of armor of a single thickness of 18 inches (the thickest ever manufactured, with the exception of the 22-inch experimental plate which was rolled at Messrs. Cammell & Co.'s works, at Sheffield, for the turrets of the Italian frigates), with backing of the same thickness, and an inner plating of 1 inch in two equal thicknesses. All experience has proved that, for many reasons, this arrangement is the best. The wood backing distributes the blow when struck, deadens the vibrations, protects the fastenings, and stops the splinters, while the inner iron is also of advantage, since it renders the backing more compact, and also assists in arresting the passage of *débris*. The height of the turret ports from the load-line is 12 feet, and a foot less from the fighting-line, and all the plating in the wake of the guns is considerably strengthened.*

OFFENSE.

A very special interest attaches to the armament of the *Inflexible*, not only because it consists of guns vastly more powerful than any yet mounted afloat, but because these guns are carried and worked on the new and remarkable hydraulic system which has hitherto only been tried in the fore turret of the *Thunderer*. Each turret weighs no less than 750 tons (including the guns), and having to deal with a moving mass of such enormous weight, and with the superadded difficulty of a float-

* Some important experiments were made by the British admiralty, in December last, on the target-ship *Nettle*, at Portsmouth, for the purpose of testing the powers of resistance of steel and compound steel and iron plates, having for their immediate object the solution of the problem as to the kind of armor of which the turrets of the *Inflexible* shall be constructed. Steel plates and compound iron and steel plates from different manufacturers were fired at; among them was one made of Whitworth's compressed-when-fluid steel. The results of these experiments were not conclusive, but they seemed to indicate that a perfect substitute for iron as a means of resisting the huge projectiles of modern warfare has yet to be produced.

ing, and therefore unstable, platform on which to revolve, it was determined to commence at this point with the adoption of the hydraulic system of Sir William Armstrong, as developed for gunnery purposes by his partner, Mr. George Rendel. The revolution of the turrets accordingly will be accomplished by hydraulic machinery, in a manner similar to that employed by the Elswick firm for turning swing-bridges and great cranes. In such cases the weights dealt with have already exceeded that of the turrets of the *Inflexible;* and so complete is the control afforded by hydraulic machinery in the movements of heavy masses in these analogous cases, that it is believed the turrets will, by this machinery, be rotated at any speed, from a complete revolution in one minute, down to a rate as slow and as uniform as desired. The advantage of the high speed is plain; that of the slow but regular rotation will be apparent when it is remembered how much delicacy of adjustment is necessary for following with the aim an object moving rapidly and at a distance. Although the 80-ton guns will be worked on a system similar to that adopted in the case of the 38-ton guns of the *Thunderer,* yet as the design of the *Inflexible* had not been completed before the decision to work the guns by hydraulic power was formed, a much more complete hydraulic gunnery arrangement has become possible. The sponging and loading apparatus are still, as in the *Thunderer,* to be placed at duplicate fixed stations outside the turrets, and under the protection of the armored deck of the vessel. The muzzles of the guns are brought to the loading mechanism by revolving the turret and slightly depressing the guns. But there is no special loading-port as in the *Thunderer.* All that is necessary is to depress the guns to the small angle required for bringing the muzzles below the level of the deck, which, still further to reduce this angle, is raised and inclined upward at the base of the turrets so as to form a sort of glacis, and to give cover to the muzzles without involving any considerable depression of the gun. By this means the objection brought against the greater depression of the guns of the *Thunderer* is avoided. A more important novelty is the manner of mounting the 80-ton guns in the turrets. Hitherto it has been the practice to place all heavy guns upon an iron structure, called the carriage, on which they rest by means of the trunnions. This carriage bears, besides the gun, the mechanism for elevating and depressing the gun, and for "tripping," and also in part the mechanism for checking recoil. Besides the carriage, again, there is the slide upon which the carriage runs. Now in the system adopted for the *Inflexible,* Mr. G. Rendel has taken the bold step of dispensing altogether with a carriage, properly so called.

The leading features of the arrangement are shown in Fig. 3. Two guns will be mounted side by side in each turret. Each gun will be mounted so as to be supported on three points. The trunnions will rest on blocks sliding on fixed beams bolted down to the floor of the turret, while the breech will rest on a third block, sliding like the others between guides upon a beam or table. Behind each of the trunnion-blocks, in the line of recoil, are two hydraulic cylinders, connected with them by piston-rods. The cylinders communicate by a pipe, on which there is a valve, which, on the recoil of the gun, opens and allows the pistons to run back slowly, checking the recoil. By reversing the apparatus, the gun can be run out again. The beam on which the breech rests is supported by a third hydraulic cylinder, fixed vertically beneath it in the turret. By this means the breech can be easily raised or lowered, thus elevating or depressing the muzzle of the gun, which pivots on its trunnions with a large preponderance toward the breech. In order to load,

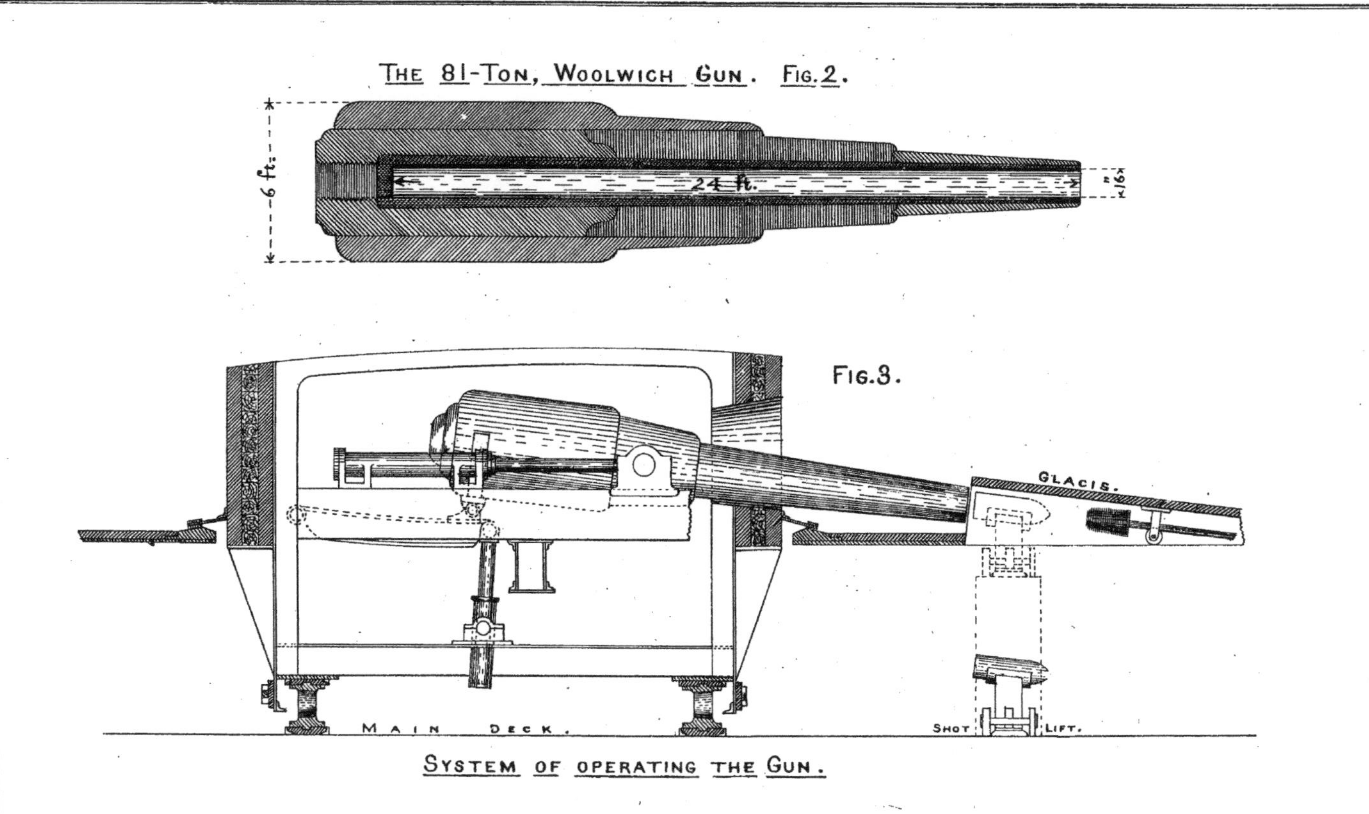
The 81-Ton, Woolwich Gun. Fig. 2.
6 ft.
24 ft.
16"
Fig. 3.
Glacis.
Main Deck.
Shot Lift.
System of operating the Gun.

the muzzle is depressed until it comes opposite to an opening made in the upper deck before the turret. A hydraulic rammer works in guides through this hole, and the rammer-head is hollow, and is so constructed that when it is driven into the recently-fired gun, and comes in contact with the sides of the powder-chamber, a valve opens, and it discharges through a number of holes small jets of water, thus acting as a sponge, and extinguishing any remnants of the charge or of the products of the explosion which may have remained smoldering in the bore. It is then withdrawn, and a hydraulic shot-lift raises up to the muzzle of the gun the charge, the projectile, and a retaining wad, and then a single stroke of the rammer drives them into the gun and home to the base of the bore. Again the rammer is withdrawn, the hydraulic piston under the breech of the gun elevates the muzzle, the turret swings around, and the shot is fired. A 9-inch gun, mounted experimentally in a turret at Elswick, and loaded on this system, was brought to the loading position, sponged, loaded, and brought back to the firing-point in forty seconds. Comparatively equally rapid loading was effected with the 38 ton gun during the experimental trial of the hydraulic gear on board the *Thunderer*. Thus, the first advantage of the system is rapidity of fire; the second is economy of labor. One man only for each gun is stationed in the turret, another works the hydraulic rammer on the main deck, six or eight others are employed in bringing up the ammunition to the shot-lift by means of a small tram way. There are two sets of loading-gear for each turret; but even if both were put out of order, the gun could still be loaded with an ordinary rammer and sponge by a number of men stationed on the main deck. The adoption of the system enables very heavy guns to be carried in comparatively small turrets. Those of the *Inflexible* are very little larger than those of the *Devastation*, so that with the old plan of having a numerous crew in the turret and running in the gun in order to load it by hand, only the 38 ton gun could be carried. As it is quite possible that the *Inflexible* will be armed with even more tremendous weapons than the 80-ton guns, this has been held in view in designing the ship; and, by a slight modification, it will be possible to mount in each of her turrets a pair of 160-ton guns, with a length of 30 feet and a caliber of 20 inches. The armament of the *Inflexible* will be composed of four of the heaviest guns (except those making for the Italian vessels) ever constructed, of which the experimental 81-ton gun completed at Woolwich and tested is the type.* Fig. 2 is a sectional sketch of the gun, showing the arrangement of the wrought-iron coils welded around the massive central steel tube. This tube, which forms the core of the gun, is bored out of a solid ingot, which cost $8,262. The bore is 24 feet long, and rifled from the muzzle to within a short distance of the base of the tube, where the unrifled portion forms the powder-chamber. The greatest external diameter of the gun is 6 feet, and at the muzzle it is 2 feet in diameter. The full caliber of the piece is 16 inches. The

* The largest rifled piece previously manufactured is the Krupp gun, exhibited at the Vienna Exposition, and subsequently exhibited at the Centennial Exhibition. It was mounted on a wrought-iron sea-coast carriage, having steel hydraulic recoil-cylinders.

This gun is a breech-loader, and is built up with steel hoops over a steel tube. Its weight, including the breech-loading apparatus, is 56½ tons. The length of the tube is 26 feet 3 inches. The length of the bore is 22 feet 6¼ inches, and the caliber 14 inches. The twist is uniform.

The projectiles consist of both steel and chilled shells, and shells 2.8 calibers in length. The heaviest projectiles, when charged, weigh, the steel, 1,124.5 pounds; the chilled, 1,157.5 pounds; and the powder-charge of the gun is 242.5 pounds for steel and chilled shells, and 275 pounds for long fuse shells.

A Krupp gun of greater weight and power will be noticed hereafter.

experimental gun was first bored out to 14½ inches and tested; for a second series of experiments it was given a caliber of 15 inches, and then bored to the full caliber of 16 inches and finally tested.

The gun is rifled with 13 grooves, each having an increasing pitch from 0 to 1 in 35 calibers. The service powder-charge is 370 pounds of 1.5-inch powder. The weight of the projectile for the service shell is 1,700 pounds, and the bursting charge about 100 pounds of powder. The details of the series of proof-trials at Woolwich, also the tests at Shoeburyness, have been widely published. Still, for reference, it is believed advisable to give here the results of the trials last made with the caliber of 16 inches, that at which the gun is to be used in actual warfare:

Number of rounds.	Size of powder.	Weight of charge.	Weight of projectile.	Muzzle velocity per second.	Mean pressure in gun.	Total energy developed.
	Cubic inches.	*Pounds.*	*Pounds.*	*Feet.*	*Tons per square inch.*	*Foot-tons.*
1	1.5	340	1,700	1,486	20.1	26,030
2	1.5	350	1,700	1,505	20.4	26,740
3	1.5	350	1,700	1,502	20.3	26,630
4	1.5	350	1,700	1,467	19.6	25,406
5	1.5	350	1,700	1,475	18.4	25,683
6	1.5	350	1,700	1,493	21	26,314
7	1.5	360	1,700	1,487	18.8	26,103
8	1.5	370	1,700	1,495	19.9	26,385
9	1.5	350	1,700	1,518	20.5	27,203
10	1.5	370	1,700	1,523	20.3	27,383
11	1.5	360	1,700	1,519	21.3	27,239
12	1.5	360	1,700	1,518	20.0	27,203
13	1.5	370	1,700	1,519	19.8	27,239
14	1.5	370	1,700	1,517	20.7	27,168

The experiments for range and accuracy were conducted at Shoeburyness, and are reported to have met the unqualified approval of the authorities. When the last experiments were concluded, viz, October 4, 1876, the gun, originally weighing 81 tons, but now reduced by reboring, &c., to 80 tons, had fired 140 rounds, and it may be interesting to summarize the amount of ammunition that has been thus expended.

With its normal caliber of 14.5 inches it fired 4,660 pounds of powder and 27,052 pounds of iron in 21 rounds. With a caliber of 15 inches it fired in 32 rounds 8,223 pounds of powder and 45,712 pounds of iron. With the same caliber, but with a powder-chamber of 16 inches, in 21 rounds it disposed of 6,020 pounds of powder and 30,810 pounds of shot. With its present uniform bore of 16 inches, and while at Woolwich, it fired 8,870 pounds of powder and 45,981 pounds of iron in 27 rounds. This gives 27,773 pounds of powder and 149,555 pounds of iron expended at Woolwich in 101 rounds. At Shoeburyness the gun has fired 39 rounds with 14,430 pounds of powder and 66,300 pounds of iron. This gives a total number of 140 rounds, 42,203 pounds of powder, and 215,855 pounds of iron.

During the experiments for range, shells were reported to have been recovered from a minimum distance of six miles; others were traced still farther, until deep water arrested the progress of the explorers.

Some idea of the amount of ammunition required for the 80-ton gun may be formed when it is estimated that, in an action, if the *Inflexible* would fire only ten shots from each of the four guns, she would expend 14,800 pounds of pebble-powder, and hurl upward of 30 tons of projectiles, at a cost of about $6,320.

The cost of the gun, exclusive of carriage and the machinery for working it, was estimated at $72,900, and the factory-plant and experi-

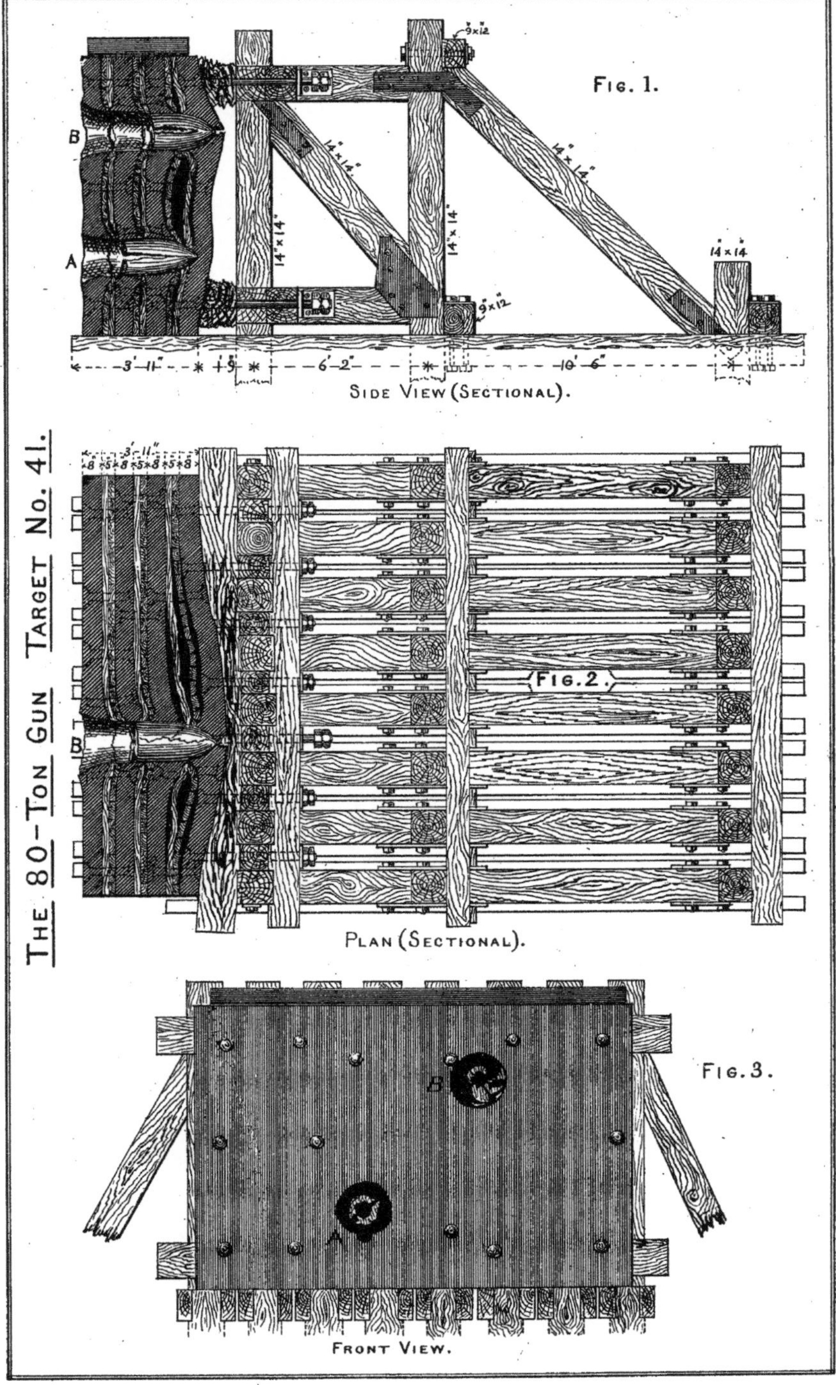
THE 80-TON GUN TARGET No. 41.
FIG. 1.
9"x12"
14"x14"
14"x14"
14"x14"
14"x14"
14"x14"
9"x12"
B
A
3'-11"
9
6'-2"
10'-6"
SIDE VIEW (SECTIONAL).
3'-11"
8
5
8
5
8
5
8
FIG. 2.
B
PLAN (SECTIONAL).
FIG. 3.
B
A
FRONT VIEW.

mental trials at $48,600. The actual cost of each of the eight guns will be best known when all are manufactured.

The target against which this gun has been proving its powers at Shoeburyness is the most formidable of any hitherto fired at. It is gigantic, even in comparison with those fired at by the 100-ton gun at Spezia, hereafter to be noticed.

Its construction is beautifully and plainly shown by illustrations giving front elevation, horizontal and vertical sections, with accompanying figures and explanations, in the *Engineer* of February 2 and 9, and from which the accompanying figure has been taken.

The target is altogether different from the Spezia targets, as may be seen by referring to the sketch. It is composed of four iron armor-plates, each 8 inches thick, sandwiched between three layers of teak each 5 inches thick, amounting in all to 32 inches of iron and 15 inches of teak. The plates are 16 feet in breadth by 10 feet in height. Each plate weighs 23 tons, the collective weight of iron being thus 92 tons. The plates are secured together in pairs by bolts; that is to say, the front plate is bolted to the second one, the second to the third, the third to the fourth, and the fourth to the horizontal beams in the rear. The bolts employed are 3 inches in diameter. The shank of the bolt has the Palliser projecting screw-thread on it, while the head is made on the ball-and-socket principle, with the hole in the plate allowing play round the neck of the shank, so that one plate may move slightly on the next without shearing the bolt.

The target is supported by a heavy frame-work of beams, mostly 14 inches square, and very firmly secured to the ground by strutted piles at the ends, and weighted with an old armor-plate on the top to keep the teak filling from escaping under the force of the blow of impact.

The cost of the armor-plates and bolts was about $24,300 in gold, and the cost of the timber and labor about $1,000 more.

The first shot was fired February 1.

The gun was charged with 370 pounds of powder, cubes of 1.5 inch, and a Palliser projectile filled with sand to 1,700 pounds' weight, including gas-check, and plugged. This projectile was of service form, having an ogival head of 1½ caliber radius.

By reference to the sketch it will be seen that the projectile buried itself into the target, having penetrated through the first three plates and the three layers of wood, also about 1 inch into the rear plate; that is, it penetrated through 24 inches of iron, 15 inches of teak-wood, and was arrested with the point 1 inch into the fourth plate, thus leaving 7 inches of iron in front of it unpierced, which iron, being cracked and bent, would offer greatly diminished resistance to further penetration—the projectile itself being split.

The horizontal beam at the base of the back of the target was crushed and split into ribbons, and the whole target structure sprung and shaken.

On the same day a blind common shell was fired from the 81-ton gun at an old 8-inch unbacked armor-plate, which was completely demolished, being split and broken across and thrown out of its position, and the shell broken up and scattered.

Doubtless the common shell from this gun would be terrible against any weak-armored ship.

After these trials, the gun being returned to Woolwich, the powder-chamber was enlarged to a diameter of 18 inches for a length of 58½ inches, and that of the bore retained at 16 inches. Early in May the trial of the gun, with the enlarged powder-chamber, took place at Shoeburyness.

The firing, as before, was against the structure known as the No. 41 target, consisting of four thicknesses of 8-inch plates and intermediate layers of 5 inches of teak, above described and here represented by illustrations taken from the *Engineer*.

These trials for penetration do not involve many rounds, which would necessitate an enormous expenditure for targets. The utmost effect of every round must be carefully considered in order to economize the splendid target against which the gun is directed.

The following are the conditions and results of the first trial with the 18-inch powder-chamber: The charge of the powder employed was 425 pounds, that previously used being 370 pounds, 1.5-inch cubes in each case. The projectile was a Palliser shell weighted to 1,700 pounds wi h sand, and plugged; it was studded, and to its base was attached a Lyon gas-check, which consists of a disk of copper having a thickened rim which is expanded into grooves of the bore by the pressure of the gases of the powder. The range, as before, was 120 yards.

The position of the shot fired on the previous round is shown at A, Figs. 1 and 3, and the position on this occasion is shown at B, Figs. 1, 2, and 3. The shot struck a spot about 6 feet from the proper left, and 7 feet from the bottom of the target. The hole in the front plate was rather larger than it should have been, either from the shot not striking quite steadily and truly, or from the setting up of the metal of the shot during penetration. The point was visible through large cracks in the back of the target, in some places open to a width of 2½ inches. There was about 5 inches of iron still in advance of the point, but this was fissured and opened in a star crack, shown in Fig. 4. The back plate was bulged or bent back nearly 14 inches, as shown in Figs. 2 and 3, and the horizontal beam behind the top of the target crushed and split from end to end. The bolts passing through this beam now protruded at the back to an extent reaching from the target's proper left to right, of ½ inch, 1½ inches, 4 inches, 4½ inches, 2 inches, and ½ inch, consecutively. There was no appearance, however, of any of the bolts having been broken.

The initial velocity of the shot was 1,600 feet per second, the highest yet attained by any gun, giving an energy of 30,180 foot-tons, or 600.3 foot-tons per inch of circumference. The striking velocity was 1,585 feet per second, giving 29,620 foot-tons of energy, or 589.2 foot-tons per inch of circumference. The average chamber-pressure was found to be 19.9 tons per square inch, a result well within the limit assigned by the Woolwich authorities, which is 25 tons. The charge had 34 cubic inches of air-space per pound of powder, and was ignited centrally through the rear or axial vent. The hollow space for central ignition in the center of the cartridge, which was formerly preserved by means of a wicker pottle or basket, is now maintained by a zinc pottle, which answers the purpose very well, and is entirely consumed or vaporized. The charge itself was contained in a silk bag. The power of penetration possessed by the 80-ton gun in its chambered condition is consequently proved to be slightly superior to that of the 100-ton gun at Spezia in its unchambered condition. It is remarked, too, that this force was generated in the gun with a defective tube. After the round was fired, a gutta-percha impression was taken, which showed that no alteration whatever had taken place in the crack, the gun still remaining in a safe condition for further work. It will thus be seen that the Woolwich gun continues to give satisfaction. It was returned to Woolwich after the last round in order to utilize its carriage for the trial of the first of the four 80-ton guns, now being constructed at the royal arsenal for the *Inflexible;* also to have its damaged tube replaced by a new tube that is being prepared for the purpose. The gun-carriage and its appendages weigh in all 40 tons, and the recoil in this last round up the railway, there being an ascending gradient, was found to be 55 feet.

THE NEW 80-TON GUN.

This is the first of the four service guns intended for the *Inflexible*, the one first constructed and described being experimental. It chiefly differs from the first in having thirty-two grooves in the bore; the projectile having no studs, but being made to take the rifling by the setting up of a copper gas check fixed on the base, and which thus enters the grooves. It adheres firmly to the base of the projectile by means of radiating saw-tooth-shaped cuts on the latter. The new gun has at present a bore of 15.5 inches diameter, without any enlargement at the chamber. The firing-charge employed was in each instance 335 pounds of powder of 1.5-inch cubes. The proof projectile weighed 1,550 pounds. The following muzzle velocities were obtained: First round, 1,603 feet per second; second round, 1,599 feet; and third, 1,598 feet. The pressure of the three rounds was registered at 22.8, 22.4, and 22.8 tons respectively. These results must be considered very good, because the bore and chamber are not yet brought to their full dimensions, and are not, therefore, in the condition to enable the powder to burn to the best advantage. It may be seen that the pressures, considering this, are not very high, and are very regular. The same carriage is employed as carried the first gun on the previous trial.

MOTIVE MACHINERY.

The machinery was constructed by Messrs. John Elder & Co., of Glasgow. Each screw will be driven by an independent set of compound engines with three vertical inverted cylinders of the collective power of 4,000 horses, giving an aggregate power of 8,000 horses (indicated) for both sets of engines. The diameter of the high-pressure cylinder is 70 inches, and the diameter of each low-pressure cylinder is 90 inches; the former is placed between the two latter. They are steam-jacketed, and are connected together by stay-bolts prolonged to bulkheads, so as to serve as ramming-chocks. The pistons have a stroke of 4 feet, and the number of revolutions expected is 65 per minute. The piston-rods are double, and are connected by crank cross-heads. They are each 7 inches in diameter, the connecting-rods having a diameter of 9 inches and a length of 7 feet 6 inches. The valves are of the piston kind. They are worked by link-motions and levers, and are reversed by an ingenious combination of steam and hydraulic power. The engines at starting are assisted by auxiliary steam-gear, the valves of which are fitted to the receiver. The steam from the low-pressure cylinders is exhausted into independent surface-condensers, having a total cooling surface of 16,000 square feet. The steam is condensed in the interior of a series of tubes of $\frac{3}{4}$-inch external diameter, of which each condenser has no less than 6,650. The condensers are constructed to be worked as common condensers. The circulating-pumps are actuated by separate engines, each having its own feed, bilge, and air pumps worked by levers from the cross-heads. The air pumps are made of gun-metal, with a diameter of 34 inches and stroke of 2 feet 3 inches, the water being discharged below the armor-deck. With respect to the centrifugal pumps, it may be mentioned that they are judiciously placed at so high a level in the vessel that in the case of leakage occurring, by which the ship's bottom may be flooded to as great a depth as 12 feet, they can be worked with perfect freedom. There are also double-acting hand-pumps, each two coupled; feed donkey-engines, each with double-acting pumps 4 inches in diameter; bilge-donkeys, each with double-acting pumps 6 inches in diameter, and fire-engines, with double-acting pumps 8½ inches in diameter. It may be mentioned that the engines which work the circulating-pumps are also made to pump out the bilge, in the event of the ship springing a leak or sustaining damage from being rammed; that the centrifugal pumps are sufficiently powerful to perform the same work in case of emergency; and that a Kingston valve is fitted through the bottom, in connection with each fire-pump.

Each cylinder is fitted with an expansion-valve, having a variable cut-off, with an extreme range of from one-sixth to one-half stroke. These valves are cylindrical gridiron-valves, of phosphor-bronze,* 3 feet in diameter, working on cast-iron gridiron seats, and giving a minimum of clearance between the expansion-valve and main slide. They are worked by an eccentric on the crank shaft and a slotted lever, and are all connected to a shaft in front of the engines, so that they may be thrown out by a single handle. Each engine is also fitted with a common injection apparatus. The crank-shaft is formed of three pieces, the diameter of the bearings being 17½ inches. The propellers will be about 20 feet in diameter, and will be worked outward, the thrust being at the after end. The shaft-tubes are of wrought iron, supported by struts, while the shafting will be made of Whitworth fluid-compressed steel, with solid couplings. It will be hollow, the inner diameter being 10 inches and the outer 16 inches. The faces of the high-pressure cylinders are formed of phosphor-bronze 2 inches thick; the liners of the cylinders are also constructed of the Whitworth compressed steel, which possesses properties rendering it not only extremely light, but at the same time much more trustworthy than the ordinary metal used for this and shafting purposes. Each engine will be fitted with a governor, to prevent racing in stormy weather; and, in addition to the hand-gear, small auxiliary engines will be erected for turning the main engines.

BOILERS.

The steam is to be furnished by twelve boilers, eight single-ended and four double-ended. They are constructed of the best Lowmoor plates,

* This alloy is now gaining favor for cylinder valve-faces where high-pressure steam is used, and for bearings where heavy pressures are applied. Its component parts consist of copper, tin, and phosphorus, and it is capable of being made tough and malleable, or hard, according to the proportions of the several ingredients. It is rendered so liquid in the molten state by the addition of the phosphorus that it forms very clean castings.

Messrs. Levi & Kingel, of the Val Benoit Nickel Works, near Liege, Belgium, have, for a number of years past, been engaged in making experiments for the purpose of improving bronzes of this kind. The results of their experiments are thus summed up by M. Dumas:

"The color, when the proportion of phosphorus exceeds ½ per cent, becomes warm and like that of gold largely mixed with copper. The grain and fracture approximate to those of steel, the elasticity is considerably increased, the absolute resistance under a fixed strain becomes more than doubled, the density is equally increased, and to such a degree that some alloys are with difficulty touched by the file. The metal when cast has great fluidity, and fills the mold perfectly. By varying the dose of phosphorus the particular characteristic of the alloy which is most desired can be varied at will."

In a series of experiments at the Royal Academy of Industry at Berlin, a bar of phosphor-bronze (proportions of components not stated) under a strain of ten tons resisted 862,980 bends, while the best gun-metal broke after 102,650 bends.

In Austria the following comparative results have been obtained:

Absolute resistance.

	Lbs. per sq. inch.
Phosphor-bronze	81,798
Krupp cast steel	72,258
Ordnance bronze	31,792

tested to 21 tons lengthwise and 18 tons crosswise, and the pressure of steam will be 60 pounds per square inch. The four double-ended boilers are 17 feet long, 9 feet 3 inches wide, and 14 feet 3 inches high, with four furnaces in each. Four of the single-ended boilers are 9 feet long, 13 feet 7 inches wide, and 15 feet 6 inches high, with three furnaces each, and the four remaining single-ended boilers are 9 feet long, 11 feet wide, and 13 feet 4 inches high, with two furnaces, each having a separate fire-box. All the boilers are to be clothed with four thicknesses of boiler-felt, and covered with galvanized sheet-iron, and are stayed to prevent their moving by concussion when the ship is engaged in ramming. They are to be supplied with water by four feed-pumps, which are attached to each engine, the pumps being 7½ inches in diameter, and having a stroke of 2 feet 3 inches. In the event of the feed-pumps receiving injury, the boilers are provided with four small auxiliary engines (one in each boiler-room), and having separate connections with the boilers. The two auxiliary engines which are used for washing the decks are also arranged to work the fire-engines in the engine-room. The safety-valves are fitted with springs upon an improved plan. The smoke-pipes, of which there will be two, are 65 feet high from the dead-plate of the lower furnaces. The bunkers, which are placed at the water-line along the unarmored sides of the ship, where the entrance of shot or water cannot injure them, are built to store 1,200 tons of coal, and are so disposed that their contents can be approached from the upper and lower compartments independently of each other.

RIG.

The *Inflexible* is also to possess sail-power, with respect to the advantages of which, however, considerable diversity of opinion exists. She will be brig-rigged, having two iron masts, but no bowsprit or stay-gear. The foremast will be 36 inches in diameter, and will measure 83 feet 6 inches from the deck to the head, while the mainmast will have a diameter of 37 inches, and a height of 96 feet. Each will have a top-mast and topgallant mast, with lower yard, topsail yard, and topgallant yard. The total area of sails will be 18,470 square feet.

In time of war it is intended that the ship will carry no masts, except for signal purposes.

The anchors, of which there are to be four, will be of Martin's self-canting pattern.

WEIGHTS.

The estimated weight of the hull is 7,300 tons.* The engines will weigh 614 tons. The propellers, shafts, and stern fittings weigh 151 tons each; the boilers, smoke-pipes, casings, &c., 522 tons, and the water in the boilers when ready for steaming is estimated at 190 tons.†

COST.

The cost is estimated as follows:

Materials	$1,307,340
Engines and appendages	486,729
Boilers	100,116
Labor	641,520

* This weight includes all armor and backing on citadel and turrets, the turrets themselves, and the deck-armor.—An English Naval Architect.

† The admiralty calculation in 1877 of the total weight of machinery was 1,405 tons; but this probably did not include coal-bunkers, stern tubes for propellers, &c.—J.W.K.

The date named in the navy estimates for the completion of this ship is March, 1878.

As a new type of a man-of-war, the leading features of the *Inflexible* may be summed up as follows: The armor is confined to the central fighting portion and to the main substructure which floats the ship. An armored deck 7 feet under water divides the vessel into two separate portions. The unarmored ends are so constructed that the vessel will float even when they are penetrated by shot. The ship has a wide beam and a comparatively light draught of water. The deck-houses give a high bow and stern, and the turrets are so arranged as to enable all four guns to be fired both ahead and astern, or on either beam.

In perusing the foregoing description of the *Inflexible*, it has been seen that her double bottom is divided and subdivided into an unusual number of spaces, and that the water-tight bulkheads have been introduced to an extent not before attempted, and in fact almost every conceivable precaution has been taken to make her secure against the ram and the torpedo. If, however, she should be fairly struck by several powerful fish-torpedoes, it is quite probable she would be crippled, water-logged, or possibly sunk. The question therefore presented is, whether two vessels of smaller dimensions, each carrying two 80-ton guns instead of four, would not have been a safer and in some respects a better investment.

STABILITY OF THE INFLEXIBLE.

During the summer of 1875, or thereabouts, Mr. E. J. Reed, C. B., M. P., made visits of observation to one or both of the great ships building in Italy. After his return to London he pronounced these ships unsafe for battle. He said: "The Italian ships *Duilio* and *Dandolo* are exposed, in my opinion, beyond all doubt or question, to speedy destruction. I fear I can only express my apprehension that the Italians are pursuing a totally wrong course, and one which is likely to result in disaster." The charge was promptly met and stoutly denied by the Italian minister of marine, from his seat in the Parliament at Rome. He said: "Mr. Reed cannot possibly prove any such statements, because no one but the designer and his confidential agents are entitled to have the particulars for making the necessary calculations; and, in case of a half-built ship, the intentions of the designer with regard to the disposition of a great mass of material not yet arranged and specified form part of these particulars."

The Italians have proceeded to complete these two great ships according to the original design, and trust for both buoyancy and stability to their unarmored ends. And in their later and far larger ships, the *Italia* and *Lepanto*, they have, in full view of Mr. Reed's criticisms, gone still further and abandoned the citadel itself.

The *Inflexible*, which is of the same type, and of the plans of which Mr. Reed no doubt had knowledge from the time she was laid down in February, 1874, was at this time building, yet nothing was said of the want of stability in this ship. The progress of construction continued rapidly to advance for three and a half years; the completed ship was promised for 1878, and the British public believed that their government would soon possess the most formidable and the most perfect war-ship ever floated, when, suddenly, surprise and alarm were created by the announcement that it had become a serious question to one or more naval architects, outside the admiralty, whether the promised and essential conditions of the safety of the *Inflexible* had been attained. It appeared to a competent critic, who had been investigating the subject, that the central citadel of the ship was too small to secure of itself the end designed, and that the added buoyancy in the unprotected ends might, in action, soon be shot away. The *Times* under date of June 18, 1876, published the result of the calculations, and Mr. Reed brought on the question, on similar lines, in the House of Commons. The first lord defended the admiralty with conspicuous gallantry, treating the criticisms as a departmental attack; the papers and letters were called for and laid before the house, when it was seen that the naval constructors asserted one thing and Mr. Reed, with no inferior authority, asserted the opposite. Mr. Reed said that in action the cork and stores might be shot away, and the unprotected ends riddled and water-logged; and that in such an event the citadel, though intact, would still capsize. The reply was, that the supposed case was too remote a possibility to be considered, and that without any unprotected ends the ship would still float. The argument became close and intricate. The various means of capsizing a ship were considered, as well as the different operations of explosive shells in the destruction of cork and stores. The varied perils to which the ship would be exposed in battle, with ends riddled—such as the action from the waves, the moving of the "free water" within the ship, the pitching and the rolling, the running out of the guns, and last, but not least, from the action of the rudder as the vessel approached its *minimum* stability—were all discussed and treated of.

The essential points in the correspondence may be summarized as follows:

First Mr. Reed said:

On visiting the *Inflexible* from time to time, I found that the unarmored ends were so very large in proportion to the citadel as to raise in my mind a doubt as to this important condition being fulfilled. Observing this, and also the introduction of cork chambers, I designed an *Inflexible* in my own office, and had the whole of the calculations made, the result showing that when these cork chambers were destroyed the vessel would have no stability whatever, but would be in a condition to capsize.

Second:

My objection is not that the *Inflexible* and other vessels do not possess that final reserve of stability, after a severe and protracted engagement, which I consider necessary, but that the cork chambers will be liable in action to speedy destruction, and that the ship will then be left without stability.

The reply of the admiralty officers, laid on the table of the House of Commons, consists of several papers quoted below—First, a letter from the director of construction, in which, besides giving the curves of stability and tables of calculations as hereafter annexed (as are also Mr. Reed's approximate curves), he also states:

But when I say that I regard this stability as being sufficient in view of the possible diminution of the stability by slow degrees by the blowing out of the cork walls and internal solid stores, I desire to add that I regard the possibility of the ship ever being reduced to this state as being infinitely remote, although not absolutely impossible. If the water be kept out of the coal-spaces by the coffer-dams, as I believe it will be, the ship will retain an amount of stability far in excess of the *Devastation*, including her wings added by us. In that case the water will not flow over her decks, as is supposed in the model; these decks will remain as high out of the water as the fore deck of the *Devastation*, and we should see no more reason for supposing the sea to wash freely from side to side in those decks than in the *Devastation*.

In order to justify Mr. Reed's objection, it is necessary to assume still further that every atom of solid material, excluding water, in the cellular store-rooms and in the cork walls has been blown out of the ship, and that only the battered iron shell remains, loading down the ship, but giving her no assistance. With regard to that, I say that no heavily-armored ship ever has been designed to comply with such a condition.

I ought, perhaps, to add that the whole of this discussion turns upon the power of the ship to resist the attacks of artillery, and I have endeavored to show that a fair balance is maintained between thickness of armor and extent of surface covered by armor. But, after all, the power of resisting attacks above water is only one element of the defense. We have also to consider the under-water attack. It would be easy, following Mr. Reed's course, to lay down some principle with regard to these attacks, and to say that no ship is well designed which is not so subdivided as to satisfy certain conditions.

The second paper is from Rear-Admiral Boys, director of naval ordnance. He said:

Looking at this as a question of naval artillery, I cannot conceive that the conditions on which Mr. Reed bases his argument as to the safety of the ship can be brought about in a naval engagement. These conditions are, practically, that the fore and after ends of the ship are to be utterly demolished. Should the *Inflexible* be made a target for continued practice, or be placed in a position similar to a fort whose walls could be breached by a battery of fixed guns, it is possible that in time the unarmored parts above water might be destroyed; but I do not think, for the following reasons, it is possible in a naval engagement to commit the havoc below the water that is presupposed by Mr. Reed:

1. The difficulty of striking a ship at or below the water-line, particularly one of the *Inflexible* type, that will scarcely ever roll.

2. The projectiles that would be fired at the *Inflexible* would certainly be armor-piercing, either chilled iron or steel; and such shell would not burst in passing through the thin iron sides of the ship, as they require the resistance of armor to ignite the bursting charge.

3. Considering the few guns that are likely to be carried by any ship engaging the *Inflexible*, and the ever-varying distance and bearing that must exist in any future naval action, it is next to impossible that any number of shells could be planted in a ship in such an exact position (even supposing them to burst) as to "blow out the cork from the chambers in which it will be fixed."

Those in charge of the ship must be devoid of all resources if during the intervals of an engagement—for intervals there must be—they could not take some steps, by the employment of stopper-mats or shot-plugs, &c., to prevent the unarmored ends of the ship from being water-logged; or supposing the water to come in, to allow it to run into the bilge, to be pumped out by the engines.

If the ship should get a list from water finding its way into the divisions at either end above the armored deck, it appears to me there are simple means at hand that can be resorted to for balancing her in an upright position.

I have no hesitation in saying I do not share, for one moment, Mr. Reed's anxiety for the safety of the *Inflexible* in action, from the effect of artillery-fire, as expressed by him.

The third paper is from Vice-Admiral Stewart, K. C. B., controller of the navy, the pith of which reads thus:

The result which has been assumed in this letter (Mr. Reed's) could, in my opinion, only be arrived at if we can suppose the ship lying perfectly helpless and immovable, and allowing herself to be attacked by an indefinite number of guns. By this means it is possible that a large portion of the unarmored structure above the water might be destroyed, but even then, I fail to see how it is possible to destroy or remove entirely all material, timber, cork, stores, coal, or other articles which, while remaining in any portion of the structure, must exclude water, or prevent water taking their place. To assume this ship placed in such a position, is, to my mind, representing an exaggerated state of circumstances which could never occur in real warfare.

Finally, a letter from their lordships indorsing their subordinates' views in full.

Tables of calculations.

TABLE No. 1.

	Condition.	Draught.		Displacement.	Deck enters water.	Angle of maximum stability.	Maximum G. Z.	Range.	Metacenter above C. G.
		Ft.	*in.*	*Tons.*	*Deg.*	*Deg.*	*Ft.*	*Deg.*	*Ft.*
b	Ship complete, cork in place	24	7	11,500	14	31.2	3.28	74.3	8.25
d	As in *b*, but in light condition	21	10	10,000	18	31.7	3.935	71.5	8.53
e	Fully equipped; ends riddled	26	8½	11,500	11	13.5	.568	30.0	2.0
f	As in *e*, but coal between decks (800 tons) removed	26	1	10,700	11½	15.4	.534	32.2	3.09
g	As in *e*, but in light condition	23	9	10,000	15	20.8	.794	36.8	2.22
k	As in *e*, but supposing the water in ship when upright locked	26	8½	12,668	11	13.9	.705	32.0	......
m	As in *k*, but supposing main deck kept free of water	26	8½	12,668	11	27.4	2.42	71.5	......

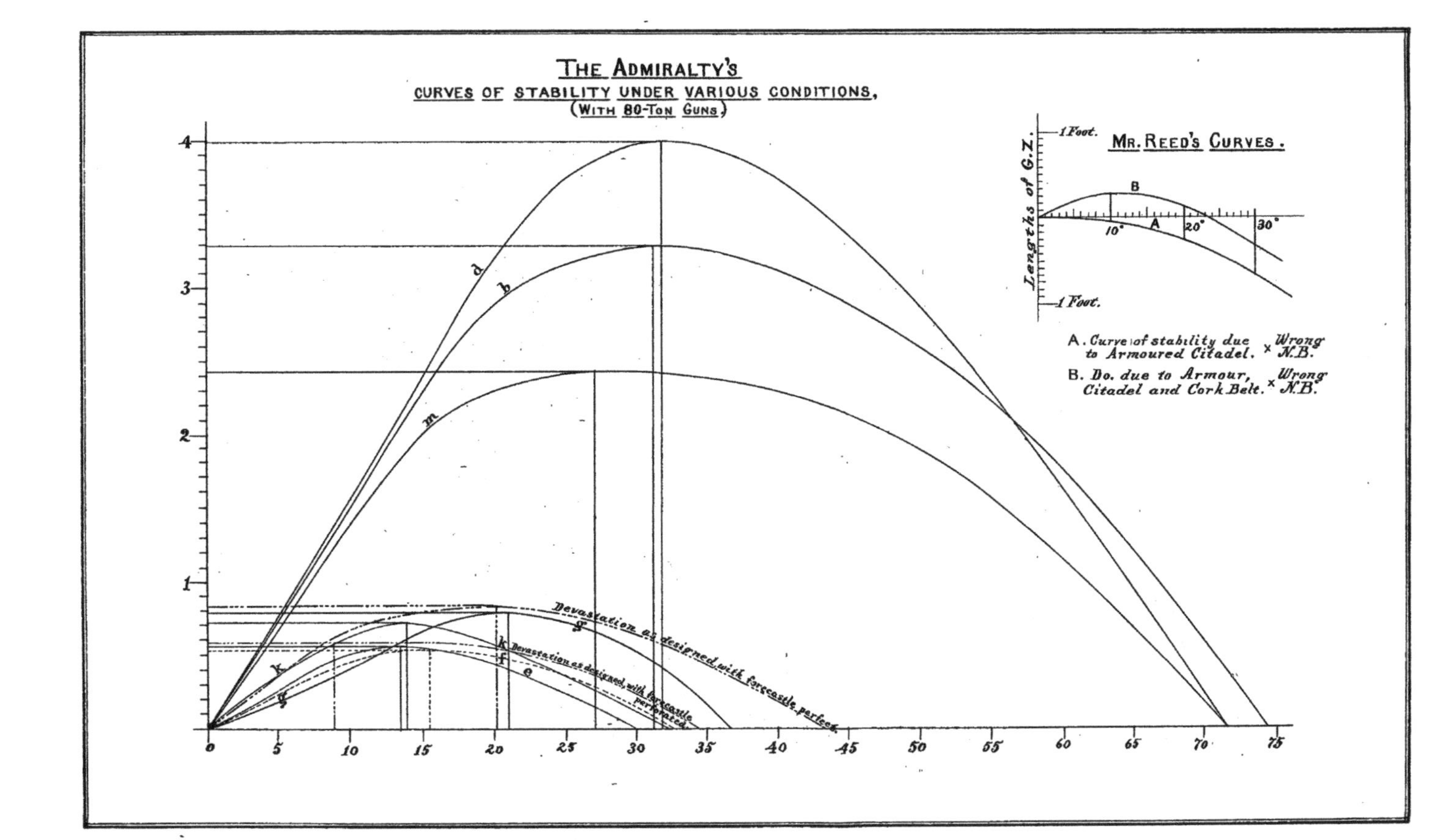
THE ADMIRALTY'S
CURVES OF STABILITY UNDER VARIOUS CONDITIONS,
(WITH 80-TON GUNS.)
MR. REED'S CURVES.
1 Foot.
Lengths of G.Z.
1 Foot.
B
A
10°
20°
30°
A. Curve of stability due to Armoured Citadel. × Wrong N.B.
B. Do. due to Armour, Citadel and Cork Belt. × Wrong N.B.
4
3
2
1
d
b
m
g
k
f
e
Devastation as designed with forecastle perfect.
Devastation as designed with forecastle perforated.
0
5
10
15
20
25
30
35
40
45
50
55
60
65
70
75

TABLE No. 2.

Condition of ship (intact draught 24 feet 7 inches.)		Draught. Ft. in.	Top of middle deck above water. Ft. in.
1. Ends riddled and one-third of buoyancy of ends clear of coals retained, but water excluded from coffer-dams and coal-spaces between decks.	Coals between decks in....	25 4	1 3
Ditto, ditto	Coals between decks out...	23 7½	2 11½
2. Same as 1, except that the one-third of buoyancy of ends referred to is neglected.	Coals between decks in	25 9	0 10
Ditto, ditto	Coals between decks out...	24 0	2 7

The following table is given on page 10 of the official report:

	Inflexible as assumed in model, unarmored ends giving no stability.	*Devastation* with forecastle riddled and giving no stability.
Maximum stability	6,532 foot-tons	5,237 foot-tons.
Angle of maximum stability	13½ degrees	9 degrees.
Range	30 degrees	34½ degrees.

In the end the government was forced to yield to public opinion and appoint a committee to investigate the subject and report their views. The committee so appointed was composed of men of the highest standing and integrity, though none of them were professional naval architects. It would be difficult to name four men in England whose opinion on the points at issue would be entitled to greater weight. They were Admiral Sir James Hope, Dr. I. Woolley, Mr. G. W. Rendel, C. E., and Mr. Froude, F. R. S. They were appointed in August, and were instructed to consider a series of questions, the investigations of which and the experiments made by them seem to have engaged their time until early in December, when their report was submitted to the admiralty. We quote the essential points from an English journal:

First. "As to the possibility or probability of the occurrence of the contingencies contemplated by Mr. Reed as being likely to happen very early in an engagement, namely, the complete penetration and water-logging of the unprotected ends of the ship, and the blowing out of the whole of the stores and the cork by the action of shell-fire."

On this point (according to the official summary of the report) the committee are of opinion that the complete penetration and water-logging of the unprotected ends of the ship, coupled with the blowing out of the whole of the stores and the cork by the action of shell-fire, is not likely to happen very early in an engagement; further, that it is in a very high degree improbable, even in an engagement protracted to any extent which can be reasonably anticipated. Nor do they think it possible, except in the event of her being attacked by enemíes of such preponderating force as to render her entering into any engagement in the highest degree imprudent.

Question two is divided into two clauses; the first is, "as to whether there would be any risk of the ship capsizing if she were placed under the conditions mentioned in the previous paragraph, supposing that the water ballast were admitted into the double bottom of the armored citadel. The committee find that under the extreme conditions assumed, the ship, even without water-ballast, would yet have stability, and would, therefore, float upright in still water, and we are of opinion that the stability that she would have in that condition, though small, is, in consequence of the remarkable effects of free internal water in extinguishing rolling, sufficient to enable her to encounter with safety waves of considerable magnitude. The ship under those circumstances, however, would require to be handled with great caution. The admission of water as ballast increases the amount of stability, and is thus of advantage as against steady inclining forces; but on account of the deeper immersion it involves it does not materially increase the range of the stability. When the immersion is such as largely to increase the depth of the water on the middle deck, it appears that the extinguishing effect of such water becomes less vigorous, so that in a seaway the ship would, in the extreme condition, be safer with a moderate than with a very large amount of water admitted as ballast. It must be clearly understood, however, that we should consider the ship in a very critical state if reduced to this condition in the presence of a still powerful enemy. Her speed and power of turning would be so limited as to prevent her being maneuvered with sufficient rapidity to insure her against being effectively rammed, or so as to avoid a well-directed torpedo, while the small residuum of stability she would possess would not avail to render such an attack other than fatal. Her guns would also have to be worked with great caution, and under restrictions imposed by the high angle to which their combined movements would in broadside firing heel the ship. We have already expressed our opinion that it is in a very high degree improbable that the ship would be reduced to this condition, even in a protracted engagement."

The second clause is "whether she would retain a sufficient amount of stability to enable such temporary repairs to be executed as would enable her to reach a port." The committee think that the destruction, implied by the extreme condition assumed, would be such that nothing effective could be done in the way of repairs at sea under any circumstances.

Question three is also divided into sections. The first is, "whether, all points considered, in so far as can be ascertained from the designs and calculations, the *Inflexible* is a safe sea-going vessel." The committee are of opinion that in the intact condition the *Inflexible* is a safe sea-going vessel. The consideration of her safety, when not in an intact condition, properly falls under the investigation involved by the clause which follows.

The second clause is, "whether, when the amount of damage to which the unprotected ends would be exposed in action is borne in mind, sufficient provision has been made to insure, in all human probability, her safety under such conditions. We have first to consider what is 'the amount of damage to which the unprotected ends would be exposed in action.' We do not hesitate to say that the complete destruction implied by riddling and gutting is so extreme an assumption that it may be regarded as a very highly improbable event even in a protracted engagement; yet recognizing the extravagance of one assumption does little toward enabling us to fix a reasonable one, and there is no sufficient basis either of actual experience or of experiment on which to decide what amount of damage to the ends is probable. Nor can we take refuge in adopting and providing for the extreme case as covering all others, because provision cannot be made for the safety of the ship in one way without prejudice to it in another, and to give undue prominence to any one provision for its security becomes a serious error where only a just balance can give the best general result. For example, any extension in the citadel in favor of the unprotected ends would necessitate a corresponding reduction of thickness of the armor on the citadel. To the best of our judgment, the condition represented under the letters *e* or *f* in the Parliamentary papers is that which might be fairly assumed to represent the greatest amount of damage the ship would be likely to suffer in any action. This condition represents the unprotected ends completely riddled and water-logged, but the materials and cork remaining and adding buoyancy. Under *e* the whole of the coal is assumed in place, under *f* it is assumed to be removed. In adopting it we include any state of partial removal of material and partial riddling which may be regarded as its equivalent. We find that the ship, if reduced to this condition, would possess both buoyancy and stability enough to enable her to face all contingencies of weather, and to exercise all her powers, subject, however, to the limitations of speed which may be imposed by the character and position of the wounds in the ends, and which might be very serious in the condition. The united movement of all her guns from the loading to the firing position would not heel her more than 2¼ degrees, and the heel due to her circling at the highest speed attainable would not be an element of danger. The actual range of her stability would be not less than 35 degrees, which is considerably below the standard provisionally laid down by the committee on designs, and referred to in the Parliamentary papers submitted to us. That standard, however, requires revision by the light of more recent investigations of the theory of rolling. It would be, at any rate, inapplicable to the present case, because the very water-logging of the ends, which so reduces the range of stability, has a most remarkable effect in preventing rolling. Should the damage to the ends go beyond what we contemplate, the ship would still be in no immediate danger of being placed *hors de combat*. The transition from the condition *e*, in which she may be said to begin to have her efficiency impaired, to that extreme in which she must be regarded as in a critical state in the presence of an enemy, is necessarily a gradual one, because it follows only the progress of destruction of the ends, and can only be completed with that destruction. It cannot be said that the armored citadel is invulnerable, or that the unarmored ends are indestructible, although the character of the risks they run is different. But in our opinion the unprotected ends are as well able as the armored citadel to bear the part assigned to them in encountering the various risks of naval warfare, and therefore we consider that a just balance has been maintained in the design, so that out of a given set of conditions a good result has been obtained."

The preceding paragraphs give a summary of the report. In the report itself the committee go into details. * * * Among other points the committee refer to the great difficulty which exists in hitting objects at sea just where the gunner wishes. "Among the chief sources of error in an action at sea are, the motion of the attacking vessel, the motion of the attacked vessel, the smoke of both vessels, the rolling and pitching of the vessel forming the gun-platform, the imperfect knowledge of the distance of the object aimed at, the action of the wind in deflecting the shot. As regards the error from imperfect knowledge of distance, the means of ascertaining which at sea are at present very rude, it is to be remembered that the high speed at which modern ships of war engage causes them to change their distance with great rapidity. For instance, two vessels approaching to or receding from each other at the rate of twelve knots vary their distance apart at the rate of 40 feet per second. Errors of range from this and other causes are, as might be expected, much in excess even of errors of direction, and a target which is low and wide, like the ends of the *Inflexible*, is much more difficult to hit than one which is high and narrow. Rifled projectiles are very devious after ricochet, so that if they fall short of the mark they have little chance of producing effect, while, if they go over, they are equally thrown away. As regards the effect of the rolling of the ship upon the accuracy of fire, the gun is generally fired at the middle of the roll, when the deck is nearly horizontal. At such time the speed of the roll is the highest and the disturbing effect greatest, rendering it a matter requiring great skill and practice to make anything like accurate firing, even at short ranges.

"It is to be regretted that there are no exact records of the results of naval firing. The custom is to record by ocular estimate made from the ships from which the practice is carried on. We are, therefore, only in a position to say that such records as we have had before us confirm, so far as they go, the conclusion we have arrived at as to the improbability of a very large number of shells being planted in the unprotected ends. The unarmored structures in question arise 9 feet above the water, and extend 7 feet below it in the fighting condition of the ship. Their length is about 110 feet in front and in rear of the central citadel respectively. The structures to be destroyed are thus about 220 feet long in broadside view, by 16 feet high, nearly one-half of which is below the water-level, and can only be reached by shells entering obliquely or when the side of the ship is partially laid bare by the action of the waves. Shells striking at or about the water-line may rip the middle deck and let water into the compartment pierced, although it is expected that the canvas and oakum with which the coffer-dam is charged will materially obstruct the inflow. Shells cannot, however, lift and blow out all the materials packed in the compartments except they enter very obliquely, which implies long range and consequent greatly increased inaccuracy of fire. The immersion of the vessel occasioned by the admission of water would in itself add to the difficulty of reaching and removing the materials below the water. Viewed obliquely or directly ahead or astern, one or other of the unarmored ends would derive a considerable amount of protection from the central battery. Shells very rarely make large breaches where they enter the side of an unarmored vessel. The process of ignition of the bursting charge, commenced on impact, takes a sensible time to complete, and the velocity of the shell being high, and but little diminished by the slight resistance offered by thin plating, it passes on at least 6 feet to 10 feet—corresponding roughly with a time of $\frac{1}{150}$th part of a second—before actual explosion takes place. It therefore enters as a shot by a hole of its own figure, and not greatly exceeding it in size, and from the point at which explosion takes place the fragments go forward in a cone of dispersion, expending themselves in indenting and cutting intervening bulkheads and the opposite side of the ship. The cork wall and coffer-dam being only 6 feet thick in all, most of the shell may be expected to pass through them and to open in the spaces inside, unless striking very obliquely. The most effective armament to bring against the *Inflexible's* ends alone would undoubtedly be one of numerous shell-guns. In an iron-clad such an armament is incompatible with armor of a thickness to be of the least avail against the *Inflexible's* guns. It must be a broadside armament, and this carried at a sufficient height above the water-level to be worked in a sea-way would involve an extended area of armor incompatible with great thickness. We cannot, therefore, conceive an enemy deliberately adopting the tactics of using or building such iron-clads with a special view to attack the unprotected ends only, nor, considering the difficulty of naval fire, do we think firing could be very successfully directed at particular portions of the ship, such as the ends, instead of against the ship as a whole. If called on to engage land forts mounting numerous shell-guns, the exceptional range of her great guns would enable the *Inflexible* to choose her distance, and to engage beyond the range at which guns of such inferior power could strike frequently or with effect. She could also, in case of need, always retire out of action, and choose her own time for renewing an engagement. Probably the most effective mode of bringing a destructive shell-fire to bear on the *Inflexible* would be by a flotilla of gunboats concentrating their fire upon her."

The committee compare the *Inflexible* as she is with a new *Inflexible*, having her armored citadel drawn out in length so as to render her much more nearly, if not absolutely, independent of the unprotected ends, the thickness of the armor being of course reduced in proportion to its extended area. It may be assumed that an addition to the citadel of at least 30 feet in length would be necessary to satisfy this condition. The thickness of armor would then be in the new ship 21 inches as compared with 24 inches in the present one. If we now suppose the actual *Inflexible* to meet in conflict the new *Inflexible*, both being armed with the most powerful guns existing, which are capable of piercing 22 inches of armor, the new ship with her 21 inches of armor would be in immediate danger of receiving a mortal wound by the penetration of her citadel, where the vital parts are so crowded together that one blow might be fatal, and would almost certainly seriously cripple her. The possibility of ultimately crippling the enemy by a multiplicity of slight wounds in his unarmored extremities would do little or nothing practically to diminish the disparity arising from the fact that one ship possessed penetrable and the other impenetrable armor. Great accuracy of fire would only render it more certain that the penetrable citadel of the supposed new ship would be struck and pierced before the destruction of the ends of the other ship could be completed. In such a case the conclusion seems inevitable that the actual *Inflexible* would be greatly the superior vessel, and if any increase in the power of existing guns takes place, the same argument would induce a shortening and thickening of the armored citadel walls rather than the reverse. Nor would there appear to be a corresponding loss of advantage to the actual *Inflexible* as compared with the supposed new one, in the event of her having to engage weak iron-clads or unarmored vessels which might be able to bring against her numerous shell-guns equally useless against 21-inch and 24-inch armor, and therefore only able to attack the unprotected ends, because, conceding to the *Inflexible* the same accuracy of fire which must be assumed for the enemy before we can contemplate the fire of the shell-guns destroying the unprotected ends, either *Inflexible* would have speedily planted among her opponents the few blows necessary to disable them.

The committee conclude with the following recommendations:

"1. Looking to the unexpectedly great demand on the ship's longitudinal stability which may possibly ensue under the circumstances referred to * * we think it deserving of careful consideration whether it will not be advisable to extend the cork chambers longitudinally to the extreme ends of the ship and upward to the upper deck.

"2. We suggest for consideration that the travel of the guns on their slides should be reduced, and that they should either be so placed in the turrets that they may range equally on each side of the center or otherwise, that a slight alteration of the distribution of weight in the turret should be made for the purpose of bringing the center of gravity of the turrets and guns over the center of evolution when the guns are at the middle of their range on the slides. At present the inclining moment due to the running out of the guns is over 1,600 foot-tons, and becomes a serious element of danger as the ship approaches the riddled and gutted condition. It might by the measures proposed be reduced to little more than one-third of that amount.

"3. We note that the total pumping power which the *Inflexible* will possess, including the use of the circulating-pumps, is capable of throwing out 4,500 tons of water per hour; and it is understood that in providing that amount of power a large increase (probably in the ratio of two to one) has been made in the proportion of pumping power to displacement hitherto adopted.

"Notwithstanding this increase, the pumping-power is very disproportioned to the enormous extent of the leakage to which a modern ship of war is subject in action. The 4,500 tons per hour might be thrown out by 200 horse-power well applied, and it appears to us to be a conclusion not to be admitted except after the most exhaustive inquiry that a ship which has at her disposal for motive purposes 8,000 horse-power should not have more than 200 available for pumping purposes when she has been struck in a vital part by ram or torpedo.

"We do not pretend to say how large a proportion of the engine-power could be made available, but we think it right to draw attention to the subject as one demanding grave consideration.

"4. Having expressed the opinion that future progress in the construction of armored ships lies in the adoption of an efficient system of armor, combined with some cellular or equivalent structure, we cannot but feel desirous that the best mode of dealing with shot and shell in the unarmored portions of the ends should be made the subject of careful and systematic experimental inquiry. Such inquiry should embrace not only the form and distribution of the shells themselves, but also the best material, if any, with which they might be wholly or partially filled. It is to be regretted that a similar recommendation of the committee on designs was so imperfectly adopted; but even the partial experiments made in 1872 added materially to our information, and, so far as they went, they justified the adoption of the cork-filled cells and oakum-and-canvas-packed coffer-dams of the *Inflexible*. If, however, as we believe, the time has come when cellular structures must form an important feature in a ship's design, the area and scope of such experiments should be greatly enlarged; and we strongly recommend this subject to the serious consideration of their lordships.

"5. Results which have been obtained in the course of the experiments at Torquay on the resistance of ships show that a considerable increase of the extreme breadth of the *Inflexible*, if accompanied by a corresponding fining of the ends so as to keep the displacement unaltered, would, if anything, diminish the resistance of the intact vessel to propulsion at full speed. Supposing the ship thus increased in beam by 10 feet and the citadel shortened so as to retain the same perimeter and thickness of armor, her transverse stability would then be about doubled in the *e* and *f* conditions, and in the riddled and gutted conditions would be more than it now is in condition *e* or *f*.

"Her longitudinal stability in the riddled and gutted condition would be reduced 10 per cent., but would not be diminished in condition *e* and scarcely appreciably so in *f*. The increase of beam would also add to the area of the citadel in a horizontal plane, and thus increase the buoyancy in the riddled condition.

"We note that the beam of the *Inflexible* was limited by the consideration of the width of the docks available for her repair, but we doubt if this consideration ought to outweigh the great advantages which a further increase of beam would give to vessels of the *Inflexible* type. We are the more inclined to doubt it because at present docks capable of accommodating vessels of any breadth can be constructed of iron rapidly, and at no serious cost in comparison with that of such vessels as the *Inflexible*.

"We therefore, in conclusion, desire to bring under the very serious consideration of their lordships the necessity, before proceeding with the construction of more vessels of the type of the *Inflexible*, of thoroughly investigating whether by more beam their safety may not be largely increased without impairing their speed and efficiency."

It must have been apparent to every one who has read the proceedings in this case and considered the subject-matter, that the real question is not so much what might possibly happen in a certain extreme supposed case, as whether this extreme condition lies within the limits of reasonable probability; for, assuming that the probability is infinitesimally small, the question of the resulting effects may be entirely disregarded. Upon this point the committee use very clear and decided language; they say that "such an extreme condition is in a very high degree improbable even in a protracted engagement." Of course this is a presumptive opinion, but it is sustained by opinions of able naval officers of high rank whose letters on the subject have been published; and as the question is not a naval architect's question, the opinions of the committee, backed by professional officers, is, in the absence of experience, the best authority attainable.

It is impossible to secure immunity from risk in battle. If this much-discussed question should ever be practically tested in actual warfare, the *Inflexible* in like manner with the *Nelson* and *Northampton* having unprotected ends, as well as other British armored ships, if engaged by a powerful enemy, will encounter greater risks of being sunk from the attacks of rams and torpedoes than from the effects of artillery-fire.

THE AJAX AND AGAMEMNON.

The *Inflexible* having been accepted as the type of the British future line-of-battle ship, two others of smaller dimensions have been put in process of construction, viz, the *Ajax*, which was laid down at the Pembroke dock-yard in 1876, and the *Agamemnon*, commenced at Chatham in the same year; besides which a third ship of the same type is provided for in the navy estimates of 1877.

After so full an account of the *Inflexible*, any detailed description of these two sister ships would be a mere repetition. By reference to the drawings it will be seen that the leading features of the design are the same. They differ only in dimensions, power of offense and defense, in motive machinery, and in minor details.

The cost to build as well as to maintain one of them will be considerably less than for the *Inflexible*, and they will be much less difficult to maneuver. The length between perpendiculars is 40 feet, and the beam 14 feet, less; the mean draught of water 23 feet 6 inches, against 24 feet 5 inches; and the displacement 8,492 tons, against 11,406. The length of the citadel is 104 feet instead of 110 feet, and the armored deck outside of the citadel is 5 feet 10 inches below the water-level, instead of 7 feet; and the free-board is 9 feet 6 inches. The cork chambers extend forward of the citadel 30 feet, and abaft it 37 feet; in depth they are 12 feet, 6 feet below water and 6 feet above. The coffer-dam is of the same length, and 2 feet wide.

The defense is considerably reduced; the armor on the water-line being, first, 10 inches of teak next to the iron hull, faced by 8 inches of iron; then 9 inches of teak faced with 10 inches of iron, making in all 18 inches of iron and 19 inches of teak, against 24 inches of iron and 25 inches of wood in the larger ship.

The armament will consist of four 38-ton guns, worked on the hydraulic system, against four 80-ton guns. The maximum indicated horse-power is to be 6,000, and the speed is expected to come up to that of any armored ship afloat.

The following are some of the dimensions and particulars:

Length between perpendiculars	280 feet.
Length over all	301 feet 9 inches.
Breadth, extreme	66 feet.
Draught of water, forward, loaded	23 feet.
Draught of water, aft, loaded	24 feet.
Depth of hold from top of citadel	21 feet 4 inches.
Area of immersed midship section	1,402 square feet.
Displacement	8,492 tons.
Free-board	9 feet 6 inches.
Length of citadel	104 feet.
Distance from stem to citadel	88 feet 6 inches.
Depth of citadel	15 feet 6 inches.
Thickness of side of citadel	3 feet 1 inch.
Distance between decks, lower	6 feet 6 inches.
Distance between decks, upper	6 feet.

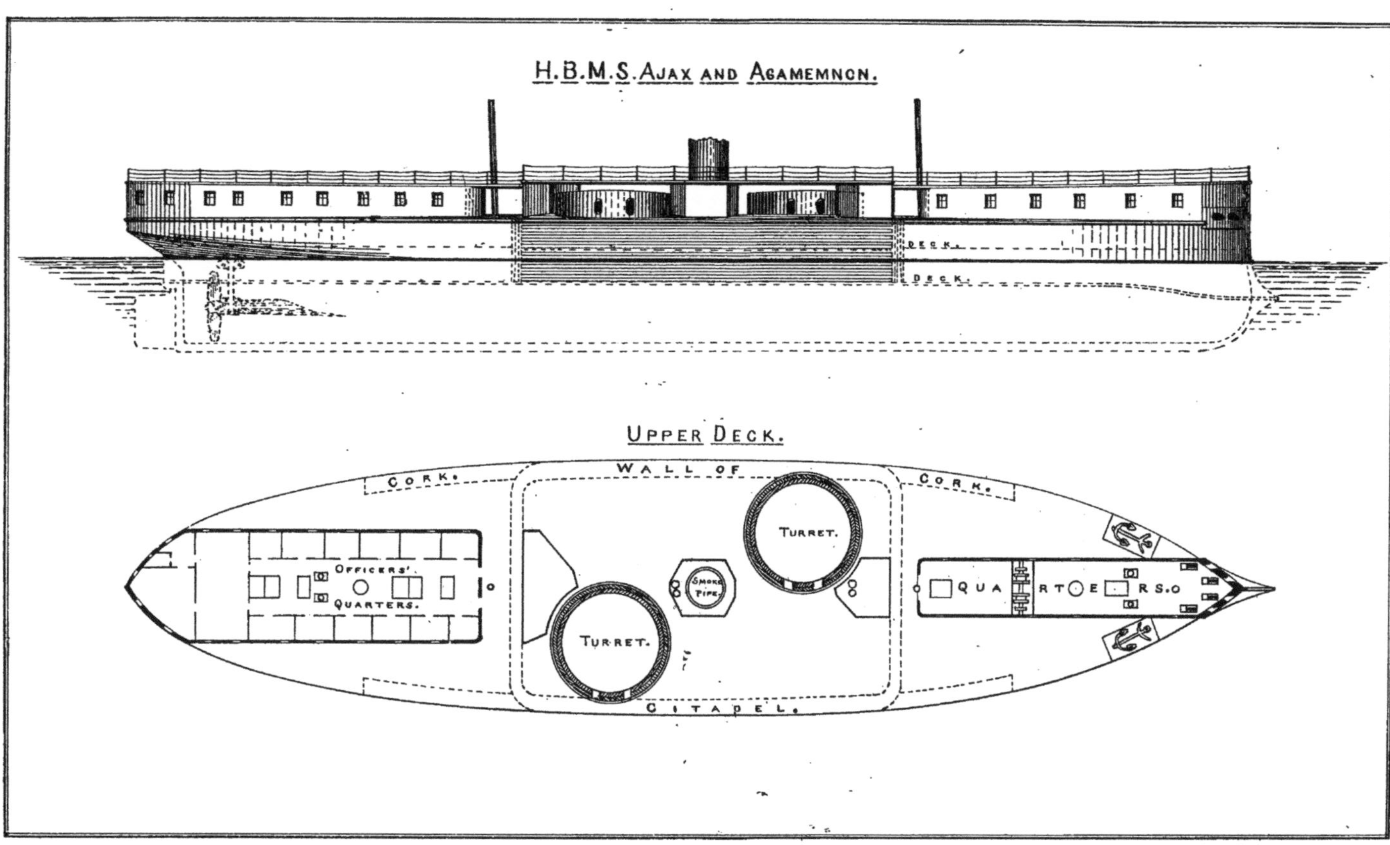
H.B.M.S. AJAX AND AGAMEMNON.
DECK.
DECK.
UPPER DECK.
WALL OF
CORK.
CORK.
CITADEL.
TURRET.
TURRET.
SMOKE PIPE.
OFFICERS' QUARTERS.
QUARTERS.

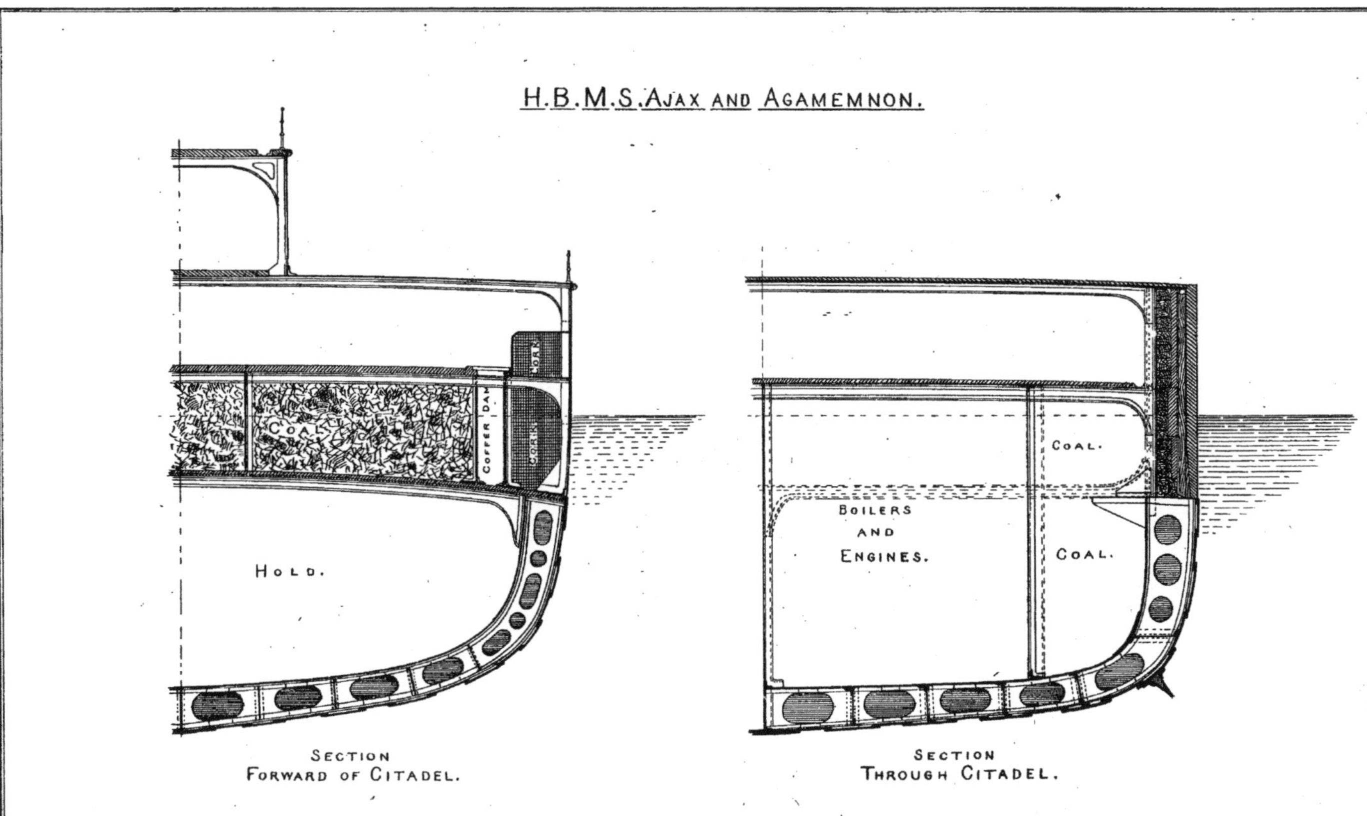

Section
Forward of Citadel.

Section
Through Citadel.

Depth of armored deck below water-line	5 feet 10 inches.
Number of turrets	2
Diameter of turrets, external	30 feet.
Height of top of turrets above water-line	17 feet 6 inches.
Projection of ram	9 feet.
Depth of point of ram below water-line	9 feet.
Width of forward superstructure	16 feet.
Length of forward superstructure	82 feet.
Width of after superstructure	29 feet.
Length of after superstructure	92 feet 6 inches.
Height of superstructure, extreme	7 feet 9 inches.
Distance between outer and inner hulls, amidships	3 feet 2 inches.
Distance between outer and inner hulls, near bilge	2 feet 8 inches.
Distance between outer and inner hulls, near water-line	3 feet 10 inches.
Citadel-armor, at water-line, 10 inches iron, 9 inches wood, 8 inches iron, 10 inches wood, and $1\frac{1}{4}$ inches iron; total thickness	3 feet $2\frac{1}{4}$ inches.
Armament	4 38-ton guns.
Number of engines, (inverted 3-cylinder)	2
Number of cylinders	6
Diameter of cylinders, high and low	54 inches.
Stroke of pistons	3 feet 3 inches.
Indicated horse-power, maximum	6,000
Diameter of crank-shafts	$14\frac{1}{2}$ inches.
Number of screw propellers	2
Diameter of screw propellers	18 feet.
Number of boilers, (return-tubular)	10
Number of furnaces	28
Total grate-surface	647 square feet.
Total heating-surface	18,062 square feet.

PART III.

THE DEVASTATION.

The *Devastation*, as designed in 1869, was a low free-board, sea-going turret-ship. She was the first of this character which it was determined to build from plans prepared at the admiralty. The great question of that day in England, "Turret *versus* Broadside" for mounting heavy guns in sea-going armored ships, will still be remembered by all persons informed in the progress of naval construction. So strong were the supporters of certain views with regard to the former system that, notwithstanding the continued opposition of the chief constructor of the navy, the order had been given for the *Captain*, a vessel embodying these views, to be designed and built by a private firm. The *Devastation* may be regarded as designed to compete with the *Captain*. She represented Mr. Reed's views of what a sea-going monitor should be. Low sides were adopted, but not in combination with rigging and sails, as was the case in the ill-fated *Captain*. As originally designed, the *Devastation* was 285 feet long between the perpendiculars, had 62 feet 3 inches beam, and 26 feet 1½ inches mean draught. Her sides, which, except right forward, arose only to a height of 4 feet 6 inches above the surface of the water, were protected by armor 12 inches thick. Her armament consisted of four 25-ton guns, mounted in pairs in two turrets, one at each end of a raised breastwork or platform, which extended about 150 feet along the middle of the upper deck. The guns were thus elevated to a height of some 14 feet above the surface of the water. The turrets were protected by armor 12 inches and 14 inches thick, and the breastwork by armor 10 inches and 12 inches thick. A forecastle extended forward from the fore end of the breastwork, at a height of 9 feet 3 inches above the water-line, but in wake of this forecastle the armor on the sides dropped to a height of only 6 inches above the surface of the water, this corresponding to the level of an armored deck. All the necessary hatchway openings, &c., into the ship were led up by iron trunks to a light flying deck, which extended between the two turrets, somewhat overlapping each.

The vessel was to be propelled by two screws, one under each counter, and each screw was to be worked by separate pairs of engines, so that the ship might be driven by their conjoint action or by either of them working singly. The total power of the engines was to be 5,600 horses, indicated, and the estimated speed was 12.5 knots. She was designed with a spur bow, the point of the ram advancing some 10 or 12 feet under water. The strength of the hull was arranged so as to give great support to the bow when ramming. She had a double bottom, the space between the two skins, some 3 feet deep, being divided, as is usual, into a number of separate water-tight cells, so that injury to the outer bottom could only result in the filling of one or more of these. The hold of the vessel, also, was divided into a number of compartments by water-tight bulkheads across the ship; so that even in the event of a clean breach being made through both bottoms, such as might be effected by a torpedo, for instance, she might still have a considerable chance for escape, from being able to confine the water to the compartment or compartments into which the breach was made.

This was the *Devastation* as first designed, and the work of building was being rapidly pushed forward at Portsmouth dock-yard. The *Cap-*

tain in the mean time was winning a high reputation. She had been launched in March, 1869, and had toward the close of 1870 made one or two successful cruises. True, when completed, it was found that a very important element in connection with the design, the weight of the ship, and consequently the draught of water and height of free-board, had been loosely calculated; but the error arising therefrom, though by no means small, was not regarded as serious; and as it did not apparently much influence her sea-going qualities, no special notice was taken of it. Her stability was never doubted by her designers; nor, indeed, was her critical state ever properly realized by any one; any doubt that may have existed was smothered by the confidence of her advocates. The chorus of praise which she elicited on all sides continued to increase, and the question, what the type of British war-ship for the future should be, was supposed to be settled in her beyond dispute. Then came the dreadful news that she had gone down during the night between the 6th and 7th of September, off Cape Finisterre. The wind had not been unusually violent; the sea had not been exceptionally heavy; there were no extenuating circumstances; she had not bravely battled with ordinarily rough weather; she was proceeding confidently under steam and sails when, in an ordinary squall, she displayed once and for all her subtle and treacherous character by slowly turning over and becoming the coffin of nearly the whole of her crew, some five hundred men, including a large number of accomplished officers. The people of England were almost panic-stricken at this terrible news. How it could have occurred with the comparatively widespread knowledge relating to the subject and the actual facts and figures of her special case before them, it was difficult to conceive. To remove the doubt which immediately arose as to the safety of the other armored ships, and particularly as to that of the *Devastation*, a special committee was appointed to examine into the designs of these vessels. This committee, which consisted of many of the highest professional and scientific authorities in England, met in January, 1871, and made their report concerning the *Devastation* class early in the following March. After numerous calculations and investigations they came to the conclusion that the stability of the *Devastation* was everything that could be desired, and reported that "ships of this class have stability amply sufficient to make them safe against the rolling and heaving action of the sea." The committee, however, agreed in recommending a plan which the constructors of the admiralty had proposed with the view of making safety doubly safe.

By this plan, which was afterward adopted, the stability of the ship has been very considerably increased; and besides this, the accommodation of the officers and men has been very largely augmented. The plan consisted in the addition of the side superstructures. They were formed by continuing the ship's side upward with light framing, as high as the level of the top of the breastwork, and continuing the breastwork deck over to the sides. The structures were extended aft on each side a considerable distance beyond the end of the breastwork, providing two spacious wings, which add largely to the cabin accommodation. Some other alterations in the design which were suggested by the committee were carried out; among them may be mentioned the introduction of athwartship armor-plated bulkheads, so as to afford additional protection to the magazines and engines. An alteration of considerable importance had been made some time before, consisting in the substitution of 35-ton guns for 25-ton guns, as originally arranged. With these and some other slight alterations the vessel was completed. Her mean draught of water is now 26 feet 8 inches. Her height of side above water-line is 10 feet 9 inches, except right forward in wake of the forecastle, where it is 8 feet 6 inches, and right aft abaft the superstructure, where it is only 4 feet.

The following tables give all the dimensions and data necessary to be known of this powerful vessel:

Statement of dimensions, weights, and other particulars of Her Majesty's ship Devastation.

Dimensions, &c.	Estimate of April, 1869.	Estimate of November, 1869.	Estimate of January, 1871.	Actual dimensions, &c., as completed in April, 1873.
Length between the perpendiculars	285 feet 0 in.	285 feet 0 in.	285 feet 0 in.	285 feet 0 in.
Length of the keel for tonnage	246 feet 3⅜ in.	246 feet 3⅜ in.	246 feet 3⅜ in.	246 feet 3⅜ in.
Breadth, extreme	62 feet 3 in.	62 feet 3 in.	62 feet 3 in.	62 feet 3 in.
Breadth, for tonnage	58 feet 0 in.	58 feet 0 in.	58 feet 0 in.	58 feet 0 in.
Depth in hold	18 feet 0 in.	18 feet 0 in.	18 feet 0 in.	18 feet 0 in.
Burden, in tons	4, 406 57-94	4, 406 57-94	4, 406 57-94	4, 406 57-94
Draught of water:				
Forward	25 feet 9 in.	25 feet 9 in.	25 feet 9 in.	26 feet 3 in.
Aft	26 feet 6 in.	26 feet 6 in.	26 feet 6 in.	27 feet 1 in.
Mean	26 feet 1½ in.	26 feet 1½ in.	26 feet 1½ in.	26 feet 8 in.
Displacement, in tons	9, 062	9, 062	9, 090	9, 298
Area of midship section, in square feet	1, 449	1, 449	1, 454	1, 487
Height of port-sills from load water-line:				
Fore turret	13 feet 6 in.	13 feet 6 in.	13 feet 6 in.	12 feet 11 in.
After turret	13 feet 2 in.	13 feet 2 in.	13 feet 2 in.	12 feet 7 in.
Height of upper deck at side water-line:				
Forward	9 feet 3 in.	9 feet 3 in.	9 feet 3 in.	8 feet 6 in.
Amidships	4 feet 6 in.	4 feet 6 in.	4 feet 6 in.	10 feet 9 in.
Engines:				
Nominal horse-power	800	800	800	800
Indicated horse-power	5, 600	5, 600	5, 600	6, 633
Speed, per hour, in knots	12. 5 (Estimated.)	12. 5 (Estimated.)	12. 5 (Estimated.)	13. 84 (Actual.)
Coals, number of tons	1, 700	1, 600	1, 600	1, 350
Water:				
Number of tons	16	16	16	30
Number of weeks' consumption	2	2	2	3
Provisions:				
Number of tons	9. 5	9. 5	9. 5	19
Number of weeks' consumption	4	4	4	6
Complement of men and officers	250	250	250	329
Armament	4 25-ton guns.	4 30-ton guns.	4 35-ton guns.	4 35-ton guns.
Total weight of armor, in tons (including fastening)	2, 307	2, 329	2, 482	2, 581
Total weight of backing, in tons (including fastening)	306	306	306	314
Depth of armor below water-line, amidships	5 feet 0 in.	5 feet 0 in.	5 feet 0 in.	5 feet 6½ in.
Height of armor above water-line:				
On sides — Amidships	4 feet 2 in.	4 feet 2 in.	4 feet 2 in.	3 feet 7½ in.
On sides — Forward	0 feet 6 in.	0 feet 6 in.	0 feet 6 in.	nil.
On breastwork — Amidships	11 feet 5 in.	11 feet 5 in.	11 feet 5 in.	10 feet 10½ in.
On breastwork — At fore end	11 feet 9 in.	11 feet 9 in.	11 feet 9 in.	11 feet 2½ in.

	Armor.	Backing.	Armor.	Backing.	Armor.	Backing.	Armor.	Backing.
	Inches.	*Inches.*	*Inches.*	*Inches.*	*Inches.*	*Inches.*	*Inches.*	*Inches.*
Thickness of armor and backing:								
On sides	12 & 10	18	12 & 10	18	12 & 10	18	12 & 10	18
On bulkheads at break of deck forward	12	16	12	16	12	16	12	16
On bulkheads in hold	......	......	......	......	......	......	4, 5 & 6	10
On breastwork	12 & 10	18 & 16	12 & 10	18 & 16	12 & 10	18 & 16	12 & 10	18 & 16
On turrets	14 & 12	15 & 17	14 & 12	15 & 17	14 & 12	15 & 17	14 & 12	15 & 17

Thickness of skin-plating behind armor:				
On sides	1½ & 1¼	1½ & 1¼	1½ & 1¼	1½ & 1¼
On bulkheads at break of deck forward	1½	1½	1½	1½
On breastwork	1¼	1¼	1¼	1¼
On turrets	1¼	1¼	1¼	1¼
Thickness of deck-plating:				
On upper deck — Amidships	2	2	3	3
On upper deck — Aft	1¾	1¾	2	2
On belt-deck	3 & 2½	3 & 2½	3	3
On deck over breastwork	1	1	1	2

Statement of dimensions, weights, and other particulars of Her Majesty's ship Devastation, as completed for sea, and as estimated at various dates.

Weights, &c.	Estimate of Apr., 1869.	Estimate of Nov., 1869.	Estimate of Jan., 1871.	Actual weights, &c. as completed in Apr., 1873.
	Tons.	*Tons.*	*Tons.*	*Tons.*
Water for two weeks; Tare of tanks	18.8	18.8	18.8	40.0
Provisions, spirits, &c., for four weeks; Tare of casks	12.5	12.5	12.5	24.0
Officers' slops and stores; Tare of casks, boxes, cases, &c	12.0	12.0	12.0	12.0
Officers, men, and effects	32.0	32.0	32.0	42.0
Mast and derrick for hoisting boats	....	....	20.0	20.0
Cables	87.5	87.5	87.5	70.0
Anchors	25.0	25.0	25.0	23.5
Boats	12.0	12.0	12.0	12.0
Warrant-officers' stores	34.0	34.0	34.0	50.0
Armament	355.0 (4 25-ton guns)	422.2 (4 30-ton guns)	512.2 (4 35-ton guns)	514.5 (4 35-ton guns)
Total weight of rigging, guns, and ship's stores	588.8	656.0	766.0	808.0
Engines, and boilers with water in, including engines for turrets, ventilating and fire service, spare gear, &c	970.0	952.0	982.0	1,064.0
Engineers' stores	15.0	15.0	15.0	23.0
Coals	1,700.0	1,600.0	1,600.0	1,350.0
Total weight of equipment to be received on board	3,273.8	3,223.0	3,363.0	3,245.0
Weight of hull	2,874.0	2,894.0	2,487.0	*2,882.0
Weight of protective deck-plating, including glacis-plates and armored skylights	413.0	413.0	522.0	556.0
Weight of armor, exclusive of turret and pilot-tower armor	1,604.0	1,626.0	1,542.0	1,629.0
Weight of backing, exclusive of turret and pilot-tower backing	266.0	256.0	256.0	254.0
Weight of turrets	590.0	581.0	592.0	622.0
Weight of pilot-tower	....	....	110.0	110.0
Weight of conning-hoods	15.0	15.0	....	....
Total displacement required	9,035.8	9,008.0	8,872.0	9,298.0
Total displacement per drawing	9,062.0	9,062.0	9,090.0	9,298.0
Difference	26.2	54.0	218.0	Nil.

*This includes the superstructure added in January, 1871, and the additions recommended by the committee on designs.

Estimated consumption of coal in the Devastation, at speeds of ten and twelve knots.

	Speed of ten knots an hour.		Speed of twelve knots an hour.	
	Total distance steaming.	Numbda consumption.	Total distance steaming.	Number of days' consumption.
	Knots.	*Days.*	*Knots.*	*Days.*
Coal carried, 1,600 to 1,700 tons: Statement in Mr. Reed's memorandum on new designs for iron-clad ships, dated 2d March, 1869, page 311, report of committee on designs of ships (printed for Parliament)	4,320	18	2,880	10
Coal carried, 1,400 tons: Calculations based on results of measured-mile trial, 31st October, 1872	4,580	19	2,890	10
Coal carried, 1,600 tons: Calculations based on results of measured-mile trial, 31st October, 1872	5,236	21.8	3,300	11.5
Coal carried, 1,400 tons: Calculations based on six hours' trial, 15th April, 1873	4,876	20.31	3,109	10.79
Coal carried, 1,600 tons: Calculations based on six hours' trial, 15th April, 1873	5,572	23.21	3,553	12.33

MOTIVE MACHINERY.

The *Devastation* is propelled by twin screws, each driven by an independent pair of engines. These engines, which have been constructed by Messrs. John Penn and Sons, of Greenwich, are of that firm's direct-acting trunk type, contracted for prior to the adoption of compound engines, and they have cylinders 88 inches in diameter, and trunks 36½ inches in diameter; the trunks reducing the effective area of the pistons to that due to a diameter of 80 inches. The cylinders are steam-jacketed, both at sides and ends, and the stroke of pistons is 3 feet 3 inches. The engines are fitted with expansion-gear, which enables the steam to be cut off at any required part of the stroke. The main slide-valves are double-ported, and fitted with an equilibrium-ring. The expansion-valves are of the gridiron form, with a variable stroke and cut-off. The admission of steam to the engines is regulated by equilibrium-valves, which are worked by screws and suitable gearing, led away to the starting-platform. To insure ready handling of the engines, small auxiliary slide-valves are fitted to each cylinder. Each pair of engines is fitted with a surface-condenser containing 5,432 ¾-inch tubes 6 feet 3½ inches long; the condensing surface for each pair of engines being thus 6,710 square feet. The tubes are packed with screwed glands and tape packing. The air-pumps and the circulating-pumps are double-acting, and are worked direct from the pistons. The condensing water is drawn through the tubes and the steam admitted to the outside. The crank-shafts are in two pieces, with solid couplings forged on. The turning-gear consists of a worm-wheel and worm, worked by hand by means of a long ratchet-lever. The disconnecting-coupling is fitted with four steel pins, which can be drawn out of gear by means of screws and ratchet-spanners. The thrust of the propellers is taken by a bearing fitted with ten movable collars. The screw-propellers are 17 feet 6 inches in diameter, and have 19 feet 6 inches pitch, and are so fitted that the pitch can be varied from 17 feet to 22 feet. The number of blades to each is four, and the propellers are formed on the Griffith principle. The boilers are eight in number, of the old kind, and contain thirty-two furnaces, the four boilers in the forward fire-room having four furnaces each, while of the four boilers in the after fire-room two have three and two have five furnaces each. The length of bars is 6 feet 6 inches, and the width of furnaces 3 feet 2 inches, the total fire-grate area being thus 779 square feet. The boilers contain in all 2,592 tubes 3½ inches in diameter and 6 feet long. The working pressure of steam is 30 pounds per square inch. The total heating surface is 17,806 square feet. A superheater is fixed in each chimney, of which there are two, the total superheating surface exposed being 1,866 square feet. The chimneys are telescopic, and are fitted with hoisting-gear and shell-proof gratings at the bottom. The length in the ship occupied by the engines is 32 feet, and that by the boilers 80 feet, this latter length being divided into two equal compartments, separated from the engines and each other by water-tight bulkheads. Telegraphs are fitted between the engine-rooms and the bridge, and, in addition to the ordinary means of ventilation, fans are fitted, driven by independent engines. A powerful fire-engine is provided with pipes leading to all parts of the ship. Engines are also fixed for moving the capstans and hoisting the ashes. The weight of engines and boilers complete, with water in boilers and condensers, and including spare gear and all the fittings above enumerated, is 985 tons, or but 388.6 pounds per indicated horse-power developed when working at full power on the official trial. The following results were obtained on the measured mile, September 2, 1872:

	Ft.	In.
Draught of water forward	26	4
Draught of water aft	26	6
Immersion of upper edge of screw	7	2

Pressure of steam in engine-room, 27 pounds.

	Full power.	Half power.
Revolutions	76.76	63
Mean pressure in cylinders, starboard	22.066	12.88
Mean pressure in cylinders, port	21.53	14.41
Indicated horse-power, starboard	3,359.21	1,566.03
Indicated horse-power, port	3,278.5	1,838.89
	Knots.	Knots.
Speed of vessel	13.839	11.909

Immersed midship section, 1,460 square feet at 26 feet 5 inches draught. Coefficient, 582.

During the full-power six-hour trial there were developed by the two pairs of engines 5,678 horse-power, and the areas of grate, heating, and condensing surface, per indicated horse-power, were as follows:

	Square feet.
Fire grate	0.137
Heating surface	3.136
Heating surface, including superheating surface	3.465
Condensing surface	2.363

The trial cruises of the *Devastation*, to ascertain the degree of success attained in the design as an engine of war, as well as her sea-going qualities, were made in the summer of 1873; and it may be said with confidence that never before did the proceedings of any single vessel elicit so large an amount of public interest as did those of this ship. The novelty of her design as an ocean-cruising man-of-war, her odd appearance, and her fighting power, formed continued topics of discussion in the scientific and other papers, but the real source of interest to the English people was doubtless to be found in the fact that the vessel was looked upon by the general public as belonging to the same type as the unfortunate *Captain*. Hence, notwithstanding the vital points of difference between the two vessels, to which attention had been repeatedly drawn, her trials were watched with an interest amounting almost to anxiety.

The preliminary trials had reference principally to the performance of the engines, boilers, turrets, and other machinery. The great importance of the first of these will be evident, since the vessel is an ocean-going cruiser, without masts and sails—*i. e.*, she is entirely dependent on her engines for propulsion. As has been seen, at the full-speed measured-mile trial a speed of 13.8 knots per hour was obtained, the engines indicating 6,637 horse-power; and from a series of continuous steaming trials at various speeds, it was shown, with her full supply of coal, what distances could be run, the result of which have been recorded previously.

The gunnery trials were made subsequently at the usual testing-ground off the Isle of Wight. The guns are capable of being raised and lowered by hydraulic pressure through a height of 20 inches, and may be thus placed for firing so as to obtain any desirable degree of elevation or depression in combination with small port-holes. The projectiles used in the trials were 700-pound Palliser cored shot, with a battering charge of 110 pounds of pebble-powder. At the trial the guns were fired first with extreme elevation, and then with extreme depression, in all directions around the ship. During two or three trials made by the vessel off Portland and Queenstown, the difficulty of judging of her behavior, with reference to the seas inducing that behavior, as compared with the behavior of ships of ordinary form under similar circumstances, suggested the desirability of prosecuting the ocean trials in company with some ship or ships of about the same dimensions but of less unusual type. Carrying out this idea, the vessel was placed in company with the *Agincourt* and *Sultan*, and thus made to form part of a division of the channel squadron. The *Agincourt* is one of the early iron-clads, having been built in 1862–'65. She is 400 feet long, and is somewhat heavily rigged with five masts. She is completely protected by armor 5½ inches thick. Her armament consists of twenty-eight pieces in two rows, after the old style of frigates; but although the thickness of her armor and the weight of her guns are now out of date, she is claimed to be one of the best sea-boats in the whole fleet. The *Sultan*, on the other hand, is one of the more modern iron-clads. She is short, not much longer than the *Devastation*, and is rigged with three masts as a ship. Her armament, consisting of twelve guns, is mounted in a central two-storied battery and protected by thick armor. The water-line also is protected by a belt of thick armor.

A scientific gentleman who was on board the *Devastation* during these sea trials wrote a highly interesting and valuable account of the proceedings, and, as a matter of interest in relation to the behavior of this class of vessels at sea, the notable points of this letter are extracted, as follows:

The squadron, consisting of these three vessels, put to sea from Plymouth Sound at the end of August, 1873; the programme laid down being to proceed to Bear Haven, on the southwest coast of Ireland, and from this point make occasional cruises into the open Atlantic, as suitable weather should occur. This programme was pretty strictly adhered to in all respects. The vessels arrived and anchored off Bear Haven on the 2d of September, after a cruise of four days, during which many points of interest came out, although no very heavy weather was met with. For purposes of comparison in pitching and lifting, &c., the *Sultan* had the height of the *Devastation's* upper deck at side painted on her in a broad white stripe, so that the behavior of the two ships might be quickly appreciated apart from the records of instruments. The lowness of the extremities of the *Devastation* gives a great deal of interest to the pitching and lifting (really the longitudinal rolling) of the vessel. Two trials were made, one on the 9th and the other on the 15th of September. On the first of these occasions, she was accompanied by the *Sultan* only, and on the second she was accompanied by the *Agincourt* only. The seas met with on the 9th of September were lumpy and irregular, the wind having shifted somewhat suddenly during the previous night. Having got well out to sea, about 40 miles off land, the wind was found to be blowing rather north of west with the force of a moderate gale, its speed varying from 40 to 45 miles per hour; and the largest of the waves were found to vary from 300 to 350 feet in length from crest to crest, occasionally reaching 400 feet—the greatest heights from hollow to crest being 15 and 16 feet. Going head to sea, at from six to seven knots, both vessels pitched considerably; the *Devastation*, however, had the best of it, pitching through smaller angles than the *Sultan*. The latter vessel was remarkably lively; at one moment she was to be seen with her forefoot completely out of water, and the next with her bow dipped down to so great an extent that it was difficult to see from the flying deck of the *Devastation*—although the ships were pretty close together—whether the sea did not really break inboard; and this notwithstanding that the bow of the *Sultan* rises forward some 30 feet above the surface of the water. On the other hand, the forecastle-deck of the *Devastation* was repeatedly swept by the seas, to each of which she rose with surprising readiness; indeed, it invariably happened that the seas broke upon her during the upward journey of the bow, and there is no doubt it is to this fact that her moderate pitching was mainly due, as the weight of the water on the forecastle-deck during the short period it remained there acted as a retarding force, preventing the bow from lifting as high as it otherwise would, and this of course limited the succeeding pitch, and so on. The maximum angle pitched through on this occasion, *i. e.*, the angle between the extreme elevation and depression of the bow, was 7½°. Each vessel behaved extremely well when placed broadside on to the sea, rolling very little. The trial of the ship on the 15th of September, in company with the *Agincourt*, was by far the most severe of any. Early in the morning the vessel got under weigh and steamed out to sea, accompanied by the *Agincourt*. The wind was blowing with considerable force from the northwest, while the sea was at times very regular, long, and undulating; just the sort to test the rolling propensities of a ship, but scarcely long enough to be most effective in doing so, either in case of the *Devastation* or *Agincourt*. The largest waves ranged from 400 to 650 feet long, and from 20 to 26 feet high. The ships were tried in almost every position with regard to the direction of the sea, and at various speeds, the result in point of comparison being extremely interesting, and, so far as the *Devastation* was concerned, very satisfactory. With the sea dead ahead, and proceeding at about seven knots, the *Devastation* pitched rather more than the *Agincourt*, although the great length of the latter compared with that of the former caused her bow to rise and fall through a much greater height, giving her the appearance of pitching through a greater angle. The usual angles pitched through by the *Devastation*, measuring the whole arc from out to out, were from 5° to 8°; the maximum angle pitched through was, however, 11¾°. The scene from the fore end of the flying deck when the vessel was thus going head to sea was very imposing. There was repeatedly a rush of water over the forecastle, the various fittings, riding-bitts, capstan, anchors, &c., churning it up into a beautiful cataract of foam; while occasionally a wall of water would appear to rise up in front of the vessel, and dashing on board in the most threatening style, as though it would carry all before it, rushed aft against the fore turret with great violence, and, after throwing a cloud of heavy spray off the turret into the air, dividing into two, pass overboard on either side. All the hatchways leading below from the upper deck were closed; it was not, however, thought necessary to close the doors in the sides of the trunks leading up from the main hatchways to the flying deck, most of the men on deck preferring to remain here under the overhang of the flying deck. It was quite the exception for the water coming over the bow to get much abaft the fore turret; but this, however, occurred occasionally. The foremost turret makes a most perfect breakwater; it receives with impunity the force of the water, which, after spending itself against it, glances off overboard, leaving two-thirds of the deck seldom wetted. There was one sea which came on board, while thus proceeding head to sea, which was much heavier than any other; it rose in front of the vessel some ten or twelve feet above the forecastle, and broke on the deck with great force, for the moment completely swamping the fore end of the vessel. A mass of broken water swept up over the top of the fore turret, and heavy volumes of spray extended the whole length of the flying deck, some small portion of it even finding its way down the funnel-hatchway—which had been left uncovered—into the fore stokehole. It should be borne in mind that the angles pitched through, given above, do not measure the inclination of the ship to the surface of the water, but only her inclination to the true vertical. Pitching and lifting are produced by the vessel endeavoring to follow the slope of the waves, or, roughly speaking, to keep her displacement the same as in still water, both as to volume and to longitudinal distribution.

As to the depressing effect of the water on the bow, a layer of water one foot deep over the entire forecastle exerts a pressure of 65 tons; this will produce a change of trim of 11 inches, together with an increase in the mean draught of 1¾ inches; *i. e.*, the draught of water forward will be increased by 7¼ inches, while that aft will be diminished by 3¾ inches. A layer two feet thick will have double this effect; one three feet thick will have treble the effect, and so on up to a considerable angle. This follows from the fact that the front slope of the longitudinal curve of stability, up to a considerable angle, is very nearly straight. Hence the effect, even of a large body of water passing over the forecastle, tending to make the vessel dive down head foremost, is small and of no importance. It modifies, however, the transverse stability. When proceeding head to sea there was no appreciable rolling motion. With the wind and sea on the bow she pitched considerably less than when going head to sea, but rolled through 5° or 6°. With the wind and sea abeam, lying passively in the trough of the waves, the maximum angle rolled through was 14° from port to starboard, 6½° to the windward, and 7½° to leeward, and this without perceptible pitching. When, however, proceeding at about 7½ knots, with the wind and sea on her quarter, she rolled through 27½° from port to starboard, 13° off the perpendicular to windward, and 14½° off the perpendicular to leeward, besides also pitching through some 4° or 5°. This is by far the greatest angle she has ever rolled through. It is the apparent period of the waves, *i. e.*, their period relatively to the ship, which operates in making a vessel roll. The motions of the vessel, both as to pitching and lifting and to rolling, were extremely easy. She indeed claims to have behaved better than her companion, the *Agincourt*. Certainly her rolling motion was somewhat slower, and she rolled less deeply; when the *Agincourt* was rolling 17° from port to starboard the *Devastation* was only rolling 14°. At to pitching, the *Devastation* may fairly claim to have had the advantage, for, as we have seen, although the *Agincourt* pitched rather less, her bow moved vertically through a greater distance, so much so that while going head to sea at 7 knots she shipped a sea over her high forecastle, showing that she could not be driven under the circumstances at a much higher speed with at least anything like comfort. The behavior of the vessel generally accorded, with considerable approximation, with what was to be expected under the circumstances from the theoretical knowledge possessed on the subject; and although on no occasion during the trials were the waves quite so long as was wished for, the data obtained have been most valuable in testing and correcting the theory, so that the behavior of this ship, or of any similar ship, in any weather, may now be foretold with considerable accuracy. The instruments measuring and recording the behavior of the vessel were most perfect in their action. They were personally attended throughout the trials by their inventor, Mr. Froude.

The *Devastation* has lately been cruising in the Mediterranean. Two winters ago, at a public meeting in London, Admiral Inglefield thus spoke of her: "I have just returned from Malta, and I saw the *Devastation*, having come into port from a long cruise. The captain spoke of the ship as being perfectly seaworthy, wholesome, and comfortable for the men and officers, and everything he could wish."

PART IV.

THE THUNDERER; THE DREADNOUGHT; THE 38-TON GUN AND EXPERIMENTAL FIRING: THE ARMSTRONG 39-TON BREECH-LOADING GUN.

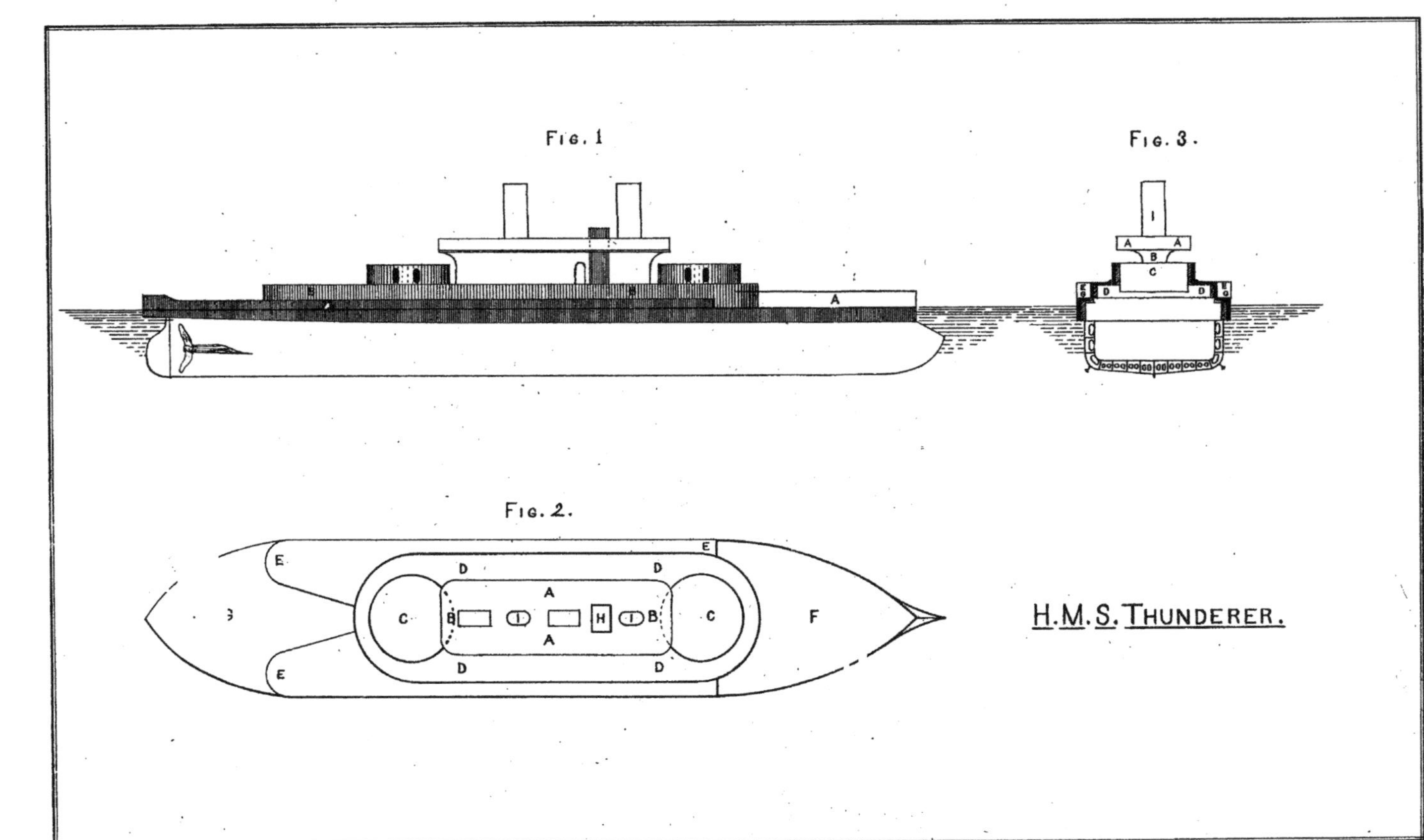
Fig. 1
A
Fig. 3.
I
A A
B
C
D D
Fig. 2.
E
E
E
D D
A
C
B
H
B
C
A
D D
F
H.M.S. Thunderer.

THE THUNDERER

The *Thunderer* is a sister ship to the *Devastation*, launched March 12, 1872, nearly one year after the *Devastation*, but was only being prepared for the first commission at the time of my visit, early in 1876. The principal dimensions of the two vessels are alike; they differ only in detail of construction, in the type of motive machinery, and in armament.

The *Thunderer*, like her sister ship, was designed to be thoroughly seagoing, and to be capable of performing every service which can be required from a first-class modern line-of-battle ship.

At the date of construction they were admitted to be the most powerful fighting-ships then laid down. The committee on designs of the ships of the royal navy gave their judgment upon them in these words: "They represent in their broad features the first-class fighting-ships of the immediate future."

A clear understanding of the general arrangement of the vessel will best be seen by the annexed drawings, Figs. 1, 2, and 3. The forecastle is shown at A, in Fig. 1, where the height of the side-armor above and below water is also shown. The position of the armored deck is indicated by the black line along the upper edge of the side-armor. In Fig. 1, the armored portions are shaded, the unarmored left plain. Where the armor is visible from the outside, as on the turrets and sides, it is shaded dark; where concealed by any unarmored structure, it is lighter. The breastwork, except at one corner where it is not screened, is shaded light in Fig. 1, being concealed by a structure to be noticed hereafter.

An end view of the deck-house, and the hurricane-deck which it supports, is seen in Fig. 3, a section through the fore turret. The broadside superstructure, as it is usually called, is shown at B B in Fig. 1, and at E E in section 3. The superstructures and other parts are clearly shown in the plan, Fig. 2, where, commencing with the highest points, I I are the smoke-pipes, H is the conning-tower, A A the hurricane-deck, B the elevated deck-house which supports it, C C are the turrets, D the breastwork, E E the broadside superstructure level with the breastwork, F the forecastle, 3 feet lower than the breastwork, and G the armored deck, about 7 feet lower than the breastwork in the only part where it comes in sight, though it extends, of course, under E and F, and, indeed, throughout the ship except inside the breastwork. The same letters apply to Fig. 3. The conning-tower will be noticed in Fig. 1. It is sufficiently high to give a view over the hurricane-deck bulwarks, and wide enough to command a view forward and aft past the smoke-pipes, which are oval.

MACHINERY.

The *Thunderer*, in common with all modern fighting-ships, is operated in every essential particular by the power of steam. The motive power of the ship is solely steam, and there are in all twenty-eight steam-engines and nine boilers. Thirteen of these engines are in pairs, having two cylinders, and the remaining fifteen are single engines, having one cylinder only. Two of the pairs are employed for driving the twin screws, and are termed the motive-engines. The others are small engines, employed for subsidiary purposes, such as revolving the turrets, working the hydraulic gun-machinery, hoisting shot and shell, working the capstans, hoisting anchors and boats, working the steering-apparatus, working pumps for circulating cold water through the surface-condensers, starting the motive engines, pumping water from the spaces between the double bottoms, feeding the boilers, hoisting ashes, and driving fans for ventilating the ship. In addition to this great responsibility, the engineer department is charged with all the water-tight doors in the ship, and all valves and pipes. In short, the interior of the ship is a vast engineering workshop, requiring skill and energy successfully to manage it.

The motive machinery of this vessel, as well as that of the *Devastation*, was contracted for previous to the introduction of the compound engine into the royal navy. It was constructed by Messrs. Humphrys, Tennant & Co., and the engines are of the horizontal, direct-acting type adopted by that firm, and built for several other ships of the navy. There is one pair to each of the two screw-propellers; the cylinders are 77 inches in diameter, and the stroke is 3 feet 6 inches. The boilers are of the old box variety, and a description of them follows, under the head of "Boiler explosions." In consequence of the explosion of one of the boilers in July, 1876, the final official trials at the measured mile were not made till the autumn following: the accompanying data of the performance are believed to be correct.

Two days after the measured-mile trials on the 4th of January, 1877, a crucial test of the working of the machinery by a six hours' continuous full-power run was made up and down the Solent in boisterous weather, the force of the wind being between seven and nine, and the sea rough; the following results were obtained as the means for the twelve half-hours:

Pressure in boilers	27.80 pounds.
Vacuum, starboard forward engine	27.3 inches.
Vacuum, starboard after engine	25.79 inches.
Vacuum, port forward engine	27.99 inches.
Vacuum, port after engine	25.52 inches.
Revolutions per minute, starboard	75.20
Revolutions per minute, port	75.03
Pressure in cylinders per square inch, starboard	19.491 pounds.
Pressure in cylinders per square inch, port	19.25 pounds.

Indicated horse-power, 5,748.97, or 149 horse-power beyond the contract.

The best quality of Nixon's steam-navigation coal was used, and the expenditure was 3.14 pounds per indicated horse-power per hour.

This ship has proved herself thoroughly seaworthy by successfully going through one of the most severe tests that can be applied. On the 18th of November, 1877, during the very height of a gale of almost unexampled fury, even in the English Channel, the *Thunderer* made the passage from Portland to Spithead. On this occasion, as reported by her officers, although having her bow immersed to a depth of 6 feet, her reserve of buoyancy was so great that she lifted readily and shook the water from her decks freely. As might have been expected, she suffered somewhat in the more perishable parts; a great quantity of glass was broken, the steam steering-gear was strained, and in some other particulars the effects of the tremendous seas encountered were made evident. The admirable behavior of the *Thunderer*, and her complete success under circumstances of an unusually severe character, will increase the confidence in the belief that mastless turret-ships may be relied upon to perform voyages in the worst weather with safety and comparative comfort to the officers and crew.

ARMAMENT.

The *Thunderer* was originally fitted, like the *Devastation*, with two 35-ton, 12-inch Woolwich rifled guns in each of the two turrets, mounted on carriages similar to those of the *Glatton*, and known as Captain Scott's design; but after Mr. Rendel brought out his system for working heavy guns by hydraulic pressure, it was decided to introduce the principle for the first time into the forward turret of this vessel. Accordingly the two 35-ton guns were removed, and guns 38 tons in weight, having a bore of 12½ inches, were substituted. The same carriages were retained. At the time of my visit all the machinery for working these guns had been fitted on board and subjected to the first tests; as, however, deficiencies usually experienced in new and untried machinery were expected to be developed, no reports were permitted to be made. Yet it was confidently stated by reliable authority that the result of this first trial on board ship was satisfactory, the proof of which may be found in the fact that the system has been adopted and ordered for the *Dreadnought*, *Inflexible*, and other vessels. All British service gun-carriages are at present mounted on their slides in such a manner as to recoil on a dead bearing, but to run out on wheels thrown into action by eccentrics. By placing the carriage permanently on wheels and trusting more to the compressor to arrest recoil, the operation of "tripping" the carriage, *i. e.*, of throwing the wheels into action for running out, is avoided. In 1867 and 1868 a partial muzzle-pivoting carriage was made at the Elswick works for an 18-ton gun on the plan of raising and lowering the trunnion-bearings of the gun in vertical grooves formed in the carriage. Captain Scott modified the arrangement by the substitution of hydraulic jacks, in combination with chocks, for the screw lifting-gear, and by the application of the jacks to act from fixed positions in the slide or turret floor on a bow-piece carrying the trunnions. In this form the system has been applied for heavy turret-guns in the British royal navy. It makes the carriage, however, high and top-heavy, a disadvantage for naval service which would become more serious with every increase in the weight of guns. The object of muzzle-pivoting is the reduction of the size of the ports. The size of ordnance, however, continues to increase with rapid strides, while the number of men employed to work the guns cannot be much further added to, and the train of mechanism required to apply the constant and limited power of men to the forces to be exerted in loading and working heavy guns becomes larger and more complicated as the weight of the guns is increased. Hence, the adoption of some inanimate power in the place of mere hand-labor for loading and working heavy ordnance has become an absolute necessity for existing guns, and for those of the immediate future. Adopting the steam-engine as the most ready and convenient source of power, it has been found that that power can be best applied through the medium of water under pressure. The simplicity and compactness of hydraulic machinery, and the perfect control it gives over the motion of heavy weights, especially adapt it for the purpose. Power sufficient for the heaviest guns may be transmitted by water through a very small pipe for long distances and by intricate ways, so that a steam-pumping engine may supply power by this means for working many guns.

HYDRAULIC MACHINERY FOR WORKING THE GUNS IN THE FORE TURRET OF THE THUNDERER.

In the turret arrangement of the *Thunderer* (which differs from that of the *Inflexible*, and is shown by Figs. 1 and 2) the carriage is placed upon rollers. In this carriage the gun is made partially muzzle-pivoting by hinging the slide at the rear horizontally, and raising and lowering the front end upon a press to three or more positions, in which it can be chocked by turning under it the bracketed supports.

The cylinder, Fig. 1, performs the double office of checking recoil and moving the gun in or out along the slide. The gun or recoil drives back the piston, and is arrested by the resistance which the valve D offers to the escape of the water from the cylinder. The valve is loaded with a spring, which may be adjusted to give any required resistance, and so meet the variations of the force of recoil. It is also partly balanced, to lessen the load required upon it. The area of the piston-rod is one-half that of the piston, and the gun is run out by admitting the water-pressure to both sides at once. For running the gun in, the pressure is admitted to the front of the piston only, the exhaust being at the same time opened to the rear. Clack-valves in connection with a waste-water tank are used to insure the cylinder being always full, and there is a relief-valve on the front for preventing any excessive strain. On the rear the recoil-valve acts as a relief-valve upon occasion. It will happen in some cases that the pressure required on the valve D to arrest recoil falls short of that necessary for running the gun in or out, in which case the water admitted to the cylinder for the purpose would lift the valve and escape to waste. This is provided for by making the act of opening the cylinder inlet-valve A place an additional load on the recoil-valve D, retaining it there so long as the inlet-valve remains open. Fig. 1 shows one method of placing the extra load on the recoil-valve, viz, by a small inverted press, having in its normal condition an open communication with the waste-water tank, which communication is closed and the press charged with water under pressure by the first movement of the lever employed to open the inlet-valve A of the recoil-cylinder. It was stated that the recoil could be regulated with precision, and that excellent control could be exercised over the movement of the gun on its slide. By the arrangements described the following advantages are claimed: The loading operation is transferred from a confined space in the port of the turret to a roomy and convenient place on the main deck. The dimensions of the turret can be reduced; one man in the turret and one outside may direct and control all the movements of a pair of the heaviest guns, and may load and fire them without other help than that involved in bringing up the ammunition, and, finally, far greater rapidity of fire is attainable than would be possible by manual labor.

An objection raised to this system is the alleged liability to premature explosions in loading, and, as a consequence, to risk of self-destruction, to which it is said the ship is thus exposed. This objection is, however, it is said, obviated by not depressing the gun for loading to such an extent as to aim a shot below the water-line, and furthermore provided for by a special arrangement for drenching the bore of the gun with

Fig. 1.

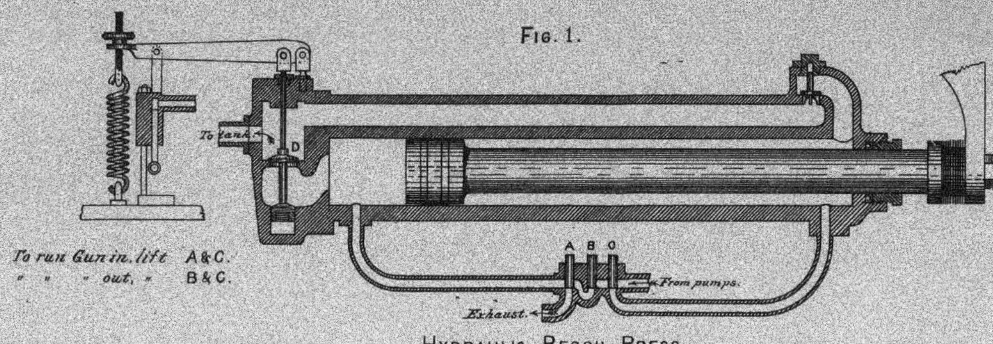

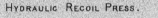

Hydraulic Recoil Press.

Fig. 2.

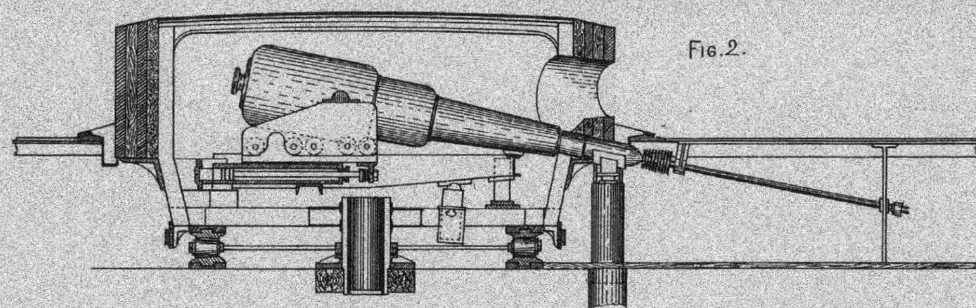

Turret with two 38-Ton Guns, showing Hydraulic Buffer, and method of loading from below, as applied to H.M.S. Thunderer.

water in sponging, and it is entirely removed by the arrangement that will be applied to the *Inflexible*, in which the loading-gear is placed so that the gun is little depressed when in the loading position.

THE RECENT BOILER EXPLOSION.

It was after my visits on board the *Thunderer* and immediately after the foregoing account of that vessel had been written that the awful disaster occurred, occasioned by the explosion of the forward starboard boiler, which produced results more terrible than any accident on board a ship of war, from the effects of steam, hitherto recorded. The report on the subject will be found under the head of "Boiler explosions."

THE DREADNOUGHT.

This ship, recently completed at the Pembroke dock-yard, South Wales, is spoken of as a modified and improved *Devastation*, on a larger scale; but the modifications, resulting from the experience of the trials of the former vessel, are such as to change the type. The *Dreadnought* cannot be called a low free-board ship, or a breastwork monitor, for the height from the load water-line to the turret-deck is nearly 12 feet, and the superstructure is more properly an armored oval citadel or tower, of which the sides are those of the vessel carried up from the broadside surface; or, in other words, instead of building a breastwork on the deck of the armored hull some 185 feet long amidships, with a passage of, say, 10 feet between it and the sides of the vessel, as was done in the *Thunderer* and other low free-board ships, the sides of the *Dreadnought* are built up flush to the top of the upper or turret deck. This armored side rises nearly 12 feet above water, and is extended in length amidships 184 feet. The design, as will be seen, requires considerably more armor than would be used if the vessel had been built on the breastwork system, but the advantages derived from having the whole width of the vessel below the upper deck unobstructed, affording light, with facilities for loading the guns on the hydraulic system, and working the turrets, is of importance; besides which, it increases the room and allows comfortable quarters for officers and crew above water.

The citadel, as before stated, is 184 feet in length, and the height between decks is 7 feet 6 inches. It is armored with solid plates, 11 inches thick, except at the ends and abreast the bases of the turrets, where the thickness is increased to 13 and 14 inches. The increased thickness at the ends is to protect more thoroughly the bases of the turrets, the machinery for working them, and for loading the guns; in short, all the working apparatus inclosed therein. The armor-belt, which is carried entirely around the vessel, is 11 inches thick on the water-line, tapering to 8 inches at 5 feet below water, where it stops. It also tapers above water, fore and aft of the citadel, as well as toward the ends. This armor-belt, fore and aft the fighting part of the ship, rises only 4 feet above water, and is intended solely to protect the vital portion of the hull; all parts above it are destructible, and may be riddled with shot without detriment to the fighting or sea-going qualities of the vessel. The turret-deck, or deck over the citadel, is plated with two courses of 1½ and 1 inch iron respectively, and the main berth-deck below is also plated with the same thickness of metal fore and aft of the citadel; of course, no armor on this deck inside of the citadel is needed.

The turrets rise through the citadel-deck to a height of 12 feet from the base or revolving deck-platform inclosed by the citadel. The diameter of each turret inside of framing is 27 feet 4 inches, the depth of the framing being 10 inches. They are built up with two courses of plates and two courses of teak in the following manner: First, the shell or wall consists of two ¾-inch plates, bolted together and riveted to the framing; on the exterior of this shell is a teak backing 6 inches thick; on this backing, armor-plates 7 inches thick are secured; next, teak backing 9 inches thick is fastened on; finally, armor-plates outside of

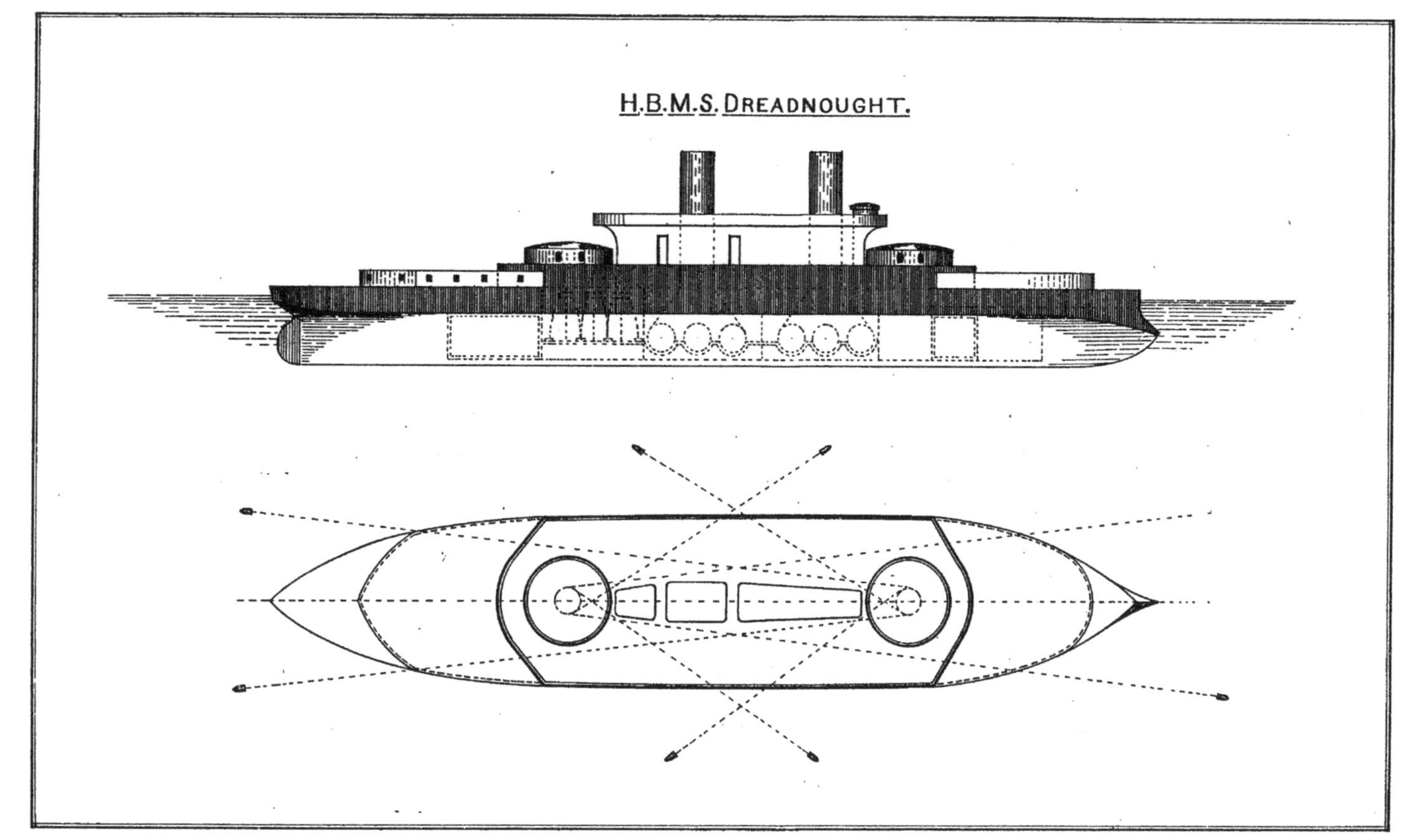
H.B.M.S. DREADNOUGHT.

all 7 inches thick; all securely bolted together. The plates were rolled at Sheffield, and curved to templates drilled and prepared for their places. The following are the dimensions of the plates for one turret:

Inside course.

No. 1, 15 feet $2\frac{1}{2}$ inches by 8 feet 1 inch, by $3\frac{1}{2}$ inches thick.
No. 2, 15 feet 2 inches by 8 feet 1 inch, by $3\frac{1}{2}$ inches thick.
No. 3, 14 feet $8\frac{1}{4}$ inches by 8 feet 1 inch, by $3\frac{1}{2}$ inches thick.
No. 4, 15 feet $2\frac{1}{2}$ inches by 8 feet 1 inch, by $3\frac{1}{2}$ inches thick.
No. 5, 19 feet $6\frac{3}{4}$ inches by 8 feet 1 inch, by $3\frac{1}{2}$ inches thick.
No. 6, 13 feet $\frac{1}{2}$ inch by 8 feet 1 inch, by $3\frac{1}{2}$ inches thick.

Outside course.

No. 1, 16 feet $10\frac{5}{8}$ inches by 8 feet $2\frac{1}{2}$ inches, by 7 inches thick.
No. 2, 18 feet $3\frac{7}{8}$ inches by 8 feet $2\frac{1}{2}$ inches, by 7 inches thick.
No. 3, 13 feet $\frac{1}{2}$ inch by 8 feet $2\frac{1}{2}$ inches, by 7 inches thick.
No. 4, 14 feet $8\frac{1}{2}$ inches by 8 feet $2\frac{1}{2}$ inches, by 7 inches thick.
No. 5, 21 feet $8\frac{1}{2}$ inches by 8 feet $2\frac{1}{2}$ inches, by 7 inches thick.
No. 6, 16 feet $7\frac{5}{8}$ inches by 8 feet $2\frac{1}{2}$ inches, by 7 inches thick.

The displacement of the *Dreadnought* when loaded will be 10,950* tons, or 1,650 more than that of the *Devastation* and *Thunderer*. The length between perpendiculars is 320 feet; breadth, extreme, 63 feet 10 inches; mean draught of water, loaded, 27 feet. The hull is constructed with the usual double bottom, and, including the spaces in it, there are sixty-one water-tight compartments in the vessel. The same general system of divisions and water-tight doors is adhered to, including the longitudinal bulkhead, which commences about 40 feet from the stem and extends to nearly the same distance of the stern, thus giving backbone and dividing the vessel in the center.

The flying deck is quite similar to the one on the *Thunderer* before described. The armored pilot-house, or conning-tower, is fitted with a steel roof as a protection from rifle-fire, and it is provided with a complete set of communicating gear for directing the movements of, and fighting the ship.

The exterior of the hull is not sheathed with wood as is usual for all cruising-ships.

There is one mast only, to be used for signal purposes and for hoisting boats, the complement of which includes three steam-launches.

The armament consists of two 38-ton guns in each of the two turrets, all of them being worked on the hydraulic system of Rendel, previously described. In addition to which the new weapons, Whitehead torpedoes, now carried by all recently-commissioned ships are provided. In this case, arrangements have been effected for ejecting them through apertures on the lower deck within the citadel. Upon this deck have also been built the magazines for storing these torpedoes; here likewise are the engine and machinery for charging them with compressed air and the appliances for handling the shot and shell.

MOTIVE MACHINERY.

The *Dreadnought* has been engined by Messrs. Humphrys & Tennant. The engines are of the compound type, very similar to those in

*The navy list of November, 1876, gives the displacement of the *Dreadnought* as 10,886 tons, instead of 10,950 tons, as in former navy lists.

the *Alexandra*, constructed by the same firm, and to be described hereafter.

There is an independent set of vertical inverted engines to each of the twin screws. Each set consists of three cylinders, the high-pressure exhausting into the two low-pressure cylinders. The diameter of the former is 66 inches; the diameter of each of the latter is 90 inches, and the stroke of pistons 4 feet 6 inches. All the cylinders are steam-jacketed. The high-pressure jacket is adjusted to the working boiler-pressure of 60 pounds, and those of the low-pressure to 30 pounds. The air-pumps are of composition, and placed fore and aft, immediately under the low-pressure cylinders, from which they are worked. The arrangement is compact and accessible, the condensers forming the midship framing, while the wing framing consists of two wrought-iron columns to each cylinder.

The surface-condensers are of the usual variety employed in the British service. They contain upward of 16,500 square feet of cooling surface, and are fitted to be worked as common jet-condensers in the event of necessity. The condensing water is supplied by two powerful centrifugal pumps, and an extra bilge-suction is provided, so that the air-pumps can receive directly from the bilge in case of an accident by which large quantities of water would enter the ship.

The engine crank-shaft is composed of three pieces, interchangeable; each piece has a length of 10 feet 7⅞ inches, and a diameter of 17½ inches. The diameter of the propelling-shafts is 16 inches, except the lengths in the stern-tubes, of which the diameter is 18 inches.

The engines are started and reversed by an auxiliary engine having cylinders 6 by 8 inches. The ship is provided with six ventilating-engines, two auxiliary fire-engines, four main fire-engine pumps, steering-engines, turret-turning engines, capstan and ash-hoist engines, besides the engine and hydraulic apparatus for working the guns; in all, there are twenty-nine steam-engines on board, and there are one hundred and eighty valves connected with the ventilating-pipes.

The screw-propellers are of Griffith's recent pattern, four-bladed, the diameter of each screw being 20 feet, with pitches adjustable from 21 to 26 feet.

Boilers.—The steam for the motive and other engines is supplied by twelve main boilers and one auxiliary boiler, having a total heating surface of 21,912 square feet. Instead of being arranged in the vessel face to face, as are those in former ships, so that the firemen have fires behind as well as in front of them, they are placed back to back against the middle-line bulkhead of the ship; the firing is thus done at the sides, convenient to the coal-bunkers. The boiler-rooms are further divided by athwartship bulkheads, whereby four rooms are formed, the forward ones being 42 feet and the after ones 40 feet in length. The length of the engine-room aft of this is also 40 feet.

The shells of the main boilers are elliptical in shape, 15 feet high by 11 feet 10 inches wide, and 9 feet 8 inches in length, with three furnaces in each boiler; the shells and tube-plates are ¾ inch thick, and the furnace and back-plates ½ inch. The shells are double-riveted, and butt-straps are placed inside and outside the longitudinal seams; the tubes are of composition, and 3 inches in diameter; each furnace is 6 feet 10 inches long, and fitted with 66 wrought-iron grate-bars, 3½ inches deep by 1¼ inches broad, and 2 feet 3 inches long.

In addition to the ordinary safety-valve chamber, containing a couple of spring safety-valves, each boiler is fitted with a supplementary test-valve placed on the front of the boiler. The stop-valves and safety-valves can be worked from the fire-room floors.

The engines and boilers are entirely surrounded by the coal-bunkers; the bunker immediately forward of the boilers is 22 feet in length, fore and aft, and the one immediately aft of the engines is 8 feet in length; there are consequently 152 feet in the length of the ship by the extreme breadth occupied by the motive machinery and coal. The bunkers contain 1,200 tons of coal only.

The ventilation of the fire-rooms is supplied by fans, supplemented by eight cowls on the hurricane-deck and four others at the side, and which are also utilized as ash-hoists.

As the ship is mastless, steam-power is to be used solely.

The total weight of the steam-machinery is given as 1,430 tons; hull and accessories, 7,350 tons; all other weights, including coal and stores, 2,170 tons.

The improvements of the *Dreadnought* over the other two mastless sea-going ships, the *Devastation* and *Thunderer*, or the difference between them, may be summed up briefly as follows: The displacement is 1,650 tons more; the length between perpendiculars is 35 feet greater, with an increase of beam of 1 foot 7 inches, and an increased depth of hold of 1 foot 2 inches.

The structural improvements consist in carrying out the breastwork, entirely to the sides of the vessel, and in bringing the forecastle and after-deck up to near the level of the breastwork, or, rather, armored citadel; by this arrangement the facilities for working the guns and ship have been greatly increased and better accommodations provided, besides which the objectionable *cul de sac* that exists in the other two ships has been obviated, and a high free-board all around is obtained, which must make her both a drier ship and a more comfortable cruiser than either of the others.

Another important structural arrangement first introduced here, and since adopted in other ships, is the longitudinal water-tight bulkhead between the respective sets of engines and boilers; so that in the event of one set being disabled by rams, torpedoes, or other causes, it can be effectually shut off, the ship being propelled by the other set.

The side-armor of the *Dreadnought* varies in thickness from 8 to 14 inches, while in the other two ships it varies from 10 to 12 inches, besides which it extends 7½ inches deeper under water than in the former-named vessel. The external diameter of the turrets is 32 feet 3 inches, and the thickness of its armor is uniformly 14 inches, while the other turrets are 31 feet 3 inches in diameter, and the thickness ranges from 12 to 14 inches.

The armament is also more formidable. It consists of four 38-ton guns, all worked on the hydraulic system, while the *Devastation* carries four 35-ton guns worked by hand-power, and the *Thunderer* two 38-ton guns worked by hydraulic power, and two 35-ton guns worked by hand.

Again, the *Dreadnought* is engined on the compound system, by which greatly increased power is obtained with reduced consumption of fuel.

The six-hour trial of the motive machinery was made in January, 1877. The London *Times* reports the results as follows:

Hours.	Vacuum.	Revolutions.	Horse-power.
1........27	ins. and 27.43	67.3 and 67.4	8,201.52
2........27	ins. and 27.18	67.6 and 67.4	8,233.78
3........27	ins. and 27.25	67.4 and 67.3	8,179.94
4........27	ins. and 27.37	66.9 and 67.0	8,177.77
5........27.5	ins. and 27.43	67.1 and 66.9	8,155.93
6........27	ins. and 27.37	66.0 and 66.8	8,207.39

The mean power developed by the engines during the six hours was 8,216.28, or 216.28 horse-power beyond the contract. * * * The blasts were not used from first to last. * * * The mean boiler-pressure was 60.3 pounds. The mean pressure of steam on the pistons was, high, 31.6 pounds; low, 9 pounds. * * * The coal consumed on the trial amounted to 50 tons 122 pounds, being equal to 2.27 pounds per indicated horse-power per hour. The consumption on the six-hour trial of the *Thunderer* was 3.14 pounds. In order to show the superiority of the *Dreadnought's* compound engines from an economic point of view, it may be mentioned that had she been fitted with engines of the common type, as those of the *Thunderer*, she would have to burn 80 tons* a day more fuel to develop the same power as on the day of trial.

The engines were stopped from full speed in 18 seconds, and from going astern were started full speed ahead in 15 seconds.

The speed of the ship is not given, but the pitch of the screw-propellers was 23½ feet, which, with 12½ per cent. slip, would give 13.8 knots per hour.

The *Dreadnought* is now the most formidable fighting-ship on the ocean, and next after the *Inflexible* will be the most powerful ship of war ever floated in British waters.

TRIALS OF THE GUNS.

What the 81-ton gun can do against an armored target has been seen, and as some important results have been achieved with the piece which comes next below it in magnitude in the British service, viz, the 38-ton gun of the pattern mounted in both turrets of the *Dreadnought*, also in the fore turret of the *Thunderer*, all being now in practical use at sea, besides which others are being manufactured for the sister ships *Agamemnon* and *Ajax*, it may be interesting for reference and comparison to state as a matter of fact that the 38-ton gun has sent its projectile nearly through an armored target of the following combination: A 12-inch plate, an 8-inch plate, 6 inches of teak, and a 5-inch plate, into which last it penetrated 2 inches, or altogether 22 inches of iron and 6 inches of teak. In another instance the projectile was sent through a target built up in the following manner: A 4-inch plate, an 8-inch plate, 6 inches of teak, a 5-inch plate, 6 inches of teak, and an inner 1-inch skin supported by angle-irons, making altogether 18 inches of iron and 12 inches of teak. This was done at close quarters. Besides, in October, 1876, at a range of 70 yards, the projectile was sent nearly through a target composed as follows: Three plates, each 10 feet wide, 8 feet high, and 6½ inches thick; between the plates were 5 inches of teak backing, making the total thickness of 19½ inches of iron and 10 inches of teak, or, in all, a target of 29½ inches. The shot, which had a striking velocity of 1,421 feet per second, punched a clean hole 13 inches by 12½ inches in the two front plates, and penetrated into the rear plate, where it broke up. The charge was 130 pounds of 1.5-inch pebble-powder, and the projectile weighed 812 pounds. The target was an exact sample of the armor of some of the English coast forts; therefore, as a prelude to experiments on a large scale, this experiment opens up the question of coast defense.

The caliber of this gun is 12½ inches. Since the firing above noted the powder-chamber has been increased to a diameter of 14 inches, and in March, 1877, the first trial against armor after chambering took place. It was calculated that after chambering the quantity of powder which could be burnt and the consequent velocity and power of the shot would be considerably increased. The powder-charge was enlarged to 200

*The *Times* is in error; the correct figures are 76 tons; but this may be instructive, in view of our very limited naval appropriations, to the officers who have been raising objections to the use of compound engines in our Navy.

The 38-Ton Gun Target No. 40.

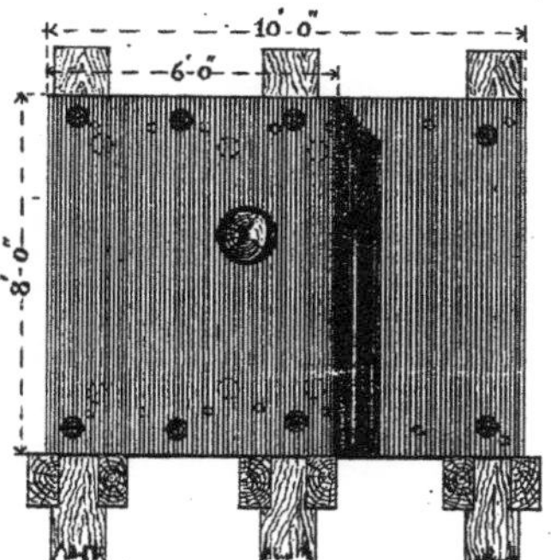

Front View.

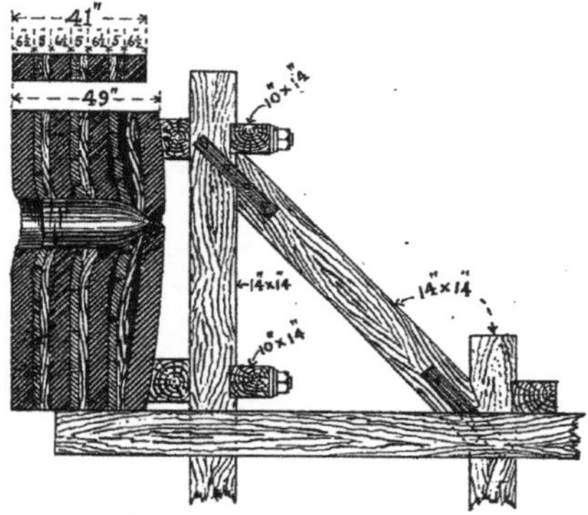

Side View.

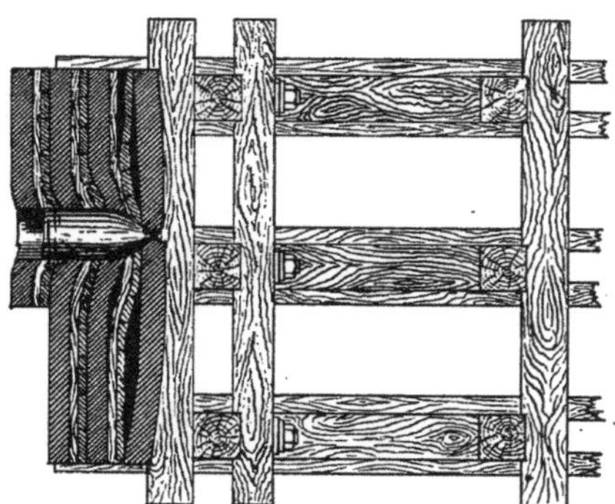

Plan.

pounds, other conditions being equal, except that for this trial the target was strengthened by the addition of another 6½-inch plate and 5 inches of teak bolted on to the front of the target, extending over the greater portion of it, making in all 26 inches of iron and 15 inches of teak. The drawing of the target, preceding this description, together with the data relating to the gun, has been taken from the *Engineer.*

The structure is shown in front view, side view, and plan. The plates are bolted together in pairs by means of the Palliser bolts in the same manner as in the target for the 81-ton gun previously described.

The shot struck a point 4 feet 11 inches from the ground and 2 feet 5½ inches from the junction-line of the thinner and thicker portions of the target. The shot buried itself deep in the target, the base, after the removal of the copper gas-check, being found to be 8.2 inches from the front face. The base was split into quarters, speaking roughly (*vide* Fig. 1). The face of the target was but slightly bulged. The rear of it was not so abruptly bulged as in the case of the target of the 81-ton gun, though it seems apparent that the general effect of the 38-ton-gun shot on its target very closely resembles that of the 81-ton gun on its target. It seems, judging from the work done by this shot compared with that of its predecessor, that the projectile penetrated to a less extent than the one in the last round before chambering. It penetrated only 21½ inches of iron and 15 inches of teak, which is not considered a fair representative of the effect of the chambered gun at 70 yards range. The deficiency is represented to have been owing to a difficulty that occurred in the setting up of the 200-pound cartridge, which prevented its being rammed down, so that in fact it had to be blown out and the gun recharged; two cartridges, each containing 100 pounds, were then employed, and they occupied more room than was intended, extending beyond the chamber. This interfered with the action of the charge, which was burned under less favorable conditions than were designed. The actual initial velocity attained, however, was 1,540 feet per second, and the striking velocity at the target was 1,525 feet per second. This, with a shot of 800 pounds weight and gas-check of 12 pounds weight, it is calculated, would give a penetration of about 23 inches of iron, or an increase of about 2 inches on the unchambered gun, results that are expected to be attained in future trials.

THE ARMSTRONG 39-TON BREECH-LOADING GUN.

This gun, the most powerful breech-loader manufactured in England, was tested at the Elswick works in March, 1877. It was described in the *Engineer*, February 23, and the details of the trials were given by the same paper March 23.

The dimensions of the gun are as follows: Caliber, 12 inches; length of gun, 282 inches; length of bore, 264 inches, or 22 calibers; weight of gun, 39 tons; weight of projectile, 700 pounds; description of rifling, polygroove; pitch of rifling, increasing from 1 turn in 100 to 1 turn in 45; number of grooves, 45.

The arrangement of the breech-closing mechanism follows the pattern adopted in the French marine as far as the form of the breech-screw and the mode of securing, releasing, and withdrawing it are concerned; but differs materially in the arrangements for closing the end of the bore against the passage of gas. The gas-check is a steel cup attached to and supported by the end of the breech-screw. The supporting surface is slightly curved, and when the body of the cup is acted upon internally, by pressure of the gas, the rim expands, and, fitting tightly upon a copper ring rolled into a groove in the cup-chamber, forms a perfect joint against any escape of gas. When the pressure of the gas is withdrawn, the gas-check cup frees itself by its own elasticity, and the breech-screw is withdrawn with ease. This arrangement has been subjected to the tests of practical working by the Italian Government, who, having fired over 500 rounds from a 12-centimeter gun of this description, manufactured for them by Sir W. G. Armstrong & Co., subsequently ordered a considerable number of guns of the same caliber. The projectiles used were fitted with a copper ring at the rear, which, being expanded into the rifling by the gas-pressure, gave the necessary rotation, and at the same time perfectly closed the bore against the passage of gas. A slightly projecting band at the front end of the projectile, carefully turned to fit easily into the bore, supported the projectile at that end. This gun is best dealt with by comparing its trials with the Woolwich 38-ton muzzle-loader just mentioned. The results are tabulated as follows:

	Armstrong breech-loader.	Woolwich muzzle-loader.
Weight, in tons	39	38
Caliber, in inches	12	12½
Weight of charge, in pounds	180 (pebble-powder).	130
Weight of projectile, in pounds	700	812
Velocity at muzzle, in feet per second	1, 615	1,450
Work stored up, in foot-tons	12, 656, or 336 per inch of circumference.	11, 838, or 301 per inch of circumference.

PART V.

BROADSIDE ARMORED SHIPS; THE AUDACIOUS CLASS; THE ALEXANDRA.

BROADSIDE ARMORED SHIPS.

France is credited with having produced the first ship clad in armor. The *Gloire* * was commenced in 1858; two others of the same type immediately followed; all of them wooden ships plated with iron. The first great English armored ship, the *Warrior*, was begun in 1859, and was followed by the *Black Prince*. These ships were ordinary single-screw war-steamers, with iron hulls 380 feet long, incompletely protected by plates 4½ inches thick. After the *Black Prince* followed the *Minotaur* and *Northumberland*, each 400 feet in length; but it was soon found that these long ships were not well adapted for maneuvering in line of battle, and hence the later armored ships were made gradually broader in the beam and shorter in length.

At the same time various improvements were introduced into the build, one of which was the change of the old oblique projecting bow into the reverse-curved, swan-breasted shape, which is substantially the same as that of the present running-down bow or ram. The armor was no longer restricted to the amidship portion of the vessels; it was extended fore and aft, until they were completely covered above water, and for a short distance below it. The weight of the guns steadily increased, and with it the thickness of armor. Leaving out of account the converted ships, we come to the *Bellerophon*, put afloat in 1865. This ship was Mr. Reed's first production. She was considered to be a long step in advance; still the central battery delivered no end-on fire. That was sought to be obtained by the contrivance of a bow-battery on the main deck. Thus, though the *Bellerophon* was known as carrying ten 12-ton guns, the bow-fire was intrusted to two 6½-ton guns. The next first-class ship, the *Hercules*, gained an improved fire from the central battery (18-ton guns) by the expedient of recesses in the ship's sides forward and aft of the battery; advantage was taken of the recesses to make four ports in the ends, or rather corners, of the battery,

* The first account we have of an armored ship is in 1530. The largest ship then known, one of the fleet of the Knights of Saint John, was sheathed entirely with lead, and is said to have successfully resisted all the shot of that day.

At the siege of Gibraltar, in 1782, the French and Spaniards employed floating batteries, made by covering the sides of the vessels with junk, rawhide, and timber, to the thickness of 7 feet, and bomb-proofing the decks.

Iron armor was suggested in the United States in 1812. In March, 1814, a bomb-proof vessel was patented by Thomas Gregg, of Pennsylvania.

In 1842 R. L. Stevens, of New York, commenced the construction of an iron-armored ship of war.

The first practical use of wrought-iron plates as a defense for the sides of vessels was made by the French during the Crimean war. The vessels used were of light draught, exposed little surface above water, and were termed floating batteries. They rendered very efficient service, especially at the bombardment of Kinburn, in 1855, and their success doubtless led to the adoption, by the French Government, of armor-plating for ships of war.

The English built armored boats at about the same time, which were also used in the Crimean war.

from which four of the guns were able to fire within a few degrees of the line of the keel. If required to fight upon the broadside, these guns, which were mounted on turn-tables, were revolved to other ports. The armament of this ship, when put on board in 1870, was considered very powerful. It consisted of fourteen Woolwich rifled guns, of which eight were of 10-inch, two of 9-inch, and four of 7-inch caliber. Her water-line is defended by a belt of 9-inch armor, believed at the time it was put on to be impenetrable at the thickest part to any of the guns at that time afloat in European waters. This armor-belt extends for upward of 3 feet above and 3½ feet below the water-line, from stem to stern of the ship. This defensive strength is, however, confined to the belt. The battery from which the largest guns are worked is only protected with 6 inches thickness of armor, and experiment has shown that armor of that thickness with the ordinary backing, can be penetrated at a distance of 1,000 yards and at an inclination of impact of 30° by the 9-inch rifled gun, and at close quarters by the 7-inch rifled gun, such as is carried by many armored ships.* But the *Hercules* has other excellencies; she is, for an armored ship, a fair sailer, though represented to be awkward in tacking or wearing. She had a speed under steam on the measured mile of 14 knots, and can probably make 12 knots steadily for a few hours at sea. She is said to be a very steady ship, and can, therefore, use her offensive powers under conditions of sea in which a less steady ship would be almost *hors de combat.*

In the *Sultan*, built subsequently, of the same general dimensions and much resembling the *Hercules*, a step in advance was made by adding an upper-deck battery. This ship carries eight 18-ton guns on the lower gun-deck, two of less weight in the upper-deck casemate, and two on this same deck forward, but they do not command an all-round fire. In most essential points the ships are the same, though the *Sultan* is reported to have the defect of excessive top weight, to counterbalance which, considerable extra ballast has been put into her. Both vessels are built with bows having projecting rams, and they are counted with the formidable sea-going fighting-ships.

---

* The *Hercules* is much more efficiently protected than the text above indicates. In evidence of this we may quote the following passage from Our Iron-Clad Ships, written in 1869 by Mr. Reed, the designer of the *Hercules*: "The thickness of armor carried has, however, for the present, reached its maximum for sea-going broadside-ships in the *Hercules*, which has 9-inch armor at the water-line, 8-inch on the most important parts of the broadside, and 6-inch on the remainder. Outside the 1½-inch skin-plating of this vessel, teak backing 12 and 10 inches thick is fitted together with longitudinal girders of the usual character. This does not, however, constitute the whole of her protection, for below the lower deck down to the lower edge of the armor, the spaces known as the 'wing-passages' are filled in solid with additional teak backing, and inside this there is an iron skin ¾ inch thick, supported by a set of vertical frames 7 inches deep. The total protection, therefore, of the most vital part of the ship, in the region of the water-line, consists of the following thicknesses of iron and wood: Outside armor, 9 inches; then 10-inch teak backing, with longitudinal girders at intervals of about 2 feet, worked upon 1½-inch skin-plating, supported by 10-inch vertical frames spaced 2 feet apart; the spaces between these frames are filled in solid with teak, and inside the frames there is a further thickness of about 19 to 20 inches of teak, the whole being bounded on the inside by ¾-inch iron plating, stiffened with 7-inch frames. The total thickness of iron (neglecting the girders and frames) is, then, 11¼ inches, and of this 9 inches are in one thickness; the teak backing has a total thickness of about 40 inches. The trial at Shoeburyness of a target constructed to represent this part of the ship's side proved that it was virtually impenetrable to the 600-pounder gun; and perhaps no better idea of the increase of the resisting power of the sides of our iron-clads can be obtained than that derived from a comparison of the 68-pounder gun which the *Warrior's* side was capable of resisting with the 600-pounder tried against the *Hercules's* target."—AN ENGLISH NAVAL ARCHITECT.

THE AUDACIOUS CLASS.

The loss of the *Vanguard*, by sinking, off the coast of Ireland, in September, 1875, from the effects of an accidental blow of the ram of a sister vessel—the *Iron Duke*—drew public attention for a time to the easy manner in which one of these powerful and costly ships could be disposed of, as well as to this particular class of vessels. In order that the department may be correctly informed on the subject, I have submitted, independently of this report, a complete set of drawings showing the construction of the vessel, the entire internal arrangements, and the point penetrated by the ram of the *Iron Duke*.

The class consisted of six broadside-vessels of similar design, viz, the *Audacious*, *Iron Duke*, *Vanguard*, *Invincible*, *Triumph*, and *Swiftsure*. The loss of the *Vanguard* leaves five. They were all built for sea-going cruising-ships, but the *Triumph* and *Swiftsure* only were sheathed in wood and coppered. A brief outline of their history may serve to show how designs for ships of war have sometimes been decided on by the admiralty.

It was in the year 1867, in the midst of the controversy between the advocates of broadside and turret systems, that the board of admiralty resolved to invite the principal private ship-builders of the kingdom to compete in designs for either a turret or a broadside ship, at the option of the designer. Certain conditions were imposed in either case: the displacement was fixed, the draught of water was to be 22½ feet, and the speed 13½ knots.

The armor-plating was to be at least 8 inches in thickness at the water-line, and 6 inches in other parts, except at the bow and stern, and it was essential that an all-round fire should be practical, or at least that some one gun behind armor-plates should command every point of the horizon.

A prize was to be awarded to the successful competitor. Seven ship-building firms responded to the invitation, and sent in designs of various degrees of merit. The London Engineering Company proposed to build a broadside-ship of 3,800 tons; the Millwall Company, a compound of broadside and turret of nearly the same tonnage; Messrs. Palmer & Co., a broadside-ship with a movable upper-deck battery; and the Thames Company, a broadside-ship; while the firms of Messrs. Napier & Son, Messrs. Samuda, and Messrs. Laird each designed a turret-ship, fulfilling the proposed conditions.

To the surprise of the competitors, all the designs were referred to Mr. Reed, then chief constructor of the navy, and he, though selecting the designs of Messrs. Laird as the best offered, referred to a turret-ship of his own, of the same dimensions, previously submitted, and the controller of the navy in reporting on the designs, which was the entire subject-matter referred to him, decided in favor of Mr. Reed's ship over all the private designs, and that the admiralty designs of the *Audacious* class of broadside-ships was superior to either.*

* The author states that the competitors were surprised to find that their designs were referred to the chief constructor of the navy for report, but it would, we think, have surprised them much more to find their designs referred to any one else, and the author omits to suggest what other or better authority existed for framing a report upon such designs. We doubt if the competitors really felt any surprise at all in the matter, for they must have known that the admiralty invariably referred competitive designs to their responsible professional officers for report. We have never heard the report which the chief constructor made in this case called in question either as regards its scientific accuracy or its fairness. It is, no doubt, always to be regretted

This view was adopted, and the result of all this competition among the naval architects of Great Britain was that six of the *Audacious* class were ordered to be built, four of them being given out to be built in the yards of the disappointed ship-builders. These ships have all been thoroughly tested at sea, and the results have not been entirely satisfactory. It is reported, in the first place, that the calculations were so defective that the ships have turned out lighter than was intended, and it has been necessary to fill in the bottoms with concrete and ballast to give moderate stability, and that it became necessary to alter the rig and largely reduce the masts and sails. The *Audacious*, when broadside on, presents an area of 6,670 square feet, and of these only 3,207 square feet, or less than one-half, are plated. Amidships for 100 feet by 3 feet, at the water-line, the armor is 8 inches in thickness, tapering to 4½ inches at the bow and stern, and at the other portions the armor is 6 inches thick, except that the ends of the main-deck battery have only 4 and 5-inch armor, while the ends of the upper-deck battery are unprotected against a raking fire, and more than half the ship's side is in the same unprotected state.

It seems, however, that some officers entertain high opinions of the sea-going qualities of this class of ship, as will be seen from the following. The *Audacious* is now the flag-ship of the China station. Admiral Ryder, in command, writes to the controller of the navy of this ship as follows: "Whatever objections may have been raised to ships of the *Audacious* class, the longer experience I have of them the more I am struck with their wonderful steadiness. I have just lately made a passage running before a heavy sea and strong wind, all my stern ports barred in, and to our great surprise the ship did not roll more than 2° to 1° each way. I half made up my mind to broach her to, to see what she would do in such a sea, but the helmsman did it for me. In giving the ship a yaw he brought her to the wind, and positively to our surprise she declined to take any notice of the sea at all. An iron-clad flag-ship of a first-class naval power accompanied me. We were both proceeding before the same sea, my flag-ship rolling 2° to 1°, the flag-ship of the other power rolling 20°."

After the above brief outline of vessels of former types, we now come to the more recent and important designs of broadside armored ships, taking up first—

---

when a competition ends without result, and especially so when the judges of the competitive designs are themselves designers, and have their designs adopted. But we venture to think that a competition for the design of a war-ship can very seldom terminate otherwise in this country, in which the admiralty staff of naval architects enjoy far better opportunities than private firms of knowing the requirements and the working of the naval service, and are in continual intercourse with the lords of the admiralty, who are the real judges in all such cases, and who are bound to build only such ships as they approve of. In the case of the Indian troop-ships a similar result of competition to the above ensued, in spite, as we happen to know, of the chief constructor's (Mr. Reed's) urgent wish to have a design of Mr. Laird's adopted. A design previously prepared by Mr. Reed was much preferred by the admiralty, and was ordered accordingly.—AN ENGLISH NAVAL ARCHITECT.

"An English Naval Architect" shows himself needlessly apprehensive for Mr. Reed's ability and honesty, neither of which I have called in question; nor is it my province to "suggest what other or better authority existed." The point lies in the questionable equity of the system which permitted one competitor to decide upon all the competitors' designs, and to pass judgment upon his own. That he preferred the designs of another, speaks well for Mr. Reed's sense of equity and his good taste; that he called attention to his own, and so influenced their adoption, speaks ill for the system.—J. W. K.

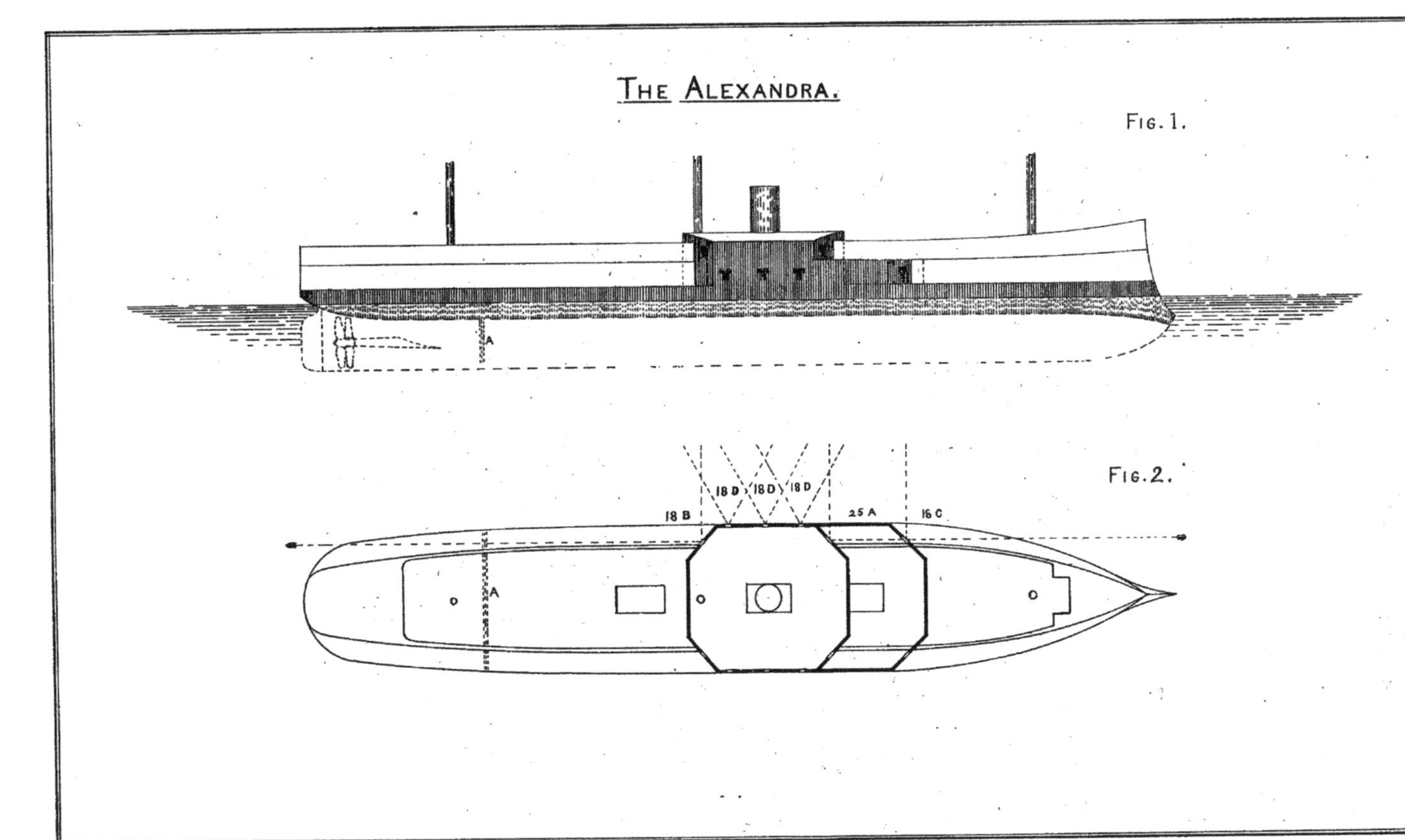
THE ALEXANDRA.
FIG. 1.
A
FIG. 2.
18 D
18 D
18 D
18 B
25 A
16 C
A

THE ALEXANDRA.

The broadside system has proved tenacious of life. For masted vessels it fairly holds its own against the turrets. The hitherto unknown perfection to which it has been brought in the *Alexandra* appears likely to give it a new lease of life, especially in combination with all-round fire from fixed turrets on the upper deck, as will be applied in the *Téméraire*. This vessel, the largest masted iron-clad heretofore designed, and now well advanced toward completion at the Chatham dock-yard, is a central-battery ship in the best sense—that is, she needs no bow or stern batteries to give her end-on fire. For the first time a broadside armored masted ship is built with satisfactory all-round fire, for, out of twelve guns, four of them, including the heaviest, can fire straight ahead and two straight astern. On each broadside from four to six guns can be fought, according to the bearing of the enemy. In other words, she has almost as perfect an all-round fire as is attainable in a broadside armored vessel, and this forms her chief claim to consideration. So far as the fighting portion of the vessel is concerned, she is a two-decker, unlike the six armored vessels of the *Audacious* class referred to above, and she may be considered a perfect example of a war-ship shadowed forth in those vessels. The battery consists of two Woolwich rifled muzzle-loading guns of 25 tons each, and ten of the same kind but of 18 tons each, the former being a size not previously attempted to be carried on a broadside-ship. In Fig. 2 of the accompanying drawing, the numbers denote the weight of the guns in tons. It will be observed that the only two 25-ton guns she possesses are located in the upper battery forward. These can be trained from 2° or 3° across the fore-and-aft line forward, to several degrees abaft the beam, as shown at A. B B are 18-ton guns with much the same training aft that the others possess forward. These four guns comprise the armament of the upper battery. To localize the effects of shells exploding between decks, the main-deck battery is divided into two by an armored bulkhead which forms a continuation downward of the forward bulkhead of the upper battery. In the portion which lies under and corresponds with the upper battery are six 18-ton guns, three on each side, for broadside-fire only. These are shown at D D in Fig. 2. In the forward and detached portion of the main battery are two other 18-ton guns for end-on fire, which they attain by means analogous to those employed to give similar fire to the upper-battery guns. Forward of the main-deck battery the whole side of the ship is set back from the level of the main deck (at the top of the armor-belt) upward. In other words, the ship forward of the battery is narrower above the main deck than below it; the two guns, C as well as A, can therefore fire right ahead past the sides. Their arc of training is about the same as that of A, or nearly 100°. The sills of the main-deck ports are 9 feet, and those of the upper-deck ports more than 17 feet, above the water. The water-line is protected by a belt having a maximum thickness of 12 inches, and it will be seen by Fig. 1 that the armor forward is carried down over the ram, both to strengthen the latter and to guard the vital parts of the ship from injury by a raking fire from ahead, at times when waves or pitching action might expose the bow. The machinery, magazines, &c., are similarly protected against a raking fire from aft by an armor bulkhead, 5 inches thick, shown at A, Figs. 1 and 2. The batteries are protected by armor only 8 inches thick below and 6 inches above, which is a deficiency of protection against guns now in common use on board armed vessels in European navies.

The total weight of armor and backing is 2,350 tons.

The principal dimensions and other data of the vessel are:

Length between perpendiculars	325 feet.
Breadth, extreme	63 feet 8 inches.
Depth of hold	18 feet 7½ inches.
Tonnage	6,050.
Displacement	9,492.
Draught forward	26 feet.
Draught aft	26 feet 6 inches.

The system of framing adopted in former armored vessels has been preserved in its main features. The great weight of armor and machinery, together with the immense power to be developed, necessitates arrangements which shall give extraordinary strength to the hull. The chief characteristics of the system, as in other vessels, consists in the adoption of an inner bottom and short angle-irons connected by bracket-plates. Increased strength longitudinally is gained by the use of much deeper longitudinal frames than employed in many former vessels; an advantage in this feature is that the space between the two bottoms (4 feet amidships) is roomy and easy of access for cleaning and painting, operations essential to the preservation of an iron structure. Facilities are also afforded by these arrangements for letting in water between the bottoms to regulate the trim of the vessel. Provision is made to pump out any compartment required, the space being in several divisions. In addition to the strength and safety proceeding from these numerous water-tight cells between the two bottoms, great increased strength is gained by the employment of a heavy longitudinal bulkhead through the center of the ship, commencing at 40 feet aft of the stem and extending to within 40 feet of the stern. Besides the wing-passages, bulkheads on either side form longitudinal divisions of the hold, while advantage is taken of transverse bulkheads to form subdivisions for the magazines, shell-rooms, chain-lockers, shaft-passages, and passages between the engines and boilers. By the bulkheads the twelve boilers are subdivided into four separate sets of three each, and the engines of the twin screws into two sets. In other words, the center longitudinal bulkhead divides the engines of each screw; it also divides the boilers, six being on either side of it, besides which there is a transverse bulkhead aft of the boilers, one immediately forward of them, and one in the center of the six. These several water-tight bulkheads are so arranged that any one or more sets of boilers can be worked independently of the others. All communication can also be shut off from either set of engines, so that if one side of the ship be damaged the engines on the opposite side can be worked independently. In the event of damage to the bottom, or accident by fire or other causes, any one of the compartments can be shut off or flooded. All the bulkheads are butted at the joints, beautifully fitted, and strongly secured as one rigid bridge. The water-tight doors on the lowermost deck are fitted with hinges having loose pins, and are secured when shut by levers placed at short intervals all around the edges of the doors, which may be worked from either side of the bulkhead. The doors in the hold are made to slide up and down, being raised or lowered by screws worked from the main deck. Flooding arrangements are fitted to the magazines, shell-rooms, and torpedo-rooms, proper stop-cocks with locks being fitted in each case to prevent the possibility of water being let in by mistake. Excellent facilities for pumping have been applied, to be worked by steam, also by hand, from the decks. Drain-pipes are placed between the two bottoms, giving control over the water in every compartment, so as to fill or empty them; the former when they are used for carrying water-ballast, the latter when they are pumped out in case of accident.

The frame spaces and the hollow masts constitute excellent ventilating tubes, the masts especially being good uptakes. To prevent the liability of the inhabited decks becoming contaminated, great attention has been given to the necessity of conveying the foul air away to the upper deck by distinct pipes from the hold and from the berth-decks. This point has not, however, been carried to the extent it deserves. The last consideration of atmospheric influence consists in providing means for closing all ventilators in event of fire. This seems the more important in the case of hollow masts used as ventilating-tubes, for if a fire should occur in the hold, the masts at once become tall chimneys creating enormous draughts to fan the flames. One case of this kind is known to have occurred in the mercantile vessel *River Boyne* within a year or two, and it is probable other unrecorded cases have happened.

MOTIVE MACHINERY.

The *Alexandra* is the first cruising armored broadside-ship of the royal navy engined on the compound system. The machinery was designed and constructed by Messrs. Humphrys & Tennant, at their works, Deptford-on-the-Thames.

As in all late armored ships of the royal navy, twin screws are applied to the *Alexandra*. Each screw is driven by an independent set of engines with three vertical inverted cylinders of the collective power of 4,000 horses, giving an aggregate indicated horse-power of 8,000 for both sets of engines. The diameter of the high-pressure cylinder is 70 inches, and the diameter of the low-pressure cylinders each 90 inches. The high-pressure cylinder is in the center; its faces are made separately, of hard, close-grained iron, 2 inches thick, and secured to the cylinder with brass countersunk screws. The linings or working-barrels of all the cylinders are made separately, and bolted to the cylinders at one end, and fitted with an expansion-joint at the other, with a width of steam-space of 1 inch between. The intermediate spaces between the high and low pressure cylinders are also steam-jacketed. The slide-valves are double-ported and fitted with packing-rings on the back, to relieve them of part of the steam-pressure. Chocks are fitted to the cylinders to stay them in the event of the ship being used for ramming. The crank-shafts are made in three pieces, which are interchangeable. The diameter of the bearings of the crank-pins is 17½ inches, and their lengths equal to the diameters. The crank-shaft brasses are lined with white-metal, and so fitted that they can be removed without necessitating the removal of the shaft, and each top brass and cap has a hole large enough to admit a man's hand for the purpose of ascertaining its temperature. The diameter of the propeller-shafting is 16 inches, except the stern-shaft through the tube, which is 18 inches exclusive of the brass casing. The tubes through the stern are of one length, and of gun-metal, and the driving-shafts within them are also cased with gun-metal cast on. The length of the lignum-vitæ bearings in the tubes are at the after end 4 feet 6 inches, at the forward end 2 feet 3 inches.

The engines are raised considerably above the inner bottom of the ship, with the view to prevent damage to them in case of accident to the ship's bottom. The propeller-shafts are in consequence inclined toward the stern. The surface-condensers, one to each set of engines, are so fitted as to be worked as common jet-condensers if necessary. They contain an aggregate of 16,500 square feet of cooling surface. The tubes are of solid drawn brass, $\frac{5}{8}$ inch in diameter, and fitted to be packed at each end in the usual manner practiced at present in Great Britain. The water is supplied by means of centrifugal pumps, worked by independent engines. The air-pumps, two to each low-pressure cylinder, are worked directly from the main pistons.

Each set of engines is fitted with two feed-pumps and corresponding bilge-pumps; and an auxiliary engine is fitted to each boiler-compartment, capable of working another feed-pump having a set of feed-pipes, feed-cocks, and overflow-valves, separate and distinct from the pipes and apparatus belonging to the main feed-pumps. Two additional auxiliary engines (one to each engine-room) are fitted as fire-engines. A double hand-pump to each set of engines is also provided; also a hand-pump to each screw-tunnel, with all necessary attachments, for the purpose of drawing water from the lowest point in the ship. The screw-propellers are of the Mangin type, 21 feet in diameter, and work outward.

The boilers are twelve in number, divided by bulkheads into four several sets. They are placed in the ship back to back against the longitudinal bulkhead. The fronts face the sides of the ship; consequently they are fired from fire-rooms convenient to the coal-bunkers. An additional advantage in this arrangement consists in keeping the sides of the ship clear of boilers, and accessible in the event of torpedoes or rams piercing holes through the sides of the ship The four sets of boilers are arranged to be used separately, in sets or singly. Each boiler contains three furnaces 40 inches in diameter and 6 feet 6 inches long. The total heating surface is 21,900 square feet, and the pressure of steam to the square inch will be 60 pounds. The smoke and gases are all carried into one chimney. Each boiler is fitted with an internal steam-pipe to obviate the effects of priming. This pipe is of brass, $\frac{1}{8}$ inch thick; it extends the whole length of the boiler, and has narrow transverse slits through which the steam must pass. "So as to prevent any catastrophe through the safety-valves getting out of order, each boiler, in addition to the ordinary safety-valve box, containing two spring safety-valves, has a supplementary test-valve loaded with a lever and weight and placed on the front of the boilers. There are also two pressure-gauges to each boiler, one being intended to act as a check to the other, the working gauge being graduated to 80 pounds and the other to 120 pounds. The stop-valves and safety-valves are all worked from the stoke-hole floors, and the main engine stop-valves are so arranged that they can be either worked from the engine-room or from the main deck. The whole of the boiler mountings, including the safety-valves and their boxes, are made of gun-metal."

Two blocks of zinc, measuring 12 by 6 by 2 inches, are suspended in each boiler with the view of preventing corrosion. The furnaces are made in short lengths, riveted together with flanges, having short distance-pieces between them. The furnaces, the tube-plates, and fire-boxes are of Lowmoor iron, all other parts of B B Staffordshire iron. The thickness of the plates is as follows: tube-plates $\frac{3}{4}$ inch; shells $\frac{7}{8}$ inch; back $\frac{5}{8}$ inch; front, above and below the tube-plates, $\frac{5}{8}$ inch; all other parts $\frac{1}{2}$ inch. The shells of the boilers are double-riveted, and longitudinal seams have butt-plates inside and outside. The grate-bars are of wrought-iron, 3½ inches deep by 1¼ inches wide, and made in three lengths. The fire-rooms are ventilated by means of circular fans, located well up, and driven by small engines.

The complement of engineer officers is one chief and ten assistants, and the complement of stokers, eighty. In addition to the care and management of the enormous machinery to propel the ship, the engineer officers have charge of the machinery for steering the ship, for hoisting the anchors, for ventilation, and all the water-tight doors, valves, and cocks in the ship. There are also two pairs of auxiliary engines, immediately aft of the main engines, for revolving the screws when disconnected from the main engine with the ship under sail. These engines work the screws through the intervention of brass bevel-wheels, working into wood cog-wheels on the couplings of the disconnected shafts. The ship has three masts, is bark-rigged, and is designed as a cruiser. Ventilators are abundantly provided, carrying fresh air throughout the ship; water-pipes are also extended along the decks, with attachments to the steam and hand fire-engines. The estimated speed of the ship is 14 knots as the maximum, and it is believed that 12½ knots under sail may be attained under favorable circumstances.

The mind of an officer who has passed his sea-life on board the old type of wooden ships of war, and become accustomed to their low, dark "between-decks," small and badly-arranged air-ports, wretched ventilation, and horrid bilge-water odor, together with the familiar cast-iron smooth-bore guns, mounted on wooden carriages, must be impressed to a degree of astonishment, when, for the first time, he enters the batteries of the *Alexandra*, and sees the great rifled guns mounted on Scott's system of wrought-iron carriage; the unusual height between decks, 10 feet 4 inches in the upper battery, and 9 feet 6 inches in the lower. But lofty and spacious as these battery-decks are, his surprise would be still greater upon entering the berth or living deck. Here will be seen the extraordinary height between the berth-deck planks and the gun-deck beams of 11 feet 6 inches, equal to the lofty ceiling of a modern dwelling-house. He would also be impressed with the large air-ports; pleasant, light, commodious state-rooms for the officers; and a wardroom centrally located, with a passage between it and the state-rooms on either side; an arrangement for convenience and comfort unknown to elderly officers. The admiral's cabin is of small dimensions and aft of the wardroom, and the captain is quartered on the upper deck, directly over the admiral.

At the official trial on the measured mile, a speed at the rate of 15½ knots per hour was attained, with satisfactory working of the machinery. The revolutions made by the screws on this trial were 67 per minute, and 8,600 indicated horse-power was developed, which was 600 horse-power above the contract. During the six-hour trial, 8,300 indicated horse-power was obtained "without difficulty, under somewhat unfavorable circumstances."

PART VI.

THE TÉMÉRAIRE, AND SYSTEM OF WORKING THE BARBETTE-GUNS; THE SHANNON.

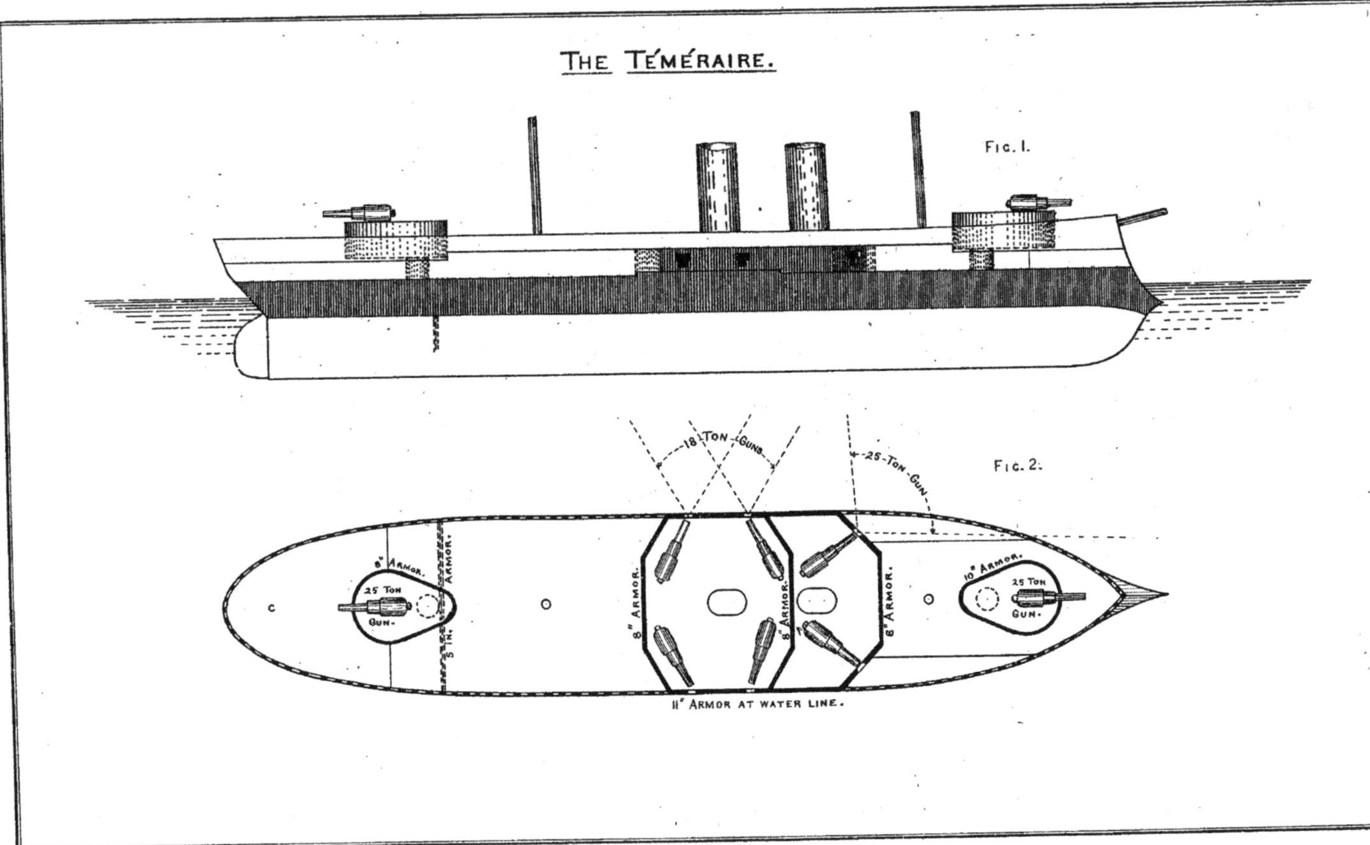
THE TÉMÉRAIRE.
FIG. I.
FIG. 2.
18 TON GUNS.
25 TON GUN
8" ARMOR.
25 TON
GUN.
ARMOR.
5 IN.
8" ARMOR.
8" ARMOR.
8" ARMOR.
10" ARMOR.
25 TON
GUN.
11" ARMOR AT WATER LINE.

THE TÉMÉRAIRE.

The *Téméraire*, building at the Chatham dock-yard, is designed for a sea-going ship. Her most important feature—the feature, in fact, which distinguishes her fundamentally from all other armored ships of the British navy—is that she carries the upper-deck armament in two fixed open-topped turrets instead of a central battery. At each end of the upper deck is a pear-shaped tower or battery, standing about 6 feet above the deck, and measuring about 33 feet on its longest axis by 21 feet 6 inches across. This contains a turn-table, on which is mounted a 25-ton gun, worked by hydraulic machinery, on Mr. Rendel's disappearing principle; that is, the gun is raised to be fired over the edge of the tower, and immediately after firing sinks under cover to be reloaded. As the sides of the vessel from the level of the upper deck to that of the main deck are, of course, not armored at the extremities, connection is maintained between the tower and the lower part of the ship by an armored trunk or tube, so placed that on the gun being revolved after firing, into the fore-and-aft line, with its muzzle toward the middle of the ship, the muzzle comes just over the opening, ready for the fresh charge from below. It must, inconveniently, always be brought to one position for loading. The French engineers have, from the first introduction of armored ships, entertained a noted antipathy to revolving turrets. Their objections were to protecting the guns by a weight equal to that of those guns and their ammunition; to a system which prevented an enemy from being clearly seen, and to the impossibility of getting an all-round fire with two turrets on a line; while the advantages they claimed for the barbette system were that an enemy could be clearly seen, and that the freedom and *morale* of the gunners were better assured in a barbette than in a turret battery.

The French naturally defend their own system against the opinions of all other European naval authorities. In the usual sense of words, the guns in the open battery of the *Téméraire* are not barbette-guns at all. They are fired *en barbette*, just as guns in Moncrieff gun-pits are, but they are not *en barbette* at any other time, which is an important distinction. They and their crews are not exposed to one-half of the risks which attend guns mounted permanently above the parapet of the battery, as is the case in all the French open-top turrets. The upper-deck guns of the *Téméraire* have much more in common with the Moncrieff system of mounting, or even with guns in ordinary turrets, than with the old and objectionable system of barbette-firing. But it remains to be seen whether this disappearing system of Rendel's, to be tested in the *Téméraire*, will be successful.

The foremost turret is protected with 10-inch armor, the after one with 8-inch. The guns have a clear sweep around the respective ends of the ship to some distance abaft or forward of the beam, as the case may be. In order not to obstruct the fire the bulwarks are kept low, about 4 feet above the deck, an arrangement hardly to be avoided, but likely to be objected to by sailors. The high bulwarks of ordinary men-of-war are liked by the men for the protection they give against wind and wet; on the other hand the *Téméraire* gains an upper deck, with no break in it except the poop and forecastle (and the fixed turrets, which are partly inclosed in them). All recent cruising iron-clads had an upper deck here, interfering more or less with the working of ropes; this is got rid of in the *Téméraire*.

Like all belted ships, the *Téméraire* has weak places at her water-line; but amidships, over the most vital parts, she has 11-inch armor (against 12-inch in the *Alexandra*), reduced slightly above and below; it is also tapered toward the bow and stern.

At the bow, to guard against exposure to raking fire in pitching, the armor is carried down over the point of the ram, and similar protection is gained for the magazines, &c., against raking fire from aft, by an armored bulkhead across the hold (shown in the sketches); this is plated with 5-inch armor. The iron deck at the level of the top of the belt outside the main-deck battery is $1\frac{1}{2}$ inches thick. The hull, which has the usual double bottom, and is divided into very numerous water-tight compartments, is built on the well-known bracket-frame system, and it is sheathed externally with wood covered with zinc. In like manner with other armored ships a ram is fitted which, in this case, projects 8 feet beyond the bow. The system of framing for the hull is quite similar to that of the ship just described. The vertical keel-plate is of steel 45 inches deep by $\frac{5}{8}$ inch thick; it is secured to the flat keel-plates by angle-irons 5 by 5 inches by $\frac{3}{4}$ inch. The inner and outer keel-plates are respectively $\frac{3}{4}$ inch and 1 inch in thickness. The longitudinals, of which there are six, have dimensions varying from 41 inches by $\frac{1}{2}$ inch to $31\frac{1}{2}$ inches by $\frac{7}{16}$ inch. The transverse frames behind the armor are 10 inches by $3\frac{1}{2}$ inches by $3\frac{1}{2}$ inches by $\frac{7}{16}$ inch thick, and are spaced 2 feet apart; those in the double bottom are 4 feet apart, while the water-tight frames are separated by a space of 20 feet. The frames above the armor-belt forward and aft of the battery are 4 feet from center to center, and of angle-irons 7 inches by 3 inches by $\frac{7}{16}$ inch; forward of the battery these frames are connected to the deck-plating by brackets $\frac{3}{8}$ inch thick and angle-irons 4 inches by 3 inches by $\frac{7}{16}$ inch. Behind the armor, and secured to the skin-plates, longitudinal girders of angle-iron 10 inches by $3\frac{1}{2}$ inches by $\frac{7}{16}$ inch are worked.

The athwartship water-tight bulkheads are seven in number. They extend from the inner bottom to the main deck, the thickness being $\frac{7}{16}$ inch below the berth-deck and $\frac{3}{4}$ inch above it; they are secured and stiffened by angle-irons, and each one is fitted with a water-tight door, so arranged as to be worked from the main deck.

The deck-beams are made with solid welded knees; those of the lower deck are bulb T-irons, 9 inches deep, with $3\frac{1}{2}$-inch top-flanges; those of the main deck under the battery are formed of plates 16 inches deep by $\frac{1}{2}$ inch thick, stiffened on the upper edge by angle-irons 3 inches by 3 inches, and on the lower edge $2\frac{1}{2}$ inches by $2\frac{1}{2}$ inches. Those of the same deck, fore and aft of the battery, are of the same kind, but 11 inches deep, and those of the upper deck are of T-iron, 10 inches deep, with a 6-inch flange. They are rounded 7 inches in their length and spaced 4 feet apart. The outside plating of the hull varies in thickness according to the strength and stiffness required, being 1 inch, $\frac{7}{8}$ inch, $\frac{3}{4}$ inch, $\frac{11}{16}$ inch, $\frac{5}{8}$ inch, and $\frac{1}{2}$ inch. All seams are double-riveted except the after hood ends, which are treble-riveted.

A bilge-keel, 112 feet long by 2 feet deep, outside of the wood sheathing, is fitted to each side of the hull; they are made of zinc plates $\frac{1}{2}$ inch thick, wedge-shaped and stiffened by wood-filling between the two plates of each bilge-keel.

The weight of the armor and backing is about 2,300 tons, or nearly the same as in the *Alexandra;* the bunkers contain only 600 tons of coal; and the guns, ordnance stores, engines, boilers, and all other equipments weigh about 2,200 tons. These weights, amounting in all to 5,100 tons, are carried by a hull weighing 3,300 tons. The spread of canvas is considerable, being 23,380 square feet, but it is carried brig-fashion, on two masts only, to avoid obstructing the end-on fire of the upper-deck guns. It is intended that the top-hamper may, if necessary, be disposed of overboard when going into action.

The following are the principal dimensions and other data, with like information added for the *Alexandra* for the sake of comparison :

	Téméraire.	*Alexandra.*
Length between perpendiculars	285 feet.	325 feet.
Breadth, extreme	62 feet.	63 feet 8 inches.
Draught aft	27 feet.	26 feet 6 inches.
Draught forward	26 feet 6 inches.	26 feet.
Displacement	8,412 tons.	9,492 tons.
Indicated horse-power (by contract)	7,000	8,000
Speed (intended)	14 knots.	14 knots.
Armor:		
Maximum thickness on belt	11 inches.	12 inches.
Thickness on batteries	10-inch, 8-inch.	8-inch, 6-inch.
Guns of 25 tons	4	2
Guns of 18 tons	4	10
Weight of broadside-fire	2,600 pounds.	2,600 pounds.
Weight of bow-fire	1,800 pounds.	2,000 pounds.
Weight of stern-fire	600 pounds.	800 pounds.
Cost, estimated	$1,817,640	$2,532,060

ARMAMENT.

The *Téméraire*, from the upper deck, fires right ahead one 25-ton gun; right astern, the same; and through a large arc on the beam, two. On the main deck, protecting also the smoke-pipe, is the *Téméraire's* double or divided battery, shown in plan in Fig. 2, and resembling, except in being shorter, the main-deck battery of the *Alexandra.* The forward part contains two 25-ton guns, whose arc of training extends from slightly abaft the beam on each side to slightly across the fore-and-aft line, so as to secure a converging fire at some distance ahead of the vessel, as already described in the case of the *Alexandra.* These guns, of course, fire from corner ports, and the sides of the ship above the main deck or top of the belt forward are set back several feet; this is shown in Fig. 2. The after part of the battery contains four 18-ton guns, for broadside-fire only. On the whole the *Téméraire* fires three 25-ton guns right ahead; on either bow, two 25-ton; right aft, one 25-ton; on either quarter, one 25 ton; on either beam (if engaged on one side at a time), two 25-ton and two 18 ton, with a third 25-ton gun available through only half the usual arc.

The guns of the *Téméraire* are better defended than those of any other broadside-ship, and this fact, coupled with a water-line defense nearly equal to that of the *Alexandra,* an armament which many will prefer to hers, and a much less size and cost, should give her the character of being the most successful masted ship, provided the system of working the barbette-guns prove successful. She is at least immeasurably superior to everything, however large, which preceded the *Alexandra.*

MOTIVE MACHINERY.

The steam-machinery has been designed and constructed by Messrs. Humphrys, Tennant & Co., of Deptford, and the contract provides that the engines shall indicate 7,000 horse-power. Though resembling in general appearance and construction the machinery which was supplied by the same eminent manufacturers to the *Dreadnought* and *Alexandra*, the engines differ from them in several important details, the principal variations being that, in consequence of want of room, the cylinders are limited to two. They are of the compound vertical inverted type. Each of the twin-screws is operated by an independent pair of engines, which, with the boilers, are separated by a longitudinal bulkhead. The diameter of the high-pressure cylinders is 70 inches, of the low-pressure 114 inches, the stroke 3 feet 10 inches, and maximum revolutions about 70. The air-pumps are worked directly from the pistons. The crank-shafts are 17½ inches in diameter, coupled in the center, and the two sections are interchangeable. The screw-propellers are of the Griffith new type, each having a diameter of 20 feet, a pitch of 23 feet 6 inches (variable from 19 to 24 feet), and an immersion of the upper edge of 4 feet 10 inches at deep load draught. A novel feature in the design of the engines, introduced here for the first time, has been the employment of wrought iron, steel, and brass to a large extent in lieu of cast iron, the cylinders, their valves and covers, being the only parts made of that material. Thus the whole of the framing is constructed of wrought iron, the bearings of the crank-shafts being also formed of heavy forgings of the same tough metal, and connected to box-girders of wrought-iron plates; while for additional security and strength the girders are riveted to the ship's framing, and are thus made to form a part of the general structure of the hull; the cylinders are also supported on wrought-iron box-girders placed vertically and strengthened by wrought-iron columns. The whole of the condensing apparatus, including the tube-cases, air-pumps, their connections, &c., are made of brass. The cases are made each in four pieces and bolted together; they contain 11,236 solid drawn brass tubes 7 feet 7½ inches in length, with an external diameter of ⅝ inch; they are tinned on both sides, and each tube is secured in its place by a stuffing-box tapped into the plate with a canvas washer behind it. The total cooling surface is 14,000 square feet. The water is circulated through the condensers by means of centrifugal pumps, which are driven by independent engines. The valve-faces of the high-pressure cylinders are of phosphor-bronze, secured in place by composition screws. Altogether the whole structure presents an appearance of lightness and beauty, composing a splendid piece of workmanship.

Boilers.—The steam is furnished by twelve boilers, elliptical in shape, containing three furnaces in each. They are placed in the ship back to back, against the longitudinal bulkhead, with the fronts facing the sides of the vessel, and consequently fired with convenient access to the side coal-bunkers. They are divided by bulkheads into four several sets, in the same manner as those in the *Alexandra*, previously described. Any one set or any one boiler can be worked independently of the others. The whole of the boiler-mountings, including the stop and safety valves and their boxes, are made of composition. The working-pressure of steam is 60 pounds per square inch, and the boilers have been tested up to 120 pounds. As in other recent twin-screw ships, the engine and boiler rooms are divided into two, longitudinally, to limit the entry of water and its ill consequences to the engine and boilers in case of injury from rams or torpedoes.

Engines exclusive of the motive power.—Besides the main engines described, the *Téméraire* is provided with thirty other steam-engines. These include two pairs of small engines placed near each screw-shaft coupling for the purpose of turning the great engines, when they are not at work, so as to bring the pistons, steam-valves, or other parts to convenient points for examination and adjustment from time to time, as required; two starting-engines, for the purpose of starting or reversing the main engines; four feed-engines, for supplying the boilers with water, or drawing it therefrom; two circulating-engines, for forcing water through the condensers; two bilge-pump engines; four pumping-engines, to free the water-bottoms, or to be used in the event of fire or accident to the hull under water; four engines for hoisting ashes, coal, or provisions; four engines for working the ventilating-fans; one capstan or anchor-hoisting engine; one engine for steering the ship; two engines for working the hydraulic gear of the guns; an engine to charge the torpedo air-reservoir, and an engine to work the electric machine which feeds the lights on the bridge.

OFFICIAL TRIALS OF THE MOTIVE MACHINERY.

The measured-mile trial in Stokes Bay was made before all the weights were placed on board, the draught of the ship then being 25 feet 4 inches forward, and 26 feet 2 inches aft.

Six runs were made over the mile, with and against the tide, with results reported as follows: first, 13.846 knots; second, 15.319 knots; third, 13.636 knots; fourth, 15.859 knots; fifth, 13.636 knots; and, sixth, 15.721 knots. The mean of the means showed a speed of 14.65 knots, with an indicated horse-power of 7,697. The amount of coal consumed during the trial was 51 tons 2 quarters, being equal to 2½ pounds per indicated horse-power per hour, a result comparatively low in consideration of the fact that the fires had to be pushed to the utmost, regardless of economy. The six-hour trial for endurance was made on the 17th of September last, after all the weights were on board and the ship ready for sea; the draught of water at this time being, forward, 26 feet 8 inches, and aft, 27 feet 4 inches, or about the same as the estimated draught of the ship. The sea was smooth, and the run was made near Cowes. The following table shows the results of each of the twelve half-hours during which observations were taken, as reported by the London *Times:*

Pressure of steam, in pounds.	Vacuum, in inches.		Revolutions.		Indicated horse-power.
	Starboard.	Port.	Starboard.	Port.	
49	28.75	28.75	71	70	6,462.98
57	28.75	28.75	74.4	74	7,538.94
57	28.5	28.5	73.6	73.9	7,470.41
57.5	28.25	28.25	74.7	74.4	7,784.19
57	28	28	74.3	74.5	7,562.12
59	28	28	74.2	74.2	7,796.38
56.5	28.25	28	73.6	74	7,447.03
58.5	28.5	28	74.5	74.5	7,517.14
60	28.25	28	73.6	73.7	7,585.61
58.5	28.25	28	74	75	7,586.19
61	28	28	73.3	74.8	7,723.17
59.5	28	28	72.4	72.7	7,644.53

The means were: Pressure of steam in boilers, 59 pounds. Vacuum in condensers—starboard, 28.20 inches; port, 28 inches. Revolutions per minute—starboard, 73.60; port, 74.13. Pressure of steam on square inch of piston—starboard, 26.6 pounds high, and 11.7 pounds low; port, 26.1 pounds high, and 11.68 pounds low. Indicated horse-power—starboard, 3,801.09; port, 3,782.95. The total collective power developed by the engines during the six hours was thus 7,584.04 horses, or 584.04 beyond the contract. * * *

Comparing the results with the measured-mile data, and taking four consecutive half-hours, counting from the third, as an equivalent for the mile-runs, we have 7,653 horses as compared with the 7,696 horses on the Maplin Sands. * * *

Subsequently the ship made a trial run at various speeds, under trying conditions of weather, between Spithead and Queenstown, and when in a fresh gale she is reported to have made the extraordinary speed of nearly 14 knots per hour, the wind doubtless being favorable. It is said that during the roughest weather she was remarkably steady, and that her barbette-guns might have been easily and effectively worked, and her after main-deck guns were available the whole time.

For comparison of the motive machinery of the three recently-constructed powerful armored ships, engined by Messrs. Humphrys & Tennant, the following table is given:

	Dreadnought.	Alexandra.	Téméraire.
Type of engines	Vertical, compound; twin screw; three cylinders driving each screw.	Vertical, compound; twin screw; three cylinders driving each screw.	Vertical, compound; twin screw; two cylinders driving each screw.
Cylinders:			
Number of	Six	Six	Four.
Diameter	Two of 66 inches; four of 90 inches.	Two of 70 inches; four of 90 inches.	Two of 70 inches; two of 114 inches.
Length of stroke	4 feet 6 inches	4 feet	3 feet 10 inches.
Diameter of crank-shaft	17½ inches	17½ inches	17½ inches.
Screws:			
Diameter	20 feet	21 feet	20 feet.
Pitch	23 feet 6 inches	22 feet 3 inches	22 feet 6 inches.
Type	Four-bladed Griffith	Mangin	Two-bladed Griffith.
Condensers:			
Number of	Two	Two	Two.
Cooling surface	16,500 square feet	16,500 square feet	16,500 square feet.
Type	Surface	Surface	Surface.
Boilers:			
Number of	Twelve	Twelve	Twelve.
Total grate surface	820 square feet	780 square feet	780 square feet.
Total heating surface.	22,025 square feet	21,912 square feet	19,824 square feet.
Six-hour trial:			
Pressure of steam	60 pounds	60 pounds	60 pounds.
Revolutions of engines.	67	64	74.
Indicated-horse-power.	8,206	8,313	7,518.
Speed of ship	14.52 knots per hour	15 knots per hour	14.65 knots per hour.

The steam steering-gear is operated by a set of Messrs. Brotherhood & Hardingham's three-cylinder engines, which is the first of its type introduced into a ship of war; it required several little alterations and adjustments after the first trial.

The *Téméraire*, in like manner with all recently-commissioned ships, is provided with the apparatus and appliances for using the Whitehead torpedo. On each side of the vessel forward, above the armor-plating, there has been fitted a tube, the diameter of which is 21 inches, for the purpose of ejecting those instruments of destruction. She is also supplied with the Harvey torpedoes, and with outrigger torpedoes, the latter to be used from steam-cutters. Gatling guns are provided for use to guard against the approach of the enemies' torpedo-boats.

The electric light which proved so successful on board the *Alexandra* has been applied to the *Téméraire*, and when tested in the river Medway, in August last, objects were distinguishable for a considerable distance in all directions around the ship.

SYSTEM OF WORKING THE BARBETTE-GUNS.

The principle of sinking guns entirely under cover from horizontal fire behind any sufficient parapet, and raising them only to deliver their fire, is quite old, and, like very many inventions introduced into European warfare, owes its origin to American genius.

It was proposed more than twenty years ago by officers of the United States Army for our fortifications, and models were made representing the principle of storing and utilizing the force of recoil; *i. e.*, the gun on delivering fire and sinking behind the wall raises a counter-weight, the fall of which again lifts the gun when required; and some years ago Captain King, Engineer Corps, United States Army, successfully applied to one of our forts a carriage of his invention on this principle.

Captain Eads, of Saint Louis, Mo., invented as early as 1861, and soon after successfully applied to the two-turreted gunboats *Winnebago* and *Milwaukee*, built at that time on the Mississippi River, a system of mounting heavy guns on a turn-table within a rotating turret. The table, with the guns and their attachments, was raised, lowered, and revolved by steam-power; the guns were also moved out to the firing positions by the same medium, and the recoil was taken on steam-pressure.

For the purpose of loading the guns, the table was lowered to the berth-deck. The work of construction was done under the government supervision of the writer. The trial tests of the machinery and firing of the guns to test rapidity and accuracy were personally executed by him, and an official report of the machinery and results of the target-firing was also made by him April 30, 1864, to the Secretary of the Navy, and published in pamphlet form.

Subsequently Captain Eads invented and patented the principle of raising and lowering guns by the elastic force of compressed air, the mechanical appliances being very similar to those afterward used by Major Moncrieff in his second invention, where he has substituted for the counter-weight, air compressed by the recoil through the medium of water; this part of the Moncrieff invention is thus described:

> The gun being supported in firing position on levers, supplemented by a ram working in a cylinder which is in communication with a vessel the upper part of which is filled with compressed air, the lower portion containing water. The air has an initial pressure given it sufficient to raise the gun. When the gun is fired the energy of the recoil drives the ram down into the cylinder, forcing the water up into the air-vessel, thus further compressing the air. A self-acting valve prevents the water from returning after the recoil has been completed. When the gun has been loaded behind the protecting parapet a valve is opened and the water allowed to flow into the cylinder. The air-pressure is thus brought to act on the ram, which at once raises the gun into the firing position. No power beyond that obtained from the discharge of the gun is required for working the gun, the air-vessel remaining always ready for use.

The Rendel system as applied to the *Téméraire* is analogous to that originated by Eads, except that the power used by the former is applied through the medium of water, that used by the latter being air. An important distinction, however, is in the fact that as here carried out no attempt has been made to store up and utilize the force of recoil of the gun, that force being taken on a hydraulic plunger working in a charged cylinder having a safety-valve loaded to about 750 pounds per square inch.

The towers in which the two 25-ton guns are mounted are 7 feet in depth, and are pear or egg shaped, the guns being placed within the broad part of the egg. The circular platform is rotated by means of hydraulic presses, which are fitted within the structure of the platform itself; the platform is arrested by a weighted pawl, which falls into notches in much the same way as may be observed in the turn-tables of railway-stations. The gun itself is raised and lowered by means of massive forged bell-crank levers, of which the heads are attached to the trunnions of the gun, and the elbows work on bearings upon the platform, the extremities being connected with hydraulic pistons, the outward or inward thrust of which imparts the upward or downward motion to the piece. The elevation or depression of the gun is accomplished by means of an elevating arc, which is actuated by a wheel and pinion after the ordinary manner, and the radial action of which, in conjunction with that of the lever, always enables the gun to be brought to the same plane—3° of inclination—for loading. The sights are fitted to the platform so that the gun may be elevated and laid while being revolved into position for firing, the gunners being at the same time protected by a bullet-proof shield. The powder and shell are brought from the magazine directly to the mouth of the gun, without the circumlocution of trolleys, by means of a hydraulic hoist working up and down an armored shaft or well, 3 feet 6 inches in diameter, in which also are placed the pipes communicating with the presses. The upper story, so to speak, of the cradle contains the cartridge, and the lower the projectile. After the former has been introduced into the gun by a push of the hydraulic rammer, the hoist is lifted a step higher and the projectile and the cartridge are forced home. The rammer, levers, and gearing are placed at the small end of the egg-shaped belt, and are protected by a splinter-proof. Indeed, the gun is the only thing which is exposed in the act of firing. The hydraulic machinery is actuated by a couple of small engines, which may be used either in combination or separately, and which, though placed within the armor-belt below the water-line, are each worked from within the turrets.

The final trial of the disappearing carriages and hydraulic apparatus for loading, training, and working the guns in the towers, was made November 13, and attracted much attention from the officials who were present. Fourteen rounds were fired from the after tower and eleven from the forward, with charges of 85 pounds of powder, the projectile weighing 530 pounds. Including the firing on former occasions, fifty rounds in all have been discharged from the barbette-guns, sufficient, it is thought, as a test of endurance in respect to the mechanism. Many of the rounds were fired at a floating target, but four were fired against time for the purpose of testing the rapidity with which the gun could be loaded, laid, and discharged, and also of proving the hydraulic gear under such conditions. From fire to fire the time was 1½ minutes. The number of men required to work the gun being one man to lay and fire electrically, two men to attend the elevating-gear, one man to take charge of the levers for lifting the gun and rotating the platform, and five men to manage the rammer and shot-hoist. It is not, however, rapidity of fire which is the most important point, for, considering the weight of the projectile, accuracy is everything, a few fair hits being probably all that will be required to disable an enemy.

Although the recoil of the gun with service charge amounts to 96 foot-tons, this enormous force is so absorbed by the water-presses that the recoil upon the cylinders did not exceed an average of twelve inches.

It has been reported that the success of the disappearing system as applied in this ship has not been such as to justify its adoption into other ships of the British service.

The whole of one day was devoted to the practical test of the torpedo-fittings of the *Téméraire*, and to a series of experiments with the Whitehead torpedo, an account of which will be found under the head of "Torpedo Warfare."

but the double bottom is only 168 feet in length; it is divided into twenty separate water-tight spaces. There are nine principal athwartship water-tight bulkheads, and fifteen water-tight coal-bunkers; but there is no longitudinal bulkhead extending through the vessel, as in all recently-constructed armored ships having twin screws. This backbone of strength and safety becomes impracticable in single-screw vessels.

The stem of the ship has fitted to it a shifting ram, the snout of which is 8 feet 3 inches above the keel, and extends 8 feet ahead of the stem. This ram is at present stowed on board the vessel, the idea being that as so many accidents have occurred in time of peace from the ram, and especially in view of the loss of the *Vanguard* from the blow by the ram of the *Iron Duke*, it would be more prudent to make the ram portable and to fit it in place only in time of war. In favor of this plan much can be urged, but it seems to suggest the questions—first, whether ships on foreign stations will be able in emergencies of war times to go into docks to have their rams secured in place; secondly, whether, if they should succeed in this, the officers, who up to that time will be deprived of all experience in guarding against accidents from it, will be able to avoid multiplying those accidents which have hitherto occasionally happened under the most ordinary circumstances, notwithstanding the experience they have acquired.

The outer hull of the *Shannon*, in common with that of all recently-constructed cruising-vessels of the royal navy, is sheathed with wood. The material is teak, put on in the usual way, in a single course, with the seams left uncalked, except above the water-line, for the purpose of admitting sea-water freely between the iron hull and the zinc with which the planking is covered.

ARMAMENT.

The armament is placed on an open deck not unlike the uncovered decks of corvettes. It consists of nine Woolwich guns, two of which are 18-ton guns, under protection of armor at the bow from raking fire ahead; six 12-ton guns (three on either broadside unprotected by armor), and one 12-ton stern-gun, which is carried on a platform amidships aft, and is intended to be fought at a port on either side of the deck. It is also unprotected by armor. The two 18-ton bow-guns can be trained to fire on a line with the keel, or to any point around at right angles with it.

One of the features noticed in the design of this belted ship is the protection by horizontal armor at the top of the belt; an important feature, since the side-armor extends only 4 feet above water.

The plated decks protect the magazines, machinery, steering-gear, &c., from plunging fire of any guns that might be carried on an enemy's upper deck, and could easily send projectiles through the unarmored side above the belt.

A second feature is the system of coal-tanks introduced for the first time at the bow of the vessel. An English writer says:

> Portions of the ship that have no armor are protected by coffer-dams, which consist of iron boxes about two feet broad, filled with old rope, canvas, &c., to resist shot. The parts so protected extend from the main to the lower deck abreast the engines and boilers, and on the fore-side of the armor-bulkhead. The engine-hatch and other hatchways on the main deck will be protected in action by 2-inch iron shutters, which at other times will remain open.

Another noticeable arrangement is the adoption of two ventilating-cowls upon the outside of the vessel—one for carrying air directly to the fire-rooms, and the other for ventilating the coal-bunkers.

MOTIVE MACHINERY.

The ship is propelled by a single screw. The machinery was constructed by Messrs. Laird Bros., of *Alabama* fame. The engines are of the compound, horizontal, return connecting-rod type, with four cylinders, two high and two low pressure, the two high-pressure cylinders being placed behind and bolted to the two low-pressure, the pistons of the former being attached directly to the latter by a single piston-rod and working simultaneously with them.

The dimensions, weights, and other important data are:

Engines:	
Diameter of high-pressure cylinders	44 inches.
Diameter of low-pressure cylinders	85 inches.
Length of stroke of pistons	4 feet.
Diameter of crank-shaft at journals	17½ inches.
Diameter of air-pumps	22½ inches.
Stroke of air-pumps	4 feet.
Cooling surface of condenser-tubes	8,000 square feet.
Diameter of screw-propeller	19 feet 6 inches.
Pitch of screw, adjustable from	18 to 22 feet.
Revolutions of engines per minute, maximum	70
Indicated horse-power, maximum	3,540
Speed of ship on six-hour trial, maximum	12.5 knots.
Boilers:	
Number of	8
Diameter	12 feet.
Length	12 feet.
Number of furnaces in each boiler	2
Number of tubes in boilers	1,700
Dimensions of tubes	6 feet 6 inches by 3 inches.
Total grate surface	380 square feet.
Total heating surface	8,500 square feet.
Pressure of steam per square inch	70 pounds.
Weights:	
Engines, appendages, and spare gear, with water in condensers	266 tons.
Boilers, including everything between boiler-room-bulkheads; also, water in boilers	310 tons.
Propeller, shafts, &c	61 tons.
Total	637 tons.

Total cost of machinery, $245,430.

The length of the boiler-space fore and aft is 56 feet, and the width, including fire-room, is about 40 feet. The boilers are placed in the ship back to back, against a longitudinal bulkhead, consequently they are divided into two sets with fire-rooms facing the side coal-bunkers; in this position they are conveniently supplied with coal. A transverse bulkhead separates the boilers from the engines, and by a second transverse bulkhead forward they are separated from the hold; hence the central position of the boilers in the ship, the division into two rooms, and protection by water-tight bulkheads, give all the security possible in event of damage to the hull by rams, torpedoes, or other causes.

The engines operate a single line of screw-shafting. The screw-propeller is of Griffith's latest pattern, and is fitted to be disengaged from the driving-shaft and lifted when the ship is to be put under sail. The air-pumps, one to each low-pressure cylinder, are worked directly from the pistons. The circulating water is supplied by means of centrifugal pumps worked by independent engines. The starting and stopping is effected by a small engine under the control of one man. The engines are designed to work the steam expansively to any desired extent, and there is fitted a special arrangement of valves admitting of the use of steam of low pressure directly into the low-pressure cylinder, and this has been tested to the very moderate figure of 2 pounds; the object of this arrangement being to meet a danger which it is apprehended by some officers may arise in going into action with steam of high pressure. The coal-bunker capacity is only 500 tons; it is therefore evident that steaming will be the exception and sailing the rule in the *Shannon.*

It is to the Russian admiralty that the credit is due for the introduction of the first belted cruiser. It has been four or five years since they built the first vessel in which the vital part, *i. e.*, the water-line only, was protected by armor, leaving the guns and crew unprotected. The *Shannon* is, however, a notable improvement on the Russian idea, and yet it has been authoritatively stated that, while she is capable of taking part in general engagements if required, she was primarily designed for distant cruising service, the rig and sail-power being above the average for armored broadside-ships.

The trials of this ship at sea have not been as satisfactory as desired. An error appears to have been made in calculating the weights entering into the vessel, and this has been aggravated by additional weights put on board unprovided for. As a consequence the ship is immersed more than was anticipated, besides which alterations became necessary in the topmasts, and the machinery when on trial did not prove satisfactory.

ART VII.

THE NELSON AND NORTHAMPTON; THE WARRIOR; THE WATERWITCH; THE GLATTON; A REMARKABLE EXPERIMENT; A TORPEDO-RAM.

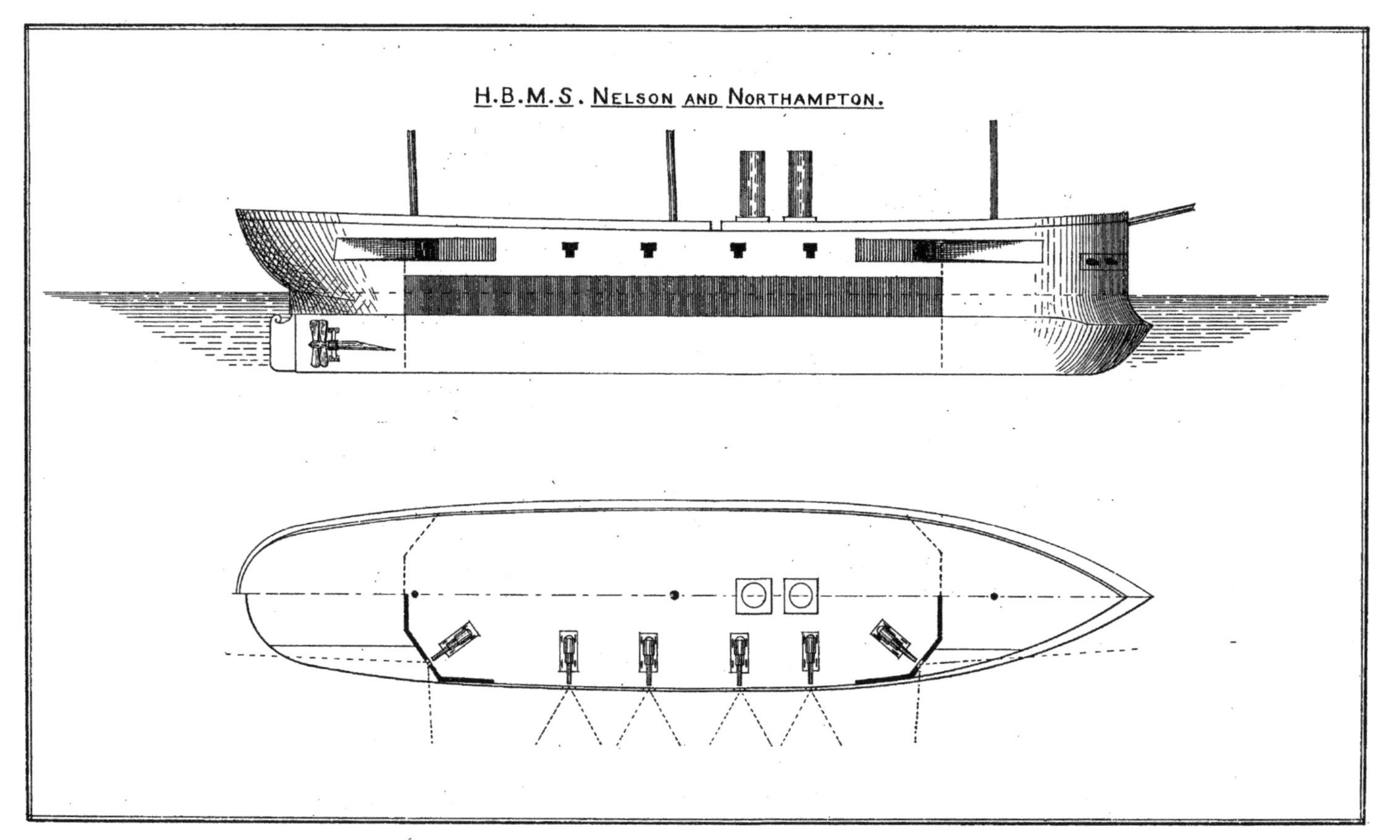
H.B.M.S. Nelson and Northampton.

THE NELSON AND NORTHAMPTON

These two sister ships, the former built by Messrs. Elder & Co., and the latter by Messrs. Napier & Sons, near Glasgow, on the Clyde, and just completed at the dock-yards, constitute a new type of ocean-cruising broadside armor-plated ships. They are the last productions of armored vessels by the chief naval architect, and before an audience in the summer of 1876, at the loan exhibition, South Kensington, he pronounced them to be his "ideal of cruising fighting-ships." A glance at the preceding longitudinal section and plan of gun-deck will convey an idea of the general design.

The length between perpendiculars is 280 feet; breadth, extreme, 60 feet; mean draught of water, loaded, 24 feet 2 inches; depth from upper deck, 42 feet; load-displacement, 7,323 tons.

The framing is on the usual longitudinal system adopted in the construction of Her Britannic Majesty's ships of war, and in this instance the longitudinal frames are made of steel, so as to combine lightness with strength. The double bottom extends for about 150 feet amidships, and the space between the inner and outer skins is divided into many water-tight compartments. According to the system recently adopted for armored ships, there is a central longitudinal bulkhead, and along the whole length of the engine and boiler spaces she is divided longitudinally by three water-tight bulkheads, besides numerous transverse bulkheads underneath the lower deck; also, wing-passage bulkheads. Altogether, including the spaces between the two skins, there are 90 water-tight compartments; all the doors leading to these compartments are likewise water-tight and are to be worked by machinery, and every conceivable precaution has been taken to provide against destruction by rams and torpedoes.

There are three principal decks: the lower, main, and upper.

The protecting armor consists of a belt on the water-line of about 181 feet in length amidships; this belt is 9 feet deep, 4 feet above water, and 5 feet under water. It is put on in two strakes; the upper plates are 9 inches thick on a 10-inch backing of teak, and the lower plates are tapered to 6 inches thick, supported by a teak backing 13 inches thick. Extending across the ship at each end of this armor-belt there is an armor-bulkhead; it starts at the bottom of the armor-belt 5 feet under water and extends to the upper deck, having in all a depth of 22 feet. Its thickness is 9 inches above water, tapering to 6 inches at the bottom. Between the main and upper decks these bulkheads are shaped to form corner ports at the fore and after ends of the battery. Between the armor-bulkheads, and at the upper level of the armor-belt, the lower deck is formed throughout of 2-inch plates, by means of which protection is afforded to the machinery, boilers, magazines, &c. A peculiar feature is the horizontal armor as here applied. For about 57 feet at the fore end there is an armor-deck. This deck is 3 inches thick, and it is 5 feet under water at the junction with the armor-bulkhead, but inclines deeper toward the stem and terminates forward in the ram. There is likewise a horizontal armor-deck of the same thickness and depth under water, extending from the after armored bulkhead to the stern. These submerged armor-decks are intended to protect the lower part of the ship fore and aft of the armored bulkheads, especially the steering gear provided against emergencies. From the above outline and reference to the annexed drawing, it will be seen that the central part of the vessel for 181 feet in length, in which all the motive machinery is contained, may be regarded as completely protected from ordinary shots of the enemy. The ends of the vessel above the submerged decks are entirely unprotected by armor, and may, it is supposed, be riddled with shot without serious injury to the flotation of the vessel.

ARMAMENT.

The armament consists of four 18-ton guns and eight 12-ton guns on the main deck, also six small guns on the upper or spar deck; the latter are designed to be used as torpedo-boat destroyers. Two of the 18-ton guns, one on either side forward and one on either side aft, are situated behind the oblique portion of the armor-bulkheads, and the ports are so cut that these guns can command a fire across the line of bow and stern. The eight 12-ton guns, four disposed equally on either side, are termed intermediate, and have in front of them the thin sides of the ship only. They are separated by a transverse bulkhead or splinter-screen 1 inch thick, intended to cut off each gun's crew from the others. This broadside of guns is designed to be loaded and laid in close engagement under the shelter of the bow or stern armor, and may be fired by electricity without exposing the crew.

The ram is a heavy plate, triangular in shape, set vertically, and terminating in a sharp point about 11 feet in advance of the stem; it is supported by two side plates 3 inches thick, which may be regarded as a continuation of the armor-deck. The rudder, which is massive, is 18 feet deep by 11 feet in breadth, and is formed by two thicknesses of teak planking set in a strong iron frame. The vessel has fitted to it bilge-keels 33 inches deep, formed of two plates riveted together and extending amidships about 100 feet; and the outer bottom of the hull below the water-line is sheathed with one course of teak planks 3 inches thick, while over that there is also a sheathing of zinc. The seams between the sheathing-planks are left uncalked, with the view of admitting free communication between the iron hull and zinc. There are to be three masts fitted, as for a full-rigged ship, and the coal-bunker accommodation is sufficient for a long voyage and cruising in distant seas. In time of war it is intended that only the lower masts shall stand.

The novelty of design worked out in the *Nelson* and *Northampton* consists in the system of armoring, and, as may be readily seen, the object contemplated is to give thicker plates to vessels of this class over the vital parts of the ship, at the expense of the exposed parts, and to increase the offensive power by carrying a heavier weight of ordnance.

MOTIVE MACHINERY.

The machinery of the *Nelson* was designed and constructed under the direction of Mr. Kirk, the manager of the engineering works of Messrs. Elder & Co. The ships are fitted with twin screws, each driven by an independent pair of compound engines, with vertical inverted cylinders, of the collective power of 3,000 horses, giving an aggregate power of 6,000 indicated horse-power for both pairs of engines. The diameter of the high-pressure cylinders is 60 inches, and that of the low-pressure cylinders 104 inches. The length of stroke is 3 feet 6 inches, and the number of revolutions to be obtained on the trial was 75 per minute. These engines are constructed entirely of wrought iron and brass, except the cylinders, cylinder-covers, steam and expansion valves. The two pairs of engines are fixed in the ship directly opposite each other. The central longitudinal bulkhead of the ship, however, divides the two pairs, and all the pipes are arranged so that, in the event of collision or other casualty to the bottom of the vessel, either pair can be worked separately. Each cylinder is supported by four wrought-iron columns, two cylindrical, the other two partly of channel section and serving as guides for the cross-head slippers. The bed-plates, or what serves in their stead, is also of wrought iron. For stiffening athwartships, there is at each end of the engines a wrought iron × frame extending across the ship from one pair of engines to the other, and in addition each front column has a small diagonal stiffener to the bed-plate, while longitudinal stays (wrought-iron bars in cast-iron tubes) connecting the upper part of the engines to the hull give stiffness fore and aft. The condensers are composed of brass plates riveted together. The valve-gear is a modification of the Allan type. The crank-shafts are in two parts, interchangeable, 16½ inches in diameter in the bearings, with an aggregate length of bearing of 9 feet 6 inches. The propeller-shafts are hollow, made of Whitworth's compressed steel. The screw-propellers are 18 feet in diameter, work outward, and they are of the Mangin type, *i. e.*, two 2-bladed screws on each hub, separated from each other.*

Boilers.—The boilers are ten in number, with three furnaces in each, set back to back against the longitudinal bulkhead, with the fronts toward the sides of the ship. They are oval in shape, 12 feet 6 inches wide by 14 feet 6 inches high, and 9 feet 6 inches long, and are divided by transverse water-tight bulkheads into four separate fire or boiler rooms. The furnaces are made in short lengths, riveted together with flanges having distance-pieces between them. The pressure of steam is to be 60 pounds per square inch. All the necessary pumps and equipments described for other vesels are here provided.

The machinery for the *Northampton* was designed and constructed by Messrs. John Penn & Son, of Greenwich. The engines of this ship are of Penn's new type, three-cylinder vertical inverted, the first design of which was made in 1875 for the Italian corvette *Cristoforo Colombo*, built at Venice. The three cylinders are of equal diameter, 54 inches, with a stroke of 39 inches. The cranks are set at equal angles, and the shafts are interchangeable. When the ship is to be driven at full speed, the engines are to be worked as simple expansive engines, cutting off the steam long or short, as desired; but under all ordinary conditions of steaming they are intended to be worked on the compound system, taking the steam first in the center cylinder and expanding it in the two outside cylinders, valves and connections being provided to effect the change from one system to the other when desired. Each cylinder rests on wrought iron columns in front, *i. e.*, on the side nearest the center of the ship, and on cast-iron columns behind; each has a main slide and an expansion-valve, each valve having its own link-motion. The starting-platforms are about midway up the engines, and placed between them, with direct communication between the two through the longi-

*The Mangin screws have given satisfaction as applied to several twin-screw vessels in England and France where the space permitted sufficient separation of the screws on the shafts; but as applied to one of our single-screw second-rates, in a short well, it proved, as might have been expected, unsuccessful.

tudinal bulkhead. The power to be developed by the engines of the *Northampton* is to be the same as provided for the *Nelson*, and the boilers are almost identical, ten in number, of the elliptical type, with three furnaces in each, and to carry a pressure of 60 pounds per square inch.

The ends of the ship are provided for coal-tanks.

On the trial trip made last autumn, with steam at 60 pounds pressure per square inch in the boilers, 27 inches of vacuum and 83½ revolutions, a mean of 6,037 horse-power was obtained. On the five-hour contractors' trial of the *Nelson*, made in February of this year, the mean indicated power developed was 6,250 horses, with a maximum speed of 15 knots.

THE WARRIOR.

This fine old ship, the first iron-armored ship built in England, was lying at Portsmouth during my visit to that place. It is worthy of note that while the first productions of all nations, of wooden ships clad in armor, have gone to decay, the iron hull of the Warrior, now seventeen years old, presents no sign of deterioration, being to all appearances as sound and as durable as when set afloat in 1860; and although large expenditures for repairs have been made on the vessel and her machinery, the hull proper has required no expenditure beyond that for its preservation by cleaning and painting. The description, dimensions, and performance of this ship have appeared in various publications many years ago. In consequence of her great length (380 feet), unhandiness, and thin armor (4½ inches), she is no longer regarded as suitable for action in great naval battles, but her success as to speed at so early a day in the race for naval supremacy seems to deserve attention still. In 1874 a new set of boilers, made at the Portsmouth dockyard, were fitted on board. They are of the old box type, having superheaters, with the usual appliances of this variety. The grate-area of the boilers is 780 square feet, and the heating surface 19,906 square feet.

The cylinders were rebored, the slide-faces and ports strengthened and braced, and the expansion-valves altered to give an earlier cut-off. After these repairs the ship was put on trial at the measured mile, to test the machinery and speed against the runs made over the same ground thirteen years previously.

The following is a comparison of the trials at full power on three several occasions:

	October, 1861.	April, 1868.	May, 1874.
Pressure of steam in boilers	22.00 pounds.	21.6 pounds.	21.5 pounds.
Revolutions	54.25	53.14	56.0
Indicated horse-power	5,471	5,270	4,811
Speed, in knots	14.354	14.079	14.158
Pitch of screw	30 feet.	30 feet.	27 feet 8¼ inches.
Immersion of screw	11 inches.	27 inches.	13 inches.
Ship by the stern	11 inches.	23 inches.	5 inches.

One curious result noted is the same speed on the last trial with 459 less horse-power than on the former trials many years previously.

THE WATERWITCH.

The *Waterwitch* was built as an experimental vessel, to test the Ruthven system of propulsion by a turbine wheel, or what is known as the water-jet engine. The vessel is built of iron; is 162 feet long, 32 feet broad, 13 feet 9 inches deep; has a load-displacement of 1,279 tons, and the indicated horse-power on the measured mile was 777. She has an excessively flat floor, is double-ended, and fitted with a rudder at each extremity. An armor-belt 4½ inches thick at the water-line extends around the hull, which rises at the middle of her length into a casemate rendered complete by athwartship bulkheads. The propelling instrument consists of a turbine wheel, or centrifugal pump, 14 feet 6 inches in diameter, made of wrought and cast iron. This wheel revolves in a chamber 19 feet in diameter, in the center of the hull, below the water-line, and the chamber is bored to a smooth surface inside, in order to reduce hydraulic friction to a minimum. The turbine has 12 radial blades or vanes, and weighs about 8 tons; it is put in motion by a set of three engines, arranged at angles of 120 degrees, the connecting-rods taking hold directly of a single crank rising vertically above the wheel-casing, an application similar to the manner in which the engines of the old *Union* and the *Alleghany* were connected to the once well-known Hunter wheels.* The engine-cylinders are 38½ inches in diameter and the stroke of pistons 3 feet 6 inches, and are supplied by steam from two ordinary box-boilers having 6 furnaces.

The wheel receives the water from a rectangular box, or tank, resting on the keelsons of the ship, and placed in free communication with the sea by means of a large number of rectangular orifices in the bottom. From the wheel-casing perimeter at opposite sides, two copper pipes, about 27 inches by 25 inches internally, lead to the discharge-nozzles at the ship's side. These are 24 inches by 18 inches, and extend about 8 feet along the side of the hull just above the water-line, so that the engines have to raise the water through a very small height. A sluice-valve is arranged at each side in such a manner that the current from the turbine may be directed ahead or astern at pleasure by simply moving a lever, the engines revolving always in one direction. The water taken in through the bottom of the ship is expelled at both sides in the line of the keel, and the reaction of the fluid issuing at high speed imparts forward motion to the hull. The movement of the vessel ahead or astern is regulated by the direction of the escape of the water. If the water escapes aft, the movement will be ahead; if it escapes toward the bow, it will be astern.

The idea is exceedingly simple and very old. As far back as 1661, Togood received a patent for propelling vessels by expelling water from

* As early as 1782, James Rumsey made a public experiment on the Potomac with a boat 80 feet long, propelled by a steam-engine working a vertical pump in the middle of the vessel, by which the water was drawn in at the bow and expelled through a horizontal tube at the stern; she went at the rate of 4 miles per hour. Benjamin Franklin and Oliver Evans suggested substantially the same mode of propulsion. Subsequently various applications of the principle were tried in the United States without success.

their sterns. In 1730, Allen secured a patent for nearly the same thing; and the proposal was also made by Bernouilli eight years subsequently. Indeed, the extreme simplicity of the system seems to have attracted many inventors, for down to the year 1857 it appears that upward of fifty persons have either proposed or patented the scheme in Europe, and many experiments were tried from time to time, but none of them received much encouragement until Mr. Ruthven entered the field, and the success, such as it has been, which attended his exertions, seems to have been mainly due to the adoption of the centrifugal pump, with equable and enormous delivery, instead of the ordinary piston-pump commonly adopted by other inventors.

Ruthven's first patent is dated in 1839. Under this, two small boats were built and exhibited on a canal at Edinburgh, Scotland. In 1849, another boat was built and exhibited on the Thames. In 1853, the *Albert* was built on this principle in Prussia by Mr. Sydel, the engines and pump being furnished by the patentee. In 1865, the *Nautilus* was built in England, embodying all of Mr. Ruthven's improvements up to that date. With this little vessel, several experiments were made in the presence of the admiralty authorities, the results of which betrayed them into the construction of the *Waterwitch*.

In consequence of the convenience of directing a vessel ahead or astern by the simple movement of a lever from the deck, this system of propulsion has been very fascinating to many officers; but unfortunately for this instrument of propulsion, in common with the Hunter wheel, the Fowler wheel,* and all such submerged water-wheels as applied to steam-vessels, an extraordinary power must be developed by the engines to obtain a small result; or, in other words, only a small amount of the power developed is utilized.

At the trial of the *Waterwitch*, a vessel of only 1,279 tons displacement,

* Fowler's steering-propeller is a submerged wheel revolving on a vertical shaft, with paddles which are feathered by an eccentric cam in such a manner that the paddles shall have a pushing and drawing action on the water while passing through the propelling arc, and present only their edges to the water while passing the dead-points. By turning the cam-wheel, which is done at the wheel on deck by a simple connection, the feathering is done at different points, and the vessel may be backed or turned on her center without reversing the engines.

The letters patent of Mr. F. G. Fowler, dated January 4, 1870, describe it as follows: "It is a submerged marine propeller, or feathering sculling-wheel. It consists of a vertical shaft, from which proceed horizontal arms, to the extremities of which are attached blades by pivots placed on their vertical central line. These blades oscillate on their pivots, as the propeller revolves, in such a manner that they exert a propelling force throughout their entire circuit except when passing two points or centers, when they are neutral. This oscillating motion is produced by an eccentric with which each blade is connected, and the propelling force is exerted in the direction in which the short radius of the eccentric extends. By suitable connections between the eccentric and helm the steersman is enabled to turn the eccentric, and thereby cast the propelling force to any point of the compass, by which means he is enabled not only to move the boat forward and backward in a direct line, but to steer it gradually to the right or left, or in a very short curve, or cause it to turn in either direction, on its own center and in its own length of water, the said arrangement serving the treble purpose of propeller, rudder, and reversing-gear."

A propeller-wheel of this description was applied to the revenue-cutter *Gallatin*, on Lake Erie, in 1872; but after several trials it was condemned, as being less efficient than the screw propeller, and, with the machinery to work it, removed from the vessel.

A Fowler wheel was also applied to the United States torpedo-vessel *Alarm* about the same time, and a board of officers has recently recommended its removal and the substitution of the Mallory steering-propeller in its stead.

Hunter's steering-propeller, patented in 1874, and intended for canal-boats, is similar to the Fowler wheel in some respects. It has two wheels on opposite sides of the stern-post, revolving in opposite directions. The blades are feathered so as to have but one dead-point.

of light draught and good lines, a power of 775 horses was developed to obtain an average speed of 6½ knots per hour.

Additional alterations and experiments were made two years ago, with the view to obtaining better results. These alterations consist in superseding the 140 small apertures through which the water is admitted by one large aperture under the wheel, and in the bottom of the ship; also in lengthening the nozzles at the sides through which the water makes its escape. The results of the trials after these alterations will, of course, be nearly the same as in previous trials. The speed of the vessel at sea has never exceeded 5 or 6 knots, and although ten years old she has never been trusted out of sight of land, and, as she is neither fit for coast defense nor harbor service, it is believed that the next move will be to break her up; and thus ends the experiment of propelling vessels by means of turbine wheels.

THE GLATTON.

COAST-DEFENSE VESSELS.

Of the twelve coast-defense vessels named in Part I., the *Glatton* is the most powerful. She is an iron double-screw turret-ship of 4,912 tons displacement. Her length is 245 feet; breadth, 54 feet; and draught of water, 19 feet.

The maximum horse-power is 2,868, and the maximum speed 12 knots. The hull is double-bottomed, and divided into water-tight compartments in the usual manner.

The armor of the hull proper consists of two strakes, the upper (above water) being 12 inches thick, and the lower (below water) 10 inches in thickness. The former has a teak backing of 18 inches, and the latter a backing of 20 inches.

The breastwork, which rises 6 feet 3 inches above the upper deck, is armored with plates 12 inches thick, having behind it a teak backing of 18 inches. The deck extending on either side of the breastwork consists of 1-inch plates covered by 2-inch plates, and over this, 6 inches of oak planking. The turret, which rises out of the center, above the breastwork chamber, is 30 feet 6 inches in external diameter, and there is a space of 6 inches between it and the surrounding glacis-belt, which is 3 feet in breadth.

The general thickness of the turret-armor is 12 inches, with 15 inches of teak backing.

All the coast-defense vessels are engined on the old system, the *Cyclops* and *Hydra* excepted, the engines of which were the first examples of Elder's compounds introduced into armored ships. Each vessel has a pair of vertical inverted cylinders to each of the two screws.

The boilers are cylindrical, and the pressure of steam is 60 pounds per square inch.

A REMARKABLE EXPERIMENT.

At the time the *Glatton* and other monitors were undergoing construction, considerable diversity of opinion existed in England as to the ability of the turret to revolve and to be worked after having been struck in action by heavy projectiles; that is, whether by the impact of a 600-pound shot, propelled by a 12-inch rifled gun at short range, a turret such as the one represented would be jammed or prevented from working. There was also to be ascertained the probable damage that might be caused to the guns and other interior fittings of the turret. With a view of arriving at a definite solution of this question, it was determined to select the *Glatton* as a target, and to cannonade her turret with projectiles from the heavy guns of the coast-defense vessel *Hotspur*. In compliance with this decision, the two vessels were moored at a distance of 200 yards from each other.

The *Glatton*, above mentioned, carries a single turret, in which were mounted at the date of the experiment, July, 1872, two 25-ton Woolwich rifles, being the heaviest guns at that time in the British navy.

The turret of the *Glatton*, against which the shots were directed, is shown in the horizontal section A on the accompanying sketch. The armor consists of plates laid on in two rings or tiers, eight plates in each ring, the upper ring or belt having six plates 12 inches thick, and two plates 14 inches thick, namely, those pierced by the port-holes. The lower ring contains seven plates 12 inches and one plate 14 inches thick; the last mentioned being that between and beneath the port-holes. The backing is of such thickness as, with the plates, to make up a total of 29 inches everywhere; that is, 15 inches of oak behind 14 inches of iron, or 17 inches of oak behind 12 inches of iron. Behind the backing comes 1½ inches of skin, consisting of two thicknesses of ¾-inch plate; then vertical girders 5 inches in depth with spaces between, and, finally, what may be termed an inner skin or mantlet skin, of ¼-inch iron, to prevent bolt-heads and splinters from flying into the interior of the turret and injuring the men working the guns on service.

The *Hotspur* is a ram, 235 feet between perpendiculars, 50 feet in extreme breadth, with a mean draught of water of 19 feet 10 inches, and her armament consists of one 25-ton Woolwich muzzle-loading rifle. Against the strongest portion of the *Glatton's* turret this gun was brought to bear, at a range of 200 yards. The projectiles used were Palliser 600-pound shot, chill-headed, and the powder-charge was 85 pounds large pebble. The results have been summarized by the *Engineer*, as follows:

The first shot struck at the spot marked B in the elevation, with effects shown in section at A and at B.

(1) The entire upper plate was forced back to a distance, at point of junction with lower plate, of 5½ inches; (2) shot penetrated to a depth of nearly 20½ inches; (3) horizontal joint between upper and lower plate was opened to a width of 2 inches; the same effect being manifest in the corner of the top plate being lifted 2 inches higher than that of the adjacent plate; (4) the lower plate was cracked in a vertical direction and otherwise contorted at the edge; (5) a bolt was driven some inches backward, the head flying into the interior of the turret; (6) the double skin was bent back and forced open to a width of about 3 inches, the wood protruding; (7) the ¼ inch or inner skin was torn open and hanging down to the extent of about 4 feet by 18 inches, a number of rivet-heads, as well as bolt-heads, being thrown into the interior of the turret.

Although a little below the spot intended, it was quite clear that this round gave a heavy contorting blow to the turret, the top of which had been so far forced back; it was, nevertheless, found that the turret revolved without the slightest difficulty, and for the object of the experiment the next round might be proceeded with. Considering the spot struck by the first blow, it seemed advisable to pass on at once to the trial of a blow at the line of junction between the turret and glacis-plate. By means of a mark painted at C, elevation, a shot was delivered, grazing the glacis-plate at a point 3 feet from the turret and glancing into the turret, which it penetrated to a depth of about 15 inches, the shot, as before, standing well up to its work and coming easily out of the hole uninjured as far as the front row of studs. The effects produced by this round are chiefly shown in section C. They are (1) penetration about 15¼ inches; (2) glacis-plate grooved to a depth of about ½ inch and cracked; (3) flange-ring covering joint of turret and glacis cut through and bent; (4) lower side of glacis-plate bent back and split open to a width of about ¾ inch; (5)—not shown in figure—a sort of binding-plate, fixed on the lower edge of the armor side beneath the deck, broken off for a length of some feet and the edge bulged downward.

This round again severely tested the working of the turret, not perhaps quite so severely as might be conceived were a similar blow to fall in a more downward direction, but quite the kind of blow intended. On trial the turret was again found to work freely and easily. The ports, which up to this time had been covered and plugged up with beams of wood, were cleared open and two rounds were fired from each gun; one a full blank charge of 70 pounds of pebble-powder, and one a battering charge of 85 pounds of pebble-powder with shot. The turret revolved easily in about a minute, and we are not aware that any effort was used to obtain speed. In short, the *Glatton* was in good fighting trim at the conclusion of the experiment. Considering how great are the chances against a second shot falling exactly on a spot already struck, it would hardly be going too far to say that the *Glatton* was in nearly as good condition to go into action as before the trial. Yet it would be difficult to put her through a more severe ordeal, except by bringing the 35-ton gun to bear on her, and as for the object of the experiment, namely, injury to the working of the turret, it may be doubted whether much more effect would, even then, have been produced. A plunging fire we are inclined to believe the most likely to jam the turret.

Sections of the Glatton's Turret.

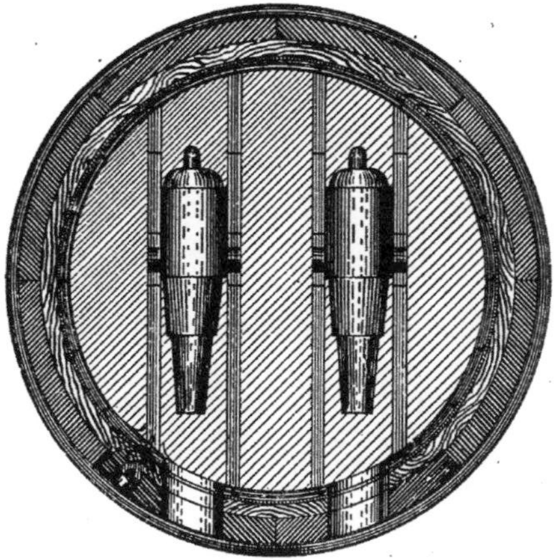

Section at A.

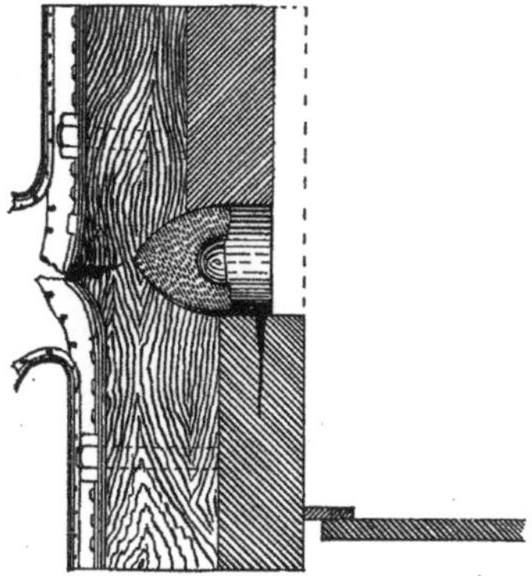

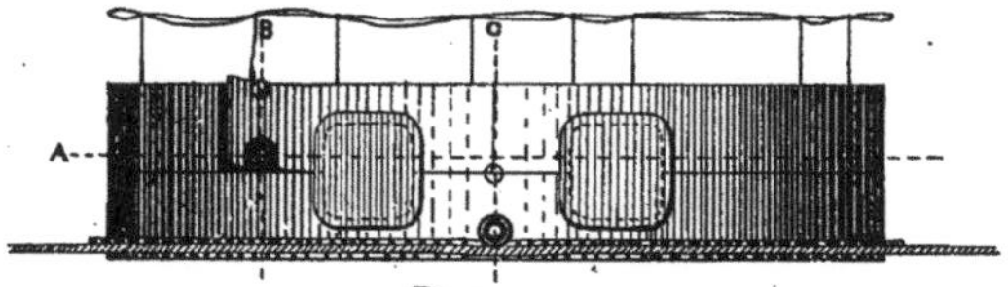

Elevation.

A

B

C

Section at B.

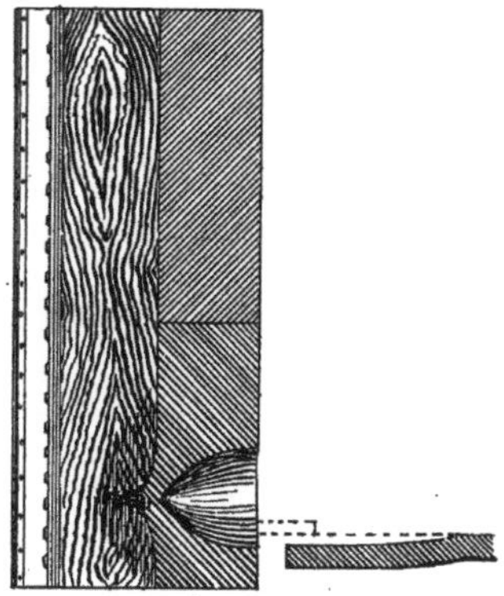

Section at C.

At the beginning of the experiment several animals and fowls were placed in the turret, and at the conclusion they were found uninjured.

The damaged plates, shown in the drawings, were removed to the Chatham dock-yard, where I had the privilege of examining them.

SARTORIUS TORPEDO-RAM.

The sum of $60,264 has been appropriated toward the construction at Chatham of a vessel now known as the Sartorius Torpedo-Ram. But as yet little definite information regarding its structural arrangement or dimensions has been made public. It has, however, been reported that the vessel will be 250 feet long, will have a draught of 20 feet and a displacement of 2,500 tons; that she will only expose about 4 feet of the hull proper out of water, and this portion will be convex and armored with steel plates; that the engine-power will be sufficient for a very high speed, while the coal-carrying capacity will also be great; and that she will have a light hurricane-deck above the cigar-shaped hull, but will not be provided with masts or guns.

This extraordinary craft is intended to ram armored ships about 5 or 6 feet below the water-line, and for this purpose she is provided with a formidable submerged ram-snout; while at the same time she is to discharge a number of torpedoes from her stem and also from her sides.

Some of these particulars resemble very much the ideas embodied in the model prepared some years ago by Commodore Ammen and still exposed to view in our Navy Department.

PART VIII.

COST OF BRITISH ARMORED SHIPS; TABLE OF DIMENSIONS, ETC., OF THE ARMORED SHIPS OF GREAT BRITAIN.

COST OF BRITISH ARMORED SHIPS.

The account rendered of the cost of the construction or repair of a vessel or its machinery in one of our navy-yards is that of the actual labor and material entering into such construction or repair, no account being taken of the cost of plant, tools, appliances, fuel, &c., used in connection therewith.

In the British navy a different system prevails. A nominal percentage is added to the actual cost of labor, materials, and stores entering into the construction of the vessel, to cover what is believed to be the ship's share of the value of maintenance of the plant, appliances, materials, &c., necessarily employed in the dock-yards as ship-building establishments. This nominal percentage has differed materially in the years prior to 1866, but since that time a bulky volume has annually been presented to Parliament, containing upward of 700 pages of figures, which gives in great detail information of a definite character as to the cost of building and repairing every vessel in the British navy. These volumes were, to some extent, in existence before 1868, but prior to that time they could not, for purposes of comparison, be regarded as trustworthy, consequent upon the results of individual judgments obtained by taking the actual cost of maintaining the dock-yards and of their plant, and distributing it ratably over the dock-yard construction and repairs of ships at the dock-yards. Thus, in comparing the cost of the *Achilles* with that of the *Bellerophon*, as the former was built when the percentage was low and the latter when it was high, it is found that these accounts make the former-named vessel apparently cost considerably less than the latter, though she really cost considerably more. Since 1866 this fluctuating percentage has been excluded from the accounts, and the details relating to the cost of ships have been limited to the actual prime cost of production; or, in other words, to the cost of the labor, materials, &c., expended on them. At the same time, the necessary information is given enabling a percentage to be added. The following extract from a carefully-prepared paper in the London *Times* will be found interesting, showing, as it does in detail, the cost of construction and maintenance of the British navy for eighteen years:

While it was difficult to assess beyond the reach of dispute the proper charge ships should bear as their share of the cost of keeping up the dock-yards and machinery necessary for construction and repair, it was not so difficult, though the difficulty must, necessarily, have been great at first, so to limit the items of charge on account of ship-building as to enable an effective comparison to be made between the estimate voted by Parliament and the actual expenditure incurred on our fleets for construction and repair; to be able, in fact, to prepare an account which should show on one side the actual votes for labor and stores as given in the navy estimates, and on the other side the actual fruit of those votes in dock-yard work of all kinds. The difficulty, which for years existed, of establishing anything like a satisfactory and direct relation between the money voted for ships by Parliament and the money actually paid for ships and expended upon ships by the admiralty was seriously felt, and was acknowledged to be a blot on naval finance; but the course taken by Mr. Childers guaranteed this important result, and on this principle these accounts since 1866 have been framed. To make this comparison more effectual, the navy estimates are now always accompanied by an analysis of the ship-building votes, or "retabulation," as it is called, which forms the basis of the dock-yard accounts. This, however, is not the only result which these accounts give. In addition to the comparatively simple process of setting off actual against estimated expenditure, an exhaustive balance-sheet, which prefaces these accounts, shows the value of land, stock, buildings, and other elements of property representing the value of our dock-yards at the commencement of each year, with the receipts from various sources on one side; on the other, the expenditure of stores and labor during the year; and, finally, the balance at the end of the year of stores in hand, and of the appreciated or depreciated value of dock-yard property. Regarded simply for the purposes they are intended to serve, these are, perhaps, the most complete government accounts to which the public have access.

In endeavoring, therefore, to ascertain what iron-clad ship-building has cost the country, it is possible to refer for eight successive years to accounts which give, readily and uniformly, this information. While it is only since 1866 we are able to accept the figures in these accounts as useful for this purpose, we are, fortunately, previously to that date, not left without trustworthy information. Mr. Reed, in his book on "Our Iron-clad Ships," gives in detail the cost of iron-clads from the earliest date, and his figures are no doubt derived from official sources.

Before the date of these accounts, then, we have Mr. Reed's figures; and since the 31st of March, 1874, the date of the last published account, an official return gives the estimated cost of each iron-clad which was then incomplete, or has been since commenced. Mr. Reed's figures give the sum of £7,338,687 as the cost of construction before 1866; the admiralty accounts, the sum of £5,961,203 as the cost between 1866 and 1874; and the estimated cost from 1874 to the present time gives the sum of £3,439,035, of iron-clads incomplete but still under construction. The result is that the cost of our iron-clad fleet which has been both completed and commenced may be calculated at the sum of £16,738,935. During the past 18 years, then, while the navy estimates have amounted to £197,000,000, the sum of £16,500,000 has been spent on the construction of iron-clads; and, if the cost of wear and tear, repair and maintenance, be added, we may raise this sum to £18,000,000. Mr. Reed's statement that our iron-clad fleet had, since it was first commenced, cost the country about £1,000,000 a year, was founded on a liberal calculation, and is beyond a doubt correct.

The two divisions of expenditure on iron-clads into construction and repair will be dealt with separately.

In turning, then, to the cost of construction, it will be found that, including four floating batteries, sixty iron-clads of all kinds have been built either by contract or at the royal dock-yards. * * * * *

The following table exhibits the total amount spent each year on iron-clad construction, and the amounts paid to contractors and spent at the royal dock-yards for this purpose:

Year.	Cost of construction.		
	At the dock-yards.	By contract.	Total.
1866–'67	£436,301	£154,840	£591,141
1867–'68	440,143	348,474	788,617
1868–'69	384,146	725,914	1,110,060
1869–'70	536,293	540,055	1,076,348
1870–'71	558,800	455,415	1,014,215
1871–'72	345,750	349,288	695,038
1872–'73	237,956	61,869	299,825
1873–'74	377,715	8,244	385,959
	3,326,104	2,644,099	5,961,203
Before 1866	3,481,843	3,856,844	7,338,687
Total	6,807,947	6,530,943	13,299,890

The following table gives the expenditure for construction and wear and tear (including repairs and maintenance) of vessels of all kinds in the effective navy:

Year.	Cost of building.		Cost of wear and tear, &c.		Total.
	Iron-clad.	Not iron-clad.	Iron-clad.	Not iron-clad.	
1866–'67	£591,141	£423,265	£109,145	£782,728	£1,906,279
1867–'68	783,617	1,103,132	159,552	568,489	2,519,790
1868–'69	1,110,060	584,302	187,699	426,084	2,308,150
1869–'70	1,076,348	310,699	130,743	468,623	1,986,413
1870–'71	1,014,215	316,599	182,065	451,880	1,964,759
1871–'72	695,038	439,134	87,595	358,388	1,630,155
1872–'73	299,825	509,262	158,933	336,259	1,304,279
1873–'74	385,959	904,069	291,381	464,911	2,046,320
Total	5,961,203	4,540,462	1,307,113	3,857,362	15,666,145

This table throws a fuller and more accurate light than the previous one on the ship-building policy and expenditure of these eight years.

* * * * * * *

But to give these figures their full value and to test them, as it were, independently, it is desirable to compare them with the amounts voted by Parliament for ship-building and for all naval services, and to compare them further with the number of artisans employed and the tonnage of shipping annually built. The result should present a complete chart of the ship-building work and a guide to the ship-building policy of these eight years, which cannot fail to be interesting. The following table gives this information:

Year.	Expenditure on building and repairs.	Estimates for ship-building (votes 6 and 10).	Total naval estimates.	Number of workmen.	Tonnage built.	
					Iron-clad.	Total (iron-clad and unarmored).
1866–'67	£1, 906, 279	£2, 718, 000	£10, 031, 000	18, 607	7, 013	15, 384
1867–'68	2, 519, 790	3, 091, 000	10, 976, 000	18, 309	12, 448	33, 701
1868–'69	2, 308, 145	3, 219, 000	11, 157, 000	15, 464	15, 045	26, 291
1869–'70	1, 986, 413	2, 655, 000	9, 996, 000	14, 124	18, 769	24, 230
1870–'71	1, 964, 759	2, 124, 000	9, 370, 000	11, 223	12, 567	19, 925
1871–'72	1, 630, 155	2, 557, 000	9, 756, 000	12, 831	10, 678	21, 137
1872–'73	1, 304, 279	2, 384, 000	9, 532, 000	12, 826	4, 798	16, 092
1873–'74	2, 046, 320	2, 797, 000	9, 872, 000	13, 485	4, 050	17, 329
Total	15, 666, 090	21, 545, 000	80, 690, 000	....	85, 368	174, 089

To return, however, to the cost of construction. The sum of sixteen and one-half millions sterling represents, as has been shown, the total cost of iron-clad construction, past and prospective, so far as naval accounts and estimates are concerned. To the 31st of March, 1874, the sum of £13,299,890 had been actually spent of this sum. What have we got for it? Forty-nine finished and seven unfinished iron-clads. These last consist of the *Alexandra*, *Téméraire*, *Dreadnought*, and *Inflexible*, whose partial cost had amounted in March, 1874, to £1,084,887. The remaining 49 vessels, whose total cost had been brought to account in March, 1874, had cost £12,215,000. Judging by the standard suggested by Mr. Reed, we may consider seven only of these 49 vessels effective; or, we may say that, of the above sum, two millions sterling represent the cost of our effective iron-clads, and ten millions of those which are obsolete or ineffective. To go still further into detail, it may be interesting to compare the cost of these two classes. In the obsolete class of broadside-vessels the *Warrior*, for instance, one of the oldest, was built at a cost of £379,154, and her sister ship, the *Black Prince*, at a cost of £378,310, each having a burden of 6,109 tons. Then the *Achilles*, with 6,121 tons, cost no less than £470,230, and the *Northumberland*, with 6,621 tons, cost £490,681. Of modern iron-clads the *Hercules* and *Sultan* are smaller, having a tonnage of 5,226 tons only, but they may each be regarded as four times as strong as the *Warrior*, while their cost in construction was respectively £377,007 and £374,777. A smaller class of vessel, which the loss of the *Vanguard* has drawn attention to, namely, the *Audacious* class, consists now of five vessels, each having a tonnage of 3,774 tons. The *Audacious* cost £256,295, the *Iron Duke* £208,763, and the *Vanguard* £272,100. In the construction of the *Bellerophon*, an attempt was first made to economize the cost of construction, and at the same time increase the strength. The vessel cost to build £364,327, or more than £100,000 less than the *Achilles*, with which it may be compared. Of turret-vessels, two, the *Monarch* and *Glatton*, are included in the class of the seven effective iron-clads already referred to. The *Monarch* cost £371,415, and the *Glatton* £223,105, to build. The unfortunate *Captain* cost £355,764. The most expensive of the four small coast-defense vessels, built in 1870, is the *Cyclops*, which cost to build £149,465; and the two rams, the *Hotspur* and *Rupert*, cost respectively £175,995 and £235,032. So far, then, as these 49 vessels are concerned, it seems clear that, as time advanced, the cost of construction diminished; it may, indeed, be said that recent invention has had the tendency to increase power with diminished size and cost. The satisfaction, however, this statement may cause must, we fear, be short-lived; for, while up to 1874 or 1875 this is true enough, since then iron-clads have become much more expensive than ever. Increased prices and more costly appliances have had an unmistakable effect on the iron-clads now under construction. The *Inflexible* is estimated to cost £521,750, the *Dreadnought* £508,395, and the *Alexandra* £521,500. Then the *Téméraire* is to cost £374,000, the *Nelson* £333,800, and the *Northampton* £349,000. These vessels, not yet finished, show a startling advance in cost as compared with some of the vessels we have noticed, or even with vessels like the *Thunderer* and *Shannon*, which are not yet complete, but are of earlier construction, and are estimated to cost £334,000 and £268,500 respectively. These points are worth noting, as they are essential to any effective criticism or comparison of past and present naval expenditure, and will help to correct many erroneous impressions on the subject. But a more serious element of disturbance in naval finance has been caused by the sudden if not entirely unsuspected demands made for the repair of iron-clads.

The repair of iron-clads is a difficulty which has only made itself felt seriously in the past three years. Mr. Goschen, in bringing forward the naval estimates in 1873, alluded to the necessity which existed for making special provision for this work by employing additional hands at the dock-yards. But Mr. Hunt, following in the same line, explained last year that extensive repairs of a costly character were inevitable. When iron-clads were first built it was considered satisfactory that, although the cost of construction was great, the ships would last longer than unarmored vessels. This, however, seems doubtful, now that we have the experience of fifteen years as a guide; and even were iron-clads exceptionally durable, it is quite certain their repair is exceptionally costly. For the first ten years the charges for repair were, comparatively speaking, light; but during the last three years the necessity of entering upon an expensive course of repair has made itself felt. The following table shows the actual cost during the eight years from 1866 to 1874 of providing for the wear and tear of iron-clads and unarmored vessels, and of keeping them in repair, in commission, and reserve:

Year.	Iron-clads.	Unarmored vessels.	Total.
1866–'67	£109, 145	£782, 728	£891, 873
1867–'68	159, 552	568, 489	728, 041
1868–'69	187, 699	426, 084	613, 783
1869–'70	130, 743	468, 623	599, 366
1870–'71	182, 065	451, 880	633, 945
1871–'72	87, 595	358, 388	445, 983
1872–'73	158, 933	336, 259	495, 192
1873–'74	291, 381	464, 911	756, 292
Total	1, 307, 113	3, 857, 362	5, 164, 475

Here it will be seen that in 1873–'74 the largest expense was incurred for the repair of iron-clads. It is also worthy of remark that a small number of vessels only was dealt with, as a reference to the accounts would prove. Thus, in that year alone the *Achilles* cost £24,907, the *Bellerophon*, which had cost nearly £30,000 in 1870, was again in 1873 charged with an expense of £40,395; the *Minotaur* cost in this year £16,681; the *Northumberland*, £10,255; the *Resistance*, £31,647; and the *Warrior* the large sum of £50,000. From these instances it will be seen that the maintenance and repair of iron-clads have introduced a new and serious element of expense into the navy which for some years is likely to make itself severely felt. A reference to the above table will show that the cost of the maintenance and repair of the entire fleet, for the eight years which have been analyzed, amounts to more than five millions sterling, of which £1,307,113 is applicable to iron-clads; and that of this sum no less than a third was incurred during the last ten years. * * * * Coming now to the consideration of what has been the cost of repairing individual vessels during this period, it appears that those which have already been classed as obsolete or old-fashioned, including serviceable and unserviceable ships, have cost during this period £870,000. Of these the *Warrior* is the most remarkable; for her cost has amounted, for maintenance and repair, to £124,245. She was built in 1860, cost £379,154 to build and fit out for sea, and has now, in the eight years since 1866, cost no less than a third of this sum to keep in good order. The *Defence* and *Resistance*, which were built at the same time as the *Warrior*, but are much smaller, have cost during the same period no less than £82,450 and £91,965, which, when compared with the cost of their construction, are large sums. The *Black Prince*, sister ship to the *Warrior*, and built at the same time, has for the same period cost only £59,193. But this difference would, in all probability, vanish if we knew what her cost had been during the past two years. The *Minotaur* has cost for the same period £55,627, and the *Achilles* £61,209, against a cost of construction of £478,885 and £470,230 respectively; the *Hector*, the cost of whose construction was about half that of either of those vessels, has cost for repair and maintenance £56,490. These, however, are all old-fashioned broadside-vessels. The turret-ships are most of them too recently built to enable an opinion to be formed of what they will cost to repair; it will be seen that the whole cost amounts to £139,845. Of this sum the *Monarch* absorbs a large share, which amounts to £53,018. Of effective and powerful sea-going broadside-vessels, the *Bellerophon*,

Hercules, and *Sultan* may be taken as good examples. The *Bellerophon* was launched in 1865, and the cost of her maintenance and repair, from 1866 to 1874, is £87,256, or about one-fourth of the cost of her construction, which was £364,327. The *Hercules* was launched in 1866, and has, similarly, cost £37,537, against a cost of construction of £377,007; and the *Sultan*, which is, comparatively speaking, a new vessel, has only cost £15,666. The *Vanguard* shows a cost, since 1869, of £12,891; the *Iron Duke*, since 1870, of £17,022; and the *Invincible*, since 1870, of £18,271. These instances prove sufficiently that the cost of the maintenance and repair of iron-clads is a growing item of naval expenditure, and an item which requires as much consideration, from a financial point of view, as building. When a million a year represents the full normal average amount available for iron-clad construction out of naval estimates amounting to ten millions sterling, it is alarming to find £300,000 is required in one year alone for their repair and maintenance.

* * * * * *

Armored ships of Great Britain.

[The principal data are from Parliamentary returns.]

Name and class of ship.	Vessel.							Armament.		
	Length between perpendiculars, in feet and inches.	Extreme breadth, in feet and inches.	Depth of hold, in feet and inches.	Immersed midship section, in square feet.	Displacement, in tons.	Draught forward, in feet and inches.	Draught aft, in feet and inches.	Number of guns and description.	Height of battery above water, in feet and inches.	Thickness of armor at water-line, in inches.
Turret-ships of the first class. (a)										
Inflexible (b)	320 0	75 0	23 3.5	1,658	11,500	23 3	25 4	4 81-ton guns	11 0	24 to 16
Dreadnought	320 0	63 10	....	1,510	10,886	26 6	27 0	4 38-ton guns	13 0	14
Thunderer	285 0	62 3	18 0	1,454	9,190	26 2	26 3	do	13 4	14 and 12
Devastation	285 0	62 3	18 0	1,463	9,157	26 2	26 3	4 35-ton guns	13 0	14 and 12
Agamemnon	260 0	66 0	....	....	8,492	23 0	24 0	4 38-ton guns	....	18
Ajax	260 0	66 0	....	....	8,492	23 0	24 0	do	....	18
Monarch (c)	330 0	57 6	21 1	1,228	8,322	22 7	26 0	4 25-ton and 2 6½-ton guns	16 0	10 and 8
Neptune (d)	300 0	63 0	....	....	8,960	24 4	25 2	4 38-ton and 2 6½-ton Whitworth guns	....	13 to 9
New Agamemnon (no dimensions)	....	....	....	....	....	....	....	....	....	....
Broadside-ships of the first class.										
Alexandra	325 0	63 8	18 7.5	1,413	9,492	26 0	26 6	2 25-ton and 10 18-ton guns	....	12 and 8
Téméraire	285 0	62 2	18 8	1,442	8,412	26 6	27 0½	4 25-ton and 4 18-ton guns	9 0	11 and 8
Sultan	325 0	59 0	21 1	1,401	9,286	24 10	27 6	8 18-ton and 4 12-ton guns	11 7	9 to 6
Hercules	325 0	59 0	21 1	1,314	8,700	23 0	26 5	8 18-ton, 2 12-ton, and 4 6½ ton guns	12 2	9 to 3
Bellerophon	300 0	56 0	17 3.5	1,218	7,540	21 0	26 0	10 12-ton and 4 6½-ton guns	7 11	6
Swiftsure	280 0	55 0	25 10.6	1,156	6,660	24 1	26 2	10 12-ton guns } Exclusive of smaller guns.	7 2	8 to 6
Triumph	280 0	55 0	25 10	1,156	6,660	24 2	26 2	do	....	8 to 6
Audacious	280 0	54 0	24 8.8	1,063	6,034	22 0	23 0	do	7 8	8 to 6
Invincible	280 0	54 0	24 9	1,063	6,034	22 0	23 0	do	....	8 to 6
Iron Duke	280 0	54 0	24 9	1,063	6,034	22 0	23 0	do	....	8 to 6
Penelope	260 0	50 0	....	....	4,394	15 9	17 4	do	....	6 and 5
Superb (e)	333 0	59 0	....	....	8,994	23 0	25 0	1 25-ton and 12 18-ton guns	10 0	12
Armor-belted. (f)										
Shannon	260 0	54 0	21 7.5	1,006	5,103	20 0	22 6	2 18-ton and 7 12-ton guns	....	9
Nelson	280 0	60 0	....	1,162	7,323	24 2	24 2	4 18-ton, 8 12-ton, and 6 Gatling guns	....	9
Northampton	280 0	60 0	....	1,162	7,323	24 2	24 2	do	....	9
Belleisle (g)	245 0	52 0	22 0	....	4,700	19 0	19 6	4 25-ton guns	....	12 and 8
Orion (h)	245 0	52 0	22 0	....	4,700	19 0	19 6	do	....	12 and 8

Breastwork-monitors. (i)										
Glatton	245 0	54 0	19 4	913	4,912	19 0	19 0	2 25-ton guns	10 5	14 to 12
Hotspur (ram)	235 0	50 0	20 1	840.5	4,012	19 0	20 8	1 25-ton gun	10 $6\frac{1}{2}$	11 to 8
Rupert (ram)	250 0	53 0	20 0.5	1,067	5,358	21 6	23 6	2 18-ton guns	10 10	14 to 9
Prince Albert	240 0	48 1	25 3	787	3,900	17 9	19 8	4 18-ton guns	9 3	$4\frac{1}{2}$
Cyclops	225 0	45 0	16 7	666	3,430	16 5	16 5	do	10 7	10 to 6
Gorgon	225 0	45 0	16 7	666	3,430	16 5	16 5	do		10 to 6
Hecate	225 0	45 0	16 7	666	3,430	16 5	16 5	do		10 to 6
Hydra	225 0	45 0	16 7	666	3,430	16 5	16 5	do		10 to 6
Scorpion	224 6	42 4		634	2,777	16 $11\frac{1}{2}$	16 $2\frac{3}{4}$	4 9-inch guns	4 9	4.5 and $0\frac{1}{2}$
Wivern	224 6	42 4			2,725	14 $6\frac{1}{2}$	15 8	do		4.5 and 5
Viper	160 0	32 0	13 3	339	1,220	10 8	11 $7\frac{1}{2}$	2 7-inch guns	5 11	$4\frac{1}{2}$
Vixen	160 0	32 0	13 3	339	1,220	10 8	11 $7\frac{1}{2}$	do	5 11	$4\frac{1}{2}$
Abyssinia	225 0	42 0	12 3	557	2,901	14 $0\frac{1}{2}$	14 $0\frac{1}{2}$	4 10-inch guns	7 6	7
Magdala	225 0	45 0	16 6	557	3,344	13 0	14 8	do	9 0	8
Cerberus	225 0	45 0	16 6	557	3,344	15 3	15 3	do	9 0	8
Iron broadside-ships of the old type.										
Agincourt	400 0	59 5	21 0	1,356	10,395	26 0	27 0	17 12-ton guns	9 0	$5\frac{1}{2}$
Minotaur	400 0	59 5	21 0	1,356	10,395	26 0	27 0	do	9 0	$5\frac{1}{2}$
Northumberland	400 0	59 5	21 0	1,356	10,395	26 0	27 0	10 12-ton and 7 $6\frac{1}{2}$-ton guns	9 0	$5\frac{1}{2}$
Achilles	383 0	58 0	21 0	1,326	9,681	26 3	27 4	14 12-ton guns	9 0	$4\frac{1}{2}$
Black Prince	380 0	58 0	21 0	1,326	9,681	26 0	27 0	10 12-ton and 16 $6\frac{1}{2}$-ton guns	9 0	$4\frac{1}{2}$
Warrior	380 0	58 0	21 0	1,326	9,681	25 6	26 5	do	9 0	$4\frac{1}{2}$
Hector	280 2	56 5	21 0		6,960	24 0	25 7	18 $6\frac{1}{2}$-ton guns		$4\frac{1}{2}$
Valiant	280 2	56 5	21 0		6,420	25 10	26 10	do		$4\frac{1}{2}$
Resistance	280 0	54 0		1,094	6,074	23 9	26 0	16 $6\frac{1}{2}$-ton guns	6 11	$4\frac{1}{2}$
Defence	280 0	54 0		1,094	6,074	24 3	25 6	do	6 11	$4\frac{1}{2}$
Wooden broadside-ships of the old type.										
Lord Warden	280 0	58 9			7,675	24 6	26 0	18 $6\frac{1}{2}$-ton guns		$5\frac{1}{2}$
Royal Alfred	273 0	58 5			6,400	24 0	25 6	do		6
Royal Oak (k)	273 0	58 5			6,416	23 8	25 8	24 $6\frac{1}{2}$-ton guns		$4\frac{1}{2}$
Prince Consort (k)	273 0	58 5			9,430	23 8	25 8	do		$4\frac{1}{2}$
Caledonia (k)	273 0	59 0			5,800	23 6	26 0	do		$4\frac{1}{2}$
Zealous (k)	252 0	58 7			6,027	24 6	26 0	20 $6\frac{1}{2}$-ton guns		$4\frac{1}{2}$
Ocean (k)	273 0	56 6		1,208	6,920	24 10	27 0	24 $6\frac{1}{2}$-ton guns	7 6	$4\frac{1}{2}$
Repulse	252 0	59 0			6,190	25 4	26 2	10 9-ton guns		6 and $4\frac{1}{2}$
Pallas	225 0	50 0	16 6	821	3,797	18 3	24 3	4 9-ton guns	8 6	$4\frac{1}{2}$
Favorite	225 0	46 9	25 0	771	3,232	19 7	22 7	8 $6\frac{1}{2}$-ton guns	8 6	$4\frac{1}{2}$
Research	195 0	38 6			1,780	13 2	15 5			$4\frac{1}{2}$
Enterprise	180 0	36 0	16 6	399	1,350	12 2	15 6	4 [illegible]-inch guns	6 11	$4\frac{1}{2}$
Waterwitch (l)	162 0	32 1	13 3	339	1,279	11 9	11 9	2 7-inch guns	5 0	5
Royal Sovereign (m)	240 7	62 0		985.5	5,079	21 6	24 11	5 9-inch guns	7 6	$5\frac{1}{2}$
Torpedo-ram	250 0				2,500	20 0	20 0	No guns		

Armored ships of Great Britain—Continued.

Name and class of ship.	Machinery.								Cost.	
	Kind of engines.	Number of cylinders.	Diameter of cylinders, in inches.	Stroke of pistons, in feet and inches.	Number of screws.	Indicated horse-power on trial.	Maximum speed, in knots, per hour.	Makers of engines.	Machinery, in dollars (gold).	Total cost of hull and machinery, in dollars (gold), exclusive of armament, rigging, outfit, &c.
Turret-ships of the first class. (a)										
Inflexible (b)	Inverted vertical compound	6	70, 90, & 90	4 0	2	8, 000	14	John Elder & Co	586, 845	*2, 535, 705
Dreadnought	do	6	66, 90, & 90	4 6	2	8, 206	14. 5	Humphrys & Tennant	520, 020	2, 470, 800
Thunderer	Horizontal	4	77	3 6	2	5, 600	12. 5	Humphrys & Co	225, 990	*1, 623, 240
Devastation	Horizontal trunk, surface-condenser, &c.	2	88	3 3	2	6, 638	13. 8	J. Penn & Sons	*306, 637	1, 717, 038
Agamemnon	Inverted vertical compound	6	54	3 3	2	†6, 000	......	do	437, 400	1, 701, 000
Ajax	do	6	54	3 3	2	†6, 000	......	do	......	1, 701, 000
Monarch (c)	Horizontal, 4 piston-rods	2	120	4 6	1	7, 842	14. 9	Humphrys & Co	*362, 906	*1, 805, 077
Neptune (d)	Horizontal trunk	2	118	4 6	1	9, 000	14	J. Penn & Sons	......	......
New Agamemnon (no dimensions)	......	...	......	......	...	......	......	......	......	......
Broadside-ships of the first class.										
Alexandra	Inverted vertical compound	6	70, 90, & 90	4 0	2	8, 313	15	Humphrys & Tennant	*537, 030	2, 539, 253
Téméraire	do	4	70 and 114	3 10	2	7, 518	14. 65	do	451, 980	1, 817, 640
Sultan	Horizontal trunk, surface-condenser, &c.	2	117. 9	4 6	1	8, 629	14. 13	J. Penn & Sons	362, 600	*1, 821, 416
Hercules	Horizontal trunk, surface-condenser, &c.	2	118	4 6	1	8, 529	14. 69	do	396, 900	*1, 832, 254
Bellerophon	Horizontal trunk, surface-condenser, &c	2	104. 2	4 0	1	6, 312	14. 05	do	425, 320	*1, 770, 629
Swiftsure	Horizontal, double piston-rod	2	98	4 0	1	4, 913	13. 75	Maudslay & Co	239, 120	1, 249, 414
Triumph	do	2	98	4 0	1	4, 892	14. 07	do	238, 140	1, 255, 450
[illegible]us	Two pairs, horizontal direct	4	77	3 0	2	4, 021	12. 8	Ravenhill & Co	254, 800	*1, 245, 594
Invincible	Two pairs, horizontal	4	72	3 0	2	4, 832	14. 09	Napier & Sons	252, 720	1, 163, 683
Iron Duke	Two pairs, horizontal direct	4	77	3 0	2	4, 268	13. 6	Ravenhill & Co	252, 720	*1, 014, 588
Penelope	Two sets, horizontal, three cylinders	6	55¼	2 6	2	4, 703	12. 76	Maudslay & Co	199, 260	908, 081
Superb (e)	Horizontal, double piston-rod	2	116	4 0	1	7, 300	13. 75	do	......	......
Armor-belted. (f)										
Shannon	Horizontal, back-acting	4	44 and 85	4 0	1	3, 500	13	Laird Brothers	245, 430	1, 304, 910
Nelson	Inverted vertical compound	4	60 and 104	3 6	2	6, 250	15	John Elder & Co	397, 548	1, 622, 268
Northampton	do	6	54	3 3	2	†6, 000	14	John Penn & Sons	*464, 130	1, 696, 140
Belleisle (g)	Two pairs, simple	4	65	2 6	2	4, 020	13	Maudslay, Sons & Field	......	......
Orion (h)	do	4	65	2 6	2	†3, 900	†12	do	......	......
Breastwork-monitors. (i)										
Glatton	Two pairs, horizontal	4	60	2 3	2	2, 868	12. 1	Laird & Co	145, 040	*1, 084, 290

Hotspur (ram)	Two pairs, horizontal direct	4	64	2	9	2	3,497	12.65	Napier & Sons	176,400	*855,336
Rupert (ram)	Two pairs, horizontal	4	80	4	0	2	4,200	13.6	......do	187,110	*1,142,255
Prince Albert	Horizontal	2	72	3	0	....	2,128	11.65	Humphrys & Co	117,600	984,957
Cyclops	Compound, tilt-hammer	4	31 and 57	2	3	2	1,660	11.03	John Elder & Co	78,400	*726,410
Gorgon	......do	4	45	2	0	2	1,670	11.14	Hall & Co	77,760	659,512
Hecate	......do	4	45	2	0	2	1,755	10.9	......do	77,760	630,483
Hydra	......do	4	31 and 57	2	3	2	1,472	11.2	John Elder & Co	77,760	645,393
Scorpion	Horizontal, double piston-rod	2	56	3	0	1	1,455	10.52	Laird Brothers	*110,104	537,385
Wivern	......do	2	56	3	0	1	1,476	10	......do	*89,405	566,258
Viper	Two pairs, horizontal	4	32	1	6	2	696	9.5	Maudslay & Co	37,240	248,477
Vixen	......do	4	32	1	6	2	696	9.5	......do	37,240	263,778
Abyssinia	Two pairs, [illegible], direct-acting	4	34	1	9	2	960	9.6	Dudgeon	....	....
Magdala	Horizontal, direct-acting	4	48	1	9	2	1,369	10.6	Ravenhill & Co	....	630,857
Cerberus	Two pairs, horizontal	4	43	2	3	2	1,369	9.7	Maudslay & Co	....	....
Iron broadside-ships of the old type.											
Agincourt	Horizontal, double piston-rod	2	101	4	6	1	6,621	15.4	Maudslay & Co	404,726	2,262,218
Minotaur	Horizontal, trunk	2	104	4	4	1	6,949	14.4	J. Penn & Sons	390,040	*2,327,381
Northumberland	......do	2	104	4	4	1	6,558	14.1	......do	*388,173	*2,381,710
Achilles	......do	2	104	4	0	1	5,722	14.3	......do	*335,909	*2,285,318
Black Prince	......do	2	104	4	0	1	5,772	13.6	......do	361,983	*1,838,587
Warrior	......do	2	104	4	0	1	5,092	13.9	......do	361,628	*1,842,688
Hector	Horizontal	2	82	4	0	1	3,256	12.3	R. Napier & Sons	222,286	1,376,362
Valiant	Horizontal, double piston-rod	2	82	4	0	1	3,560	12.6	Maudslay & Co	*195,975	1,556,560
Resistance	Horizontal, trunk	2	70¾	3	6	1	2,428	11.8	J. Penn & Sons	168,560	1,176,193
Defence	......do	2	70¾	3	6	1	2,537	11.8	......do	168,560	1,128,667
Wooden broadside-ships of the old type.											
Lord Warden	Horizontal	3	91	4	6	1	6,706	13.5	Maudslay & Co	331,836	1,533,772
Royal Alfred	Horizontal, double piston-rod	2	82	4	0	1	3,434	12.3	......do	230,908	1,295,977
Royal Oak (k)	......do	2	82	4	0	1	3,704	12	......do	220,207	1,141,629
Prince Consort (k)	......do	2	92	4	0	1	3,953	12.7	......do	*255,762	1,083,493
Caledonia (k)	......do	2	92	4	0	1	4,538	13	......do	252,204	1,261,194
Zealous (k)	......do	2	82	4	0	1	3,623	11.7	......do	263,091	1,069,584
Ocean (k)	......do	2	92	4	0	1	4,244	12.8	......do	262,640	1,213,435
Repulse	Horizontal, trunk	2	82	4	0	1	3,347	12.2	J. Penn & Sons	133,134	892,490
Pallas	Horizontal, Wolf's. Trunk in large cyl.	4	2 51 & 2 91½	3	3	1	3,581	13	Humphrys & Co	193,477	889,234
Favorite	Horizontal	2	64	2	8	1	1,773	11.8	......do	117,600	704,355
Research	Horizontal, surface-condenser	2	50	2	0	1	1,042	10.3	James Watt & Co	49,308	317,091
Enterprise	Horizontal, surface-condenser	2	45	1	6	1	710	9.7	Ravenhill & Co	42,532	*243,092
Waterwitch (l)	Horizontal	3	38.5	3	6	....	777	6.5	Dudgeon	66,300	*294,127
Royal Sovereign (m)	......do	2	82	4	0	1	2,463	11	Maudslay & Co	....	1,204,648
Torpedo-ram	....	...	....	...	...	...	....	....	....	....	‡60,064

Four armored vessels of a new type are provided for in the navy estimates, to be commenced during the present financial year.

* These figures have been changed to conform to those of "An English naval architect," his information in regard to English vessels being probably late and correct. † Estimated. ‡ Appropriation for commencing. *a* Mastless, turret, modern line-of-battle ships. *b* Estimated cost of guns and incidentals, $600,000. *c* Masted, turret line-of-battle ship. *d* Late Independencia. *e* Late Memdoohiyeh. *f* Cruising-ships with lifting screws. *g* Late Poyk Sheref. *h* Late Burdj. Sheref. *i* Coast-defense vessels. *k* These converted vessels, built about 1863, are not likely to be sent to sea again, as they are fast decaying, but are employed in harbor service. *l* Hydraulic gunboat, Ruthven's water-jet. *m* Wooden turret-ship. NOTE.—The Lord Clyde having had her engines removed, is omitted from this list.

PART IX.

MODERN UNARMORED SHIPS OF GREAT BRITAIN; TABLE OF DIMENSIONS, &c.; THE INCONSTANT, SHAH, RALEIGH, BOADICEA, BACCHANTE, ROVER, EURYALUS, AND SMALLER VESSELS.

9 K

UNARMORED SHIPS OF GREAT BRITAIN.

It has been noted already in this report that all the *old types* of British naval vessels are gradually disappearing from the navy list for fighting or cruising purposes; sailing-craft, paddle-wheel vessels, and auxiliary or low-speed screw-vessels are alike obsolete. Besides which, wooden vessels of all classes are docmed, and at no distant day will be counted out or relegated to harbor service. The modern cruising-fleet consists of full-powered screw-ships, having iron or steel hulls cased in wood; and vessels of the smaller class called composite, in which the materials are either iron or steel, except the outside planking, put on in two courses, as is hereafter described.

The corvettes *Sapphire* and *Diamond*, built in 1874, were the last wooden war-vessels that will probably ever be added to the British navy, for it has been authoritatively made public that no more wooden ship-frames, knees, or beams will be required at the dock-yards.

The British modern unarmored cruising-fleet may be divided into seven or eight classes or types. The frigates, corvettes, and sloops of this new fleet, constructed and in process of building, are as follows:

MODERN UNARMORED SHIPS.

	Length between perpendiculars, in feet and inches.	Extreme breadth, in feet and inches.	Mean draught of water, in feet and inches.	Load displacement, in tons.	Kind of engines.	Maximum horse-power.	Speed* at the measured mile, in knots.	Number of guns.	Cost of hull and machinery, in dollars (gold).	Remarks.
IRON FRIGATES.										
Shah	334 8	52 0	23 0	6,040	Simple	7,477	16.6	26	1,119,861	In Pacific; first commission.
Inconstant	333 0	50 1	23 7	5,782	do	7,361	16.5	16	1,036,756	Second commission.
Raleigh	298 0	49 0	22 0	5,200	do	6,158	15.5	22	939,856	Do.
IRON CORVETTES.										
Boadicea	280 0	45 0	22 0	4,027	Compound	5,130	14.8	†16	1,040,040	Fitting for sea.
Bacchante	280 0	45 0	21 7	3,932	do	5,250	15	†16	1,020,600	Do.
Euryalus	280 0	45 0	21 7	3,932	do	5,250	‡	†16	1,015,740	Do.
Rover	280 0	43 6	20 2	3,494	do	4,964	14.5	18	782,460	First commission.
Volage	270 0	42 0	....	3,078	Simple	4,532	15	18	617,706	Built in 1870.
Active	270 0	42 0	.....	3,078	do	4,015	15	10	613,104	Do.
One of new type, dimensions not given.										
STEEL DISPATCH-VESSELS.										
Iris	300 0	46 0	20 0	3,735	Compound	7,000	16.5	10	§889,380	Fitting for sea.
Mercury	300 0	46 0	20 0	3,735	do	7,000	§17	10	....	Building at Pembroke.

* The speeds of the ships were taken when the engines were being driven to the utmost for a short period only.

† Under covered deck. ‡ Trial trip not made. § Estimated.

Modern unarmored ships of Great Britain—Continued.

	Length between perpendiculars, in feet and inches.		Extreme breadth, in feet and inches.		Mean draught of water, in feet and inches.		Load displacement, in tons.	Kind of engines.	Maximum horse-power.	Speed* at the measured mile, in knots.	Number of guns.	Cost of hull and machinery, in dollars (gold).	Remarks.
STEEL AND IRON CORVETTES.													
Cleopatra........	225	0	44	6	20	0	2,383	Compound	2,300	†13	14	400,000	Building by Elder Co., Glasgow.
Carysfort........	225	0	44	6	20	0	2,383	....do.....	2,300	†13	14	400,000	Do.
Champion.......	225	0	44	6	20	0	2,383	....do.....	2,300	†13	14	400,000	Do.
Comus...........	225	0	44	6	20	0	2,383	....do.....	2,300	†13	14	400,000	Do.
Conquest.........	225	0	44	6	20	0	2,383	....do.....	2,300	†13	14	400,000	Do.
Curaçoa..........	225	0	44	6	20	0	2,383	....do.....	2,300	†13	14	400,000	Do.
COMPOSITE CORVETTES.													
Opal.............	220	0	40	0	16	3	1,864	Compound	2,116	12.5	14	420,390	In Pacific; first co mission.
Tourmaline......	220	0	40	0	16	3	1,864	....do.....	1,972	12.5	12	397,908	First commission.
Turquoise.......	220	0	40	0	16	3	1,864	....do.....	1,990	12.5	12	....	In Pacific; first co mission.
Garnet...........	220	0	40	0	16	3	1,864	....do.....	2,100	12.5	12	....	Fitting for sea.
Emerald..........	220	0	40	0	16	3	1,864	....do.....	2,100	12.5	12	....	Do.
Ruby.............	220	0	40	0	16	3	1,864	....do.....	1,830	.....	12	....	In Mediterranean first commission.
COMPOSITE SLOOPS.													
1st class.													
Pelican..........	170	0	36	0	14	6	1,124	Compound	987	12	6	220,678	Fitting for sea.
Penguin.........	170	0	36	0	14	6	1,124	....do.....	720	.....	6	220,678	In Pacific; first co mission.
Cormorant.......	170	0	36	0	14	6	1,124	....do.....	900	.....	6	220,678	Building at Chatha
Pegasus.........	170	0	36	0	14	6	1,124	....do.....	900	.....	6	220,678	Building at Dev port.
Wild Swan......	170	0	36	0	14	6	1,124	....do.....	800	.....	6	220,678	In East Indies; fi commission.
Osprey...........	170	0	36	0	14	6	1,124	....do.....	1,010	.....	6	220,678	In Pacific; first co mission.
Dragon..........	170	0	36	0	14	6	1,124	....do.....	900	.....	6	220,678	Building at Devo port.
Gannet..........	170	0	36	0	14	6	1,124	....do.....	900	.....	6	220,678	Building at Sheernes
New Gannet.....	......		....		....		......	....	.....	.....	...	....	
2d class.													
Daring...........	160	0	31	4	13	0	894	Compound	916	.....	4	171,558	In commission; Pa cific.
Sappho..........	160	0	31	4	13	0	894	....do.....	884	.....	4	171,558	In commission; Aus tralia.
Flying Fish......	160	0	31	4	13	0	894	....do.....	836	.....	4	171,558	In commission; Eas Indies.
Egeria...........	160	0	31	4	13	0	894	....do....	1,011	.....	4	171,558	In commission; China
Albatross........	160	0	31	4	13	0	894	....do.....	838	.....	4	171,558	In commission; Pa cific.
Fantôme.........	160	0	31	4	13	0	894	....do.....	975	.....	4	171,558	Do.
SINGLE-SCREW, COMPOSITE GUN-VESSELS.													
Six of Arab class.	150	0	29	0	12	0	700	Compound	656	.....	3	145,800	Displacement an horse-power vary fo different vessels.
Twenty-one of Coquette class.	125	0	23	6	9	0	430	....do....	406	8	4	70,837	Do.

* The speeds of the ships were taken when the engines were being driven to the utmost for a shor period only.

† Estimated.

In addition to the above there are eleven twin-screw iron gun-vessels of 774 tons displacement and 811 indicated horse-power; twenty-one twin-screw composite gun-vessels of 584 tons displacement and 587 indicated horse-power; thirty-eight twin-screw iron gun-vessels of 254 tons displacement and 262 indicated horse-power; and five single-screw gun-vessels of 570 tons displacement and 336 indicated horse-power, all of modern build.

Several other unarmored corvettes are projected. It is intended to construct them with an armored deck three feet below water, and with a ram, and to fit them to use the Whitehead torpedo. The speed is to be 13 knots per hour.

The first essential element of power for an unarmored cruising man-of-war of the present day is speed; a speed sufficiently high to overtake any vessel on the ocean desirable to capture, and to escape from any powerful fighting-ship desirable to evade.

The above-named vessels constitute the present fleet of fast unarmored cruising-ships of the British navy. The hulls of these vessels, larger than the *Opal*, are of iron and steel, and they are built with improved structural arrangements for securing great strength to withstand the immense engine-power put into them and to endure it for any desirable length of time. The bottoms are sheathed with wood, and coppered or zinked to prevent fouling. They carry a large spread of canvas, are provided with lifting screw-propellers, and are in all respects fitted to keep the sea. They have not the speed aimed at by the naval authorities, *i. e.*, such speed as will be attained in the class of vessels now building called rapid cruisers, and in which vessels requirements for keeping the sea for lengthened periods must be sacrificed; but they have speed superior to that of the cruising-ships of any other navy, besides which they are reputed to be excellent sea-boats, fast under sail, and are armed with rifled guns, some of which are of heavy caliber.

The *Inconstant* is the only one of the number to my knowledge that has as yet been driven under full steam-power alone for upward of twenty-four consecutive hours. This was in the emergency when carrying the news of the loss of the unfortunate *Captain* from Cape Finisterre to England, on the 7th of September, 1870. On this occasion a speed of very nearly 15 knots per hour was averaged for the whole distance.* The speed of each of these cruising-vessels on the measured mile has been shown in the column above, and it is fair to conclude that the maximum performance of any one of them at sea, in smooth water, for a period of, say, twenty-four hours, would be one and a half knots less than was recorded on the measured-mile trials.

THE INCONSTANT.

Every officer familiar with the progress and advancement in naval science of late years is acquainted with the fact that when the *Wampanoag* and that class of vessels were under construction at New York and Boston, reports went abroad of the extraordinary speed and terrible power they were designed to possess, and of the fearful destruction which would follow their path. These reports, published in American journals, and copied and commented on in the London papers, alarmed the British government; and as a consequence the chief constructor of the admiralty was directed as early as 1866, before the *Wampanoag* was completed, or her defects known in Europe, to prepare the plans

* It has been understood that the *Inconstant's* speed on this occasion was nearly $15\frac{1}{2}$ knots.—AN ENGLISH NAVAL ARCHITECT.

for competing vessels. Thus originated the *Inconstant.* She was designed for full-sail power, and provided with a lifting screw, and to give sufficient scope for the combination of high speed with what was then a heavy armament, good sea-going qualities, a large carrying capacity, and provision for all requirements; she was made to have a displacement of 5,782 tons, was 333 feet long, had 50 feet beam and 23 feet mean draught of water. Iron sheathed with wood, and coppered, was the material chosen to stand the immense strain of the power to be developed by the engines. She was launched November 12, 1868, and the trial cruises were made upward of a year afterward. Subsequently a sister ship, the *Shah*, was constructed, and was launched in 1873. These two ships represent the largest class of unarmored frigates, and are in all essential features of construction the same. The latter has, however, an increased length of 20 inches over the former, and, in order to confer greater stability than is possessed by the *Inconstant*, an increase of beam to 52 feet was given.

The armaments of the *Inconstant* and *Shah* are as follows:

	Main deck.	Upper deck.	
		Pivot-guns.	Broadside.
Inconstant	10 12½-ton.	2 6½-ton.	4 6½-ton.
Shah	{ 16 6½-ton. 2 64-pdrs. }	2 18-ton.	6 64-pdrs.

The particulars of the *Shah's* guns are taken from a paper published by the committee on designs. Several alterations, tending to greater efficiency, have been made, and one—the substitution of 18 for 12 ton pivot-guns—is embodied above.

The bow-guns fire on a line with the keel. The 18-ton 10-inch guns mounted on the *Shah* are similar to the broadside-guns of the *Sultan* and the *Hercules.*

The *Inconstant* has been refitted during the last autumn and again commissioned. It is worthy of consideration that this iron ship, sheathed in wood, now nearly ten years old, is still sound in all parts of the hull and iron work, and notwithstanding the alterations necessarily made to keep pace with the times—such as fitting tubes and gear for Whitehead torpedoes, and apparatus for charging the torpedoes with air, building magazines for these, also for the Harvey torpedoes, rearranging the armament, scraping, cleaning, and painting the iron skin, repairing the machinery and fitting the ship for sea—the cost of the refit, exclusive of boilers, is reported at only $58,320.

The heavy armament formerly carried remains unchanged, but the power of the guns is increased by using battering charges of 50 pounds of pebble-powder for the Palliser shell, instead of 43 pounds as previously used, and the weight of the projectile is increased to 250 pounds.

THE RALEIGH.

This ship, launched also in 1873, is of the same general design and construction, but of reduced dimensions. The cost of her construction was $180,005 less than the cost of the *Shah*, and $96,947 less than that of the *Inconstant.* Although the *Raleigh* is not possessed of as powerful a battery as the larger ships, she is by far more serviceable, more easily handled, and less costly to maintain. The comparison between the *Raleigh* and the *Inconstant* stands thus:

		Raleigh.	Inconstant.	
			As designed.	As completed.
Tonnage		3,210	3,978	4,066
Displacement		5,200	5,495	5,782
Length between perpendiculars		298 feet.	333 feet.	
Breadth, extreme		49 feet.	50 ft. 1 in.	
Draught	forward	20 feet.	22 feet.	
	aft	23 feet.	24 feet.	
Horse-power.	nominal	800	1,000	
	indicated	5,639	7,361	
Guns	upper deck	2 12½-ton. 4 64-pdrs.	4 6½-ton.	
	main deck	14 90-cwt. 2 64-pdrs.	10 12½-ton.	

These ships, as previously stated, are built of iron; externally they are entirely sheathed with wood, the chief object of which is to admit of copper being applied to the bottom. There are two thicknesses of sheathing, except over parts of the top-sides, where there is one thickness only; the first is secured to the skin of the ship by galvanized iron screw-bolts, which are tapped into the skin, but also bear lock-nuts on the inside. The bolts are of course screwed in through holes prepared in the wood, for the holes become unduly enlarged if there is the slightest want of concentricity between the threaded and the plain parts of the bolt, thus occasioning leakage through the wood to the skin. The second layer of planking is secured to the first by metal wood-screws, which stop short, naturally, of the iron skin, and avoid contact with the galvanized bolts. The two courses of planks break joint, and are carefully put on the iron skin, the planks being properly painted and their joints made tight by calking. The wood is teak, and the thickness as a rule is, for the first course 3 inches, and for the second 2½ inches, tapering as the top is approached. This system has been tested for some years on the *Inconstant*, and has proved to be so successful that its adoption for all naval iron cruising-vessels, both of English and Continental build, has become general. It may be well to mention that the courses of planks are both applied horizontally, except in the *Shah*, where one course is vertical and the other over it horizontal.*

The motive steam-machinery of all these ships was designed and contracted for prior to the date of the adoption of the compound engine by the admiralty. They are, therefore, of obsolete types, and the consumption of fuel in each of them is upward of 30 per cent. greater than in the ships of recent construction. Moreover, the anticipated speeds have not been realized.†

A brief description of the machinery of the *Raleigh* and of the trials may not be out of place. The engines, constructed by Humphrys, Tennant & Co., are of the horizontal, direct-acting, return-connecting-rod

* It is the *Inconstant*, the first built of the type, in which one thickness of sheathing is worked vertically and the other horizontally. In the other both thicknesses are horizontal.—AN ENGLISH NAVAL ARCHITECT.

† The speed the *Inconstant* was designed to have over the measured mile was 16 knots, and the actual speed 16.52. The *Shah* was designed for 16 to 16½ knots, and made 16.65, and the *Raleigh* was designed for 15 to 15½ knots, and made 15½.—AN ENGLISH NAVAL ARCHITECT.

type, and intended to indicate 6,000 horse-power. A few of their dimensions are as follows:

Number of cylinders	2
Diameter of cylinders	8 feet 4 inches.
Stroke	4 feet 6 inches.
Diameter of piston-rods, 4 to each piston	7 inches.
Diameter of back trunk, 1 to each piston	1 foot 6 inches.
Travel of slide-valves (vertical, double-ported)	10¼ inches.
Length of valve	5 feet 10 inches.
Width of valve	5 feet 1½ inches.
Length of ports	5 feet $2\frac{7}{16}$ inches.
Width of bars in extreme ports, 2 in each	2 inches.
Width of bars in other ports, 2 in each	1 inch.

The expansion-valves are horizontal and of the gridiron type, placed above the slide-casing, and worked from rocking-shafts by means of eccentrics on the main shaft. The cut-off ranges from 3½ inches of the stroke onward, and is varied by moving the end of the eccentric-rod along a quadrant arm on the rocking-shaft. There is one surface-condenser to each cylinder. The air-pumps are 25½ inches in diameter, double-acting, and worked direct from the pistons. The circulating-pumps are centrifugal, and driven by separate engines with 12-inch cylinders having 12 inches stroke, exhausting into the condensers; the water is forced through the condensers outside the tubes. The condenser-tubes are vertical, and there are 6,740 in each condenser; their dimensions are as follows:

Diameter inside	½ inch.
Diameter outside	⅝ inch.
Length between tube-plates	6 feet 6½ inches.
Cooling surface	12,000 square feet.

The starting-gear can be worked by hand or steam. A Silver's governor is fitted. The propeller is by Hirsch, two-bladed, has 26 feet 8 inches pitch, and is fitted for being raised when the ship is under sail alone.* The number of boilers is nine; but two of them are small and intended for auxiliary purposes; these are together equal to one of the large ones. The chief data relative to the boilers are as follows:

Number of furnaces	32
Length of fire-grate	6 feet 10 inches.
Breadth	3 feet 3¼ inches.
Area of fire-grate	720 square feet.
Number of tubes	2,880
Length between tube-plates	6 feet 4 inches.
Diameter of tubes outside	3 inches.
Tube-heating surface	14,300 square feet.
Number of superheaters (steam passing through the tubes)	2
Number of tubes	248
Diameter of tubes, inside	2 inches.
Length of tubes	9 feet.
Superheating tube surface	1,170 square feet.
Number of chimneys	2

* The screw-propeller invented by Hermann Hirsch, of Paris, patented in France December 16, 1865, and in the United States April 26, 1870, has been successfully applied to many ships in Europe. The claims set forth in his letters patent are, "1st. Constructing screw-propellers with blades the faces of which are formed of concave lines in cross-section, in combination with an increasing pitch or helicoidal inclination from the axis to the circumference. 2d. In combination with said transverse curvature and graduated helical form, the recession of the forward terminal edges of the blades."

A *two-bladed* screw, made from the drawings of Mr. Hirsch, was applied to one of our second-rates in 1872, but after a trial at sea in rough weather it was reported on unfavorably, and removed from the vessel. The Hirsch screw was also applied to the United States vessels *Trenton*, *Tennessee*, and *Huron*. It was removed from the latter, but remains on the former and has given satisfaction.

The boilers are of the ordinary box kind, and carry 30 pounds pressure on the square inch. The dimensions quoted above give the following relative areas per indicated horse-power:

	Square feet.
Grate surface	.117
Tube-heating surface	2.320
Superheating tube surface	.190
Condenser-tube cooling surface	1.950

Trials of Her Britannic Majesty's ship Raleigh.

Date of trial	April 1, 1874		September 2	September 2.
Kind of trial	Measured mile		6 hours	Measured mile.
Where tried	Maplin Sands		Off Portsmouth	Stokes Bay.
Draught:				
Forward	19 feet 6 inches		21 feet 10 inches	21 feet 11 inches.
Aft	23 feet 6 inches		25 feet 2 inches	24 feet 10 inches.
Mean	21 feet 6 inches		23 feet 6 inches	23 feet 4½ inches.
Ship by the stern	4 feet.		3 feet 4 inches	2 feet 11 inches.
Screw:				
Diameter	21 feet.		21 feet 1 inch	21 feet 1 inch.
Length, greatest	3 feet 2½ inches		3 feet 3¾ inches	3 feet 3¾ inches.
Upper edge, immersed	2 feet 10 inches		2 feet 5½ inches	2 feet 1½ inches.
Force of wind	4 to 5		4 to 5	3
State of sea	Smooth		Moderate	Smooth.
Steam-pressure and temperature:	Full power.	Half power.		
Boilers	32.7 pounds.	29.1 pounds	27.4 pounds	28 3 pounds.
Superheaters	Mean, 310°.	370°	Mean, 325°	Mean, 320°.
Engines	31.2 pounds, 274°.	27.6 pounds, 282°	25.9 pounds, 289°.	26.8 pounds, 281°.
Vacuum:				
Forward	27.6 inches.	27.4 inches	27.0 inches	26.8 inches.
Aft	26.9 inches.	26.4 inches	26.5 inches	26.8 inches.
Revolutions per minute	73.9	58.5	68.8	69.6
Indicated mean pressure	19.8	13.8	19.1	19.2
Indicated horse-power	6157	3413	5541	5639
Speed of vessel	15.503	13.455	15.139, calculated	15.320
Revolutions per knot	286.0	260.8	272.5, calculated	272.5
Slip, per cent	20.28	12.6	16.35, calculated	16.35
Temperatures:				
Deck	56 degrees		65 degrees	64 degrees.
Engine-room	77 degrees		87 degrees	83 degrees.
Stoke-hole, aft	105 degrees		118 degrees	108 degrees.
Stoke-hole, fore	118 degrees		110 degrees	92 degrees.
Kind of coal used on trial.	Nixon's navigation		Nixon's navigation.	Nixon's navigation.

The *Inconstant* has seen considerable service; the *Raleigh* made a cruise to India and returned; but the *Shah* was just put on the last measured-mile trials at the time of my visit on board that vessel in April, 1876, and she is now in the Pacific on her first commission.

It is now freely admitted by the authorities that both the *Inconstant* and the *Shah* are undesirable property. They were too costly to build, are too costly to maintain, and too unwieldy to handle.* It is said by Mr. Brassey, M. P., that "the designers of these vessels were betrayed into an exaggeration of size from over-anxiety to combine in a single ship every quality with which an unarmored vessel can possibly be endowed. They were to possess unrivaled speed, both under steam and

* We have never seen any reason for believing that this is a correct statement of the estimation in which these ships are held by our admiralty. The placing of the *Shah* on the South Pacific station is a proof that their qualities are highly estimated, while the action between the *Shah* and the Peruvian iron-clad *Huascar* showed that instead of the *Shah* being too unwieldy to handle, as Mr. King says, she was maneuvered with such ability and success as to avoid being struck by the enemy. The power of the *Shah*, which might have been made much greater by the introduction of the heavier armament she is able to carry, combined with her handiness, was sufficient to enable her to come successfully out of a contest with an unquestionably handy vessel protected by armor-plating. It should be remembered, however, that, as Mr. King has said, these ships were built to compete with the American *Wampanoag* class, about which some alarm was felt in this country while they were being built, and that in this competition the English ships have been successful. There have been no threats expressed since the

under sail, and to be armed with such batteries of armor-piercing guns that it was hoped engagements might be fought even against armored ships with some prospect of success. The attempt was ambitious and not altogether unsuccessful, but they are now found to be too expensive for mere protection of commerce, and their guns would be useless against armored ships of the present day." "A perfect ship of war," as it was very prudently observed by the admiralty committee on designs, "is a desideratum which has never yet been obtained; any near approach to perfection in one direction inevitably brings with it disadvantages in the other."

We now come to recent productions, ships set afloat two years ago, and ships not yet completed. They are the *Boadicea*, *Bacchante*, *Euryalus*, and *Rover*. These vessels are of smaller dimensions than the *Raleigh;* they are engined on the new system with three-cylinder compound engines, and promise, with modifications which experience may prove advantages, to have permanence as types. They are all rated as corvettes, although the main batteries of the first three named are carried under covered decks.

THE BOADICEA.

This ship has 95 tons more displacement than the other two just described; she has a brass stem, and the bottom is sheathed with two courses of planks and coppered, while the other two ships are built to utilize the power of the ram; and with this object in view, they are formed with upright bows, have iron stems, and are sheathed with only one course of planks which is covered with zinc instead of copper.

When such vessels are fitted for ramming, it not only becomes necessary to make the stems of iron or steel, but also to build them of additional strength, and to avoid coppered bottoms, for the reason that any damage to the bow in ramming which should expose the iron skin to the action of the copper sheathing would cause serious galvanic action.

The *Boadicea* has been completed and equipped for sea. Her engines were tested at the measured mile in October, 1876, and in December a special run of six hours' duration was made, when the following results were attained, the draught of ship forward being 20 feet 6 inches, and aft, 22 feet 4 inches:

Pressure of steam in boilers, 70 pounds; horse-power in the first hour, 4,893, with 73 revolutions; in the second hour, 5,406 horse-power and 74.6 revolutions; in the third hour, 5,414 horse-power and 75.3 revolutions; in the fourth hour, 5,227 horse-power and 74.5 revolutions; in the

---

building of the English ships, of sweeping our commerce off the seas in the event of war, by fast American cruisers. This is a complete justification of the *Inconstant* and *Shah* if any were required. Mr. Brassey's criticism was obviously based upon an imperfect appreciation of the objects with which these ships were first designed, and of the manner in which they have been realized.—AN ENGLISH NAVAL ARCHITECT.

In selecting the action between the *Shah* and *Huascar* to prove the handiness of the former, "An English Naval Architect" is unfortunate, for, whereas the *Shah* possessed this quality only in a small degree, the *Huascar* possessed it in a still smaller, and moreover was by far the slower vessel.

I must concede that the *Wampanoag* attained her speed, viz, 16.97 nautical miles per hour, for only twenty-four hours during her trial trip at sea, in one hour of which she achieved 17.75 nautical miles, and that she never performed any service thereafter. This class of vessel was built during our war, at a time when all the iron-mills were working to their full capacity to fill orders. As a consequence, the hulls being unfortunately built of wood, were not capable of sustaining the weight and great power of the engines. The war having ceased before the vessels were completed, no necessity existed for their use.—J. W. K.

COMPOUND ENGINES OF H.B.M.S. ROVER.

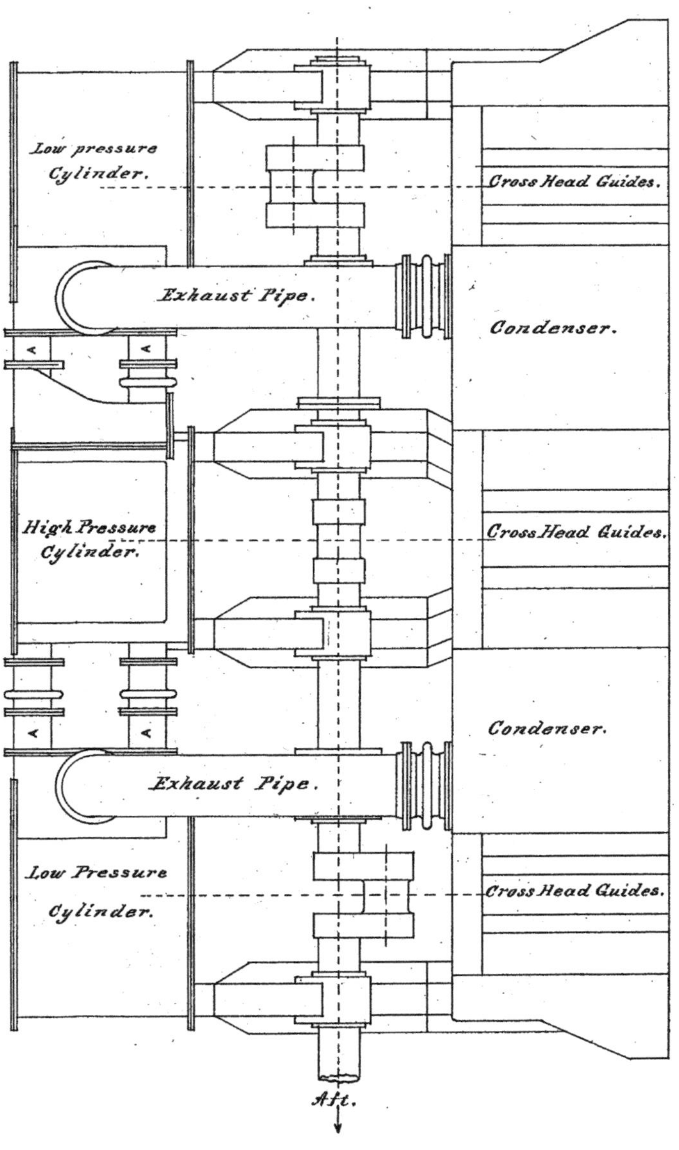

fifth hour, 5,523 horse-power and 75 revolutions; in the sixth hour, 5,320 horse-power and 74.4 revolutions; the mean being nearly 5,300 horse-power, 74.5 revolutions, and speed 14.5 knots.

The engines are reported to have worked satisfactorily, and one feature of the trial was working them with a pressure of only 10.5 pounds in the boilers.

THE BACCHANTE.

The *Bacchante*, built at the Portsmouth dock-yard, and recently fitted for the first commission, is of the same length and breadth as the *Boadicea*, but the draught of water is 5 inches less, and the displacement 3,932 tons against 4,027 tons. Unlike her sister, she has been built with a ram-bow and running-in bowsprit, to enable her to ram wooden and unarmored ships. A second difference is in the wood sheathing applied to the outside hull; this is formed of a single strake covered with zinc instead of copper, the seams between the planks being uncalked, with the view that as the water gains access to the iron skin, and galvanic action is set up, the hull is then supposed to be preserved at the expense of the zinc. With the exceptions named, she is similar to the *Boadicea*, even as regards the armament, in regard to which there exists considerable diversity of opinion among officers, since the *fiasco* between the *Shah* and *Huascar*.

The engines, which have been constructed by Messrs. Rennie Bros., are of the compound, horizontal, return-connecting-rod type of three cylinders; the high-pressure cylinder being 84 inches and the low-pressure 92 inches in diameter, with a stroke of 4 feet.

The steam is supplied by ten cylindrical boilers; the initial pressure of steam is 70 pounds per square inch, and the contract indicated horse-power is 5,250.

Late in the summer of 1877 a trial of the machinery was made with the ship light, the draught at that time being 15 feet 4 inches forward, and 21 feet 6 inches aft. A six-hour run was made, and the following results were recorded:

Pressure of steam, in pounds.	Vacuum, in inches.	Revolutions.	Horse-power.
67	27 and 26½	72.60	5,293.78
70½	27 and 26½	73.36	5,432.38
71	26½ and 25	73.33	5,246.34
69	26½ and 25	72.50	5,154.51

The mean results gave 5,282 horses, or 32 beyond the stipulated power, the mean revolutions per minute being 73.

The means for the entire six hours were as follows: Pressure of steam in boilers, 68.33 pounds; vacuum forward, 26.75, aft, 26.37 inches; revolutions, 72.68; mean pressure of steam in cylinder, high-pressure, 31.80 pounds, low-pressure, 11.08 pounds; horse-power, 5,164.

The speed through the water attained during the day never fell below 15, and, measured by the ship's log, a speed of 15½ knots per hour was occasionally realized.

THE ROVER.

This vessel was built by contract; she has 493 tons less displacement than the *Bacchante* and *Euryalus*. She was launched August 12, 1874, put on the trial trips in November, 1875, and sent to the West India squadron immediately thereafter. She approaches the size and type of a class of vessels very greatly needed in our own Navy. The length between perpendiculars is 280 feet; breadth, extreme, 43 feet 6 inches; draught of water forward, 17 feet 2 inches; aft, 23 feet 2 inches; displacement, 3,494 tons.

The contract price for hull and machinery was $782,460. She was made large enough to be sea-going, to act as a ram, and to be fast; and no ship having structural strength to endure the engine-power necessary to drive her 15 knots per hour, to act as a ram, to have all the requirements necessary for a war-ship, and to keep the sea as a cruiser, is likely to be made very much smaller until some further advancement is made in engineering.

The engines are the first of their kind completed and used in the British navy, and special interest has been taken in their performance. They are compound and surface-condensing, and belong to the type known as the horizontal return connecting-rod, and have their cross-heads working in slipper-guides. According to the contract, they were to indicate 4,750 horse-power at the measured-mile trial.

The preceding diagram shows the main parts of the engines in plan. The diameter of the high-pressure cylinder is 72 inches, equivalent to an area of 4,071.51 square inches of piston. The diameter of each low-pressure cylinder is 88 inches, and the area of piston 6,082.13 square inches. The two low-pressure pistons combined giving an aggregate area of 12,164.26 square inches, the high and low pressure pistons bear the ratio to each other of 1 to 2.987 in effective area. The high-pressure cylinder, which is fitted with a working barrel made of Sir Joseph Whitworth's patent compressed steel,* is placed between the two low-pressure ones, and each cylinder stands separately and distinctly by itself, securely attached to its own two main frames—these latter, six in all, carrying the crank-shaft. The slide-valve of the high-pressure cylinder is placed on the upper side of it, and lies on its face; the valve-face of the cylinder is cast separately, and bolted to its place, with a view to its ready repair or removal in case of need. The slide-valves of the low-pressure cylinders are placed on the sides of the latter, the valve-faces of these cylinders being also bolted on; and all the slide-valves are fitted with the usual metallic packing-rings on their backs to relieve the pressure. Steam starting-gear, in addition to the ordinary hand-gear, is fitted for facility of handling them, and the easy way in which these large engines can be handled is best told by describing the results obtained. While running full speed ahead, they were stopped

---

* This metal, first introduced by Sir Joseph Whitworth, about twenty years ago, and since used for his ordnance, is now gradually gaining favor for such constructions as demand a material of exceptional strength and toughness, or where a combination of strength and lightness is essential, as in many parts of marine steam-engines, especially in cylinder-linings and valve-faces, where hardness is also necessary to resist abrasion.

Steels, as ordinarily cast by pouring into ingot-molds, are found to be porous and comparatively heterogeneous in texture, and brittle in consequence of gas and air-holes. Whitworth adopted the simple but effective expedient of subjecting the ingot or casting, while still molten, to the tremendous action of a heavy hydraulic press while solidifying; by this means the pores and bubble-holes are effectually closed, and the compressed steel is given a strength and homogeneity unequaled by any other known metal. Although a simple expedient, it has required a considerable amount of experiment, ingenuity, and cost to produce this compressed steel with uniform success; this was, however, accomplished, and the material is now used in the British navy, in France, and in Austria.

The inventor has stated that by a careful selection and treatment of metals, a steel can be produced capable of resisting a tensile stress of 45 tons per square inch of section, and of elongating 25 per cent. before breaking.

First cost is the only objection raised against its general use.

in nine seconds, went astern at full speed in six seconds more, and were reversed from that position to full speed ahead again in eight seconds. The casing of the high-pressure slide-valve, to which the expansion-valve casing is attached, is connected with the low-pressure slide-valve casings by four copper pipes, A, A, A, A, which are fitted as shown, and through which the steam passes after leaving the high-pressure cylinder. These pipes and the passages connected with them form the only steam-reservoirs between the cylinders. The surface-condensers are on Hall's system, the steam passing through the tubes; they stand on the side of the connecting-shaft opposite to the cylinders, and each condenser is connected with its own low-pressure cylinder by an eduction-pipe, through which the steam passes on leaving the latter. The condensers have a total number of 7,224 brass tubes, tinned inside and outside, with a total cooling surface of 9,500 square feet. The tubes are fitted in the tube-plates with screwed glands and stuffing-boxes, and the condensers are so arranged that they may be worked as common condensers, if required. The air-pumps, double-acting, are two in number, 23 inches in diameter, with a stroke of 4 feet, and the circulating water is driven through the condensers by two centrifugal pumps worked by an independent pair of small engines, and the main receiving-pipes to these pumps are fitted with branches leading into the bilges of the vessel, so that the pumps may become large bilge-pumps in the event of any leak arising in the vessel. There is the usual arrangement of feed and bilge pumps fitted on the main engines, two of each; they are of gun-metal, and are 5 inches in diameter, with a stroke of 4 feet. In addition to these pumps there are two feed auxiliaries, a bilge auxiliary, hand-pump, and fire-engine, all fitted in accordance with the admiralty specification.

The length of the stroke of the main engines is 4 feet, and the length of the connecting-rod is twice the length of the stroke. The diameter of the connecting-shafts is 18 inches, and of the crank-pins 20 inches. The shaft is made in three pieces, having couplings forged on the ends, by which they are bolted securely together. The diameter of the line-shafting is 16½ inches, and that of the stern-shaft is 18½ inches, the latter running in the usual *lignum-vitæ* bearings fitted in the stern-tube. The screw-propeller is of gun-metal, is on the Hirsch principle, has a diameter of 21 feet, and is driven by an ordinary cheese-coupling keyed on the stern-shaft, and is fitted to lift in a banjo-frame by means of sheer legs and tackles on deck.

BOILERS.

The boilers, which are ten in number, stand athwartships in two groups. Each group of four and six, respectively, has its own separate chimney, fitted on the telescopic principle, and the boilers carry a working pressure of 70 pounds to the square inch. They are about 11 feet 10 inches in diameter, 9 feet 6 inches long, fitted with brass tubes 3 inches in outside diameter, and with wrought-iron stay-tubes. Each boiler has two furnaces 3 feet 10 inches in diameter and 6 feet 8 inches long. The total heating-surface of the boilers is 12,700 square feet, and the grate-bar surface is 510 square feet.

Among other fittings of the engines it may be mentioned that both low-pressure cylinders have separate starting-valves; a double-beat regulator-valve is placed in the main steam-pipe close to the expansion-valve, and beside this a steam-separator. There is also a throttle-valve to be used with a Silver's governor. The high-pressure piston is provided with a 20-inch trunk at the back, which is inclosed in a casing bolted upon the cylinder-cover, and works always upon an adjustable composition block; the latter thus takes the principal weight of the piston, by which means it is hoped that excessive wear of the cylinder may be prevented. This adjustable block is regulated by set-screws, and although at the time when made it was thought to be an improvement, it has since been found to be a disadvantage, because there is no means of ascertaining just how much to raise or lower the piston. The starting and reversing gear referred to is simply a vertical steam-cylinder with the necessary valve-gear; the lower end of its piston-rod is connected with the reversing-gear by a link, the upper end of the same rod is formed as a very coarsely-pitched screw; this is so proportioned that it does not move of itself when steam is turned on, but is then so far balanced that scarcely an effort is required at the hand-wheels (of which two are provided and connected with the piston-rod by bevel-gearing) in order to move the reversing-links in either direction.

The following results were obtained in November, 1875, on the measured-mile trial: The mean speed on the six full-power runs was then 14.533 knots, the mean revolutions being 68.51; steam, 68 pounds to 70 pounds; vacuum, 27½ inches. The mean indicated horse-power on these runs was 2,476.6 in the high-pressure cylinder, 1,343.2 in the forward low-pressure cylinder, and 1,143.7 in the after low-pressure cylinder; the total being thus 4,963.5 indicated horse-power. The mean speed on the four half-power runs was 11.714 knots with 54.26 revolutions, 66 pounds of steam, and 28 inches of vacuum. The indicated horse-power was 1,240 in the high-pressure cylinder, 580.6 in the forward low-pressure cylinder, and 501.7 in the after low-pressure cylinder, the total being 2,322.3. The propeller was set at a mean pitch of 24 feet 11 inches.

As previously stated, the engines of the *Rover* are the first of the kind put on trial in the royal navy. The most marked feature about them is the use of two low-pressure cylinders instead of one. The diameter of one cylinder, equal in area to the two low-pressure cylinders of the *Rover*, would be nearly 124½ inches. Cylinders of about this size have been used in Her Britannic Majesty's navy, working with pressures greater than those at which the low-pressure cylinder of a compound engine works, and they have been used in the merchant service in compound engines; but the experience with these large cylinders has been very unsatisfactory, consequent upon the number that have been cracked by unequal expansion and contraction in these large castings, besides which the inconveniences of handling and working with such large and heavy parts in the confined space of an engine-room are very great. For these reasons the low-pressure cylinders have been kept within reasonable limits in all recently-designed engines for the naval service.

THE EURYALUS.

The engines and boilers of this ship are similar to those of the *Rover* except that the high-pressure cylinder has a diameter 2 inches greater than that of the *Rover;* the low-pressures are correspondingly increased in diameter, and also the working parts. The power of the boilers is also increased by an addition of one furnace in each, thus making 30 furnaces in the ten boilers, each having a diameter of 3½ feet and a 6-foot length of grate.

During the time these and the several other sets of three-cylinder engines have been building, the subject of the position in which the cranks should be set relatively to each other has received considerable

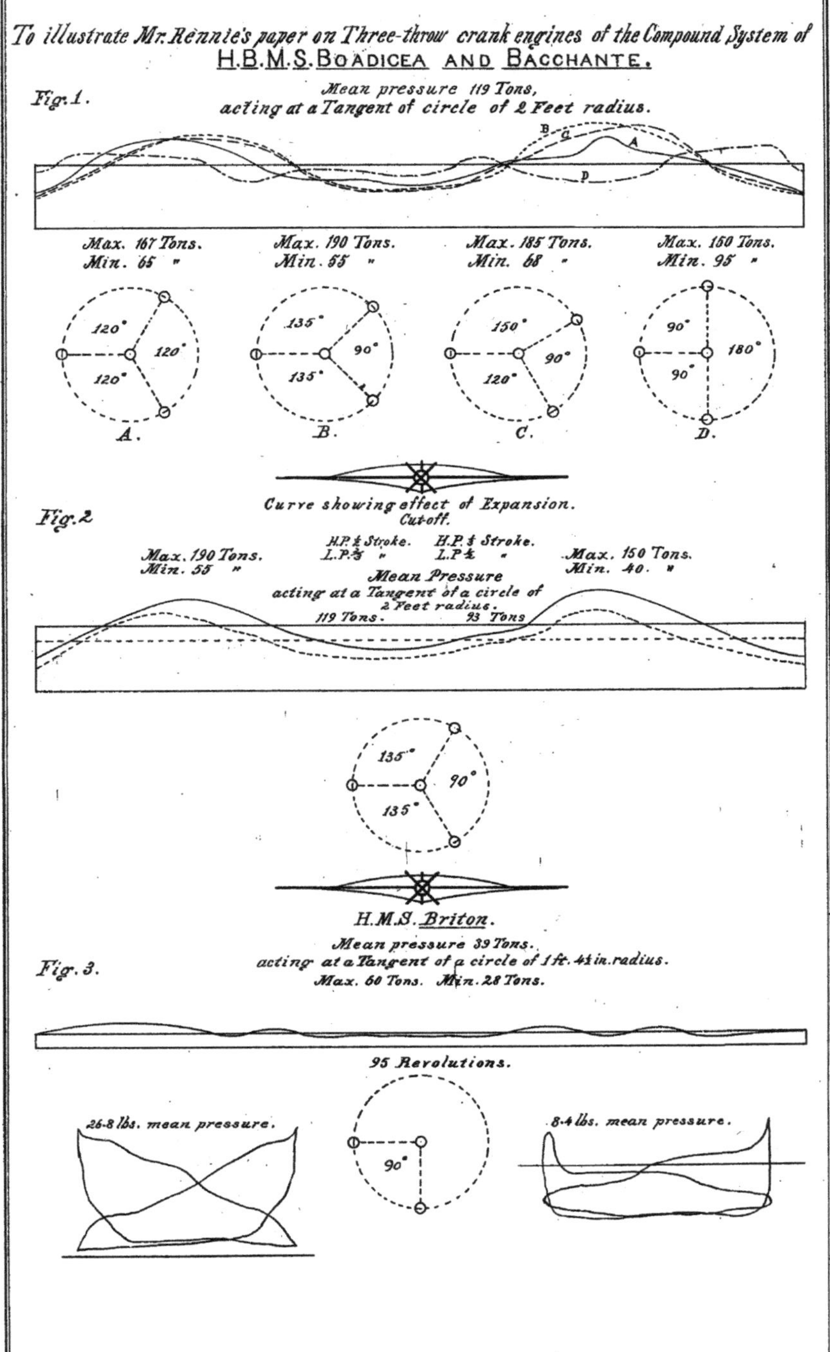

To illustrate Mr. Rennie's paper on Three-throw crank engines of the Compound System of
H.B.M.S. Boadicea and Bacchante.
Fig. 1.
Mean pressure 119 Tons,
acting at a Tangent of circle of 2 Feet radius.
B
C
A
D
Max. 167 Tons.
Min. 65 "
Max. 190 Tons.
Min. 55 "
Max. 185 Tons.
Min. 68 "
Max. 150 Tons.
Min. 95 "
120°
120°
120°
A.
135°
90°
135°
B.
150°
90°
120°
C.
90°
180°
90°
D.
Curve showing effect of Expansion.
Cut-off.
Fig. 2
H.P. ½ Stroke.
L.P. ⅔ "
H.P. ¼ Stroke.
L.P. ¼ "
Max. 190 Tons.
Min. 55 "
Max. 150 Tons.
Min. 40. "
Mean Pressure
acting at a Tangent of a circle of
2 Feet radius.
119 Tons.
93 Tons
135°
90°
135°
H.M.S. Briton.
Mean pressure 39 Tons.
acting at a Tangent of a circle of 1 ft. 4½ in. radius.
Fig. 3.
Max. 60 Tons. Min. 28 Tons.
95 Revolutions.
26·8 lbs. mean pressure.
90°
8·4 lbs. mean pressure.

attention and discussion. At the fifteenth session of the Institution of Naval Architects, Mr. G. B. Rennie, a distinguished mechanical engineer and marine-engine builder, read a paper on the subject (to be noticed directly), and at the last session of the same institution Mr. John R. Ravenhill, also well known as a mechanical engineer, made the following statement:

In the month of March, 1874, a paper was read by Mr. G. B. Rennie, on engines at that time under course of construction by his firm for Her Majesty's steamships *Boadicea* and *Bacchante*, in which he brought under our notice a series of diagrams showing the theoretical force exerted by the three cylinders at all parts of the path of the crank-pin's centers, in four different ways of setting the relative angles of the cranks, which he described as follows: "No. 1. With equal angles of 120° with each other. No. 2. The two low-pressure cranks with 90° between them, and 135° between these and the high-pressure crank. No. 3. The low-pressure cranks with the same angle between them, but at angles of 150° and 120° with the other one. No. 4. The low-pressure cranks placed opposite to each other, and at right angles to the high pressure crank." And he proceeded to give his reasons for concluding to adopt the position described in No. 4, for the angles of the cranks of the two above-named ships. (See Figs. 1 to 4.) In the discussion that followed, I stated that in the three-cylinder compound engines then making by my late firm for the *Rover*, I had adopted the plan of placing the cranks at equal angles of 120° with each other (see Fig. 1), and promised to furnish the institution with the practical result obtained. Recent experience has demonstrated most clearly that we have closely approached, if we have not actually reached, the point at which the crank-pins of large high-speed engines will work satisfactorily in consequence of the very limited amount of bearing surface that of necessity can be allotted to them; and unless phosphor-bronze should come to the assistance of the marine engineers, as white-metal and *lignum-vitæ* did in days gone by, it may be necessary to reduce the load on the high-pressure crank-pins. Rather than depart from the angle of 120° for my cranks I would alter the multiple of my cylinders, for with the three cranks set at equal angles to each other you possess the great practical advantage of being able to work on with any two cylinders out of the three, under many circumstances, in the event of temporary derangement with the third, an advantage that might prove to be the salvation of a ship and her crew on a lee shore or in the time of war.

It will be seen that the cranks of the engines of the *Rover* are set as represented by the position at Figure No. 1. The cranks of the engines of the *Boadicea* and the *Bacchante* are set according to the positions represented at No. 4. In the armored ships *Alexandra* and *Dreadnought*, the position adopted for the engine-cranks is that of No. 2.

The following is the paper by Mr. Rennie, and the preceding are the diagrams produced by him:

ON THREE-THROW CRANK ENGINES OF THE COMPOUND SYSTEM—HER MAJESTY'S SHIPS BOADICEA AND BACCHANTE.

By G. B. Rennie, Esq., *Member of Council.*

[Read at the fifteenth session of the Institution of Naval Architects, 27th March, 1874; the Right Hon. Lord Hampton, G. C. B., D. C. L., president, in the chair.]

Since I had the honor of reading a paper before you on the subject of compound engines of Her Majesty's ship *Briton*, in 1871, the system of engine adopted in the navy has been almost entirely compound, even to the largest size. The engines for the ships *Thetis*, *Encounter*, and *Amethyst* closely followed those of the *Briton*, of similar size and construction, besides others of the same size and many smaller. All of these were made with one high-pressure and one low-pressure cylinder; but when a much larger power than that developed in the *Briton* was required, it was considered advisable by Mr. Wright, chief engineer of the admiralty, to have two low-pressure cylinders and one high-pressure cylinder, on account of the risk involved in making good castings of the size of cylinders that would be required if made with one low-pressure cylinder, especially as casualties had taken place in some of the larger cylinders in Her Majesty's ships.

Last year my firm entered into a contract for two sets of compound engines, each 5,250 horse-power, to be fitted on board Her Majesty's ships *Boadicea* and *Bacchante*; each set of engines has three cylinders, one high-pressure of 73 inches diameter, and two low-pressure of 93 inches diameter, the stroke being 4 feet.

Wishing to ascertain the most advantageous angles to place the cranks in relation to each other, so as to develop the required power with the greatest regularity of motion, with the least strain on the shaft, I had some carefully-made diagrams constructed, taking into account the cut-off to which the valves were made to [work], as well as making allowance for the capacity of the ports, passages, &c., between one cylinder and the other at each successive point of the travel of the pistons. The steam being cut off at *half-stroke* in the high-pressure cylinder by a separate expansion-valve, and at three-fifths of the stroke in the two low-pressure cylinders; the initial pressure being 82 pounds, and the back pressure, from imperfect vacuum, at 4 pounds. It was impossible to make allowance for the "wire-drawing" of the steam through the ports, pipes, &c., with any degree of accuracy; this, therefore, was not taken into account, so that the total horse-power indicated on the diagrams is in excess of that which would be actually realized in practice; but as each case is relatively the same in this respect, it would not affect the general comparison.

The diagrams are made for four different positions of cranks:

No. 1. With equal angles of 120° with each other.

No. 2. The two low-pressure cranks, with 90° between them, and 135° between these and the high-pressure crank.

No. 3. The low-pressure cranks, with the same angle between them, but at angles of 150° and 120° with the other one.

No. 4. The low-pressure cranks placed opposite to each other, and at right angles to the high-pressure crank.

Having thus ascertained the pressures at different points, in each case, when the cranks were placed at the above-named angles, the diagrams of the tangential forces were constructed, taking into account the length of the connecting-rod, to be four times that of the crank.

These diagrams show the force exerted by the three cylinders at all parts of the path of the crank-pin center; and the greater regularity of the curve, and the nearest approach to a straight line, the more uniform the rotation of the shaft and screw-propeller.

It is seen clearly by the diagrams that the most uniform motion is derived by placing the cranks at right angles, as in No. 4; and the next best position is where the angles are equal (No. 1).

The more uniform motion also gives the least maximum strain on the shaft, and thus allowing for the same margin of safety in each case. And, supposing in one case the diameter of the shaft to be 18 inches, in the other it would have to be 19½ inches diameter.

As regards the better propelling-machine, I can fancy there can be little doubt among naval engineers that a steady, continuous pressure will be a far better propelling-machine than one subject to a series of jumps and variation of strain during the rotation of the propeller.

In order to ascertain the effect of an earlier cut-off than half-stroke, I had a further diagram (No. 5) made, supposing the steam cut off at one-third in the small and one-half in the large cylinders, but this does not appear to affect the irregularity of the motion, but merely shows a gradual depression throughout the circle. How far a greater cut-off in the low-pressure cylinders affects the curves I have not yet gone into, the engines in question not being fitted with separate expansion-valves on the low-pressure cylinders.

The sixth set of diagrams are those of the *Briton*, with two cylinders, taken on her trial trip, and reduced to a curve, showing the tangential force to turn the shaft round.

From these examples, it seems to me that in compound engines—whether with two or three cylinders—the best position to place the cranks, both for uniformity of motion as well as strain on the shaft, is where the low-pressure cylinder cranks are placed at right angles to the crank of the high-pressure cylinder.

SMALLER VESSELS.

There is yet a smaller class of modern corvettes of a type known formerly as the *Magicienne*, now as the *Opal* class, that promises permanence; it consists of the *Opal*, *Tourmaline*, *Turquoise*, *Ruby*, *Emerald*, and *Garnet*. The two first-named were launched in 1875, the next three in 1876, and the last on June 30, 1877. They have all been completed and are in commission except the *Garnet*, which vessel is just receiving the finishing strokes.

These vessels are of composite build, and, as may be seen from the annexed drawings of one, the *Garnet*, inspected when in process of

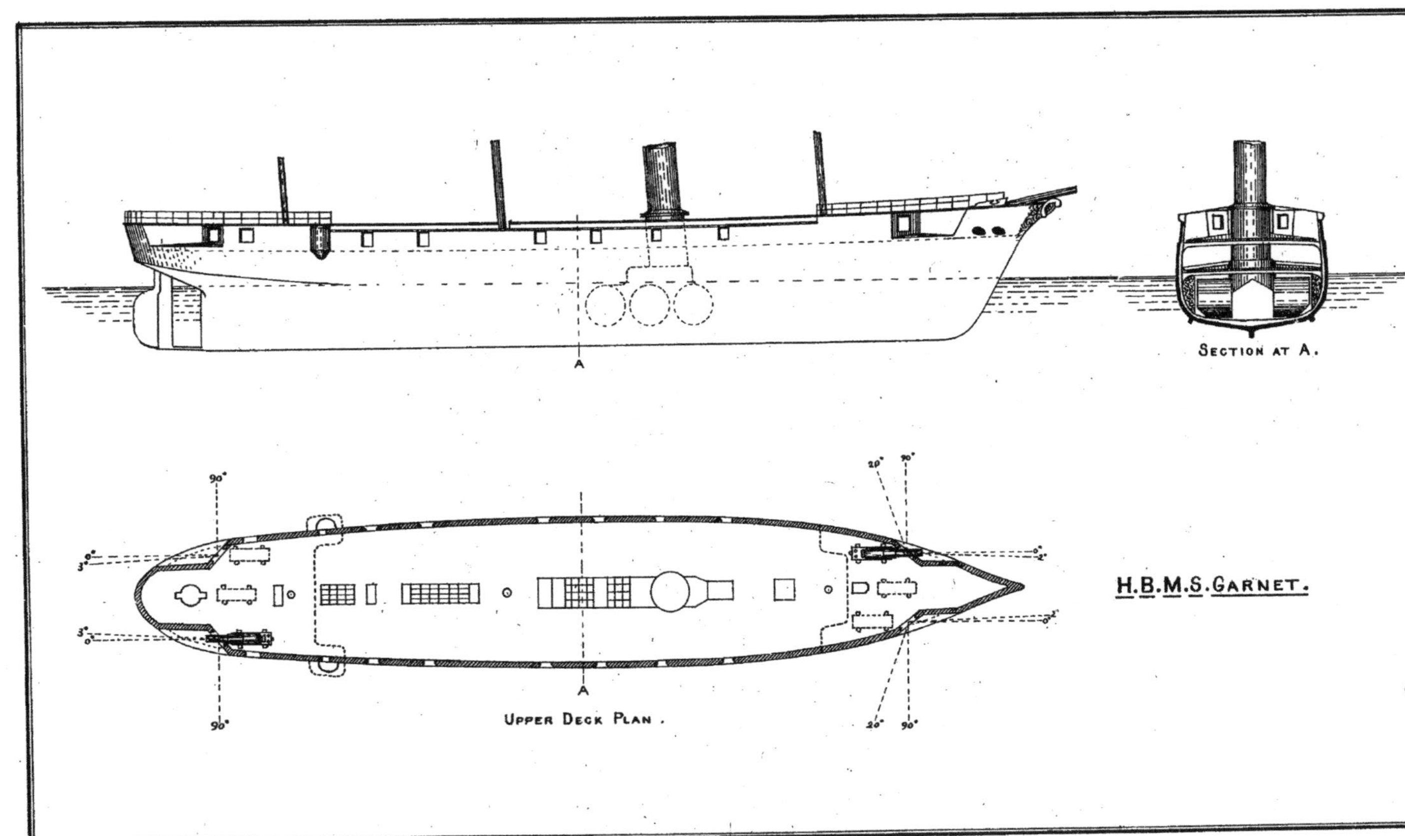
A
Section at A.
90°
3°
0°
90°
20°
A
Upper Deck Plan.
H.B.M.S. Garnet.

construction at the Chatham dock-yard, they are of a type desirable to possess, having a light draught of water, fair speed, considerable battery, full sail-power, and being easily handled. Indeed, they are greatly superior to our third rates, especially in strength and durability, there being no wooden frames, beams, or knees to *rot*, and the outside planking of the hull being mostly of a material—teak—of great durability; besides, very considerable strength of hull is gained by the transverse and longitudinal bulkheads, in securely tying the frames and deck-beams of the vessel together.

The iron keel-plate is 30 inches wide by $\frac{5}{8}$ inch thick, and is continuous from end to end, turning up at the stem and stern and extending to the bowsprit forward and to the main deck aft. The keelson is intercostal, 24 inches deep, and is secured to the flat plate-keel by angle-irons 3 by 3 inches and $\frac{1}{2}$ inch thick. The intercostal plates extend above the floors 3 inches, and are fastened to each other and to the reverse frames by double angle-irons from end to end of the vessel. There is a wood keel of English oak secured to the flat keel-plate by malleable bolts, also a false keel of oak 4 inches thick. To the iron stem there is secured first a course of teak, then a course of elm. This wood stem is molded 24 inches inside of the rabbet and sided 12 inches. The stern-posts are of teak and oak, and the two are connected together by a composition shoe. The rudder is of wood cased with brass in the usual way. The bottom of the vessel under the engines and boilers is strengthened by two intercostal longitudinals on each side, composed of plates 3 inches by $\frac{7}{16}$ inch, connected to each other and to the reverse angle-irons of the floors; in the wake of these and connected to the intercostal plates are additional plates. The frames of the vessels are spaced 20 inches from center to center, and are composed of angle-irons 9 inches by $3\frac{1}{2}$ inches and $\frac{7}{16}$ inch thick, and 3 inches by 3 inches by $\frac{1}{2}$ inch thick; the former extend to the rails and the latter are continuous from bilge to bilge. The beams are of bulb T-iron with knees welded on; those of the gun-deck are 9 inches deep by $\frac{9}{16}$ thick; those of the berth-deck are 7 inches deep, and at the poop and forecastle they are 5 inches deep. The transverse water-tight bulkheads are seven in number; they are $\frac{5}{16}$ inch thick at the bottom, $\frac{3}{8}$ inch at the top, are stiffened by angle-irons, and they extend from the floor to the gun-deck, thus tying the bottom, sides, and decks of the vessel together, and thereby adding strength to the hull and a rigid foundation for the machinery. Each bulkhead is provided with a water-tight door that can be opened and shut from the deck. The coal-bunker bulkheads are also water-tight; and fitted with water-tight doors; all other bulkheads are of the usual kind. The deck-stanchions are wrought-iron tubes with solid heads and feet welded on. Stringer-plates are riveted over the ends of the deck-beams; those of the poop and forecastle are 15 inches wide and secured to the sheer-strake by angle-irons; on the gun-deck the width is 36 inches by $\frac{5}{8}$ inch thick, and they are secured in a similar manner to the gunwale sheer-strake; on the berth-deck the stringer-plate is 28 inches wide and secured to the outside iron strake by angle-irons. The deck-planks are of Dantzic oak 4 inches and 2 inches in thickness, and fastened to the beams by galvanized iron bolts having nuts on the points under the decks. On each side of the vessel just forward of the poop and partly projecting over the sides, there is an iron pilot-house $\frac{3}{4}$ inch in thickness, intended to protect the commanding officer from rifle-bullets. The method of applying the outside planks to the iron hull will be noticed farther on.

The lower masts are of iron, the topmasts and yards are of wood; the main, fore, and mizzen masts have lengths and diameters respectively

53 feet by 22 inches, 49 feet 6 inches by 22 inches, and 43 feet by 15 inches.

The *Garnet* carries one steam-cutter 28 feet long by 7 feet 3 inches wide, and six other boats.

The principal dimensions of the *Garnet* and class are: length between perpendiculars 220 feet, breadth 40 feet, draught of water forward 15 feet 6 inches, aft 17 feet, and the displacement 1,864 tons. They are fully rigged as sailing-vessels, spread about 13,228 square feet of canvas, and are fitted with lifting screw-propellers.

MOTIVE MACHINERY.

The engines, in common with those of all other cruisers of late design, are of the compound type. They are not from the same patterns for each vessel, yet it is scarcely necessary to give more than the general dimensions of those for the *Garnet*, as specified by the admiralty. They are as follows: one pair of horizontal engines, the diameter of the high-pressure of which is 57 inches, and of the low-pressure 90 inches; the length of the stroke is 2 feet 9 inches, and the maximum revolutions per minute about 90. The high-pressure-cylinder face is made separately of phosphor-bronze, secured to the cylinder with brass screws. The diameter of the crank-shaft and crank-pin bearings is 13 inches; the aggregate length of the former bearings is 7 feet 6 inches, and that of each crank-pin 13 inches. The diameter of the propeller-shaft is 12 inches, and the section which passes through the tube is 13½ inches in diameter. The screw-propeller is 15 feet in diameter, and has a pitch of 15 feet 6 inches. The boilers, six in number, 10 feet in diameter and 9 feet long, are cylindrical, with horizontal tubes, and the pressure of steam 60 pounds per square inch. The grate-surface is 245 square feet. The indicated horse-power on the measured mile is to be 2,100, and the maximum speed with this power is estimated at 13 knots. The space occupied in the length of the vessel by machinery and coal is: for the engines, 26 feet 8 inches; for the boilers, 33 feet 4 inches; passage between engines and boilers, 30 feet; coal-space forward of boilers, 6 feet 8 inches; total, 96 feet by the whole width of the vessel.

The armament consists of fourteen 64-pounder guns, one of which is a bow-chaser, working under cover of a forecastle, and another is a stern-chaser, under a poop; the remainder of the guns are uncovered.

SLOOPS.

These vessels, of modern date, are also of composite build; they are rated next after corvettes, and may be divided into two classes. Those of the first class consist of the *Wild Swan* and *Penguin*, launched in 1876, the *Osprey* and *Pelican*, in 1877, completed and in commission; also the *Cormorant*, *Pegasus*, *Dragon*, and *Gannet*, not yet completed. This class have a displacement of 1,124 tons. The length between perpendiculars is 170 feet, the breadth, extreme, is 36 feet, the draught of water forward is 13 feet, and aft, 16 feet, and the midship section 389 square feet. They are bark-rigged, and spread 10,138 square feet of canvas. The cylinders have diameters of 38 and 66 inches, with a stroke of 2 feet. The propeller has a diameter of 13 feet and a pitch of 12 feet 6 inches. The estimated revolutions are 100. There are three cylindrical boilers, 7 feet 10 inches in diameter and 15 feet long; six furnaces with a total grate surface of 108 square feet. The coal-bunkers have a capacity of 150 tons, the daily consumption being 22 tons. They carry four guns on the broadside and two bow or stern chasers. The estimated cost of the hull, &c., is $169,891, and of the machinery $50,787, in gold. The second class are all completed and in commission, and consist of the *Daring*, *Albatross*, *Flying Fish*, *Egeria*, *Fantôme*, and *Sappho*. These have a displacement of 894 tons; the length between perpendiculars is 160 feet; breadth, extreme, 31 feet 4 inches; draught of water forward, 12 feet; aft, 14 feet; the average indicated horse-power being 900 and the number of guns four, two of which are chasers. Total cost of hull and machinery, $171,558 in gold.

A still smaller class, rated according to Parliamentary returns as sloops, and on the navy list as gun-vessels, consists of the *Arab* and *Lily*, both in commission. They are each of 700 tons displacement, with a length of 150 feet; breadth 28 feet 6 inches, and draught of water forward 10 feet, and aft 12 feet; the indicated horse-power being respectively 656 and 829. Also of the *Flamingo*, *Condor*, *Griffon*, and *Falcon*, completed in 1877. These last named have a displacement of 774 tons; length, 150 feet; breadth, 29 feet; draught forward, 11 feet; aft, 13 feet; and the indicated horse-power is to be 750. Total cost of hull and machinery, $160,380 in gold.

All of these vessels, as well as all other classes of recent construction, are engined on the compound system. The largest of these sloops, *i. e.*, those having a displacement of 1,124 tons, are fitted with engines capable of developing an indicated horse-power of 900 on the measured-mile trials. The diameter of the high-pressure cylinder is 38 inches, of the low-pressure cylinder 66 inches, and the stroke of piston 2 feet. The maximum revolutions are to be 100 per minute, and the speed under steam, in smooth water, between 9 and 10 knots per hour. The steam is supplied by cylindrical boilers, and the pressure is to be 60 pounds per square inch. The screw-propeller is 13 feet in diameter, and like those of other cruising-vessels it is two-bladed and lifting. A new arrangement for feathering the blades, patented by Mr. R. R. Bevis, has been fitted to the *Griffon* and *Falcon*, which, it is said, is not open to the objections made to feathering screws previously tried. The levers and other gear for moving the blades are inclosed in the boss of the propeller, are attached to a rod passing through the center of the shaft, and are worked in the screw-shaft tunnel. During the trials of one of these vessels the gear was put into requisition, as after completing one series of runs it was considered desirable to alter the pitch of the screw, which was done in the course of a few minutes, when she was again put on the mile to complete another series; this operation, with the ordinary lifting screw, would have involved considerable delay, or, if not fitted with lifting gear, would have necessitated putting the vessel into dry-dock.

These two are the only vessels in the royal navy actually at work with this gear, but the *Garnet* and the *Cormorant* are now being fitted with it. The same arrangement of gear has been fitted in ships belonging to several other governments, and in a large number of yachts, notably to Mr. Brassey's *Sunbeam*, which has recently circumnavigated the globe in forty-six weeks, sailing or steaming at intervals as circumstances required, during which time this feathering arrangement was found to be useful and handy, and is reported as successful, and as not having been out of order.

All the sloops mentioned are fully rigged as cruisers, and are intended to keep the sea under sail. The *Arab* class are also three-masted; they are square-rigged on the mainmast as well as on the foremast.

There is also a large number of what is known as composite gun-vessels fitted as cruisers, of which the *Coquette* class are the smallest sea-going vessels in the British navy. The displacement of this class is 430 tons; the length is 125 feet; breadth, extreme, 23 feet 6 inches; draught forward, 8 feet; aft, 10 feet. They also have lifting screws, are three-masted and square-rigged on the foremast only. The armament consists of two 64 and two 20 pounder rifles.* The speed under steam is only about 8 knots in smooth water at the best, but under favorable conditions of wind and weather a run of twenty-four hours can be made with 3 tons of coal.

In addition to the composite gun-vessels and gunboats, there is a large number of iron double-screw gunboats, ten of which were building by the Palmer Ship-Building Company at Jarrow-on-Tyne at the time of my visit to that place in 1876. These vessels have a displacement of 363 tons; the indicated horse-power is to be 310, and the number of guns three. Quite a number are of still smaller size, and some of them are fitted to carry a single gun mounted on a rising and lowering platform forward, so arranged that the gun can be lowered into the hold when the vessel goes to sea.

This peculiar type of vessel was invented by Mr. Rendel, the *Staunch*, built in 1867, being the original. She cost $64,481 in gold. Since then about thirty have been added to the navy.

Recently, Sir W. Armstrong & Co. have been constructing four gunboats for the Chinese government, in which several important improvements over the *Staunch* type have been introduced. Two of these boats, each mounting a 26½-ton gun, have already been delivered at Tien-tsin, and two others, each mounting a 38-ton gun, are probably by this time completed and ready to sail for China.

These last two little vessels, built of iron, measure 126 feet over all; their extreme breadth is 30 feet; draught of water 8 feet, and displacement 430 tons. They are schooner-rigged, with tripod-masts; carry 50 tons of coal, 50 rounds of ammunition, and, in addition to the heavy gun, they carry two 12-pounders and a Gatling gun. They are propelled by twin screws, and are intended to have a speed of 9 knots.

The main peculiarities consist in the great gun mounted on so small a vessel, and the system of working it; the piece being much heavier than those used in the English boats, and the little vessel is herself made to act as the gun-carriage. The gun is worked by hydraulic power, and the entire arrangement of the mechanism is similar to that employed in working the 100-ton gun at Spezia, to be noticed hereafter. Two heavy iron beams in the fore part of the vessel are placed side by side on a level with the deck and parallel with the keel; on these beams are bolted frames analogous to the cross-head guides of a horizontal engine, and the trunnions of the gun are fitted in slide-blocks, these last taking the place of the cross-head. Thus arranged, the gun can slide back and forth through a range of about 3 feet. The preponderance at the breech-end is supported by two secondary parallel bars inside the main gun-beams. These are hinged at the rear end, while at the forward end they are carried on the cross-head of a vertical hydraulic ram fixed beneath the deck. The breech-end of the gun is supplied with a hoop and lugs; the lugs rest on the two secondary bars near their hinged ends, and thus, by causing the hydraulic ram to rise or fall, the gun can be elevated or depressed at will. No turning-gear is provided, the lateral training of the gun being effected by turning the whole boat through

* One feature to be noticed is that the bows and sterns of all recently-built British unarmored vessels are constructed so as to carry guns for fore-and-aft fire on the line of the keel; a second is, that the cabin accommodations are made secondary to the working of the guns.

the required arc by the use of the rudder and twin screws. To run the gun in and out, two hydraulic cylinders are used, one of which is fixed horizontally on each side-beam, the cross-heads of the rams taking hold of the trunnion slide-blocks. The recoil is taken up by these rams, or more properly pistons, delivering water under a weighted valve.

The gun is loaded by a hydraulic rammer, the shot being brought to the muzzle by a trolley or carriage, off which it is pushed into the bore.

During the trials of the *Gamma*, one of the vessels just tested, the 38-ton gun was fired with charges consisting of 130 pounds of powder behind an 800-pound projectile, the elevation being 3½ degrees.

These boats are regarded as the nucleus of a Chinese hornet-fleet, which fleet, if properly manned, will no doubt give a deal of trouble to Japan or any other country which may dare to invade the sanctity of Celestial waters. Extravagant estimates of their merits have, however, been formed, and some of the English papers have been urging the adoption of the type into the British navy. They are a great improvement on the English boats, but not free from objections, one of which is the want of lateral movement of the gun without moving the vessel, especially in rivers, where such craft would otherwise be particularly serviceable; and in order that they may operate with maximum efficiency the water must be tolerably smooth; and at such a time it would not be difficult to hit one of them by a shot from the small rifled guns carried by armored ships and send her to the bottom. Therefore, except under peculiar circumstances, they are useful for defensive purposes only, and service on board of them must in many cases during war be attended with extreme risk.

THE COMPOSITE SYSTEM.

All the modern cruising-vessels of the British navy (unarmored), below the rate of the *Active*, are now built on the composite system. In this method of construction the frame-work inside of the skin, including frames, beams, keelsons, stringers, shelf-pieces, water-ways, transoms, bulkheads, &c., are of iron, and arranged nearly as they would be in an ordinary iron-built ship, the frames and beams being of the same kind and dimensions and spaced the same distances apart, with bulkheads of the usual number and construction. The keel, stem, outside planking, and decks are of wood. The planks are put on in two courses laid fore and aft. The first course is secured to the iron frames by ⅝-inch Muntz's-metal bolts tapped into the iron, having also lock-nuts on the points inside. The bolts have screw-driver heads, and are screwed home against a shoulder so as to leave the head below the surface of the plank, the cavity over the head of the bolt being filled with white and red lead so as to prevent leakage. The planks on both sides, as well as the iron, are carefully painted; after the first course of planks has been carefully calked between the joints with oakum, the second course is laid on it, breaking joints with the planks of the first course. The planks of this second course are fastened to those of the first course by copper bolts driven through both thicknesses, and riveted inside the vessel. The joints between the planks of the last course are then likewise calked, and the surface below water coppered over. The kind of wood used is teak; the thickness of the first course of planks is about 3½ inches, and of the second course about 3 inches; the width is about 12 inches, and there are two bolts to every frame through each plank of the first course. The work is required to be carefully done, and special attention is directed to see that the iron is completely insulated or cut off from electrical communication with the copper sheathing and bolts used in the structure.

In this system of ship-construction the same strength of hull is not expected to be attained as when the hull is composed entirely of iron, having each skin-plate riveted to the next, also to the frames and the bottom, and decks tied together by bulkheads; the constructors employed in building them, however, estimate the strength of one of these well-built composite vessels to be from 40 to 50 per cent. greater than the strength of a wood-built vessel of the same dimensions, besides which the durability is infinitely greater, for there is no wear-out of the interior parts, and the skins are of teak, which possesses durability equal to our live-oak.

PART X.

CRUISERS OF THE RAPID TYPE, THE IRIS AND MERCURY; STEEL CORVETTES; MATERIALS FOR SHIPS OF WAR; TABLE OF COST OF REPAIRS FOR A FEW UNITED STATES VESSELS OF WOOD IN FIVE YEARS UNDER A SINGLE BUREAU; SHIPS OF THE MERCANTILE MARINE SUITABLE FOR WAR PURPOSES.

THE IRIS AND MERCURY.

These two sister ships, put afloat at the Pembroke dock-yard last year, and to be completed early in 1878, are built of steel, and are termed armed steel dispatch-vessels. They are the first of a new type designed for high speed as the pre eminent requisite; all other requirements are subordinated to this important element.

The model from which they have been built presents a beautiful, sharp bow; a long, exceptionally clean run, and altogether exquisitely fine lines for a swift and lightly-sparred vessel.

The principal dimensions and other data of the *Iris* are:

Ship:	
Length between perpendiculars	300 feet.
Breadth, extreme	46 feet.
Depth of hold to top of double bottom	16 feet 3 inches.
Draught of water, forward	17 feet 6 inches.
Draught of water, aft	22 feet.
Area of midship section	777 square feet.
Displacement	3,735 tons.
Speed per hour, maximum	16½ knots.
Estimated cost of hull (gold)	$437,400
Engines:	
Two four-cylinder horizontal compound engines.	
Diameter of high-pressure cylinders	41 inches.
Diameter of low-pressure cylinders	75 inches.
Stroke of pistons	3 feet.
Revolutions, per minute	95
Indicated horse-power (contract)	7,000
Diameter of crank-shaft	16½ inches.
Number of screws	2
Diameter of screws	18 feet 6 inches.
Pitch of screws, alterable, now fixed at	17 feet 6 inches.
Boilers:	
Number of boilers { elliptical	8 } 12
Number of boilers { cylindrical	4 }
Total grate surface	700 square feet.
Total heating surface	15,960 square feet.
Total weight of machinery, including water in boilers and condensers, about	1,000 tons.
Estimated cost of machinery (gold)	$451,980
Total cost of hull and machinery, estimated	$889,380
(Both estimates as to cost will be exceeded.)	

Armament:

Ten 64-pounder rifled guns; four on either side, and two revolving, the latter being mounted on the poop and forecastle.

The official trial test of six hours' duration was made December 14, 1877, and the results obtained were as follows: The engines maintained a noteworthy uniformity in the vacuum and the number of revolutions; the starboard engines made 91, and the port engines 89½, revolutions per minute; the mean vacuum in the after condenser was 28.33 inches, the vacuum in the forward condenser remaining at 28 inches throughout the day. The mean pressure in the boilers was 62 pounds; in the high-pressure cylinders 41.29 pounds, and in the low-pressure cylinders 11.24 pounds; the mean total horse-power developed being 7,088.5. The speed of the ship, as recorded by the electric log and confirmed by cross-bearings, was 16 knots per hour. The consumption of coal during the trial amounted to 2.7 pounds per indicated horse-power per hour, no attempt, however, being made to economize fuel. Although the horse-power developed was 88½ beyond the contract, the speed realized fell one knot short of the estimate of the admiralty.

Viewing the *Iris* with the frames in place and the bottoms plated, the scantlings appeared very light. To strengthen the ship as much as possible, and to compensate for her extreme lightness of build, the perpendicular frames are only 3 feet 6 inches apart, and they are crossed by longitudinal Z-shaped girders, thereby leaving no part of the ship's side more than 4 feet square without support. The thickness of the plates below the gun-deck is ½ inch, garboard strakes excepted, which are ⅝ inch. Above the deck the plates are ⅜ inch. The joints are double-riveted, and the rivets used are of wrought iron ¾ inch in diameter.

The ship is divided into twelve water-tight compartments by eleven transverse bulkheads, and water-tight iron flats are built in the three foremost and four aftermost compartments below the lower deck, thereby confining the water which might come into one compartment to its bottom, that is, between the bottom of the ship and the flat so built and described. These water-tight flats are also built on each side of the ship in the coal-bunkers, which are connected by small water-tight sliding doors. The ship has an inner and outer bottom for the distance occupied by the engines and boilers, the space between the two skins being about 4 feet in the middle of its length with reduced depth toward either end. The transverse bulkheads referred to, by which the bottoms and decks are tied together, add strength and stiffness to the hull, but the limited beam—46 feet—has prohibited the introduction of the central longitudinal bulkhead, extending from stem to stern, a backbone of strength so important for a vessel comparatively light-built and intended to sustain the great strain of her immense engine-power.

MOTIVE MACHINERY.

The chief requirement sought is great speed; to obtain this, engines guaranteed to develop 7,000 indicated horse-power have been placed in the ship, which, compared with the displacement, 3,735 tons, gives 1.87 horse-power per ton—a higher proportion of power than is possessed by any naval ship afloat. The space occupied by the engines, boilers, and coal is the whole width of the vessel and 150 feet fore and aft, just one-half of the length of the vessel in the most important part. The machinery was constructed by Messrs. Maudslay, Sons & Field. The ship is propelled by twin screws, each screw being worked by one pair of two-cylinder compound engines, laid horizontally, making eight cylinders in all, four high and four low pressure. Each high-pressure cylinder is secured to the front of the low-pressure cylinder with which it works, and is partly recessed into it to save length, and one piston-rod carries both the high and low pressure pistons. The piston-rod is connected to a wrought-iron cross-head, which works on guide-rods forming the connection between the main crank-shaft bearings and the cylinder, to each of which they are bolted by T-ends. The high-pressure cylinders are lined with working barrels of Whitworth steel. The air-pumps are vertical, and are worked from the low-pressure pistons by means of bell-crank levers. The surface-condensers are of the usual variety used in the royal navy, and have about 14,000 square feet of cooling surface. The crank-shafts are of wrought iron, and have a diameter of 16½ inches at the bearings. The propeller-shafts (aftermost lengths excepted) are of Whitworth compressed steel, with solid couplings; their diameters are 11½ inches inside and 17 inches outside. A considerable length of the screw-propeller-shafts extends outside of the vessel; this is of wrought iron, solid, and 16½ inches in diameter. *Lignum-vitæ* is used at both ends of the stern-tubes, as well as in the tubes through which the shafts are worked. There are also annular thrust-rings with *lignum-vitæ* facings fitted at the after ends of the stern-tubes to assist in taking the thrust of the screws. The screw-propellers are each 18 feet 6 inches in diameter, and in common with other twin screws they revolve outward when going ahead.

Boilers.—The boilers have been made of the same material—Landore steel—as the hull of the ship, and, compared with the use of iron, it effects a considerable saving in weight, even if it were not a much better material for their construction. They are twelve in number, placed in two separate water-tight compartments, six boilers in each compartment, so that if one set of boilers should become useless, by reason of the compartment being flooded, the engines can be operated by steam from the other set, and with this contingency in view the steam-pipes connecting the boilers with the engines have an outer metal covering to prevent condensation of the steam within them in the event of being surrounded with water. There are, also, two separate engine-rooms. The boilers differ in dimensions, on account of the varying shape of the ship; eight of them are elliptical and four cylindrical.

The coal-carrying capacity is estimated at 500 tons in her ordinary bunkers, and 250 tons in the midship reserve bunkers, which total amount, calculated roughly, is intended to carry her through five days' full steaming, or to enable her to steam 6,200 miles, at the rate of 10 knots per hour, or 8,600 miles, at 8 knots per hour.

One important feature introduced in this vessel, also into several other recently-constructed ships, and comforting to many officers, is an arrangement for working the steam at the low pressure of four pounds per square inch, when going into action, if so desired.

The system of pumps and fire-mains is arranged by having one large main running the length of the ship, from fore compartment to after compartment, and the whole being connected with two 40 horse-power engines, one in each engine-room, and with six 9 inch pumps which can be worked either below or on deck.

The steering is performed by the ordinary wheel, fitted with Rapson's patent connecting-gear and Frayne's patent brake, for use in turning at a high rate of speed or in heavy weather.

The wardroom and the cabins of the captain and wardroom officers are all under the poop, in the front of which is a water-tight bulkhead reaching to the ship's bottom. The engineers' berths and the warrant officers' cabins and mess-place are all on the lower deck, aft; and the sick-bay, dispensary, and engine-room artificers' berths are also on the same deck, one compartment further forward. The magazines, shell-rooms, and store-rooms are on the iron flat below the lower deck. The whole of the lower deck forward is devoted to quarters for the men. The forecastle is fitted up for supernumeraries, the after part being closed in by a water-tight bulkhead extending from the bottom of the ship up to it.

The ventilation of the magazines and decks is effected by means of air-pumps.

The *Iris* is three-masted, lightly sparred, and bark-rigged. The armament is also necessarily light, consisting as it does of ten 64-pounder rifles, one at the bow and one astern, with fore-and-aft fire, and eight broadside-guns. Provision is also made for using the Whitehead and other torpedoes. The crew will consist of 250 officers and men.

The engine-power of the *Iris* is considerably more than was ever put into a naval hull of her dimensions, and while no doubts exist as to the speed being eventually obtained, it will remain to be seen whether the hull is sufficient in strength to sustain for a lengthened time the immense strain of 7,000 horse-power.

The *Iris* and *Mercury* are the first war-vessels constructed of steel. The stems, stern-posts, and keels are of wrought iron, otherwise the hulls, including frames, beams, plating, bulkheads, &c., are built wholly of Landore steel.

In consideration of the uncertainty in regard to steel, hitherto employed in large quantities for ship-construction, unusual care was exercised in the selection of the works to manufacture it. Some months were devoted to testing specimens of steel supplied by the most eminent firms of the kingdom before the decision was arrived at, which resulted in giving the contract to the Landore Steel Company, of South Wales, some forty or fifty miles distant from the dock-yard. This company has the contract for the whole supply, under very rigid requirements. The steel is made according to the Siemens process. Its chief characteristics, and those which give it its special value for ship-building purposes, are its extreme ductility when cold, its ability to stand punching without appreciable loss of strength between the rivet-holes, its not requiring to be annealed after shearing or punching, its having a smoother surface than iron, its capability of being easily forged and worked hot, its uniformity of quality being absolutely insured and its strength along the grain. The most remarkable point about this steel is the fact that punching produces scarcely any injury to the material in the neighborhood of the hole, and that annealing the plates and angles after punching and shearing is not necessary; without this quality, which has not been hitherto obtained, steel cannot possibly make headway with ship-builders, for annealing cannot be intrusted to an ordinary ship-building yard. Therefore, where the problem is to combine the greatest speed of vessel with the smallest dimensions, Landore steel is a material admirably suited for the purpose. It is stronger than iron by from 25 to 30 per cent., and equal in ductility to iron of the best quality. The requirements of the admiralty as to the materials supplied under their contract with the Landore Steel Company are as follows:

From every plate made a strip is to be cut, which, after being heated to a "cherry-red" color, shall be plunged into water having a temperature of 82° Fahrenheit. After being thus cooled, the strip is to be bent, without fracture, until the radius of the inner curve equals not more than 1½ times the thickness of the strip. This is known as the "tempering test." Further, from each lot of 50 plates or angles a piece is to be taken, and the edges having been planed parallel, its tensile strength is to be proved. To be satisfactory, this must not exceed 30 tons nor be less than 26 tons on the square inch, and before fracture takes place there must be an elongation of not less than 20 per cent. on 8 inches of its original length. These tests are applied in the presence of a resident representative of the government. * * * *

Up to April [1876], 101 samples, representing over 5,000 plates or angles, have been subjected to the tensile test, with the following results:

Number of test pieces.	Breaking strain in tons per square inch.	Average elongation in 8 inches.
		Inches.
1	25 to 26	2.00
20	26 to 27	2.00
24	27 to 28	2.03
28	28 to 29	1.93
24	29 to 30	1.89
2	30 to 31	1.96
2	31	1.78
101	Mean, 28.16 tons.	1.94 inches, or 24.25 per cent.

* * * The tensile strenth of best best iron is 22 tons when tested with the grain and 18 tons when tested crosswise. In certain experiments by Mr. Kirkaldy on two very fine plates manufactured at Borsig's Works, Berlin, and, I believe, stated to be homogeneous, the tensile strength was found to be lengthwise 23.84 tons, and crosswise 22.6 tons. When similarly tested the Landóre steel gave lengthwise 28.85 tons, crosswise 28 2 tons; the comparison, therefore, stands thus:

	B. B. iron.	Borsig's plates.	Landore plates.
	Tons.	*Tons.*	*Tons.*
Lengthwise	22	23. 84	28. 85
Crosswise	18	22. 60	28. 20
Difference	4 tons, or 18. 18 per cent.	1. 24 tons, or 5. 20 per cent.	. 65 tóns. or 2. 59 per cent.

Therefore, taking the lowest tensile strength of best best iron and of Landore steer there is a difference in favor of the latter of 10.20 tons, or the Landore steel is 56.6 pe cent. stronger than the best best iron.

I was granted the privilege of inspecting the tested specimens at the Pembroke dockyard. For convenience they are divided into cold tests and hot-forge tests, as follows:

COLD TESTS.

(1) A piece of 6-inch by 3-inch by $\frac{7}{16}$-inch angle, with the corner bent over and flattened close by blows from a 40-cwt. steam-hammer. No fracture.

(2) A piece of 3-inch by 3-inch by $\frac{3}{8}$-inch angle, 1 foot long, flattened out by two blows of steam-hammer. No fracture.

(3) A piece of 3-inch by 3-inch by $\frac{3}{8}$-inch angle, 1 foot long, closed by two blows of steam-hammer. No fracture.

(4) A piece of 3-inch by 3-inch by $\frac{3}{8}$-inch angle, 1 foot long, flattened out same as No. 2, after which the wings were turned back over the outside angle. Slight fracture at the ends only.

(5) A piece of 3-inch by 3-inch by $\frac{3}{8}$-inch angle, 18 inches long. After being flattened out as in the previous cases, the two ends were turned over in opposite directions; the piece was then doubled up in the middle, and closed by hard blows from a steam-hammer. There is no fracture, but there are cuts from the hammer-tool in one place, and from its own angle in another.

(6) A piece of 3-inch by 3-inch by $\frac{3}{8}$-inch angle, 18 inches long, dealt with as in the specimen last mentioned, except that before bending one end over, the wings were bent back as in No. 4. There is a slight fracture, as shown, caused by its own angle.

(7) A piece of $\frac{5}{8}$ inch square bar, formed into a knot, occupying only $4\frac{1}{2}$ inches by 3 inches space. No fracture.

(8) A piece of $\frac{3}{4}$-inch round bar, formed into a knot, occupying only 3 inches by 2 inches space.

(9) A piece of $\frac{1}{2}$-inch round bar, formed into a knot, occupying $2\frac{1}{2}$ inches by $1\frac{1}{2}$ inches. This is a very fine specimen.

(10) A ring 5 inches in diameter, made from two pieces of plate welded together, and closed by blows from a sledge-hammer. No fracture.

(11) A similar ring made from Lowmoor iron, closed by one blow from a 60-cwt. steam-hammer. There are slight fractures where bent, as shown.

(12) A similar ring, made from Landore steel, and closed in the same manner. Fractured in the weld, as shown.

(13) A piece of $\frac{1}{2}$-inch steel plate, 12 inches diameter, dished to $3\frac{1}{2}$ inches deep by five blows from a 60-cwt. steam-hammer. Without fracture.

A piece of steel from an eminent firm, similarly treated, broke into three pieces at the fifth blow.

(14) A piece of 6-inch by 3-inch by $\frac{7}{16}$-inch angle, had the wings closed in, as shown; it was then placed edgewise, on bearings 5 feet apart, and bent by hydraulic pressure to a deflection of $12\frac{1}{2}$ inches. A piece of best angle-iron, similarly treated, broke at a deflection of 6 inches.

(15) A piece of plate, 3 feet by 3 feet by $\frac{1}{2}$ inch, had an iron block placed under each corner, 9 inches from the edge, and an iron ball weighing 1,291 pounds was dropped on the center from a height of 30 feet. Bent, as shown, without a sign of fracture.

(16) A piece of steel, from an eminent firm, similarly treated, was bent and fractured as shown.

(17) A piece of best best iron, similarly treated, was bent and fractured as shown.

(18) A beam piece, 4 feet 6 inches long, made of $\frac{5}{16}$-inch plate, 5 inches deep, with

double angles 2½ inches by 2½ inches by $\frac{5}{16}$ inch on the upper and lower edges, riveted together with ⅝-inch rivets (3½ diameters apart), was bent under hydraulic pressure to the form shown, without fracture, when the fixings failed. It was removed before the test could be completed.

(19) A piece of ¾-inch plate, 12 inches diameter, forced by hydraulic pressure through a ring 10 inches diameter, and dished to the form shown, that is, 3½ inches deep, by a pressure of 145½ tons, without fracture.

(20) A piece of ⅝-inch plate rested during experiment upon a perfectly horizontal surface of an annular iron anvil, a central circular portion of the plate, 12 inches in diameter, being unsupported. The anvil was firmly imbedded in the ground. A charge of 18 ounces of compressed gun-cotton was suspended over the center of the plate, an air-space of three inches intervening between the upper surface of the plate and the base of the charge. The charge was exploded by detonation.

The result is that the plate is dished down to the extent of 1½ inches, but there is no sign of fracture.

(20*a*) A piece of ⅝-inch best best iron, similarly tested, is almost shattered to pieces.

(21) A piece of plate $\frac{11}{16}$ inch thick, supported on anvil as above. Charge of gun-cotton 10 ounces, and 3 inches in diameter, placed upon the upper surface of the plate in a central position, and exploded by detonation.

The result is that a hole 1½ inches diameter is made in the center of the plate, and that for ¾ inch all round this hole the plate is beautifully cupped or countersunk, as if it had been done by a cutting-tool. There are no lateral fractures in the plate.

(21*a*) A piece of best best iron plate $\frac{11}{16}$ inch thick, similarly treated, was fractured as shown.

(22) A piece of plate doubled close up fourfold, as shown, without fracture.

* * * * * * *

HOT FORGE TESTS.

(1) A piece of ½-inch plate subjected to the ram's-horn test, the lower end being doubled close while cold, without fracture.

(2) Shearings of ½-inch plate about 1½ inches wide, welded together and bent in the weld to a radius of ¾ inch, without fracture.

(3) Two pieces of plate welded together and bent in the parts welded, one to an angle of 90°, and the other to an angle of 105°, without fracture.

(4) A box-end made from ½-inch plate to form angle-steel, 3 inches by 3½ inches by ½ inch. It is soundly welded, and is in every respect as sound as one made from angle-iron.

(5) An intercostal, made from ½ inch plate, to form angle-steel, 3½ inches by 3 inches by ½ inch, soundly welded, and considered a good job.

(6) An outside corner, made from ½-inch plate, to angle-steel. This is a fair average weld, these corners being much more difficult to make than inside corners, as the gusset piece has to be welded in after being fitted.

(7) Two pieces of plate, 2½ inches by ½ inch, welded together straight, then turned (upon) its edge to form a circle 6 inches in the clear. It is a sound weld, and there is no fracture in the turning.

(8) Two pieces of plate, 3 inches by ½ inch, welded together and bent off at right angles. It is soundly welded, and bent to the form shown without fracture.

(10) A piece of plate, 21 inches by 10 inches by ½ inch, turned to the form of a tube (not welded), then placed in a flanged socket, and a flange-pin forced into it by blows from a 60-cwt. steam-hammer. It assumed the form shown after eight blows. Stood well.

(11) A piece of ½-inch plate, 12 inches diameter, forced into a socket by blows from a steam-hammer, to the form shown. Stood well.

(12) A piece of ½-inch plate dished to form shown, No. 11, then flattened back nearly to its original form. Stood well.

(13) A piece of ½-inch plate, 12 inches in diameter, forced into a tube, then flanged back on the anvil to form shown. Stood well.

(14) An outside and inside corner, made from angle-steel, 6 inches by 3½ inches by $\frac{7}{16}$ inch, welded sound, but with more difficulty than in welding iron. * * * *

The above tests establish the fact that the material can be easily manipulated, and I was informed by the officers in charge that welding was successfully accomplished.* In addition to the qualities claimed for this steel, it is represented that a series of experiments extending over about three years, carefully conducted at the Terre Noire works, had established the fact that when exposed to the action of sea-water this soft steel

* The account of Landore steel is chiefly from a paper by Mr. Reilly, the manager of the works.

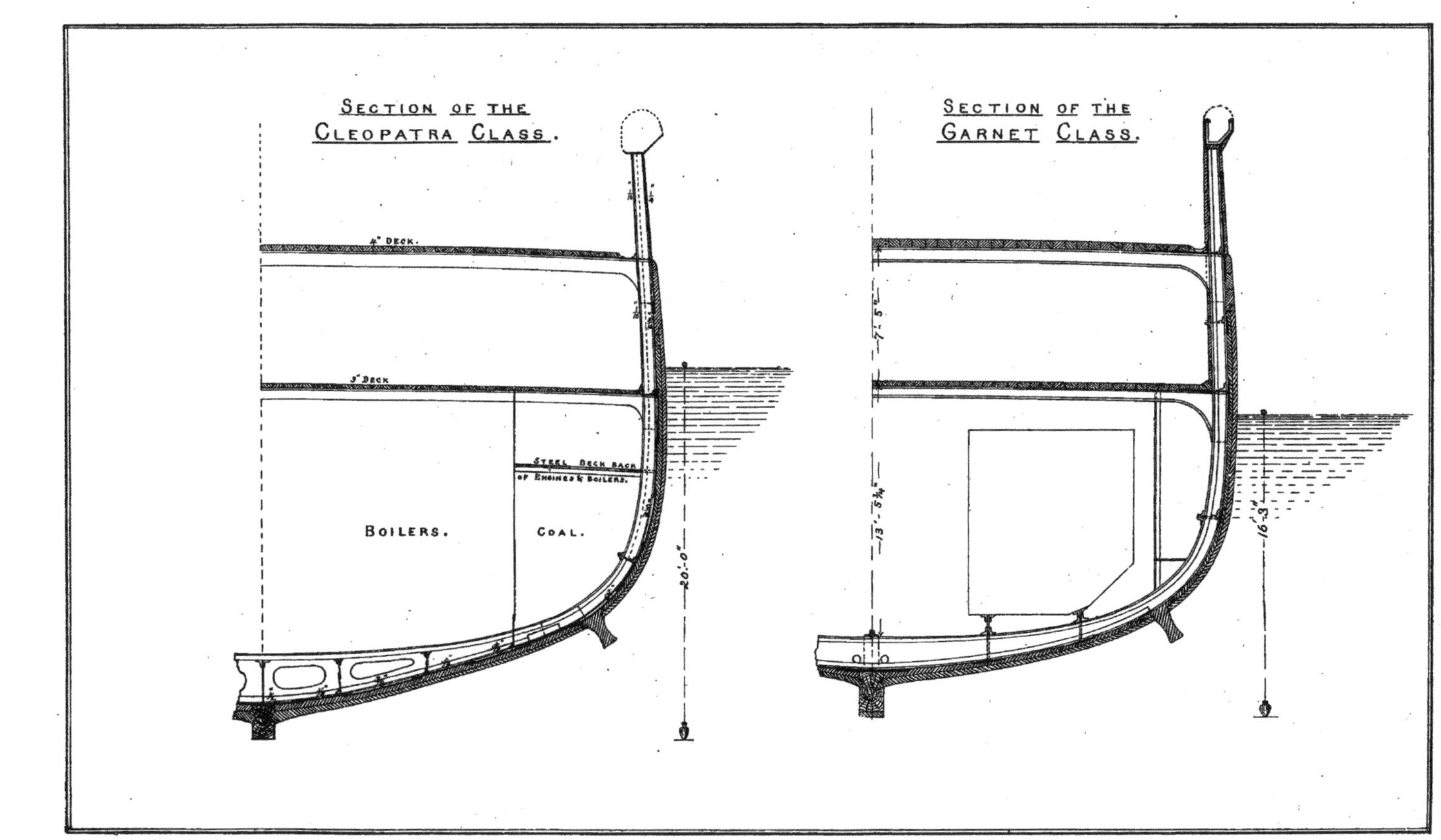

Section of the Cleopatra Class.
4" Deck.
3" Deck
Steel Deck back of Engines & Boilers.
Boilers.
Coal.
20'-0"
Section of the Garnet Class.
7'-5"
13'-5¾"
16'-3"

suffered by corrosion only in the proportion of 60 to 140, when compared with the effect of similar treatment upon iron plates. I do not know of any boilers made from this steel; it would seem, however, in consideration of the rapid decay of the plates of iron boilers consequent upon the use of redistilled sea-water, that a liberal expenditure in this direction would be wise.

PURPOSE OF THE IRIS AND MERCURY.

At a date soon after these two ships were ordered, a distinguished English writer in a London magazine enunciated the following:

> It is forgotten that by the declaration of Paris the field of operations is much restricted as regards either the necessity for protecting our own commercial marine or the possibility of injuring an enemy's commerce. This declaration, to which most European maritime powers adhered, but which the United States did not join in, is a contract to respect private property, not being contraband of war, if carried in ships bearing a neutral flag. * * * Many advantages to this country result from the declaration, although serious disadvantages press upon ship-owners, and our position as the chief maritime carriers must suffer. On the other hand, it must be noted that the United States and England are under no such mutual obligations, and with the rankling recollection of the mischief done by the *Alabama* and her consorts to their own mercantile marine, Americans never miss an opportunity of expressing their intention to make similar attacks on British commerce in case of war, *so that England is bound always to maintain an unarmored fleet more powerful than that of the United States, and not to allow individual unarmored ships in her navy to be surpassed in speed or power by vessels of the American Navy.*

STEEL CORVETTES.

This type of vessel, of which six are in process of construction at the works of Messrs. Elder & Co., Glasgow, are to be full-rigged cruisers, sheathed with wood, and fitted with single screw-propellers arranged for lifting. They are named the *Cleopatra*, *Curaçoa*, *Conquest*, *Champion*, *Carysfort*, and *Comus*. The principal dimensions are:

Length between perpendiculars	225 feet.
Breadth, extreme	44 feet 6 inches.
Depth of hold	21 feet 6 inches.
Tonnage	2,020 tons.
Displacement	2,377 tons.

The three first named are being engined by Messrs. Humphrys, Tennant & Co., the dimensions of the engines being as follows:

Type of engines: horizontal, compound, four-cylinder, with return connecting-rods.

Number of cylinders (one pair of engines)	4
Diameter of high-pressure cylinders	36 inches.
Diameter of low-pressure cylinders	64 inches.
Stroke of pistons	2 feet 6 inches.
Diameter of crank-shaft	13 inches.
Revolutions per minute (estimated)	100
Indicated horse-power, maximum	2,300
Diameter of screw	16 feet 6 inches.
Pitch of screw	14 feet 6 inches.
Type of screw	Griffith's two-bladed.
Type of boilers	Cylindrical, single-ended.
Number of boilers	6
Number of furnaces in each boiler	2
Total grate surface	266 square feet.
Total heating surface	6,714 square feet.
Pressure of steam	60 pounds.
Speed of vessel (estimated maximum)	13 knots.

Armament.—This is to consist of two 7-inch revolving guns and twelve 64-pounders on the broadside, all rifles.

The description of the method of applying the wood sheathing to iron or steel hulls has been seen a few pages back; reference to the foregoing sketch of the midship section of the vessels named, will, by illustration, make plain this system, now in common use in European navies. A section is also here given of the *Garnet*, showing the composite system of construction previously explained.

The *Cleopatra* class are built with two decks; the principal water-tight bulkheads are four in number; and for the purpose of protecting the machinery, the lower deck, for the distance extending over the engines and boilers, is armored with 1½-inch steel plates.

The materials entering into the construction of the hulls is, in the main, steel, about half of which has been made by the Siemens process and the other half by the Bessemer; both kinds furnished for the corvettes have been represented to be of good and uniform quality, having a tenacity of 25 to 26 tons per square inch, and sufficiently soft to double cold without fracture. The rivets used are of iron.

Steels are now rapidly gaining favor, both in the navies of Europe and the mercantile marine, for use in the construction of vessels, of boilers, and many parts of the engines. In consideration of the greater tenacity and ductility of this material over the iron usually entering into ship-construction, Lloyd's Register has agreed to accept for mercantile vessels a reduction of 20 per cent. from their standard for iron, and 25 per cent. in the thickness of boiler-plates, while the Board of Trade has allowed boiler-steel to be accepted at its tested strength of 28 to 30 tons, the saving in weight here being just what corresponds to the ratio in which this tenacity exceeds that of ordinary boiler-plates.

MATERIALS FOR SHIPS OF WAR.

In a paper by Mr. John Vernon, on the construction of iron ships (*Trans. I. M. E.*), it is stated: "The main points of superiority of iron ships over those built of wood consist in the superior strength, greater durability, and consequently less cost of iron ships, together with their larger carrying capability and greater facility of construction."

The greater strength of iron ships is shown in daily practice in numerous ways, and it is also shown by the fact that in many modern wooden ships it has been found necessary to introduce diagonal iron straps inside the framing, and in many cases the use of iron bulkheads, knees, beams, and stringers, and even the frame-work itself for the whole structure; but this arrangement, it is admitted, falls far short in point of strength of a vessel built entirely of iron. Again, the introduction of iron affords great facility for obtaining the necessary strength in keels, stem and stern posts, screw-port frames, and other parts, by the application of large forgings, also by tying the bottoms, decks, and sides of a vessel together with bulkheads.

The greater comparative durability of iron arises mainly from its freedom from the decay to which wood is always liable, in consequence of being unavoidably subject to constant and extreme variations of temperature and moisture. Another important source of this greater durability is to be found in the firm and substantial union of the several parts of an iron ship by means of riveting, which effectually prevents that *working* under heavy strains to which all wooden ships are more or less liable, and which is a source of great difficulty with the engines, caused by the screw-shafts being forced out of alignment, and thereby strained.

Abundant proof of the superiority of metal over wood for ship-construction may be found in the records of our Navy Department, as shown by the yearly sale or breaking up of vessels, not by reason of their being of obsolete types, but consequent upon the timber of which the hulls were composed being decayed to such an extent as to render them unfit even for repairs. Quite a number of vessels have been found so decayed after the first commission as to condemn them for further service, and others having been several years on the stocks have rotted and been condemned before being launched. Some vessels have been rebuilt, while others, retained in the service and kept in repair, have cost in the course of ten years, more than the sum necessary to build the same number of new ships of iron, of similar dimensions.

In corroboration of this statement the following table from the testimony taken by the House Committee on Naval Affairs (H. R. Mis. Doc. 170, p. 450), officially furnished by the Navy Department, July 6, 1876, showing the cost of repairs, &c., is given:

Name of ship.	Rate.	Cost of repairs between 1870 and 1875 by Bureau of Construction and Repair.	Name of ship.	Rate.	Cost of repairs between 1870 and 1875, by Bureau of Construction and Repair.
Franklin	First	$103,703 57	Omaha	Second	$179,762 51
Minnesota	do	303,899 45	Pensacola	do	550,294 27
Wabash	do	512,222 62	Richmond	do	153,905 85
Brooklyn	Second	*203,653 52	Ticonderoga	do	124,293 71
Canandaigua	do	143,766 96	Iroquois	Third	301,100 86
Hartford	do	396,690 06	Kearsarge	do	501,104 11
Plymouth	do	84,002 21	Mohican	do	143,218 31
Lackawanna	do	360,804 65	Narragansett	do	104,152 65
Lancaster	do	600,000 00	Ossipee	do	54,410 87
Tennessee	do	445,792 94			

* The estimated additional cost of repairs for this vessel in February, 1878, was $187,000.

The foregoing list * does not include all the wooden vessels of the Navy of which the cost for repairs was reported to the committee. It includes only those on which the greatest amounts had been expended in the five years between 1870 and 1875, and it does not include any vessels rebuilt under the head of "repairs."

It is proper to state that the large sums expended on repairs, as here reported, have not been due entirely to the fact of the hulls being of wood, but the largest portion of it, at least 75 per cent., is so due; that is to say, the money was expended in substituting sound timber in lieu of rotten or defective wood; almost the whole of which money would have been saved if the hulls had been of iron or steel. Evidence of this may be found by a comparison of the bills for repairs of the United States iron steamer *Michigan* during the thirty-five years she has been in continuous service with those for the above-named vessels.

It is fair to assume that the difference between the cost of the repairs of our entire fleet and one of the same number and dimensions of vessels, all having iron hulls cased in wood, would be a sufficient sum to add to the Navy yearly at least one good-sized cruising-vessel.

The preceding figures are sufficiently appalling to cause any Congressional committee to pause before recommending the construction of

* The amounts given in the table as having been expended for repairs include only those directly under the cognizance of the Bureau of Construction, and for labor and materials only; a large additional amount is also to be charged for repairs, &c., by the Bureaus of Steam Engineering, Equipment, and Ordnance.

more wooden ships. Yet that able ordnance officer, Captain Simpson, U. S. N., in a letter to the *Army and Navy Journal* of March 2 of this year, writes "A word of caution" against building iron ships of war; and, to sustain his views, quotes the antiquated reports of experiments made in England twenty-eight and thirty-eight years ago, showing that 32-pound shot, fired with charges of from 2½ to 10 pounds of powder at a distance of 450 yards, were shivered to pieces in passing through the thin sides of an iron hull; and that if the plates be of ⅝-inch thickness, the shots will be broken by impact. He also states, "In the matter of material for building ships England has no choice; with her it is a necessity that forces her to build of iron. Had she the forests of America at her back we should have never heard of her fleet of composite light cruisers." Persons well informed in the history and progress of naval construction are acquainted with the reports quoted by Captain Simpson, also with the fact that, consequent upon said reports, the introduction of iron ships for general service in the British navy was long deferred until better experience was acquired. We have knowledge of, and have referred on another page to, a report made about the same time (1840), by officers of the same royal navy, that the screw-propeller was not suitable for ships of war. It is scarcely necessary to mention the fact that when the experiments referred to by Captain Simpson were made, ships of all navies were propelled by the wind; that the guns they carried were of cast iron or bronze, of smooth bore and small caliber, and that the projectiles used were cast iron, spherical, solid shot.

The types of ships, the materials of which they are composed, the guns mounted on them, the projectiles and instruments of all kinds employed in modern warfare, differ from those of former times so radically that better authority would be the numerous and expensive iron-target-firing experiments of recent times, rather than going back twenty-eight years for illustrations. Surely no such conditions as are named in the experiments quoted will ever concur in modern warfare. The art of artillery-fire has now reached such perfection that it may be confidently asserted that any unarmored vessel, of whatever type, should be sunk in a brief period by the common shells of the present day, and whether the materials of which her hull is composed be wood, iron, or steel, will not materially alter the time necessary to send her to the bottom: the *Alabama*, *Congress*, and *Oneida* deserve the title "slaughter-pens" as well as any iron vessels Captain Simpson refers to.

The first requisite in a modern ship of war is high speed, and the strength of hull necessary for this cannot be obtained by any combination of wood bolted or spiked to wood; the *Iowa* and class, and numerous other examples, are proofs of this.

The British admiralty have abandoned wood ship-building for the reason that the requisite strength of hull and endurance of materials composing it could only be obtained by the use of metals, not for want of timber, for the markets of the world are open to them now as in former times, and wooden ships can be built in their dock-yards at less cost than in our yards. The naval authorities of all other European countries are following the example of the British, and are now building their vessels of iron and steel, the hulls of which are cased in wood, or as in the composite build, with the planking of wood and all other parts of metal. It may be added that steel is rapidly gaining favor, and will, probably, soon be most generally employed.

To say that we are right in adhering to wooden ships, is to say that the vast experience of the skilled naval architects, and that of the naval authorities of the great navies of the world, is all wrong—a position which, in view of our comparatively limited experience, we are no more justified in assuming than we would be in saying that our cast-iron smooth-bore guns are superior to the steel or wrought-iron and steel rifles of European navies.

SHIPS OF THE MERCANTILE MARINE SUITABLE FOR WAR PURPOSES.

According to the returns of the British Board of Trade, at the beginning of 1876 the number of registered sailing-vessels having a tonnage of 50 and upward was 18,696, and of registered steam-vessels of 50 tons and upward there were 3,436; total number of merchant-vessels sailing under the British flag, 22,132. Out of the number of steam-vessels thus registered about 300 are recorded as having a speed of upward of 12 knots per hour, regular steaming, at sea; quite a number of them have a speed of from 14 to 15 knots, and several of the Atlantic steamers have made upward of 16 knots per hour, under favorable conditions, for several days consecutively. In August, 1875, the writer was a passenger on board the *Germanic*, of the White Star Line, which made the passage from Sandy Hook to Queenstown in seven days and twenty-one hours, unaided by sail. This time has since been beaten by the same ship in a number of trips, also by the sister ship *Britannic*, which vessel made the run in August, 1877, from Queenstown to Sandy Hook in seven days and eleven hours.

All of these mercantile ships are iron-built, and have great strength of hull; they are engined on the compound system, and are provided with sufficient coal-storage for long runs.

But objections are raised against such vessels being fitted as fighting-ships; these objections consist, first, in the position of the motive machinery. In unarmored ships of war the engine, boilers, and steam-pipes are kept below the water-level; moreover, in recently-built British unarmored ships of war coal is stored between the sides of the hull and the engines and boilers from a few feet below the water-line to some distance above it,* while in the merchant-ship the steam-cylinders, steam-pipes, and tops of the boilers are usually above the water-level, consequently exposed to shot and shell traversing the ship, and any serious injury to these vital parts would disable her.

The second objection is to the material of which the ship is built. Unarmored ships of war of modern build almost invariably have wooden skins, even if the ribs, beams, and interior parts are of iron. In large ships of high steam-power, where the necessary strength and rigidity of the structure cannot be secured without iron or steel, wood is used as an outer covering for the iron skin. The reason for the objection to iron plates without wooden sheathing in such structures is, that if the iron skin exists alone, projectiles passing out at low velocity at the side opposite to that at which they entered are likely to drive away from the frames a considerable area of plating, and if this should happen on or below the water-line the consequences might be serious. A thick, well-fitted wood sheathing secured on the plates tends to prevent this. For this reason, and also to prevent the fouling to which the iron skins are subject, a wood sheathing is now applied to nearly all ships of war having iron or steel hulls.

* Experiments were made at Portsmouth, in October, 1877, to test the effect of firing shells, containing bursting charges, into coal-bunkers. The vessel used for the experiment was the *Oberon*, and the guns were 64-pounders, fired from the gunboat *Bloodhound*. The range was 200 yards, and the bursting charge of the shell was 7 pounds. The result of the experiment showed the coal to be a protection against this kind of projectile, but did not show whether the coal, which was bituminous, would be ignited by the explosion.

The third objection is, that there is no effectual division of the ship by water-tight bulkheads, while there is no unarmored ship of war, built of iron, which is not so divided as to be secure against foundering in ordinary weather with any one compartment in free communication with the sea, which is a condition held to be necessary.

This bulkhead objection will, however, be removed from mercantile steamers constructed in the future, as all the ship-owners in the United Kingdom have accepted the admiralty condition, and it is now believed that every first-class steamship hereafter built will float in smooth water with any one of its compartments in free communication with the outside water, and bulkheads can be readily introduced into any vessels now afloat that may be selected as suitable for naval purposes.

The first and second objections are not so easily and quickly remedied; but, assuming that such ships would be no match for unarmored ships properly built for war, they would evidently be the equals, and probably, as a consequence of their speed, the superiors of merchant-ships employed by the enemy; besides, they would be numerous and formidable against sailing-vessels and slow steamers, and they would have reasonable security against capture by *Alabamas*.

The subject of arming mercantile vessels in the event of war has long been under consideration by the admiralty, and their views have recently been foreshadowed by the director of naval construction, who read a paper before the eighteenth session of the Institution of Naval Architects, the notable parts of which are extracted, as follows:

* * *. I believe the ships may be so defended and armed as to become not only quite capable of defending themselves and of destroying armed ships not regularly built for war, but also most useful auxiliaries in all important naval operations. It is quite certain that they can at a few hours' notice be efficiently defended by a shot-proof screen across the deck before the machinery, and can as a rule be quickly and inexpensively armed with rams and with two 64-pounder guns in the bow. So long, therefore, as they can be fought bow on, they will compare favorably for ordnance, for the ram, and for the torpedo, with unarmored ships of war. The same holds good as to the defense and armament of the stern. About the broadside I am not so clear, but I should not despair, in view of structure and stability, of giving at short notice to many of our fast ocean-going ships an efficient broadside of 64-pounder guns, and 6-inch armored screens between decks, if that were ever found to be desirable. I think, however, we may be content with an armament which would be an absolute guarantee for their own preservation; for the equally fast unarmored ship of war or privateer would not, and the slower armored ship could not, attack them; an armament which would, moreover, make them more than a match for some of the slow wooden frigates and corvettes of old type in foreign navies, and more than a match also for rovers not regularly built for war. The extent to which, with the protection and armament I have indicated, they could be employed in naval warfare may be thought out, if we consider what those operations will be. I think they may be summarized as follows:

NAVAL OPERATIONS IN WARFARE.—*Defensive.*—(1) Self-protection by merchants or travelers on the high seas against rovers, whether men-of-war or armed merchant-ships. (2) The patrol of the highways of commerce by vessels in the employment of the government, for the destruction or capture of rovers. (3) Clearing the offing of important harbors, at home and in the colonies, of hostile vessels, including breaking the attempted blockades of ports. (4) Convoying merchant-ships. (5) Protecting harbors, naval stations, and coasts, at home and in the colonies, against violation.

Offensive.—(1) The capture of trading-ships belonging to the enemy, or liable to capture on his account. (2) The infliction of injury upon harbors, naval stations, and coast towns, and landing military forces on the enemy's territory. (3) Disabling or destroying the armed ships of the enemy. (4) Blockading the principal ports of the enemy to prevent the passage of merchandise inward or outward, and to lock up his armed ships. (5) Transporting troops, stores, and munitions of war, and keeping up communications by dispatch-vessels.

Of the five classes of work placed under the head of defensive warfare, a fast merchant-ship, armed, could perform two, in independence of the regular ships of war, and could take part in all the rest as auxiliaries to the iron-clads. And a precisely similar statement holds good with regard to the five classes of work placed under the head of offensive warfare. I do not stop to particularize these, as a little study of the question will, I believe, insure acceptance of this view.

There are certain general principles which may be accepted as arising out of the relation between the several types of fighting-ship. (1) The iron-clad ship will, as a rule, be slower and have less coal endurance than the first-class unarmored or lightly-armored ship. The iron-clad ship will therefore be unable to force the first-class unarmored or lightly-armored ship to engage her. (2) In duels between fast unarmored or lightly-armored steamships, the ship with most guns—supposing them to be equally good and equally well served—will generally be the victor, whatever the relative speeds or turning powers of the ships may be, because such actions will generally be determined at long ranges. (3) Since the merchant-ships cannot mount numerous guns, they will, even when armed, find the modern regular ship of war almost always their victor in single combat, and fast unarmored or lightly-armored ships will be more effective against armed merchant-ships than iron-clads would be. It follows from this that fast unarmored or lightly-armored ships of war must be of great consequence to a navy against which armed merchant-ships may be employed by an enemy. (4) The speed with which fast steamships can in any weather bear down at night upon slower steamships and sailing-ships, and the terrible nature of the attack they can make upon such ships with shells, the ram, and the spar-torpedo, will make it impossible to convoy successfully sailing-ships and slow steamships in face of the attack of even unarmored ships, provided they are fast and efficiently armed.

ENGLAND AND THE DECLARATION OF PARIS.

By the provisions of the declaration of Paris, privateering is abolished. The result of this is that Great Britain, being at war with any power possessing a navy, immunity from capture on the high seas can only be secured for British merchandise by carrying the same in ships sailing under a neutral flag and registered in a sea-port of a neutral power.

Some eminent Englishmen concluded that, as a consequence of this international law, a protracted war with a naval power would cause the transfer of a great portion of the carrying trade of England to some other country. The agitation upon this question first assumed a definite shape in the well-known pamphlet of Mr. Brassey, published nearly two years ago. Being precluded from granting letters of marque to merchantmen, they now propose to overcome the difficulty by commissioning, in time of war, as many merchant-steamers—

as may be built in accordance with certain requirements laid down by them, provided their owners consent to the arrangement upon the receipt of a certain subsidy. An admiralty officer in the controller's department has, during the past twelve months, visited every sea-port in Great Britain and Ireland, and surveyed or obtained particulars of the iron steamships sailing therefrom. The result of this inspection has been highly satisfactory, and their lordships are informed, upon the authority of this officer, that a very large number of these steamships are already built in accordance with the admiralty requirements for the purpose in view. With the addition of a little strengthening under the guns, and the construction of a magazine and shell-room, these vessels will be most formidable cruisers, fitted not only to defend themselves, but to act as policemen in the tracks which will be pursued by other merchantmen on the principal ocean voyages. These tracks will be marked out by the hydrographical department, in order that vessels may pass along the protected routes to their several destinations. The admiralty stipulate that vessels selected and subsidized for this service shall be capable of steaming at least twelve consecutive hours at not less than twelve knots an hour, at sea; also that they shall be so divided by bulkheads that, with a hole of any size in any one compartment, they shall continue to float in smooth water. It is proposed to arm these vessels with two 64-pounder guns, one forward and the other aft. * * * * It will be a question for their owners to consider whether the advantages offered are sufficient to compensate them for the cost of the alterations and the inconvenience resulting from the original intentions in the ship's design being departed from. Extensive water-tight subdivision does not meet with much favor in certain trades, as it interferes seriously with the stowage of the cargo and the operations of loading and unloading. Underwriters do not at present make that difference in the premiums of insurance which is represented by the additional outlay and the inconvenience endured, notwithstanding the greater chance of safety to vessel and cargo which these bulkheads provide. The whole question is purely a commercial one, and by commercial principles it will be tested by the shipping portion of the community.

PART XI.

THE PERKINS HIGH-PRESSURE COMPOUND SYSTEM; METHOD OF CONTRACTING FOR STEAM-MACHINERY FOR THE NAVY; TRIALS OF SHIPS AT THE MEASURED MILE; PERSONNEL OF THE BRITISH NAVY; COST OF MAINTAINING THE NAVY.

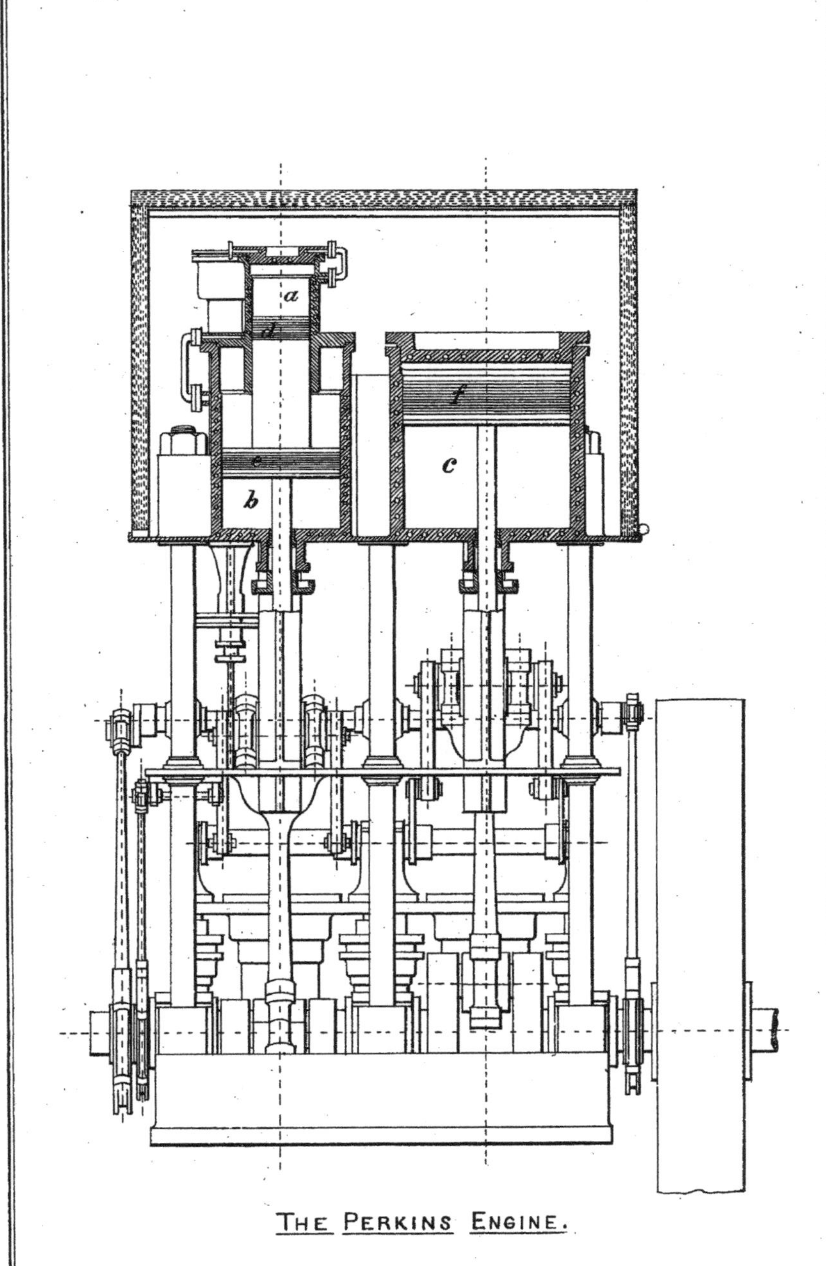

THE PERKINS ENGINE.

THE PERKINS HIGH-PRESSURE COMPOUND SYSTEM.

The machinery about to be described was under construction for Her Britannic Majesty's sloop *Pelican*, and considerable progress had been made when a suit at law between the constructors of the engines and the patentee unfortunately caused the work to cease, and as a consequence the naval authorities canceled the contract, declined to enter into any other engagements relating thereto, and placed engines in the *Pelican* similar to those in the *Fantôme*.

The British admiralty have long been noted for the careful investigation given untried plans of any kind proposed for ships of the royal navy. As a rule, they adopt useful inventions only after such have been successfully established in the mercantile marine; a case in point was the caution exercised in the introduction of the compound engine. It must therefore have been unusually strong proof which decided that august board to make the test, on a large scale, of a system so far in advance of the present day, and to pass from 70 pounds pressure per square inch as the highest used in boilers of the navy to 300 pounds pressure per square inch; it was a step beyond anything previously attempted.

The *Pelican* is a composite sloop, of 1,124 tons displacement, built at Devonport. The steam-machinery originally intended for this vessel was designed by the Yorkshire Engine Company, under letters patent granted Mr. Loftus Perkins, an enterprising American, long established as a manufacturer in London. The novelty of the design and the principles involved are so unlike those which have influenced the construction heretofore of machinery for marine purposes, that the subject excited no small degree of attention among parties interested in naval and mercantile vessels. The contract provided that the vessel should on the trials at the measured mile, and on the six-hour runs, develop 900 indicated horse-power and consume not more than 1½ pounds of coal per indicated horse-power per hour. The engines have 5 cylinders of three different diameters, two high-pressure of 16 inches diameter, two medium-pressure of 32 inches diameter, and one low-pressure of 56 inches diameter. One high and one medium pressure cylinder are bolted together end to end; these cylinders are single-acting, the steam being admitted first to the high-pressure, thence on a return stroke to the opposite end of the medium-pressure cylinder, and thence it escapes into a receiver. The other high and medium pressure cylinders are bolted together in the same manner and are acted upon by the steam in the same way. The low-pressure cylinder is double-acting, and draws its supply of steam from the receiver into which the medium-pressure cylinders exhaust, and itself exhausts into the condenser. There are three cranks, placed 120° apart, and these are coupled, the after crank to the common piston-rod of one pair of cylinders of high and medium pressure; the center crank to the piston-rod of the low-pressure cylinder, and the forward crank to that of the other pair of high and medium pressure cylinders. The stroke is 2 feet, and the revolutions per minute were to be about 100. The valves employed throughout are of the piston variety; the advantages claimed for the method of construction in these valves is the almost entire absence of friction, wear, or leakage, combined with a balanced valve. The surface-condenser, also patented by Mr. Perkins, is constructed in such a way as to prevent leakage through the tubes when they are correctly fitted.

The tubes are galvanized, and arranged as illustrated by the annexed sketch.* The cooling-tubes, which are surrounded by steam, and of which one, C D, is shown, are closed by welding up one end, D, and are accurately and permanently screwed into a strong plate at the other end, C. Each has within it a circulating-tube, B, secured to a tube-plate placed at a little distance from the former. The circulating water passes from the sea into and through the internal tube and returns by the annular space between the two tubes, passing out to the pumps.

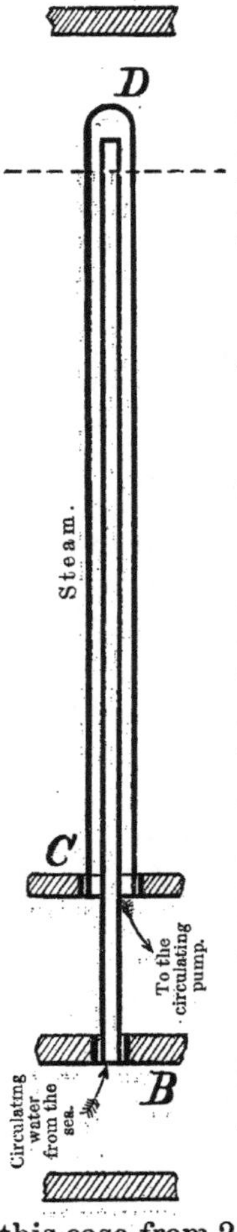

D

Steam.

C

To the circulating pump.

Circulating water from the sea.

B

The pistons of the steam-cylinders are made tight by having several rings, or sets of rings, say six, of hard metal; each being separated from the others by what the inventor terms intermediate junk-rings. These rings, of Mr. Perkins's patent metal, are composed of five parts of tin and sixteen parts of copper.

All hot surfaces of the cylinders and pipes are surrounded with a jacket of sheet-iron, packed with vegetable black. Independent air, circulating, and feed pumps were to be provided, driven by a separate pair of engines. The boilers contain a nest of built-up tubes, placed horizontally, close to each other, and having the flame and gases passing around and among them. These tubes are 3 inches in diameter and $\frac{3}{8}$ inch thick; when put together in the boilers, they are proved to 2,500 pounds hydrostatic pressure to the square inch, and worked at a pressure of 300 pounds per square inch. There is consequently an enormous margin of safety, and every precaution is taken in building the tubes together to insure tightness, and their connections are to be beyond doubt of danger or waste. So also with the engines: they are made of the best materials, and involve the best workmanship. The water-gauges employed are made of mica, and to secure a tight joint between the two plates of mica and the two opposite faces of the body of the gauges there is formed a narrow raised edge, against which the mica is firmly pressed by a metal plate applied to its front face.

The first peculiarity of the Perkins system is the extremely high pressure at which he works the steam—in this case from 250 to 300 pounds per square inch; and he claims to have quite overcome all the difficulties hitherto experienced in using steam at sea above ordinary working pressures. A second peculiarity of the system is the absence of internal lubrication with either oil or tallow, thereby avoiding the possibility of corrosion by fatty acids; while, to prevent the wear of cylinders and slides, which is experienced under ordinary circumstances, he uses a metal of his own, whose composition and working are reported as unobjectionable. Another peculiarity and important feature of the Perkins system consists in the use, over and over again, of soft

* From *Engineering*.

fresh water or rain-water. The continually recurrent use of pure water, *not distilled sea-water*, is claimed as the only remedy against the internal corrosion of marine boilers supplied with water from surface-condensers. So much importance is attached to this point, that the committee appointed by the admiralty to investigate the subject of boiler corrosion were urgent in their recommendation to test the system, without loss of time, on a scale sufficiently large to arrive at correct conclusions on the subject. Of course this involves the necessity of carrying at sea a supply of fresh water sufficient to make up for all waste accruing during a voyage, and objection has been urged to the system on this account; but, on the other hand, it is alleged that all joints are to be made mathematically correct and tight, as shown by the boilers and engines now in operation on the system, hence the leakage will be reduced to a minimum.

This would have been the first attempt to use steam of 300 pounds pressure at sea; and while an expression of opinion has become needless, it is proper to state that Mr. Perkins has been entirely successful in his land-engines built on the same system, one of which has been continuously employed at his works for fourteen years, using steam at 500 pounds pressure per square inch. The tubes of this boiler, cut out after thirteen years' service, as inspected by me, were found to be in a remarkable state of preservation, as were also the piston-packing and valve-rings, which had been at work eighteen months without lubrication.

Besides the land-engines manufactured by Mr. Perkins, he has built and kept employed during the last few years, for his own use, a small yacht, the *Emily*, engined on his own system, the greater portion of the time running under a boiler-pressure of 500 pounds per square inch. The number of cylinders in this boat is six; two high-pressure, two intermediate, and two low-pressure. The high and intermediate pressure cylinders are single-acting, the latter exhausting into a chamber from which the low-pressure cylinder is supplied. The steam is expanded 24 times, and the expenditure of fuel is a little above 1 pound of coal per horse-power per hour.

Mr. Perkins has also engined one of the Thames tug-boats on his high-pressure double-compound system. There is in this case a pair of engines working a screw 8 feet 6 inches in diameter, and of 11 feet 6 inches pitch. The cylinders are four in number, with diameters of 15 and 30 inches, and the working-pressure is 250 pounds per square inch. It was after the inspection of this boat, the yacht *Emily*, and Mr. Perkins's land-engines that the committee appointed by the admiralty recommended the system to be tested on a considerable scale in one of Her Britannic Majesty's vessels. The system of working the steam in the cylinders is the same as in the land-engines of Mr. Perkins, which the preceding drawing and his explanation following will make clear.

a is a single-acting high-pressure cylinder.
b is a single-acting cylinder of four times the capacity of *a*.
c is a double-acting cylinder of four times the capacity of *b*.
d and *e* are two pistons on the same rod, working in the cylinders *a* and *b*.

The course of the steam is as follows: The steam enters *a* at 250 pounds pressure and cuts off at half stroke; at the termination of the stroke it expands into the bottom end of cylinder *b*, making the return stroke; it then expands into the opposite end of cylinder *b*, which is in direct communication with the valve-box of cylinder *c*. This latter is double-acting, and is arranged to cut off at about quarter stroke and exhaust into the condenser after the steam has expanded 32 times.

SYSTEM OF CONTRACTING FOR STEAM-MACHINERY FOR THE BRITISH NAVY.

The policy of England has been steadily to encourage the responsible engineering-works of the kingdom with orders, which have in no small degree tended to expand and develop their special resources for marine-engine construction, and, besides, to obtain through these establishments the best constructive and mechanical engineering ability. No steam-machinery for ships of war has at any time been manufactured in the excellent works of the dock-yards. These works are employed solely on the repairs of the steam fleet. All new machinery is built by contract from the designs of responsible bidders, and the establishment to which the contract is awarded guarantees the entire work and the indicated horse-power on the measured mile. It is also held responsible for the efficiency of the machinery for a period of one year after it has been accepted by the admiralty, and any parts which during that period may be found defective, or may show symptoms of weakness owing to faulty design, materials, or workmanship, must be removed and others substituted for them by the contractors at their own expense. When the machinery for a new ship or class of vessels is to be furnished, the specifications are drawn up by the engineer-in-chief and printed. They are headed "Specifications of certain particulars to be strictly observed in the construction of ——— engines, with ——— screws, for the ship ———, of ——— indicated horse-power." These specifications give all the principal dimensions and many of the details of the engines, boilers, and appendages. They also limit the weight and space to be occupied. They are then sent out to the following-named firms: Messrs. John Penn & Son, Greenwich; Messrs. Maudslay, Field & Son, London; Messrs. Humphrys & Tennant, Deptford; Messrs. Rennie & Bros., London; Messrs. Napier & Son, Glasgow; Messrs. John Elder & Co., Glasgow; Messrs. Laird Bros., Birkenhead; Messrs. Easton & Anderson, Leith, and sometimes to other works.

In tendering for the supply of the machinery, the contractors are requested to forward a design in accordance with the specifications and tracings of the ship, &c., supplied. They are also requested to furnish a design and tender for any other plan of engines and boilers not based on the specifications, adhering to the total weight and space to be occupied by the machinery. The tender accepted by the admiralty is in most cases in accordance with the specifications, but not always so, as may be seen in the case of the sister ships *Nelson* and *Northampton*. The specifications for the motive machinery of these two ships were printed as for a class, and while the contract for the former is based on them, the latter has engines of an entirely different design, as has been seen in the brief description of these vessels elsewhere. There are other similar cases of contracts where tenders have been received on the specifications of a class, but it is unnecessary to refer to them here, as it is believed that sufficient has been said to show the system pursued in obtaining machinery for the ships of the British navy.

TRIALS OF BRITISH NAVAL VESSELS AT THE MEASURED MILE.

The speed of all steam-vessels belonging to the British navy is tested by running them over a measured mile in Stokes Bay, near Portsmouth. The indicated horse-power of new engines, as stipulated by contract, is also determined on the runs over this course. The noteworthy points in the regulations for these trials are as follows:

Her Majesty's ships are to be tried on the measured mile on the following occasions:
1st. After having new engines erected.
2d. After every extensive repair of engines.
3d. On being commissioned when all their weights are on board.
4th. When trials may be ordered on special occasions. * * *

Ships in the first division of the reserve are to be tried under way once every year during summer for not less than six hours; but it is not required that they be taken to the measured mile for the purpose, and their trial at full speed need not be of longer duration than four hours. * * * The trials of ships at the measured mile are to be conducted under the charge of the captain or commander of the reserve, who is to be accompanied by the constructor, an engineer officer from the yard, and the chief inspector of machinery afloat. * * * As a rule, the trials should not take place when the force of wind exceeds 3. * * * No other coal is to be used while running the mile than the coals specially ordered by the admiralty. * * * The best-trained stokers that can be selected from the reserve are to be employed.

On the full-power trials of ships at the measured mile, the engines and boilers are to be worked to the utmost extent of their capabilities, not only when running the mile, but the whole of the intervals between the several runs.

In the intervals between the runs, ships are to be run well away from the marks, so as to insure the attainment of full speed on their return to them.

Steam must not be partially shut off when the ship is not on the mile in order to obtain a higher result when she is on the mile. * * * * Indicator-diagrams are to be taken, as nearly as possible, at equal intervals of time during each run, in order to obtain the real mean pressure in the cylinders. * * * The revolutions of the engines during each run are to be taken by the counter.

If, during the trial, any defects should occur to the machinery or the ship, the trial is to cease; * * * a new series of trials being commenced when the defects shall have been made good.

If the boilers should prime so that they cannot be worked at full power, even at the close of the trial, it is to be considered as unsatisfactory and the trial is to be repeated.

Whenever irregularities occur in running the measured mile, either at full or half power, the officers are to repeat the runs until the results agree one with the other so nearly as to leave no reasonable doubt of their substantial accuracy.

When ships which have new engines not yet received from the contractor are under trial at the measured mile, the engines and boilers during the trial are to be under the charge of the contractor or his agent, who is to have the whole responsibility and management; * * * but the trial is to be conducted in strict accordance with the regulations laid down for all trials at the measured mile, and the engineer officers will be responsible that the regulations are never deviated from. * * * Full and detailed reports of the results of the trial of each ship at the measured mile are to be made in the prescribed forms and forwarded to the admiralty. A tracing of the screw-propeller is to be annexed thereto, also a set of original indicator-diagrams for one mile only. * * * In all reports of trials it is to be stated whether the engines and boilers did or did not work in a satisfactory manner, and are or are not in all respects fit for service at sea. If the trial should not have been satisfactory, the reason is to be stated and the time it will take to render the machinery fit for service at sea.

When the ship tried is in commission, the reports are to be signed by the captain and chief engineer of the ship, as well as by the officers of the reserve and the dock-yard.

* * * * * * *

After every trial of a ship in commission, and before proceeding to sea, the ship is to run into the harbor for at least 24 hours, and be carefully examined by a shipwright and engineer officers of the yard, to ascertain that there are no defects, and a report made. These reports are to be signed by the dock-yard officers who made the examination, the chief constructor, and the chief engineer of the dock-yard; also by the officer in command of the ship and the chief engineer of the ship, and are to be accompanied by any remarks which the superintendent may think proper to offer.

* * * Besides the above trials, it is directed that the engines of all ships in commission for service at sea are to be worked at full power twice during each year by their own complement. The first full-power trial is to be made when the ship is commissioned ready for sea, after the dock-yard trials to test the machinery have been made, and it is to be of six hours' duration. * * *

In consequence of the explosion of the boiler of the *Thunderer* and other mishaps, new instructions were issued in the autumn of 1877, which embody very important alterations, the following synopsis of which is from an English journal:

* * * With reference to the measured-mile runs, the object is to make the trial a test not only of what the engines are capable of doing when at their best and pushed to the utmost of their power, but also of what they are likely to perform in the ordinary circumstances of sea service. Great attention has been given to the importance of securing perfect observations by a series of checks and counter-checks, and of obtaining uniform results throughout the trials. Their lordships have also, profiting from recent experience, seen the necessity of rigidly defining the responsibility of every one concerned in the trials, and of taking effective measures for the prevention of accidents. Full-power trials are intended to test the capability of the machinery and boilers in Her Majesty's ships to maintain the required horse-power for lengthened periods of service at full speed, and the officers who are authorized to test the machinery are to satisfy themselves that no faults exist in the construction of the engines which may render the ship inefficient for such service. In all the old contracts the manufacturers were required to develop the guaranteed power of the engines at a trial on the measured mile. Experience, however, has shown that this stipulation was objectionable in many respects. As a test of endurance the measured-mile trial was not so conclusive as a six hours' full-power run under way; and as a trial of the performances of the ship as regards speed it was, in consequence of the difference of trim between the ship on the mile and when commissioned, positively misleading. Besides, as the contractors are in no wise concerned in the behavior of the ship apart from the working of the machinery, it was unfair to them that they should be put to the expense of keeping the engines in order until the ship was ready for the mile.

In all the new contracts a six hours' run has been substituted for the mile trial, and if the machinery works well and indicates the covenanted horse-power, it will be "taken over" at once by the admiralty. Before the official trials are held the contractors are to be allowed such preliminary trials under way as they may consider necessary to prepare the engines. During these trials the engines are to have a run of at least an hour with the expansion-valves in full gear, and such trials at lower grades as may be deemed necessary to test the efficiency of the expansion-gear, the steam in the boilers being at the same time maintained at its maximum pressure. If the engines are fitted with a surface-condenser, the jet-injection is also to be submitted to a trial of one or two hours, at full speed, for the purpose of determining the capability of the machinery and the power which it is possible to develop with the common injection. The engines are also to be stopped, started, and reversed at full speed. Before going on the official trial the contractor is required to notify the captain of the steam reserve that the steam-gauges, stop-valves, safety-valves, and sentinel-valves of all the boilers under which fires are lighted are in good working order. The stop-valves are, moreover, to be opened and the safety-valves worked in the presence of the chief engineer of the ship. At the trial the power is to be maintained for at least the time specified; the water service is to be confined to the regular engine-room service, and the engines are to be in all respects ready and fit to continue running after the trial. Rangoon oil, which has been found to cake and choke the lubricating apertures, * * * is to be discarded, and Gallipoli oil exclusively used for the lubrication of the bearings. The mean power deduced from the half-hourly records of steam-pressures, revolutions, and vacuum, will be taken as the indicated horse-power developed on the trial; and it is requested that great care be observed that the diagrams represent the mean power as nearly as possible. * * * "It is to be distinctly understood that the presence of engineer officers for the charge or observation of stokers, or the presence of other officers appointed by or on behalf of the admiralty to observe the results of the trials, does not in any way relieve the contractors, or the persons who have charge at the trials, from any responsibility in the care and management of the machinery and boilers." No banquets or entertainments are allowed to be given on board by contractors at the trials, "or on any other occasion."

Full instructions are also given for a thorough examination of the machinery after the contractor's trial, so that the engines may be ready for the runs on the mile. This trial, which is called the constructor's (as distinguished from the contractor's) or measured-mile trial, will not be held until after the ships are commissioned, and when all the legend weights are on board, so that the ship to be tried will have been brought down to her estimated draught. * * * Under the old arrangements for new ships it was customary to make six full-power runs and four half-boiler runs with and against the tide. By the new regulations the half-power runs have been abolished and two new features are to be introduced. Four runs are to be made at full power, four at two-thirds power, and four at one-third power, steam being obtained from the whole or part of the boilers as may be considered advisable. * * * Steam-pressure is to be maintained at a maximum in the boilers, the reduction in power being made with the expansion-valve only; and while the reduced-speed runs are being made, the revolutions, pressures, &c., are to be kept as constant as possible so as to develop nearly the same indicated horse-power throughout. Each trial at full and reduced power should be completed while the tide is running in one direction. * * * Stringent regulations are to be enforced with reference to the indicator-diagrams and the noting of the revolutions, which are to be taken by two counters and independent observers. * * * The turning capabilities of the ship are, as formerly, to be tried on the same day as the measured-mile trial is made. * * * * *

In reviewing the above, it may be questioned whether the system thus prescribed for testing the strength and power of motive machinery and the speed of a vessel be not seriously objectionable: it has been seen that the coal used on the trials is of the very best quality; that the firemen employed are the best trained and most expert in the service; that the machinery is placed in the most complete order, and that under these conditions the engines and boilers are forced to the utmost extent of their capabilities. Now, it may reasonably be asked whether the excessive strain to which the machinery is subjected while undergoing this severe test does not leave weakness in some hidden parts, and is not the cause of defects subsequently exposed. Furthermore, as the engines can never under any possible conditions be worked at sea for any extended time up to the measured-mile speed, it may be asked what practical purpose is served by it except that of recording a speed never afterward attainable under any other conditions.* As proof of

* Trials at the measured mile, when the greatest possible power is got out of a vessel's engines, furnish the best means that can be obtained for determining the relative speeds of different ships. The average speed that will be afterward obtained on long runs is less than this, but the percentage of reduction, so long as the engines and boilers remain in good working order, and the bottom of the ship remains clean, is pretty uniform, and is not difficult to estimate. Indeed, Mr. King himself * * * estimates the average speed of a vessel over a run of 24 hours at 1½ knots per hour less than is given on the measured-mile trials. We should put the reduction down at less than this, but, whatever it is, the average speed of a ship, when doing her best over a long run, can be pretty correctly inferred from the measured-mile speed. So far, however, from the engines being worked on the measured-mile trials beyond what is necessary, it is sometimes found that greater power is developed after the engines have been working for some time and have got thoroughly to their bearings, than is done on the measured mile. If trials were made under less stringent conditions than the usual measured-mile trials, the sources of error would be more numerous than now exist, as it would be extremely difficult to insure only a certain proportion of the maximum power being developed. Of course any falling off in a ship's speed from fouling, or from defects or deterioration in the engines and boilers, would be left out of account as much in one case as in the other.

The measured-mile trials, by putting as great a strain on the machinery as possible, is most useful as a proof-test. At any rate, it is one of the chief objects of these trials to test the machinery as much as possible; and by so doing the same plan is adopted as is usual in testing any other machinery or boilers, or any mechanical structures. Mr. King's objection would apply as strongly to testing any other material or mechanism beyond its ordinary working strain as to marine engines. But, as we have said, the power developed by the engines on the measured mile is sometimes exceeded under ordinary working conditions, so that as a test the measured-mile trials are certainly not unduly severe.—AN ENGLISH NAVAL ARCHITECT.

In the foregoing "An English Naval Architect" has, perhaps, stated all that can be said in favor of the trial at the measured mile, but he makes a comparison fatal to his own argument when he adds that in testing the machinery as much as possible "the same plan is adopted as is usual in testing * * * boilers." No fact is better recognized an that many boilers have been destructively weakened by over testing.—J. W. K.

this, it may be instructive in regard to the speed of armored ships, to refer to the proceedings of the court-martial held at Devonport in September, 1875, to inquire into the loss of Her Majesty's ship *Vanguard* by collision with a sister ship, the *Iron Duke*, off the coast of Ireland, September 1 of that year.

The commanding officer of the *Iron Duke*, Captain Hickley, being sworn and examined, testified as follows to questions:

768. It is stated in the steam-register that at 12.30 *Iron Duke* was going fifty-four revolutions; what speed would that produce?—Answer. Eight knots.

769. And at 12.40, sixty revolutions; what speed would that produce?—Answer. Eight and one-half knots, or under. And I may refer to the log of the 23d of August, that on leaving Loch Swilly under full speed, to pick the squadron up, she was going eight and one-half knots at sixty revolutions.

770. What was the state of the weather on that occasion?—Answer. Very fine.

* * * * * * *

816. How many revolutions do *Iron Duke's* engines go at their utmost speed—all boilers?—Answer. The only opportunity I had of judging was on the 23d of August, and on a short trial we had to test the engines; sixty-three revolutions was the utmost we could get, with most indifferent stokers.

* * * * * * *

J. D. Charter, engineer in Her Majesty's ship *Iron Duke*, being sworn and examined, testified as follows to questions:

1076. What is the pitch of your screws?—Answer. Twenty-one feet.

1077. What is the slip percentage in calm weather?—Answer. The slip varies according to the number of revolutions we are making.

1078. Fixing your revolutions at sixty, what is the slip?—Answer. About 15 per cent.

1079. Going at sixty revolutions with a screw of 21 feet pitch, what speed would that give?—Answer. About ten and one-half knots.

Taking the pitch of the screws and the revolutions as above given, it appears from the testimony of the captain of the ship that about 8.92 knots was the maximum speed that could be obtained, by the *Iron Duke*, while from the testimony of Engineer Charter it would appear that a speed of about 11.02 knots was possible when working the engines to the utmost.

Now, by referring to the tables of armored ships in this report it will be seen that the speed of the *Iron Duke* on the measured mile was 13.6 knots. This great discrepancy could not be owing even in a small degree to foulness of the iron bottom (the bottom is not sheathed with wood), for, if so, the slip of the screw would have very largely exceeded 15 per cent.

It appears also, from the testimony given before the same court by the commanding officer of Her Majesty's ship *Vanguard*, Captain Dawkins, that the maximum speed of that ship was nine knots or thereabouts.

It may probably not be fair to make an abatement for the speed of all British armored ships in accordance with those sworn to for the *Iron Duke* and the *Vanguard;* but from the testimony in these two cases, we may reasonably conclude that a wide margin exists between the speeds of the ships at the measured-mile trials and the maximum speed that can be obtained when cruising at sea.

The single case of the *Volage* has been instanced as proof that a greater speed may be attained at sea than on the measured mile, that vessel having logged 15.38 knots on the six-hour trial, against 15 at the mile; but every engineer will recollect conditions any one of which might have brought about the exceptional action on the *Volage*.

The real speed of a ship is that recorded under full power at sea in ordinary weather during 12 to 24 hours consecutively, and of this we have but little from any ships of the British navy.*

In considering this subject it may be well to remember that the "indifferent stokers" referred to as being on board the *Iron Duke*, although inferior to those employed on the measured-mile trials, are far superior to the firemen that have been employed in our naval vessels within seven or eight years.†

* We have no record of the *Sultan-Bellerophon* relative-speed trial of 24 hours in 1873.

† These statements respecting the speeds of the *Iron Duke* and *Vanguard* are very untrustworthy, as the discrepancy between them shows; and it must be remembered that not only were the stokers "most indifferent," but that the boilers of these ships were, at the time referred to, six years old.—AN ENGLISH NAVAL ARCHITECT.

These are two excellent reasons given by "An English Naval Architect" why the speed at the measured mile cannot be maintained in general service: difference in stokers and difference in age of boilers; many others could be given if it was considered necessary.—J. W. K.

PERSONNEL OF THE BRITISH NAVY.

The *personnel* of the British navy at this time, 1877, including all officers on the active list, enlisted men, boys, marines, and others, of all ranks and grades, and exclusive of retired officers, consists in the aggregate of 60,536 persons.

The number of officers whose names are borne on the active list of the navy register, July, 1877, is as follows:

EXECUTIVE OR LINE OFFICERS.

Admirals of the fleet	3	
Admirals	10	
Vice-admirals	16	
Rear-admirals	27	
Captains	174	
Commanders	205	
Lieutenants	800	
Sub-lieutenants	275	
Midshipmen	228	
Naval cadets	185	
Staff captains	11	
Staff commanders	93	
Navigating lieutenants	146	
Navigating sub-lieutenants	72	
Navigating midshipmen	4	
Total executive or line officers		2,249

ENGINEER OFFICERS.

Chief inspectors of machinery	5	
Inspectors of machinery	5	
Chief engineers	164	
Engineers	562	
Assistant engineers	137	
Total engineer officers		873

MEDICAL OFFICERS.

Directors-general	1	
Inspectors-general	5	
Deputy inspectors-general	12	
Fleet-surgeons	80	
Staff surgeons	125	
Surgeons	192	
Total medical officers		415

PAY OFFICERS.

Paymasters	263	
Assistant paymasters	233	
Total pay officers		496
Chaplains		97
Naval instructors		69

WARRANT OFFICERS.

Chief gunners	11	
Gunners	272	
Chief boatswains	24	
Boatswains	395	
Chief carpenters	10	
Carpenters	196	
Total warrant officers		908
Total number of officers		5,107

The petty officers, seamen, &c., are distributed as follows:

Effective, for general service:		
Seamen and others	20,478	
Boys	3,131	
First reserve ships, including tenders:		
Seamen and others	1,989	
Boys	269	
Gunnery and training-ships:		
Seamen and others	2,467	
Boys	2,789	
Stationary ships:		
Seamen and boys	3,422	
Surveying-vessels:		
Seamen and boys	334	
Troop-ships (imperial service):		
Seamen and boys	747	
Store-ships and drill-ships:		
Seamen and boys	414	
Coast-guard service, *on shore:*		
Officers (warrant)	251	
Seamen and others	3,983	
Troop-ships (Indian service):		
Seamen and others	1,155	
Total number of petty officers, seamen, &c		35,124
Total number of boys		6,305
Total number of marines (including officers)		14,000
Total		55,429
Officers on active list		5,107
Grand total		60,536

The number of ships in commission, officered and manned December 1, 1877, consisted of 27 armored and 222 unarmored, of all classes; total number, 249, representing an aggregate tonnage displacement of probably not less than 575,000, and an aggregate indicated horse-power of fully 385,000.

COST OF MAINTAINING THE NAVY.

The total amount appropriated for purposes of all kinds for the financial year 1875–'76 was $52,964,400. For the financial year 1876–'77 the total amount was $54,862,918, and the estimates for which the votes were given for 1877–'78 amounted to the sum total of $53,361,970, distributed as follows:

No.		Pay.	Allowances.	Total.
	PAY OF OFFICERS ON THE ACTIVE LIST.			
18	Flag officers			
5	Flag officers, superintendents of dock-yards, &c	$211, 881	$122, 112	
14	Flag-lieutenants			
153	Enlisted men and others, retinue to above			
2, 994	Commissioned and other officers			
463	Subordinate officers	3, 804, 272	274, 541	
858	Warrant officers			
317	Officers of coast-guard, *on shore*	250, 689		
560	Unemployed officers (half-pay)	550, 949		
	Total			$5, 214, 444
	PAY OF MEN.			
30, 890	Enlisted men	4, 960, 665	299, 182	
6, 305	Boys	305, 854	2, 076	
3, 983	Enlisted men of coast-guard, *on shore*	644, 946	29, 296	
	Total			6, 242, 019
166	Officers on reserved list			193, 375
2, 095	Officers on retired list			2, 673, 345
	Victuals and clothing for men and boys			4, 704, 733
	Total of pay of officers, also pay, victuals, and clothing of men			19, 027, 916
	MARINE CORPS.			
185	Staff officers	253, 070		
320	Commissioned officers	113, 418	191, 027	
13, 495	Enlisted men	1, 593, 691		
	Total			2, 151, 206
204	Officers on the retired list			292, 169
	Victuals and clothing for men			1, 023, 312
	Total of pay of officers, also pay, victuals, and clothing of men			3, 466, 687
	Extra pay and allowances to officers and men for special services, good conduct, &c			558, 584
1.	Grand total for pay, &c			23, 053, 187
2.	Pensions and allowances			5, 054, 016
3.	Admiralty office			942, 305
4.	Coast-guard service		$241, 838	
	Royal naval reserves, &c		768, 556	1, 010, 394
5.	Scientific branch			529, 750
6.	Dock-yards		$4, 596, 078	
	Cost of labor for building and completing ships at dock-yards		1, 924, 487	6, 520, 565
7.	Victualing yards			373, 880
8.	Medical establishments			321, 489
9.	Medicines and medical stores			379, 129
10.	Marine divisions (barracks, &c.)			103, 596
11.	Naval stores			5, 867, 478
12.	Steam-machinery and ships building by contract:			
	Ships and machinery building by contract		$1, 916, 274	
	Steam-machinery for vessels at dock-yards		2, 092, 011	
	Hydraulic and steam machinery, fittings for turrets, and torpedo machinery		316, 386	
	Repairs of ships other than at yards		145, 800	
	Purchase of torpedoes		388, 800	
	Boats to be ordered		34, 506	
	Experimental purposes		73, 143	
	Breaking up ships, boilers, and machinery		29, 160	
	Traveling and other expenses of officers and others superintending the building of ships by contract, and other works		68, 040	5, 064, 120
13.	New works, buildings, machinery, and repairs			2, 652, 175
14.	Martial law and law charges			39, 594
15.	Miscellaneous services			632, 451
16.	Army department (conveyance of troops)			817, 841
	Grand total			53, 361, 970

PART XII.

BRITISH NAVAL DOCK-YARDS.

PORTSMOUTH; CHATHAM; DEVONPORT AND KEYHAM; SHEERNESS; PEMBROKE.

ADMINISTRATION OF DOCK-YARDS.

THE SUPERINTENDENT; MASTER ATTENDANT; CHIEF CONSTRUCTOR; CHIEF ENGINEER; STOREKEEPER; ACCOUNTANT; CASHIER; CIVIL ENGINEER; WORKMEN AND BOYS; POLITICAL; PAY; POLICE; PENSIONS; REMARKS.

BRITISH NAVAL DOCK-YARDS.

The naval dock-yards of Great Britain are five in number, viz: Portsmouth, near the Channel; Chatham, on the river Medway; Devonport and Keyham, on the southwest coast; Sheerness, at the junction of the Thames and Medway; and Pembroke, on the coast of South Wales. These yards have been described by the writer in a former report, printed by order of Congress; but since that time the Woolwich yard has been abandoned for naval purposes, the yard at Deptford has been turned over to the victualing department, and Portsmouth and Chatham yards have been largely extended and the facilities greatly increased. A brief description, therefore, of the five first named will be given.

PORTSMOUTH.

The Portsmouth dock-yard is located near the English Channel. It is fortified, and the approaches to it are well covered. The series of forts which have been years in building for its defense have recently been completed, and last summer a number of 38-ton, 12½-inch, Woolwich guns were mounted in them. These guns are worked and loaded entirely by the hydraulic system of Rendel, and, it is said, may be fired at the rate of a round in a minute and a half. It possesses a capacious natural harbor, and, in respect to position, extent of ground covered, and number and capacity of basins, docks, buildings, and appliances of all kinds for building, repairing, and equipping ships of war, has occupied the foremost place among English dock-yards. In 1864 the first appropriation was made by Parliament for its extension and improvement, and, with the aid of liberal votes, since that time the work has been steadily progressing. The plan, which has been omitted here, but may be found in the first edition of this report, will show the extent and character of the new works since their commencement in 1865; also, a portion of the old dock-yard. That portion of the old yard not shown on the plan, but containing buildings, lies south of the steam-basin; and south of dock No. 6 is the ship-basin with its four stone docks. There is also another stone dock south of the ship-basin, with its inlet from the harbor. The old yard contains two basins and eleven stone dry-docks, and, as it has been one of the chief seats of manufactures for the royal navy, a brief description of its principal workshops may be found instructive.

On the northeast side of the old yard is the steam-basin, the basin in which steam-machinery is put into or taken out of vessels, or where general machinery repairs are carried on. The entrance to the steam-basin is from the harbor, through the camber, or dry-dock No. 7, hereafter to be mentioned. On the eastern end of the basin is located a pair of wrought-iron steam-operating box-shears, 145 feet high, designed chiefly for hoisting and lowering boilers out of or into vessels. The shears have two main purchases and one crab, with engines to each; these, together with the boiler, are located at a convenient distance from the basin. On the same side of the basin there is, in addition, one large Fairbairn steam-crane, having the boiler and engines attached. Besides these appliances for raising and lowering weights, there are on the borders of the basin five Fairbairn cranes, operated by manual labor. On the east side of the basin are the dry-docks, No. 8 and No. 11; the first 300 feet, and the second nearly 400 feet long, measured on the bottoms. On the west side of the basin, and parallel with its greatest length, are the machine, boiler, erecting, and copper shops, in one building, under one roof. The building, like others in Portsmouth dock-yard, is of brick. It is 720 feet long by 40 feet wide, measured inside, and two stories high. Entering the first or lower story, the height of which is 28 feet, it is found to be divided by cross-walls into five equal compartments. The boiler-shop and boiler-making machines occupy three of the north divisions. The divisions are conveniently arranged and well stocked with a variety of superior machinery, tools, and appliances of the present day. The division adjoining the boiler-shops is occupied exclusively as a machine-erecting shop. It has but one traveling crane, a few heavy machines, and the usual mechanical appliances. The next or north division, on the ground floor, is the machine-shop for heavy work. It is provided with traveling cranes, and stocked with heavy machinery and tools. Ascending to the second floor of the building, the machine-shop is first reached, which is provided with the usual machinery and tools for fitting and finishing light work; and afterward the coppersmith-shop, the chemical and store rooms. It may be added that the building containing the various shops just named was built originally for a store-house, and is therefore not altogether such a building as would have been designed for the various purposes.

On the south side of the basin are the iron and brass founderies, the pattern-shops, offices, and drawing-rooms. The iron-foundery is 200 feet long by 45 feet wide, with 30 feet height from the floor to the bottom of the overhead cranes. The building is of brick, with an iron-frame roof, slated. It is well lighted, and the internal arrangements and appliances are complete. It contains overhead travelers; also wall-cranes. The cupolas are conveniently placed, and there are hydraulic lifts for raising stock to the platforms. At right angles with this building is another, of equal dimensions, two stories high. On the lower floor of this building are the foundery cleaning-rooms and foundery store-rooms. On the upper floor are the pattern-shop and pattern store-room in separate divisions. The pattern-shop proper occupies the larger portion of the floor, and is excellent in its detail of arrangement. A complete number of wood-working machines are arranged on one side of the room, and the work-benches on the opposite side. The pattern-shop storage-room has a complete distribution of patterns. They are contained in cases with iron frames and supports, each case having its special patterns so marked, and each pattern its special mark in plain letters. The brass-foundery is also excellent. It is one story high, 105 feet long by 85 feet wide, divided longitudinally by a wall with a center opening, in which are located the cupolas and air-furnaces, arranged to discharge the metal into the division prepared for making the heaviest castings. In the opposite division are the crucible furnaces. This foundery is the largest and most complete of its kind in any of the dock-yards, and in it are made heavy brass castings for other yards. Drawings of both founderies have been submitted to the department.

On the west side of the machine and boiler shops are the smithery and forge. These occupy a large brick building, with a metal roof, and well lighted from the sides. It contains about 120 smithy fires, ranged in a quadrangle, with separate pipes from each pair of fires to carry off the smoke and gases. The forge is amply supplied with steam-hammers, furnaces, cranes, and appliances for ordinary work. Heavy shafts are not made in dock-yards. Besides this smithery, there is a smaller one on the south side of the basin.

This concludes notice of the works denominated steam factories, and no more need be said of them beyond the fact that they are of insufficient capacity, and the location of some of the buildings is objectionable. An additional building is, however, being erected next adjoining the machine and boiler shops as an extension, and a new factory is projected to be built in the not distant future.

The next building that claims attention is the one containing the block-making machinery invented by the celebrated engineer, I. K. Brunel. This machinery is the most complete in detail of the several machines for making different parts of blocks; it is perfect in combination, and up to the present day has not been superseded or surpassed; by it the wooden blocks for all vessels of the royal navy have been made for more than forty years. The only other building for mechanical purposes to be noticed is the saw-mill, which is two stories high, the lower floors containing all the usual sawing-machines, while the upper floors are provided with machines for light work.

Returning to the notice of the dry-docks, it is observed that leading from the steam-basin to the harbor, in a southwesterly direction, is the camber, the large dry-dock spoken of before. It is 644 feet long on the bottom, and has three gates: one at the harbor entrance, one in the center, and one at the basin outlet. This arrangement admits of its being used as two docks or one, as may be desired. The other is the ship-basin, south of the camber, with an outlet directly from the harbor. It is used chiefly for placing vessels in to be fitted for sea or to be docked. Attached to it are four dry-docks, namely, Nos. 2′, 3′, 4′, and 5′. These docks were constructed in the days when ships of war were of short dimension. They are of the following lengths, all measured on the bottoms: No. 2′ is 220 feet 1 inch; No. 3′, 270 feet 8 inches; No. 4′, 285 feet 4 inches; and No. 5′, 208 feet 9 inches. South of the basin is dry-dock No. 1′, with inlet from the harbor; its length on the bottom is 229 feet 4 inches. North of the basin is dry-dock No. 6′, also with inlet from the harbor; its length on the bottom is 192 feet 8 inches. Dry-dock No. 9′ is in the northwest corner of the dock-yard; its length is 253 feet 4 inches, measured on the bottom. All the docks, building-slips, and basins are formed of granite blocks, and are constructed in the most substantial manner.

Between dry-dock No. 9′ and the camber are the building-slips, which are five in number. They are in a group, covered with substantial and well-lighted ship-houses; between some of them, in a line with the vessels building, are overhead traveling cranes and steam-hoisting appiiances for raising materials to the vessels under construction.

Having now briefly described the buildings and appliances in which the principal mechanical operations are carried on, it will only be necessary to name some of the other buildings and localities, and to add that all of the former have heavy walls and are capacious and sufficiently substantial. These places may be best mentioned in order. No. 1, boiler-house; and No. 3, office; No. 7, engine-house, containing engines and boilers for pumping out the docks; No. 10, guard and store houses; and No. 12, police-station; Nos. 13 and 14, store-houses; and Nos. 15, 16, and 17, timber-sheds and wood-work shops; No. 18, officers' houses, in which reside all the principal officers of the yard; and No. 19, admiral-superintendent's house; Nos. 20 and 21, store-houses; and No. 22, building for officers; No. 23, tar-making house; No. 24, water-tank;

and No. 25, rope-making house; No. 26, mast-making house; No. 27, boat-pond, or water area, in which ships' boats are floated; No. 28, store-houses; No. 29, rigging and sail loft; No. 30, chain and cable store; No. 33, gate entrance to the dock-yard.

Such is a brief description of the Portsmouth dock-yard of the present day. What its future is to be may be best seen from the plan mentioned, showing the proposed extension and works completed thereon.

CHATHAM.

This dock-yard is situated on the river Medway, about twelve miles from the mouth of the Thames. It has now grown in importance to be second only to Portsmouth among the naval dock-yards of the world. The position is advantageous, because, not being on the coast, but far up the river, to be reached by a hostile fleet the defenses of Sheerness must first be passed; secondly, the Medway, from Sheerness to Chatham, has the natural protection of high land on both sides, readily convertible into positions of defense. A chain of forts is to be constructed for the defense of the dock-yard, the Medway, and the London approaches; one of the forts is already in process of construction at Borsted, near Rochester, and another will be built at Cookham Hill. In all there are to be seven forts of a massive character, and to be armed with heavy artillery.

The old dock-yard proper has a considerable river frontage, and the several docks, ship-houses, buildings, and necessary avenues cover an area of 89 acres. By a vote of Parliament in 1863, St. Mary's Island, adjoining the yard, was appropriated, and in the following year the works of extension on it were commenced. This island contains about 300 acres, and the place having been somewhat analogous to League Island before the work of extension was commenced, lessons may be drawn from a study of its hydrographical features.

The work of extension was commenced in 1864, and the plan in the first edition of this report will show the progress made up to January, 1875. The total estimated cost of the work was £1,750,000 sterling, but it is now seen by the memoranda of the director of works of 1875–'76 that the total ultimate cost of the whole will be not less than £1,950,000, or $9,477,000 gold. A very large part of the work has been executed by contract and a small portion by convict labor. The above does not include any sum for the erection of buildings. The director gives as reasons for the ultimate cost exceeding the estimates, the extremely treacherous character of the soil, a large portion of the work having been executed in mud, water, and alluvial deposits, of very varying tenacity, at a great depth below the surface. The plan mentioned does not show docks or buildings of the old dock-yard. In describing them, therefore, as at present existing, it may be premised that there are no basins or works denominated steam factories, but for the construction of vessels there is ample provision. Fronting the river are four dry-docks and seven building-slips. The first and largest is of sufficient capacity to take in a vessel 390 feet long, and the second, one of 295 feet. The third and fourth are comparatively small. The dry-docks and slips are excavated from soft ground, but piled, and then built throughout with granite blocks, in a similar manner to those in the other British dock-yards.

The building-slips are all covered with substantial ship-houses, four of which are grouped side by side, having the appearance of one building roofed in four spans. They are constructed solely of iron and glass, afford ample light, and will endure, there being no danger of their destruction by fire.

In the intermediate spaces from slip to slip is a sufficient area to prepare materials for vessels building, the whole neatly paved or laid with wooden blocks, making a dry, hard, and comfortable floor, both to work on and travel over.

The ship-houses are provided with overhead traveling carriages, sufficiently elevated to be above the largest vessels constructed. These carriages travel from end to end of the building, and are adapted to pick up material from the ground and carry it to and place it in any part of the vessel. There is also a traveling-carriage between two of the slips adjoining that last mentioned, used for the purpose of moving timber which has been or is to be prepared. There are, in addition, minor facilities for the purpose of carrying materials to or from the vessels under construction. The want of basins in the Chatham dock-yard has always been a great inconvenience, the fitting and repairing of vessels taking place in the river, where the rise and fall of the tides is considerable and the room insufficient. This want is now supplied at the new yard. As before stated, there are as yet no steam factories in the yard, but there is an old ordinary machine-shop, two stories high; there is also a small ordinary iron-foundery, containing two medium-sized and one small cupola, a small brass-foundery, pattern-shop, and boiler-repairing shop. All these shops are used merely for jobbing purposes. The ship-smithery, constructed many years ago, is a brick building, in the form of a quadrangle, with an open middle court, each of its sides being about 250 feet in length; the plan of the building is good, and if properly lighted and ventilated, with a simple plan of carrying off the smoke and gas, the shop would be unobjectionable. The various other buildings and workshops in this dock-yard are, with two exceptions, nearly identical with those named or described as existing in the other royal dock-yards, the exceptions being a copper-rolling mill and saw-mill; attached to the latter is an oar-making machine, which turns out all the oars for the boats of the British navy. The copper-rolling mill contains the furnaces, machinery, and appliances for the manufacture of sheathing and bolts; it has turned out all the sheathing for the bottoms of the vessels of the navy. In addition to the copper-rolling machinery, &c., there are rolls and facilities for the manufacture of small-sized angle and bar iron.

DEVONPORT AND KEYHAM.

The dock-yards of Devonport and Keyham are situated on the southwest coast of England, and are employed principally as repairing yards.

Devonport dock-yard has existed for more than a century, and it was supposed to answer all the requirements of the line-of-battle sailing-ship period. But after the introduction of steam, it was thought desirable to create an entirely new dock-yard, namely, Keyham, which is three-quarters of a mile distant from Devonport. Both dock-yards are connected under ground by a tunnel, and one general superintendence controls the two. Devonport has been chiefly the department of wooden-ship building, storage, &c., while Keyham is the department of steam-machinery repairs and manufactures; the latter alone invites attention. Besides, as Keyham is a modern dock-yard, established not more than twenty years, its factories are greater in extent and more complete in detail than those of any other dock-yard in England or France.

Keyham dock-yard contains two basins and four dry-docks. The north basin is of sufficient depth to take in vessels of any size afloat.

Around the basin are all the needful mechanical appliances for raising weights to or from vessels. The south basin is entered from the harbor and also from the north basin. It possesses three attached dry-docks, one of which is 600 feet in length. As a matter of convenience, to facilitate the movement of weights to and from the vessels in dock, rail-tracks extend parallel with either side of the dock, while traveling steam-cranes, with boilers and engines attached, pass up and down the tracks and perform the work of carrying material to or from the vessels under repair or fitting out. One of the rails of this either side track is laid directly on the edge of the top surface of the dock, and the other on beams projecting over the edge and supported by uprights; the cranes are thus brought within reaching distance of the vessels. The other two dry-docks differ only in length and in the absence of mechanical facilities for raising weights.

The machine-works, founderies, smitheries, boiler-making and pattern shops, copper-making and tin shops (which in Great Britain are all denominated steam factories), are the chief features of Keyham dock-yard. They are grouped together in a quadrangle 890 feet long by 580 feet wide. The center area of the group is a space substantially covered with a slate roof, supported on cast-iron columns and lighted from the top. The covered center area is formed into divisions by seven longitudinal rows of columns. Rail-tracks extend through two of the divisions of the buildings, and provision exists for steam traveling cranes. The whole of the covered center area admits of being appropriated for workshops, and some of it has been partitioned off and so appropriated. Within these appropriations are the smithery and forge, the coppersmiths' shop and the boiler-making smiths' shop, the smithery and forge occupying a division at the south end.

SHEERNESS.

This dock-yard is rated as second class, and is employed as a repairing and fitting-out yard. There are not any extensive steam factories in it, and it contains only one building-slip for the construction of small vessels. There are, however, three basins and five dry-docks, and all necessary facilities for repairing and equipping vessels for sea.

PEMBROKE.

This dock-yard, located on the coast of South Wales, is employed exclusively as a building-yard for hulls and accessories of hulls. Some of the heaviest armored ships have been built here.

Briefly described, such are the dock-yards of the British navy. Any one of them is worthy of an inspection, and in some the *tout ensemble* is imposing. The number of stone dry-docks at present completed in the five yards is thirty-seven. The number of building-slips having ship-houses over them is thirty-one, and the several basins have an aggregate water-area of about 130 acres.

Neither ordnance nor provisions are supplied from the dock-yards. The armaments for all vessels of the navy, as well as for fortifications and the army, are manufactured and supplied from the great arsenal at Woolwich. Provisions for the ships at home are supplied from the victualing-yards at Deptford, Gosport, Plymouth, and Haulbowline.

In addition to the large naval dock-yard facilities, the British government can in emergencies command any of the private iron-ship-building yards on the Clyde, the Tyne, the Mersey, the Thames, and other rivers, numbering in the aggregate about fifty; also as many engine-factories in the kingdom as may be desired, besides any number of stone dry-docks belonging to private companies.

ADMINISTRATION OF BRITISH NAVAL DOCK-YARDS.

The administration of the dock-yards is conducted under the system of civil employment. The officer in charge of each yard is called the superintendent. The term "command" or "commanding officer" is on no occasion used, and there is no "aid or executive officer" to the superintendent.

After the superintendent, the principal officers of the yard are named in the order of precedence as follows:

The master attendant.
The chief constructor.
The chief engineer.
The assistant master attendant.
The storekeeper.
The accountant.
The cashier.
The superintending civil engineer.
The chaplain.
The surgeon.

THE SUPERINTENDENT

in the largest yards is a rear-admiral, and in the small yards he is a captain. He has full and complete authority over all officers and other persons whomsoever, employed in the dock-yard, and control over every part of the business carried on therein. In his absence the captain of the reserve is charged with the duties, and in the absence of both these officers, "the other officers are authorized to take charge of the yard, according to precedence on the list above named." The duties of every officer, foreman, and other person having authority are defined in great detail in the admiralty dock-yard instructions.

The superintendent is required to hold at his office every working day, at 9½ o'clock a. m., a meeting of the officers above named (chaplain and surgeon excepted), and to have read to them all orders and dispatches, except such as may be marked confidential. At this meeting the captain of the steam reserve is also to be present.

Besides this method of promptly making all the principal officers daily acquainted with the business to be done, each one is furnished when necessary with copies, or extracts therefrom, of the orders pertaining to his respective department.

MASTER ATTENDANT.

The duties of the master attendant are defined. He is to attend to the docking, grounding, and graving of all ships, by day or night. He has charge of the moorings at the port, of all the yard craft, consisting of tugs and vessels of all kinds, including coal-hulks. He attends to the loading of stores, he superintends the sail-loft, rigging-house, and all persons employed therein, besides having other duties connected with vessels; but he has nothing whatever to do with any mechanical department, or persons employed in them. His rank is staff captain, and in the largest yards he has one and sometimes two assistants.

CHIEF CONSTRUCTOR.

He is to cause all ships and vessels to be built and all work to be executed in strict accordance with the approved drawings and specifications. He has charge of the shipwrights and other workmen belonging to the ship-building branch. He has the direction of the boatswain and laborers under him. He has charge of the ship-smithery, mast-house, boat-house, joiners' shop, and all other workshops appertaining to the ship-building branch, and all foremen and workmen employed therein. His assistant is a constructor; but in the absence of the chief constructor, the chief engineer is charged with his duties.

THE CHIEF ENGINEER

has the superintendence of all the engines and boilers, cranes, shears, capstans, leading-blocks, turn-tables, weigh-bridges, and pumps in the dock-yard, also the machinery at the smitheries and saw-mills, and all other machinery in the yard. He is charged with the superintendence of the ropery, and all persons employed therein, and is responsible for the steam fire-engines used for supplying water in case of fire.

The foremen of the engineering workshops have, under their directions and those of their assistants, charge of the men and all the work carried on therein and on board vessels under repairs. These foremen are answerable for the conduct of the leading-men and workmen employed under them, and responsible for all work intrusted to their supervision.

In the absence of the chief engineer, his first assistant is charged with the duties.

STOREKEEPER.

There is but one storekeeping officer to a dock-yard. He has charge of all store-houses, receiving-rooms, plank-sheds, and other inclosures used for the storage of timber or goods. He gives receipts for all stores received, and takes receipts for all stores delivered, and is responsible for all articles in store, and that the accounts in relation thereto be correctly kept.

ACCOUNTANT.

He is to audit the amount paid for wages to the workmen and all other employés, such audit to comprise a verification of the accuracy of the authorities for the entry of each person on the muster and pay books, the rates of pay, the amounts cast up as due, and a comparison of the casualties of attendance recorded in the muster and pay books with those in the weekly returns of lost and extra time.

CASHIER.

Under his directions all workmen and laborers employed in the dock-yard are mustered by persons attached to his department, and the several timekeepers are placed in charge of the muster-galleries according to his assignment. Besides which, all other responsible duties necessarily belonging to such an office are discharged by him.

CIVIL ENGINEER.

The dry-docks, basins, and other important public works are constructed from plans prepared at Whitehall by the director, who is a colonel of the royal engineers. The civil engineer attached to the yard has charge of the architectural works, buildings, and houses, alterations and repairs of existing buildings and works, also the supply of water and gas, and the maintenance and management of the lines of railways in the dock-yard to which he is attached.

WORKMEN AND BOYS.

Two classes of workmen are employed, distinct from the various classes and grades in such establishments. One of these is termed "established men," and the other "hired men." The former are permanent, are not discharged except in cases of great emergency, and are pensioned after long and faithful service.

Great care is exercised in the selection of all workmen and boys. "The superintendent is strictly enjoined to give his active personal attention to the subject of the entry of men when wanted, with a view to prevent any abuse or irregularity; he is to use his best endeavors to check and prevent any attempts being made to support applications for employment in the yard by aid of interest either political or personal." The best men are to be taken, and they are entered on probation for a fortnight, with the understanding that they will be retained if at the end of the prescribed period they shall have proved themselves competent, but if otherwise, they are discharged.

Established men are received between the ages of twenty-one and thirty-five only. They must be passed by the medical officers, and thoroughly tested as to their qualifications as workmen before their appointments are confirmed. These men are eligible to promotion as leading-men, foremen, &c.

Boys are selected by competitive examination, and thoroughly tested before being made permanent. Hired men are also selected with much care, but they are liable to discharge in the event of the reduction of force.

POLITICAL.

Under the heading of "Rules for workmen" (dock-yard instructions) is the following paragraph:

> No interference, direct or indirect, is to be exerted over any person, whatever be his rank or station, in the matter of the elective franchise; and in the event of elections for members of Parliament, the men who may be qualified to vote are to be left to the exercise of their unbiased judgment, free from all influence, inquiry, let or hinderance; and no canvassing by or on the part of any candidate is to be permitted within the dock-yard upon any pretense whatever.

The pay of all employés is regulated by the admiralty; it is the same for each class or grade of men in every dock-yard, and it is not changed to suit varied conditions. The working-hours are regulated by Greenwich time, and are also the same in every yard. The muster of men is by ticket, both when entering and leaving the yard.

POLICE.

No marines or watchmen are employed in any dock-yard. The commissioner of the metropolitan police of London details a force of intelligent and well-trained men, sufficient for each yard. These men are under the orders of the superintendent, and are stationed at the gate and at such other judiciously-chosen places in the yard as the superintendent may direct. They perform all the duties of watching the public property in the dock-yard, and on board vessels building, and at the wharves, both by day and night. They wear the uniform of the London police, perform their duties thoroughly, and are highly respected.

PENSIONS.

All officers, clerks, foremen, leading-men, established workmen, and crews of the yard craft are granted pensions, retiring allowances, or gratuities, according to pay and length of service. Ten years' service will entitle a man to a pension or superannuation for injuries received, and after forty years' service he is allowed to retire with forty-sixtieths of his full pay, intermediate rates being graduated between these accordingly.

All the principal officers and clerks must be superannuated on attaining the age of seventy years, and all inferior officers and workmen are superannuated on attaining the age of sixty years.

All the principal officers of each yard are provided with most comfortable houses inside the walls, taking precedence therefor as they are named on the foregoing list.

The superintendent wears the uniform of his rank, but no professional officer wears a uniform on any occasion whatever.

The number of persons employed in each of the principal dock-yards averages from 7,000 to 8,000; about 1,500 of whom are under the supervision of the engineer department.

No foreigner is permitted to enter any of Her Britannic Majesty's dock-yards without authority given by the admiralty through application of the embassador or minister of the country to which the person belongs, made to the secretary of state for foreign affairs, naval *attachés* of the foreign embassies in London excepted. Foreigners, including officers, whether singly or in parties, are always accompanied throughout the periods of their visits by an officer detailed for the purpose, and while all necessary attention is given, and every proper facility is afforded for viewing the works and ships under construction, sketches or written notes are not allowed to be taken except by special permission.

The system of dock-yard administration very briefly sketched in the foregoing paragraphs is the best in existence.

The superintendent and principal officers are selected especially for their fitness, and not because they may have claims for shore-duty or otherwise. The inferior officers and workmen are employed solely on their merits, and are mainly well-trained, efficient, and faithful; besides which, the permanence of employment adds stability and gives character to the whole staff of employés.

One familiar and experienced as I am with dock-yard duties and with workshops generally, at home and abroad, cannot but feel impressed on observing the prompt and business-like way in which the duties are carried on in these great British naval establishments No idle officers or loitering persons are to be seen in any place within one of these yards at any time during working-hours.

The contrast between these admirably-conducted establishments and those of France and other continental countries, Germany excepted, is noticeable soon after entering the gates; there, and especially in France, numerous idle officers in uniform and loitering workmen may be seen in every department entered.

PART XIII.

THE FRENCH NAVY.

TABLE OF ARMORED SHIPS OF FRANCE; TABLE OF VESSELS BUILDING AND PROPOSED FOR THE FRENCH NAVY IN 1877; DOCK-YARDS OF CHERBOURG, BREST, L'ORIENT, ROCHEFORT, AND TOULON; PERSONNEL OF THE NAVY.

THE FRENCH NAVY.

Frenchmen have of late been compelled to assume the attitude of critical observers, if not of careful imitators of other naval powers, and particularly of England.

Driven by stress of circumstances and by force of competition from the proud position so long occupied and so eagerly contended for, of pioneers in armored-ship construction, they have had the wisdom and courage fairly to face their position, and to endeavor to make the best of the condition in which they are placed. With their earlier armored ships growing obsolete, and rapidly becoming worn out or unserviceable, and with very limited means at command, the naval authorities in France had no easy problem to solve when they endeavored to give the best direction to their expenditure in the new programme of the fleets, which was published with their estimates soon after the worst pressure of the Franco-German war had passed away. In arranging this new building programme, English types, more or less modified, but in the main reproduced, have been taken as models, and English systems of construction have been adopted in lieu of the discarded wood-ship building which previously was almost exclusively in use. Breastwork-monitors for coast defense, central-battery armored ships with iron hulls (with added barbette-batteries), fast unarmored ships with iron hulls, sheathed in wood and coppered, constitute the last additions to the French navy.

In reviewing their fleets, it is seen that they have perpetuated not merely the forms and approximate dimensions of their unarmored steam-fleet in their armored ships, but have also clung pertinaciously to the old broadside system of armament. Not a sea-going turret-ship has been built.

Their coast-defense vessels (*garde-côtes*) include six turret-vessels, each with a single turret; but all the other vessels in the armored fleet are on the broadside principle, and all sea-going ships are rigged. There are no representatives of the mastless sea-going type to match the English *Dreadnought*, *Devastation*, *Thunderer*, and *Inflexible*, or the Russian *Peter the Great*, or the Italian *Duilio* or *Dandolo*, or analogous vessels of other nations. The revolving turret has found no favor with the French for sea-going ships, although they have adopted, in many cases in association with a broadside armament, the plan of mounting the gun on a revolving platform, and allowing it to fire *en barbette* over a fixed armor wall or turret. There are in the large ships two of these open-topped turrets on either side of the vessel, and, true to their national instinct of systematic classification, they laid down simultaneously many ships of one design, or differing but in minor details. First of all, in 1858 three of the *Gloire* class were begun, corresponding to the English converted ship *Prince Consort* class, begun three years later; then, after the iron ship *Couronne*, no less than ten vessels of the *Flandre* class, modified from the *Gloire* and not much better, were commenced; while some slight variety was introduced by building a few coast-defense vessels.

In the design of their smaller vessels (*corvettes cuirassées*) for foreign service, similar uniformity was maintained. After building the first, the *Belliqueuse*, some few changes were made, and then on the amended design (of which the *Alma* is the type) no less than ten vessels were built.

The *Océan* class of frigates, again, contains no less than five, which may be fairly termed sister ships. So, taking the three classes represented by the *Flandre*, *Alma*, and *Océan*, more than twenty-five ships, or nearly one-half the entire French fleet, will be found grouped therein.

The English admiralty have pursued a different and a wiser course. Instead of spending their great resources on the construction of ships which were mere *fac-similes* of each other, they have, with rare exceptions, in successive designs made onward steps in offensive and defensive power; from the very first the English designers struck out an independent course.

The *Warrior* was designed in 1858, with an iron hull one-half as long again and 2,000 tons heavier than the largest screw three-decker that had been built before her; and she was considerably faster than any war ship previously built.

In some respects this venture proved a success, for the *Warrior* is still, after seventeen years' service, sound and in good order; while all wooden armored vessels, both of previous and subsequent date, are counted out or relegated to harbor service. The precedent established in the design of the *Warrior* has been followed; novelties being daringly introduced and new designs constantly brought forward, until it is difficult to place more than three or four vessels in any one class, by far the greater number of designs having been given shape in not more than one vessel.

The most powerful fighting-ships of the French are the armored vessels recently put afloat and those they have yet to complete. Of these the *Redoutable*, launched September 18, 1876, is the most formidable of the number completed. She has a length of 330 feet; beam, 66 feet 5 inches; draught of water forward, 23 feet 7 inches; aft, 25 feet; and a displacement of nearly 9,000 tons. She is built of iron and steel, has a ram bow, is armored with iron from 8 to 10 inches in thickness on the water-line, and horizontal armor is used of sufficient thickness to make the decks proof against projectiles ordinarily used. She is armed with heavy rifled guns of large caliber, and the speed is represented to be equal to that of the heavy ships of the British navy.

Two other ships of still greater power, viz, the *Foudroyant* and *Dévastation*, are rapidly advancing toward completion, the first at the Toulon dock-yard and the other at L'Orient; also two others of the same type, but not yet named, have been laid down, one at Brest and the other at Industrie, making in all five powerful armored ships of the first class.

Some features of more than ordinary importance are to be noticed in these vessels. The French naval authorities have broken fresh ground, and are in some respects making steps in advance of other powers.

The *Dévastation*, about being completed, is an example of their recently-constructed armored sea-going ship. Her length is 371 feet 7 inches; beam, 66 feet 5 inches; mean draught of water, 23 feet 10 inches; displacement, 9,630 tons; indicated horse-power, 6,000; estimated maximum speed, 14 knots per hour; and armor on the water-line 15 inches thick.

The features of most interest are the great guns to be carried on the broadside and the system of working them. There is a central square citadel, with the corners truncated. In these corners are to be placed four guns, each 46 tons in weight, to be worked by hydraulic machinery, the carriages and gear having already been delivered by the Elswick firm. These guns are rifle breech-loaders of the French pattern, and are the heaviest by 15 tons yet mounted on the broadside of a ship, besides which it is the first attempt to use hydraulic power for working guns mounted in this way. It was in fact only by the determination to apply machinery that the French have been enabled to work guns 46 tons in weight on the broadside.

The firing projectile to be used in these guns will weigh about 915 pounds and the charge of powder about 200 pounds. The estimated muzzle-velocity will be 1,475 feet per second.

When completed, these ships will carry heavier metal than any vessels afloat, the *Inflexible* and Italian monsters excepted.

The sister ship *Foudroyant* will also soon be completed, and it is intended that her armament shall be of the same character and weight.

Next in rank to these, built for aggressive warfare, are the *Richelieu*, *Colbert*, *Trident*, and *Friedland*. The first two were launched in 1875, and the other two upward of a year ago. These ships have a length of 314 feet; beam, 57 feet 3 inches; draught of water, forward, 24 feet 7 inches; aft, 27 feet 10 inches. The first-named two were laid down in 1869, so that the time occupied in building them was six years, and as a consequence they are not as heavily armored as later-designed ships; but they are, nevertheless, formidable vessels, carrying on the main decks heavy rifled guns, besides guns of considerable power in the two side turrets; also lighter guns under the armored forecastle; and the speed, as represented by the officers of the *Richelieu* during my visit to that ship, at the time the trial trip was made—February, 1876—was about 13 knots per hour.

The next class of modern armored vessels, known as the *garde côte* type, of which six have been ordered, are represented by the *Tonnerre*, recently completed, and put on her sea trials in February of this year. The *Tonnerre* is 241 feet 6 inches in length, 57 feet 9 inches beam; mean load draught 21 feet, and displacement 5,495 tons. She is armored with iron 11[illegible] inches thick on the water-line, and fitted with a snout-ram. The single turret is 27 feet 6 inches in diameter, plated with iron 12 inches thick, except at the gun-ports, where it is increased to 14 inches, and revolves around a central shaft 4 feet 6 inches in diameter. The armament at present mounted in the turret consists of two rifle guns of the French breech-loading pattern, each 23 tons in weight, with a caliber of nearly 11 inches.

The leading feature in this vessel is the introduction of the Rendel hydraulic system of machinery for rotating the turret; also for working the guns. The advantage claimed for this system in the rotation of the turret consists in the fact that the speed can be regulated and controlled with a nicety hitherto found unattainable. As the revolving of the turret is the method by which the guns are trained laterally for aim, the power of nice adjustment in the operation is of much importance. The guns are also mounted and worked by hydraulic machinery which runs them in and out, arrests them in recoil, and lifts or lowers them bodily.

During the trials twenty-nine rounds were fired, ten of them over the bow and the remainder over the stern, except two rounds the guns were fired together. The revolving action of the turret was also tested when the vessel was rolling and found to be under complete control, and the working of all the machinery satisfactory.

It is intended to mount on the same carriages guns weighing 26½ tons to fire projectiles of 551 pounds, having a muzzle velocity of 1,476 feet per second.

Hydraulic power is further employed in the *Tonnerre* for hoisting the anchors by means of a capstan-engine. The reported maximum speed of the vessel on the measured mile is 15 knots per hour.

The next size of the modern armored vessels is rated second class, of which two, the *Triomphante* and *Turenne*, are under construction, the first at Rochefort and the second at Cherbourg, and three others have been laid down at the dock-yards. Besides, for coast defense, two of the first class, viz, the *Fulminant* and *Furieux*, are building at Cherbourg, and three others of the same class have been laid down or ordered to be built; also, three of the second class are building, viz, the *Tempête*, *Tonnant*, and *Vengeur;* besides two first-class gunboats.

When all of these vessels shall have been completed the French will possess a formidable modern fighting fleet, ready, perhaps, to meet their traditional foe, the German.

The dimensions and particulars of the armored ships will be found in the first of the following tables.

The French naval authorities are also progressing with the reconstruction of their unarmored fleet. The *Duquesne* and *Tourville*, of the rapid type, have been completed, also other vessels; while the *Duguay-Trouin* is nearly completed, and the *La Pérouse* and *Villars* are in process of construction; besides, five others of the second class have been ordered, and two of the third class are building.

Prior to 1873 the only French naval sea-going armored ships in service, built of other materials than wood, were three in number, viz, the *Friedland*, *Héroïne*, and *Couronne*. Neither iron nor steel had hitherto been used in the navy for ship construction to any considerable extent; the use of steel was limited to masts, boats, and very small vessels; but recently both iron and steel as materials for ship construction have rapidly gained favor. In the large new armored ship *Redoutable*, built at L'Orient, steel has been employed for the frames, beams, deck-plating, bulkheads, the plating behind the armor, and the inner bottom; consequently only the outer bottom and the rivets throughout the ship are of iron. Two other large armored ships, viz, the *Tempête* and *Tonnerre*, are building at Brest and L'Orient with the same distribution of material. The steel has been produced mainly at Creuzot and Terre Noire. All the steel employed in France for naval vessels is made under government inspection at the various manufactories, and it is not tested in the dock-yards upon receipt. The quality is defined by the conditions of official regulations lately issued by the minister of marine, as follows:

Steel plates.—Steel plates are classified under five heads. For plates of the first class the lowest ruling rates will be accepted; for those in the four other classes 2, 4, 6, and 8 francs per hundred kilograms ($3.88, $7.76, $11.64, and $15.52 per ton) are allowed. The thicknesses of the plates thus classified range from .059 inch to 1.18 inches, and the dimensions from 12 feet 4 inches by 3 feet 11¼ inches to 19 feet 8½ inches by 4 feet 7½ inches.

Testing.—The following tests will be required to ascertain the extension of the metal, and its ultimate strength, both longitudinally and transversely, the recorded results in all cases being the mean of at least five independent tests. Test pieces are to be cut from a certain percentage of plates taken at random, which are to be subjected alike to each class of test.

These sample bars are to have in each case a width $1\frac{3}{16}$ inches, excepting those taken from plates less than .197 inch thick, in which cases the width will be reduced to .787 inch, and for plates .708 inch and under, the width will be reduced to the thickness of the plate. The length of the portion submitted to test will be in all cases 7.875 inches, and the test bars will always be annealed. The initial test load will be such as to produce a strain equal to four-fifths of the load required to rupture the plate. This initial load will be kept on the test piece for a time of five minutes. Additional weights will then be added at equal intervals of time, probably half-minute periods. The corresponding extension for each increment of load will be carefully noted, and measured on the original length of 7.875 inches. The ultimate extension will be that produced at the moment of rupture.

These test bars to be passed should not break under the initial load, nor give any ultimate extension less than eight-tenths of the maximum final extension mentioned above.

The minimum loads in tons per square inch of the original section and the minimum percentages of extension are given in the following tables. For plates the mean results which should be compared with the table are those which have been obtained in the direction of least resistance.

TABLE I.—*Steel plates.*

Thickness of plate in inches.	For constructive purposes.		For boilers.	
	Average maximum load.	Average final extension.	Average maximum load.	Average final extension.
	Tons per sq. in.	*Per cent.*	*Tons per sq. in.*	*Per cent.*
.019	30.3	10		
.07 to .118	30.3	12	....	....
.118 to .157	30.3	14	....	....
.157 to .197	29.5	16	....	....
.197 to .236	29.5	18	....	....
.236 to .315	29.0	20	27.0	25
.315 to .787	29.0	20	27.0	26
.787 to 1.181	28.4	20	25.8	25

TABLE II.—*Strips and cover-plates.*

Thickness of plate in inches.	Longitudinal.		Transverse.	
	Maximum load.	Maximum extension.	Maximum load.	Maximum extension.
	Tons per sq. in.	*Per cent.*	*Tons per sq. in.*	*Per cent.*
.157 to .236	31	18	28.4	16
.236 to .630	31	22	28.4	18
.630 to 1.18	31	22	27.0	17

Hot tests.—These tests will be made with sample plates of suitable dimensions, and consist in stamping a dished cavity, the side of the plates preserving its original plane. The diameter of this cavity is to be equal to forty times the thickness of the plate, and the depth will be ten times this thickness; the flat edge to be joined to the cavity by a curve the radius of which is not to be greater than the thickness of the plate. Moreover, plates more than .197 inch thick will be stamped with a flat-bottomed depression with square angles and straight sides, the diameter of the bottom to be thirty times the thickness of the plate, and the depth ten times the same thickness. The bottom of this cavity will be pierced with a round hole with the metal forced perpendicularly beyond the bottom of the recess. The diameter of the hole to be twenty times the thickness of the plate, and the height of the sides five times the same thickness. All of the corners will be rounded with a curve not of greater radius than the thickness of the plate. The pieces thus tested with every precaution which the working of steel requires, must show no signs of yielding or cracking even when cooled in a brisk current of air.

Tempering tests.—For these tests bars 10.24 inches long by 1.58 inches wide will be cut from the plate longitudinally as well as transversely. These strips will be heated uniformly to a slightly dull cherry-red, and then tempered in water at a temperature of 82°. Thus treated they must be bent in the testing-machine to a curve of which the minimum radius is not greater than the thickness of the bars. These same bars, when the corresponding plates are to be used for boilers, will be bent double in the press without showing any traces of fracture, and in such a way that the halves of the plate may be in contact. The sides of the bars thus tested, if rounded, can be squared up with a soft file. Plates not coming up to these tests will be rejected.

Angle, profile, **T** *and* **I** *irons.*—To ascertain the qualities of different classes of profile bars three series of tests will be imposed; 1, cold tests; 2, tempering; and 3, hot tests.

1. *Cold tests.*—These have for their object to ascertain the ultimate strength and properties of extension of the metal. A certain number of pieces will be cut from the webs of bars taken at random, and care will be taken that the cross-sections are almost rectangular, the thickness being that of the web and the width 1.18 inches, except for sections less than .197 inch thick, in which cases the width will be reduced to .787 inch, and for those more than .71 inch thick, in which the width will be reduced to the thickness of the plate. The length of the samples tested to be 7.87 inches. The bars will be subjected to tensile strains either by direct weights or by levers, the load increasing up to the point of rupture. The initial load will be such as to produce a strain equal to eight-tenths the breaking strain calculated upon the basis in the following table. The first load will be kept on during five minutes, and the additional loads will be added at half-minute intervals. The ultimate extension is that produced at the moment of rupture, and no samples will be passed which show an extension less than the required amount.

The lowest average loads per square inch of original section under which the bars should break when tested, and the minimum corresponding extensions, are as follows:

Thickness of web in inches.	Angle and profile sections.		T-sections.		I-sections.	
	Average minimum load.	Average ultimate extension.	Average minimum load.	Average ultimate extension.	Average minimum load.	Average ultimate extension.
	Tons per sq. in.	*Per cent.*	*Tons per sq. in.*	*Per cent.*	*Tons per sq. in.*	*Per cent.*
.118 to .157	31	18	29.23	18	29.5	16
.157 to .236	31	20	29.23	20	29.5	16
.236 to .630	31	22	29.23	20	29.5	18
.630 to .984	31	20	29.23	20	29.5	18

2. *Temper tests.*—For these tests bars will be cut from the webs of the various sections 10.24 inches long and 1.575 inches wide. The sides of these pieces must not be rounded, but they may be squared up with a soft file. The bar will be heated uniformly to a dull cherry-red, and then cooled in water at 82°. Thus tempered they ought to take under the action of the press a permanent curvature, the inner radius of which must not be greater than one and a half times the thickness of the plate.

3. *Hot tests.*—The angle-bars will be subjected to the following tests: With one piece cut from the end of a bar taken at random from each parcel, a ring shall be made, so that while one web preserves its original plane, the other shall be bent into a circle the inner diameter of which shall not be more than three times the width of the flat web. A piece cut from another bar shall be opened until the inner faces of the webs shall be practically in the same plane, and a third piece shall be closed until the webs come into contact. The samples thus treated must show no cracks or other imperfections. Plain **T**-bars shall be subjected to the following tests: A piece cut from the end of a bar taken at random from a parcel shall be bent into a semicircle, while the vertical web preserves its original plane. The interior diameter of this curve shall not exceed four times the height of the bar. In a piece cut from another bar taken from the same parcel, a horizontal slot will be cut in the middle of the web, of a length equal to the depth of the bar; a hole will be drilled at each end of the slot, in order to prevent the plate from tearing under the test, and that part of the web beneath the slot will then be bent until it forms an angle of 45° with the remainder. Care will be taken to keep the bent portion in the same line with the rest of the web, to which it will be connected by a bend of small radius. No cracks or other imperfections must be developed under this test. Bulb and double **T**-bars will be tested as follows: At the end of a bar taken at random from the parcel, a horizontal slot will be cut in the center of the web, equal in length to three times the depth of the bar, holes being drilled at the ends as before; then the bar will be bent at one or several heats, until one part thus opened forms an angle of 45° with the other, the portion thus bent being kept in the same plane. In bulb-sections the portion bent will be that carrying the bulb. In all cases the angles must be connected with the straight parts by curves of small radius. All samples failing under these tests will lead to the rejection of the parcel from which they were selected.

Armored ships of France.

Name and class of ship.	Vessel.							Armament.			Machinery.				Cost.	
	Length between perpendiculars, in feet and inches.	Extreme breadth, in feet and inches.	Depth of hold, in feet and inches.	Immersed midship section, in square feet.	Displacement, in tons.	Draught forward, in feet and inches.	Draught aft, in feet and inches.	Number of guns and description.	Height of battery above water, in feet and inches.	Thickness of armor at water-line, in inches.	Number of screws.	Indicated horse-power on trial.	Maximum speed, in knots per hour.	Makers of engines.	Machinery, in dollars.	Ship, completely equipped, in dollars, gold.
Line-of-battle cruisers:																
Redoutable	330 0	66 5	43 2	1,353	8,658	23 7	24 11	8 10.63-inch; 4 5.5 inch	13 7	10	1	6,000	14	Creuzot	....	....
Foudroyant (*a*)	371 7	66 5	....	....	9,630	23 0	24 8	4 12.6-inch; 2 10.6-inch; 8 5.5-inch.	....	....	2	6,000	14	....	....	....
Dévastation (*a*)	371 7	66 5	....	....	9,630	23 0	24 8	......do	2 11	14.9	2	6,000	14	Creuzot	....	....
Richelieu (*b*)	314 0	57 3	28 1	1,249	8,269	24 7	27 10.6	6 10.63-inch; 7 9.45-inch; 6 5.5-inch.	13 4	8.66	2	....	14	Indret	....	....
Colbert	314 1	57 2½	28 1	1,259	8,167	21 7	27 10.6	8 10.63-inch; 1 9.45-inch	13 4	8.66	1	....	....	....	....	....
Trident	314 1	57 3	28 1	1,259	8,164	24 7	27 10½	6 5.5-inch; 4 mitrailleuses	13 4	8.5	1	4,800	....	Indret	....	....
Friedland	314 1	57 3	28 1	1,259	8,164	24 7	27 10½	8 18-ton; 2 12-ton	13 4	8.5	1	....	....	....	....	....
Marengo (*c*)	282 10	57 7	34 8	1,267	7,360	24 10	29 1½	4 10.63-inch; 4 7.48-inch.	11 9	7.75	1	3,673	13.5	Forges et Chantiers	260,680	1,302,000
Océan (*c*)	282 10	57 7	34 8	1,267	7,360	24 10	29 1½	4 18-ton; 4 7-ton	11 9	7.75	1	3,673	13.5	....do	260,680	1,302,000
Suffren (*c*)	282 10	57 7	34 8	1,267	7,360	24 10	29 1½	......do	11 9	7.75	1	3,673	13.5	....do	260,680	1,302,000
Solferino (*d*)	284 0	57 3	27 3	1,225	6,786	24 1	28 0	52 6.3-inch	6 3	4.72	1	4,019	12.86	....do	278,320	1,162,280
Flandre (*e*)	258 10	55 9	27 10	1,058	5,703	23 6½	26 10	8 12-ton; 4 7-ton	7 3	5.9	1	3,500	13.9	....do	299,880	1,066,240
Gauloise (*e*)	258 10	55 9	27 10	1,058	5,703	23 6½	26 10	......do	7 3	5.9	1	3,500	13.9	....do	299,880	1,066,240
Guyenne (*e*)	258 10	55 9	27 10	1,058	5,703	23 6½	26 10	......do	7 3	5.9	1	3,500	13.9	....do	299,880	1,066,240
Magnanime (*e*)	258 10	55 9	27 10	1,058	5,703	23 6½	26 10	......do	7 3	5.9	1	3,500	13.9	....do	299,880	1,066,240
Provence (*e*)	258 10	55 9	27 10	1,058	5,703	23 6½	26 10	......do	7 3	5.9	1	3,500	13.9	....do	299,880	1,066,240
Revanche (*e*)	258 10	55 9	27 10	1,058	5,703	23 6½	26 10	......do	7 3	5.9	1	3,500	13.9	....do	299,880	1,066,240
Savoie (*e*)	258 10	55 9	27 10	1,058	5,703	23 6½	26 10	......do	7 3	5.9	1	3,500	13.9	....do	299,880	1,066,240
Surveillante (*e*)	258 10	55 9	27 10	1,058	5,703	23 6½	26 10	......do	7 3	5.9	1	3,500	13.9	....do	299,880	1,066,240
Valeureuse (*e*)	258 10	55 9	27 10	1,058	5,703	23 6½	26 10	......do	7 3	5.9	1	3,500	13.9	....do	299,880	1,066,240
Héroïne (*e*)	258 10	55 9	27 10	1,058	5,703	23 6½	26 10	......do	7 3	5.9	1	3,500	13.9	....do	299,880	1,066,240
Gloire (*d*)	252 3	55 9	26 9	1,068	5,530	23 1	27 10	36 6.3-inch	6 2	4.72	1	2,537	12.85	....do	246,960	919,240
Couronne	260 0	54 3	....	....	....	23 0	25 0	10 7-ton; 2 12-ton	....	....	..	....	....	....	....	....
For ordinary station service: (*c*)																
La Galissonnière	230 0	45 9	....	....	3,445	17 5	21 0	4 7½-inch; 4 6¼-inch	....	6	1	3,963	12.8	Mazeline	274,912	....
Triomphante (*a*)	230 0	45 9	....	....	3,445	17 5	21 0	......do	....	6	1	3,963	12.8	....do	274,912	....
Turenne (*a*)	230 0	45 9	....	....	3,445	17 5	21 0	......do	....	6	1	3,963	12.8	....do	274,912	....
Victorieuse	230 0	45 9	....	....	3,445	17 5	21 0	......do	....	6	1	3,963	12.8	....do	274,912	....

a In process of construction.
b A central battery and two open-top fixed turrets; four cylinders to each screw.
c Central battery and four fixed turrets, with guns *en barbette*; ram bow.
d Broadside-battery; ram bow.
e A central battery and four fixed turrets, with guns *en barbette*.

Armored ships of France—Continued.

Name and class of ship.	Vessel.							Armament.			Machinery.				Cost.	
	Length between perpendiculars, in feet and inches.	Extreme breadth, in feet and inches.	Depth of hold, in feet and inches.	Immersed midship section, in square feet.	Displacement, in tons.	Draught forward, in feet and inches.	Draught aft, in feet and inches.	Number of guns and description.	Height of battery above water, in feet and inches.	Thickness of armor at water-line, in inches.	Number of screws.	Indicated horse-power on trial.	Maximum speed, in knots per hour.	Makers of engines.	Machinery, in dollars.	Ship completely equipped, in dollars, gold.
For ordinary station service—Cont'd.																
Alma	230 0	46 3	22 10	781	3,675	18 10	23 9	6 7-ton	6 3	6	1	1,897	11.8	Mazeline	127,776	495,880
Ar[illegible]de	230 0	46 3	22 10	781	3,675	18 10	23 9	6 7-ton	6 3	6	1	1,897	11.8	do	127,776	495,880
Atalante	230 0	46 3	22 10	781	3,675	18 10	23 9	do	6 3	6	1	1,897	11.8	do	127,776	495,880
Belliqueuse	230 0	46 3	22 10	781	3,675	18 10	23 9	do	6 3	6	1	1,897	11.8	do	127,776	495,880
Jeanne d'Arc	230 0	46 3	22 10	781	3,675	18 10	23 9	do	6 3	6	1	1,897	11.8	do	127,776	495.880
Montcalm	230 0	46 3	22 10	781	3,675	18 10	23 9	do	6 3	6	1	1,897	11.8	do	127,776	495,880
Reine B[illegible]he	230 0	46 3	22 10	781	3,675	18 10	23 9	do	6 3	6	1	1,897	11.8	do	127,776	495,880
Thétis	230 0	46 3	22 10	781	3,675	18 10	23 9	do	6 3	6	1	1,897	11.8	do	127,776	495,880
Coast-defenders:																
[illegible]erre (*a*)	241 6	57 9	23 7	1,098	5,495	20 8	21 4	2 12.5-inch	13 5½	11¾	1	3,400	15			
Fulminant (*b*)	241 6	57 9	23 7	1,098	5,495	20 8	21 4	do	13 5½	11¾	1					
Furieux (*b*)	241 6	57 9	23 7	1,098	5,495	20 8	21 4	do	13 5½	11¾	1					
T[illegible]te (*b*)	241 6	57 9	19 2	966	4,452	16 9	16 9	do	13 1½	11¾	1	3,400	13.5			
Vengeur (*b*)	241 6	57 9	19 2	966	4,452	16 9	16 9	do	13 1½	11¾	1					
Tonnant (*b*)	241 6	57 9	19 2	966	4,452	16 9	16 9	do	13 1½	11¾	1					
Bélier (*c*)	216 6	52 6	17 7	831	3,313	17 0	17 0	2 1[illegible]ton	17 5	8	2	1,508	12.5	Creuzot	121,968	
Bouledogue (*c*)	216 6	52 6	17 7	831	3,313	17 0	17 0	do	17 5	8	2	1,508	12.5	do	121,968	
Cerbère (*c*)	216 6	53 1	17 7	831	3,702	18 7½	19 7½	do	17 5	9	2	1,508	11.3	do	121,968	668,736
Tigre (*c*)	216 6	53 1	17 7	831	3,702	18 7½	19 7½	do	17 5	9	2	1,508	11.3	do	121,968	668,736
Taureau	216 6	53 1	17 7	831		18 7	19 7	1 18-ton					13			
Onondaga (*d*)	228 7½	51 2	13 2	549	2,551	12 3½	12 7½	2 9.45-inch	1 2½	5½	2	613	7.07	Quintard, N. Y.	124,269	978,040
Arrogante	144 4	48 5	10 3	385	1,338	8 8	8 8	9 6.3-inch	4 9	4¾	2	484	6.75	Creuzot	39,690	225,400
Implacable	144 4	48 5	10 3	385	1,338	8 8	8 8	do	4 9	4¾	2	484	6.75	do	39,690	225,400
Opiniâtre	144 4	48 5	10 3	385	1,338	8 8	8 8	do	4 9	4¾	2	484	6.75	do	39,690	225,400
Embuscade	129 7	51 10	13 0	513	1,422	10 2	10 10	4 7-inch	3 1	5½	2	433	8.5	do	39,104	329,280
Imprenable	129 7	51 10	13 0	513	1,422	10 2	10 10	do	3 1	5½	2	433	8.5	do	39,104	329,280
Protectrice	129 7	51 10	13 0	513	1,422	10 2	10 10	do	3 1	5½	2	433	8.5	do	39,104	329,280
Refuge	129 7	51 10	13 0	513	1,422	10 2	10 10	do	3 1	5½	2	433	8.5	do	39,104	329,280
Floating battery No. 5	72 0	25 3	5 7	77	140	3 3½	3 3½	2 6-inch	1 1¾	3	2	94	4.21	Forges et Chantiers	8,131	29,160
Batteries Nos. 8, 9, 10, 11	88 7	29 6	5 7	156	280	5 7	5 7	2 5½-inch	2 0	3	2	99	5.37	do	11,228	52,488

a Turret-ship; ram bow. *b* In process of construction; turret-ships. *c* Single turret-ships, ram bow. *d* Turret-ship.

NOTE.—The Rochambeau is no longer borne on the list.

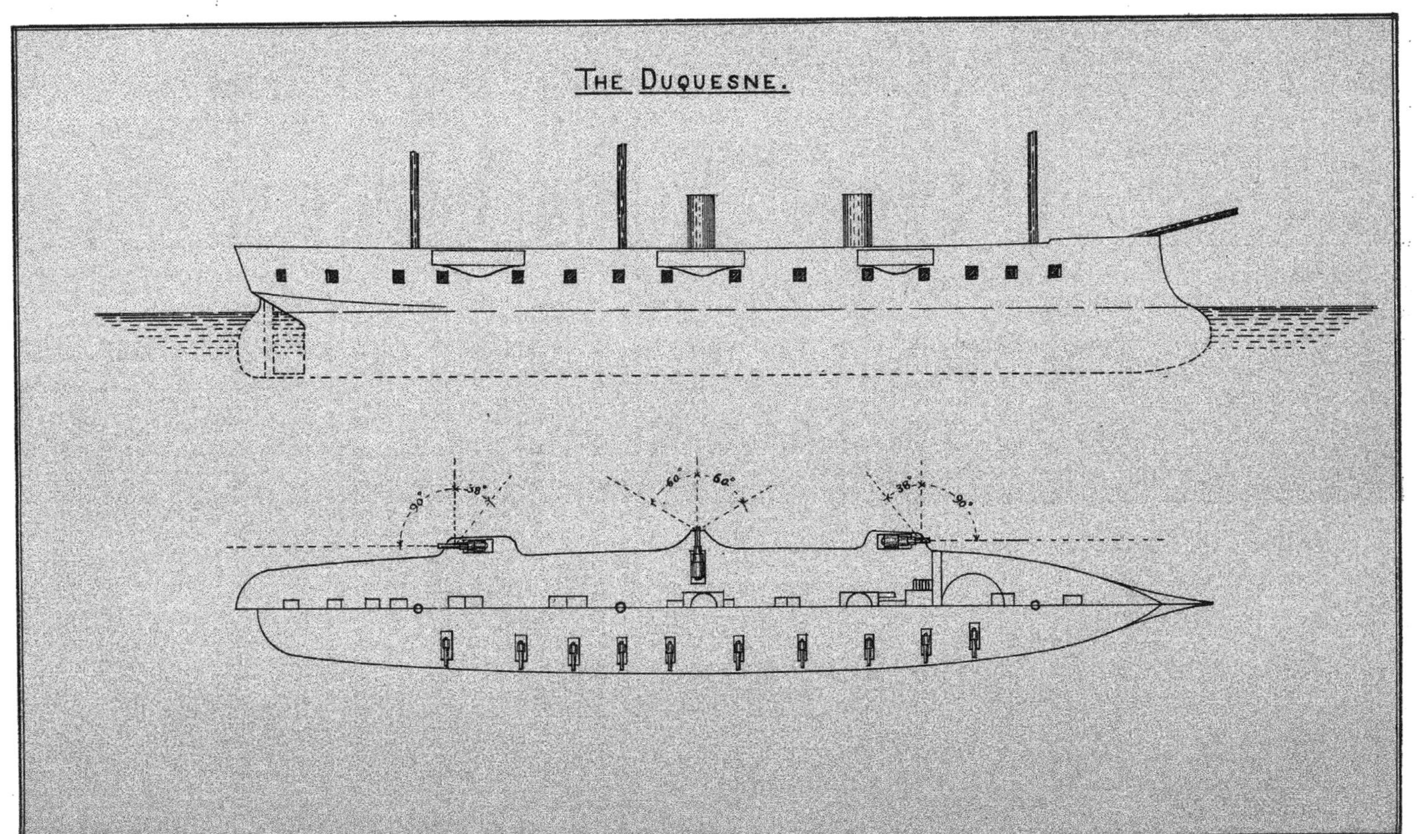
THE DUQUESNE.
90°
38°
60°
60°
38°
90°

THE DUQUESNE AND TOURVILLE.

The preceding outlines represent the modern French frigate *Duquesne*, recently set afloat at the Rochefort dock-yard. A sister ship, the *Tourville*, has also recently been constructed, under contract, at a yard near Toulon. Both ships are termed cruisers of the rapid type, are designed for seventeen knots per hour at sea, and are furnished with formidable rams. The frames, bulkheads, beams, and all interior parts, also the masts, are composed of steel; but the outside plating of the hulls is entirely of iron, and the bottoms are sheathed with two layers of teak planks, in all 7 inches thick, and coppered; put on in a similar manner to the system of the English, except that, to insulate the iron from the copper, thick layers of marine glue have been placed between the iron hulls and the teak planks, also between the teak and copper.

A few of the dimensions and other data are as follows:

Length between perpendiculars	325 feet 8 inches.
Length, total	350 feet.
Breadth at load water-line	50 feet.
Draught of water forward	20 feet 3 inches.
Draught of water aft	25 feet 2 inches.
Displacement, loaded	5,436 tons.
Area of sail-surface	21,000 square feet.

WEIGHTS.

Weight of hull	2,650 tons.
Weight of guns and ammunition	260 tons.
Weight of engines and boilers	1,280 tons.
Coal	655 tons.

ARMAMENT.

This consists of twenty-seven guns; twenty of these have a caliber of 14 centimeters, and seven of 16 centimeters, about 5½ and 6¼ inches, respectively.

MOTIVE MACHINERY.

The engines and boilers for the *Tourville* were both designed and constructed at the extensive *Forges et Chantiers de la Méditerranée*, at Marseilles. The engines, which are to develop 6,000 horse-power, are of the horizontal compound type, and consist of two sets, of four cylinders to the set, making, in all, eight cylinders, having four piston-rods connected to one crank-shaft, to work a single screw; *i. e.*, the four low pressure cylinders are laid horizontally on one side of the shaft, each having its high-pressure cylinder bolted to the outboard end, the steam being admitted directly from the high to the low pressure cylinders without passing through a receiver, and being conducted finally to four tubular condensers kept free by two rotary pumps. The boilers are of the old box type, braced to carry a pressure of 45 pounds per square inch, and built after the systematic style of the French.

The grate surface is 950 square feet, and the heating surface 24,500 square feet. The diameter of the screw-propeller is 20 feet 3 inches. The estimated number of revolutions per minute is 76. The estimated indicated horse-power is 7,500, and the speed at these figures 17 knots per hour. The space occupied by these engines is greater than that occupied by the engines of any single-screw ship I have knowledge of, and the numerous parts and multitudinous connections make the complication great, and will cause serious difficulty with the operation of the machinery, if it does not result in other evils; besides which, the boilers are of the discarded type, and cannot be worked to the pressure necessary to obtain the economy resulting from the compound system; and as the *Tourville* has been built for high speed as the first requirement, it may be regretted that a more simple type of machinery has not been adopted. The French naval authorities, in building this, their first rapid cruiser, have, in many features of the hull, followed the system of the English, and have accepted the errors admitted by the British authorities to exist in their modern frigate the *Shah.*

Besides the original great cost, about $1,470,000, large cost for maintenance, and unwieldy bulk to handle, the motive power is decidedly objectionable, and the coal supply is only 655 tons.

The motive machinery for the *Duquesne* is from a different design. It is being manufactured at the *Usine d'Indret.*

SMALLER VESSELS.

The corvette *Duguay-Trouin*, building at Cherbourg, is also designed as a rapid cruiser. She is to have 3,180 tons displacement, to make 16 knots per hour on the measured mile, and is intended to compete with the English *Rover* class.

There is also a third class building, of which the *Rigault de Genouilly* is the type. They are to compete with the English *Opal* class. These vessels are 240 feet long, have 44 feet extreme beam, 14 feet 7 inches mean draught, a displacement of 1,640 tons, 8 guns of 5.51-inch caliber, engines of 1,900 horse-power, and an intended speed of 15 knots.

The list appended contains the names of the vessels building and proposed December, 1877. It will give a very good idea of the progress made at this time in reconstructing the French navy.

List of vessels for the French navy, building and proposed, 1877.

Names of vessels.	Class.	Kind.	Where built or building.	Hulls.			Machinery.			Total appropriation for machinery and hulls for the year 1877, in dollars, gold.
				Percentage of work on the hulls, completed January 1, 1877.	Percentage of work to be done during 1877.	Amount appropriated for each class for 1877, in dollars, gold.	Percentage of work on machinery January 1, 1877.	Percentage of work to be done during 1877.	Amount appropriated for work on machinery during 1877, each class, in dollars, gold.	
Trident	First class	Iron-clad	Toulon	79	21		100	0		
Dévastation	do	do	L'Orient	17	30		0	25		
Foudroyant	do	do	Toulon	17	25	946, 580	0	25	203, 040	1, 149, 620
N	do	do	Brest	0	7		0	0		
N	do	do	Industrie	20	25		0	0		
Triomphante	Second class	do	Rochefort	49	29		100	0		
Turenne	do	do	Cherbourg	8	35		0	0		
N	do	do	do	0	3	935, 676	0	0	40, 608	976, 284
N	do	do	Brest	11	19		0	20		
N	do	do	L'Orient	0	21		0	0		
Fulminant	First class	Iron-clad coast-defender	Cherbourg	37	25		70	30		
Furieux	do	do	do	11	20		0	20		
N	do	do	Rochefort	0	7	777, 192	0	0	100, 768	877, 960
N	do	do	Toulon	0	12		0	0		
N	do	do	Industrie	0	22		0	0		
Tempête	Second class	do	Brest	92	8		100	0		
Tonnant	do	do	Rochefort	19	40	608, 368	0	16	67, 680	676, 048
Vengeur	do	do	Brest	36	35		0	50		
Lutin	First class	Iron g obt	Cherbourg	71	29	22, 560	70	30	16, 920	39, 480
Lynx	do	do		70	30		70	30		
Duguay-Trouin	Second class	Cruising-ship	Cherbourg	85	15		100	0		
La Pérouse	do	do	do	32	30		0	30		
Villars	do	do	Brest	17	24		40	60		
N	do	do	Cherbourg	6	10	446, 124	0	30	195, 520	641, 644
N	do	do	do	0	10		0	0		
N	do	do	L'Orient	0	40		0	0		
N	do	do	Toulon	3	50		0	0		
N	do	do	do	0	20		0	0		
Rigault de Genouilly	Third class	do	Brest	70	30	58, 656	100	0	0	58, 656
Eclaireur	do	do	Toulon	63	37		100	0		
Six screw dispatch-boats				47	35	202, 288	100	60	105, 280	307, 568
[illegible]e transports				50	40	627, 356	0	0	144, 760	772, 116
Six small vessels			Industrie	0	100	57, 904	0	100	42, 864	100, 768
Six [illegible] obts			do	0	100	45, 120	0	100	45, 120	90, 240

DOCK-YARDS OF FRANCE.

France is separated into five naval divisions, known as *arrondissements maritimes*, each of which is presided over by a *préfet maritime.* The names of the arrondissements are the same as those of their chief ports, viz, Cherbourg, Brest, L'Orient, Rochefort, and Toulon.

Each dock-yard is immediately under the command of a rear-admiral, *major-général de la majorité*, who controls the *personnel.*

The work in the yard is divided under seven official heads, viz, a major-general, major of the fleet, director of movements in the port, director of naval constructions, director of naval artillery, naval commissary-general, and medical director.

The director of constructions is an engineer officer; he supervises the construction and repair of all ships and vessels and all machinery built or repaired at the dock-yard. His assistants also belong to the engineer corps.

CHERBOURG.

This important French naval port is located on the English Channel, opposite to Portsmouth. The dock-yard is an artificial one, with a roadstead formed by the construction of a grand breakwater, and it is defended by works believed at the time of construction to have been the most successful effort of engineering skill; numerous forts and batteries not only encircle the yard, but also surmount the breakwater; and every other commanding point, both to the seaward and in the rear, has been turned to the best account.

The works within the yard and the fortifications without were constructed at an enormous cost and by long and patient labor, having occupied the French fifty-five years, viz, from 1803 until 1858, when the last stone was laid and the late Emperor opened the basin, which then bore his name.

Plans showing the position of the dock-yards, basins, dry-docks, building-slips, and buildings were furnished to the department some years ago by the author; particular localities mentioned in this report may be found upon reference to said plans.

The works within the yard consist of three great basins and two minor ones, eight dry-docks, eleven building-slips, and twenty-eight substantial stone buildings. The docks and basins have been excavated from slate-quartz of the same formation as that from which the Keyham docks (noticed previously) have been excavated. The first or outer basin, known as the *avant-port militaire*, or Dock Napoleon I., with entrance from the roadstead, and outlets to the other basins, covers an area of 741,185 square feet. The second or inner basin, known as the *basin de flôt*, has an area of 686,662 square feet; and the third, known as the *arrière basin de flôt*, or Dock Napoleon III., covers an area of 904,201 square feet. The three are connected by locks and gates. Contiguous to the basins are the dry-docks; the outer port possessing one, and the inner basin seven.

Around the basins are distributed the building-slips, work shops, and store-houses. After the excavations from the slate-quartz, they are formed throughout, batteries and walls, with blocks of granite laid in a similar manner to the blocks in the dry-docks and the building-slips, and the whole work presents beautiful specimens of masonry.

The outer port has no gates or caissons at the entrance from the roadstead; into it all the vessels first enter, passing afterward to the inner docks or basins at pleasure. On the south side of the entrance to this basin are the smitheries, the forges, and the old machine-works, an arrangement found to be inconvenient for the transportation of material to the shops, or of machinery to and from the vessels in the basin or docks. On the south side of the basin are four building-slips, having ship-houses of old date over them, and a dry-dock of less capacity than those attached to the inner basin. The ship-houses are 380 feet in length by 87 feet high, covered with ordinary roofs which are supported by granite columns 7 feet square, and they are spaced 130 feet from center to center. Neither on the west nor on the southwest are there any shops in close proximity to the outer port.

The building No. 6, shown on the plan mentioned, is 883 feet long and 73 feet wide; it is divided in the center by a cross-wall. The south end contains a large number of smithery fires, which are arranged at either of the side walls, with the steam-hammers, furnaces, and cranes in convenient reach. The north end of the building contains the machinery and tools for repairing and fitting the steam-machinery of the vessels in port. The appliances, tools, &c., embrace the usual kinds found in such shops, and are the productions of Whitworth and Rigby, of England, and of Mazeline and other tool-builders of France. With few exceptions, it may be said that they are of a date anterior to the improved machines in the best English engine-factories. Adjoining this building, and at the north end, are the iron and brass founderies, under one roof, containing ordinary foundery appliances. At a short distance from the machine-works, No. 14, is the machine-erecting shop. Here all the heavy machinery for the vessels is fitted and put together previous to being placed on board. This building is one story high and 203 feet in length by 83 feet wide. It contains one traveling crane movable on wheels over a rail-track, one very large lathe, several large boring, slotting, and drilling machines, besides the erecting attachments; but, like the machines and tools in all the other shops, they are from patterns made previous to recent improvements. On the east side of the ship-basin is the ordnance building, containing the small-arms and ordnance stores, and on the south side are the mast-houses, spar-houses, mold-lofts, sail-lofts, and galleys. The inner ship-basin is the great work of the dock-yard; its construction was commenced in 1836, and it was finished in 1858; its dimensions in round numbers are, length 1,365 feet, breadth 650 feet.

On the east side of this basin, No. 3, are the store-houses and offices. On the north side are four dry-docks, two of them being of the largest class, each of which, measured at the bottom, is 600 feet by 70 feet.

To the east of the dry-dock No. 15, are the boiler-making, copper, and tin shops, all under one roof and separated by longitudinal and cross walls. The building is one story high, 320 feet long by 118 wide; one portion, divided into two longitudinally, is devoted to boiler-making and has an overhead traveling crane in each division to facilitate the movement of heavy weights. None of the boiler-making tools are of the best varieties, besides which the shops are deficient in many improved machines of the time, found in English dock-yards and the best English engine-factories. In the center of the building at the extreme south end of the yard are the pumping-engines and their boilers, for removing the water from the inner ship-basin and the docks belonging to it. At both ends of the same building are workshops. The hydraulic works and model-rooms are marked No. 5 on the plan referred to. In the former all details appertaining to pumps and pump-gear are manufactured. In the model-rooms is a plan of the dock-yard and a complete model of it on a large scale, showing all the buildings, basins, docks, and building-slips. A little to the west, No. 32 is the reservoir for supplying the yard and the vessels of war in the port with fresh water. The water is received from the same source as that from which the supply of the town is drawn. The west side of the inner basin is occupied with seven building-slips and one small dry-dock. The slips are formed like those previously named, and are of sufficient capacity to allow the building of any class of vessels, but they have no ship-houses over them.

The vessels built on these slips are launched into the basin, and are brought to rest before reaching the opposite walls by means of rafts pushed before them.

The south side of the basin is occupied by two dry-docks of the largest class, capable of taking in any vessel now afloat; timber, store-houses, and the steam-factories are also here. The buildings of the steam-factories embrace those for the machine-works, the smithery, and forge. The smithery building is in the form of a right angle, with legs of equal length, each of which is 175 feet long by 70 feet in width. It is one story high, and, like all the other Cherbourg dock-yard buildings, is constructed of a peculiar soft stone of the locality. In the center longitudinal line of the building are judiciously arranged in iron frames the smithery fires, so that the anvils, on either side of the building, are brought directly under the light of the windows. In one angle of the building are the steam-hammers, furnaces, and other appliances.

The building for the machine-works forms an opposite right angle to the smithery, and is of the same dimensions. At the extreme south-west end of the dock-yard are the sailor, marine, and gendarme barracks. The joiner shops (No. 9) are west of the ship-houses. The steam saw-mills are marked No. 10, and have proved to be one of the most useful and economical branches of employment belonging to the dock-yard. The dimensions of this building are 910 feet long by 113 feet wide, divided into two parts by a cross-wall—one end containing the stores of timber and the other all the machinery necessary for sawing and working wood into any shape required in wood-ship building, including the formation of the frames of ships.

At the extreme south end of the yard are the bakery and provision stores. The bakery is extensive; the building is two stories high, 685 feet long by 64 feet wide. One of the peculiarities of the French system is that all the bread for the use of their vessels of war is baked in the dock-yards. No. 33 is a "hydrometer"; it measures the rise and fall of the tides continuously, and there is a complete daily record kept of it from beginning to end of each year. Conveniently arranged on either side of the entrance to the outer port are coal-storages, admitting of coaling vessels without interfering with other work. This can be seen at No. 26, on the plan aforesaid.

The streets and avenues of the Cherbourg dock-yards are paved throughout with blocks of stone, and rail-tracks traverse all avenues leading to or from the workshops, basins, docks, and building-slips.

Cherbourg, in short, is a dock-yard displaying great engineering skill, and shows what may be accomplished in those cases where nature has done little more than trace the outline and furnish an abundant supply of water. The basin accommodation is very extensive, and the largest basin—the inner one—with all the docks attached to it, is designed to accommodate a small fleet.

The land approaches of Cherbourg, in addition to the redoubts and forts shown on the plan to which reference has been made, possess features of an interesting kind. An enemy landing near the place for the purpose of its reduction by siege would have no easy task; his every step for miles would be impeded, even by numerically smaller force, from behind successive natural field-works. Cherbourg, when created, was regarded by the first engineers of France as a stronghold, both to seaward and in the rear, frowning defiance to the world; but in those days there were no rifled guns, no heavy ordnance, and none of the destructive projectiles of the present day; nor were there any armored ships of war. Modern improvements in naval warfare place Cherbourg dock-yard, at the present period, in the greatest danger when approached by an enemy from the sea.

TOULON,

in name and tradition, is ranked as one of the most extensive and important dock-yards of Europe. Unlike Cherbourg, it is formed on a natural harbor and in a more secure position from bombardment. It is fortified, and the approaches to it are well covered. The port is familiar to all navigators, and known as one of the best in the world. Entering it from the Mediterranean, an outer roadstead leads by a narrow channel into an inner one of large dimensions. On the shore side of this there is a deep bay, the water-front of which is occupied by the dock-yard. This bay is about one mile and a quarter across, but, measured by the curve of the shore, the water-front is nearly two miles and a quarter, most of which is devoted to government purposes.

After passing the Grosse Tour on the right, the first point reached is the construction yard, where the ship-houses are erected and all the operations of ship-building carried on.

At a distance of some hundred yards from this is the commencement of the yard proper, which is inclosed by a wall. In front of it, as now completed, there are two large basins, known respectively as the Old and the New wet-dock. These basins are formed by broad dikes or breakwaters constructed in the water of the bay, on which are erected large shops and store-houses, and the water-front on the inside of the breakwater more than doubles the accommodation for vessels which is permitted by the wharves on the front on the main.

Beyond the precincts of the building now constructed is the Missiessy wet-dock, which extends to the left point of the bay in front of the city. It is more commodious than the other basins, the water-front on the main being about 203 rods. Adjoining it is the pyrotechnic school, at the village of Brégaillon, on what would be an island but for a narrow isthmus connecting it with the mainland, and immediately south of the Grosse Tour is located the great naval hospital of St. Mandrier.

The dry-docks are seven in number, one of them being in two communicating parts, which, when used as one, admit a length of 500 feet. All are convenient to the Old and the New basins.

The disposition of basins, docks, workshops, and other buildings may be seen from the drawings transmitted to the department.

BREST.

The dock-yard at Brest is admirably situated for defense, and the harbor is well protected from the weather. The entire length of the channel is about 15,000 feet, and its average width is about 525 feet. Both sides of the inlet are devoted to the use of the dock-yard, and are lined with fine quays.

On either side is a basin with docks, and a third basin, larger than either of the others, is situated at the head of the slip and has four stone dry-docks capable of taking in the largest vessels. The construction of the yard was difficult in consequence of the formation of the land, which rises almost perpendicularly from the shores of the inlet on which it is constructed.

In order to provide space for the accommodation of the numerous workshops, store-houses, and other buildings required, the land and rock have been terraced, and the buildings on the left side toward the town have been constructed in three ranges, one above the other. The lowest tier is devoted to workshops, store-houses, offices, &c., the second tier to the *bagne*, formerly used as a prison for convicts, also to the accommodation of a large ropewalk 1,600 feet in length. On the right side are the works for the construction and repair of steam-machinery, also a barrack for the accommodation of sailors. Its capacity is sufficient for five thousand men.

L'ORIENT

is easily approached from the Bay of Biscay, and its position on the projecting land between the two rivers Scorff and Blavet gives it natural advantages which have been recognized and turned to account. The sea defenses have been strengthened in the last few years by heavier guns from the government establishment at Ruelle, and additional factory works erected. As a construction yard its capabilities are large, both for wood and iron ship building, there being two dry-docks, one of large size, and four building-slips. It was here that the first two iron armored ships of the French were built, the *Couronne* and *Héroïne*.

ROCHEFORT

is also a construction yard of large capabilities, containing a number of building-slips, several stone dry-docks, and all the usual appliances for iron and wood ship building. The five national dock-yards of France, great as are their capabilities, do not comprise all the establishments for building and repairing ships of war under the orders of the French Government.

The ship-yards at Bordeaux, at Nantes, at La Seyne, Havre, and other ports, the extensive iron-ship-building and engineering works, and iron and steel manufacturing works at Creuzot (the largest in Europe), and all other works in France, can be placed under the orders of the minister of marine on conditions to suit the government, or men can be conscripted from them into the national dock-yards.

PERSONNEL OF THE FRENCH NAVY.

The line officers of the French navy are recruited chiefly from the naval school, from the polytechnic school, and by the admission of officers from the merchant service who have passed certain examinations. Beginning at the lowest step, the grades are as follows: Cadet of the second class; cadet of the first class having the relative rank of second lieutenant in the army; ensign (*enseigne de vaisseau*), ranking with a first lieutenant; lieutenant, ranking with captain in the army; captain of frigate, ranking with lieutenant-colonel; commodore (*capitaine de vaisseau*), ranking with colonel; rear-admiral (*contre-amiral*), ranking with brigadier-general; vice-admiral, ranking with general of division; and, finally, admiral, ranking with field-marshal. Of these, the three superior grades are distinguished by the common name of "general officers of the navy"; captains of both grades are entitled "superior officers," all others being classed as "officers." The active list of "general officers of the navy," as at present established, contains the names of no admirals.

Vice-admirals, when they attain the age of sixty-five, and rear-admirals, when they complete their sixty-second year, are removed from the active list and placed *en disponibilité*; that is to say, they are no longer eligible in time of peace, but may be called upon to serve in the event of war.

Admirals, however, who have commanded a fleet, or who have specially distinguished themselves in action with the enemy, may be retained on the active list, though they have passed the prescribed limit of age. Promotion, from the lowest rank up to that of captain of frigate inclusive, is given in some few instances by selection, but for the most part by seniority. From the rank of captain of frigate upward, advancement takes place exclusively by selection, but, in any case, an officer must have served a certain prescribed time in each grade before he is eligible for promotion to the next higher. For instance, a rear-admiral must have served two years either in a squadron or naval division before he can become a vice-admiral; a *capitaine de vaisseau* must have four years' seniority and must have served three years at sea before he can be promoted; a captain of frigate must have similar service before he may be advanced; a lieutenant must have four and an ensign two years' service before becoming eligible for promotion. All officers below the rank of rear-admiral are attached to one or other of the naval ports of France, at which those below the rank of captain of frigate must reside when not embarked.

A roster, called the *liste d'embarquement*, is kept at each station by the maritime prefect of the district, and officers are embarked, as may be required, when they arrive at the top of this list, while officers required for duty on shore in connection with naval matters are taken from among those at the foot, a method which insures that, as nearly as may be, every officer shall have a fair share of sea and harbor employment.

The number of officers borne on the Navy Register for 1877 is as follows:

Admirals		0
Vice-admirals		29
Rear-admirals		51
Captains		110
Captains of frigates		230
Lieutenants of the first class	348	748
Lieutenants of the second class	400	
Ensigns		498
Midshipmen of the first class	140	182
Midshipmen of the second class	42	

Naval engineer corps:

Inspector-general, ranking after rear-admiral	1
Directors of naval construction, ranking before *capitaine de vaisseau*	12
Engineers of the first class, ranking with *capitaine de vaisseau*	22
Engineers of the second class, ranking with captain of frigate	22
Sub-engineers, ranking with lieutenant of the first or second class, or with ensign	74
Mechanicians-in-chief, ranking with captain of corvette	2
Principal mechanicians of the first class, ranking with lieutenant of the first class	8
Principal mechanicians of the second class, ranking with *enseigne de vaisseau*	39

Hydrographic engineers:

Chief engineer hydrographer, ranking after rear-admiral	1
Engineer hydrographers of the first class, ranking with *capitaine de vaisseau*	4
Engineer hydrographers of the second class, ranking with captain of frigate	4
Sub-engineer hydrographers, ranking with lieutenant and ensign	6

Pay corps or commissariat:

Rank		Number
Commissary-generals of the first class	ranking after rear-admiral	7
Commissary-generals of the second class		4
Commissaries, ranking with *capitaine de vaisseau*		28
Deputy commissaries, ranking with captain of corvette		52
Sub-commissaries of the first class	ranking with lieutenants of the first and second class	94
Sub-commissaries of the second class		93
Assistant commissaries, ranking with *enseigne de vaisseau*		113
Commissary clerks	first class 51	404
	second class 51	
	third class 151	
	fourth class 151	

Commissariat of the colonies:

Rank	Number
Commissary-generals of the first class	2
Commissary-generals of the second class	3
Commissaries	17
Deputy commissaries	34
Sub-commissaries of the first class	27
Sub-commissaries of the second class	43
Assistant commissaries	93
Commissary clerks	80

Medical corps:

Rank		Number
Inspector-general, ranking after rear-admiral		1
Medical directors of the first class	ranking before captain	2
Medical directors of the second class		3
Medical and pharmaceutical inspectors		2
Surgeons-in-chief, ranking with *capitaine de vaisseau*		21
Surgeon-professors	ranking with captain of corvette	9
Principal surgeons		40
Surgeons of the first class, ranking with lieutenant		202
Surgeons of the second class, ranking with ensign		204
Surgeons of the third class		2
Assistant surgeons, ranking with cadet of the first class		157
Apothecaries-in-chief, ranking with captain		4
Professors of pharmacy	ranking with captain of corvette	6
Principal apothecaries		4
Apothecaries of the first class, ranking with lieutenant		22
Apothecaries of the second class, ranking with ensign		25
Assistant apothecaries, ranking with cadet of the first class		26

Chaplains:

Rank	Number
Chaplain-in-chief, ranking after rear-admiral	1
Superior chaplains, ranking with captain of corvette	4
Chaplains of the first class, ranking with lieutenant of the first class	24
Chaplains of the second class, ranking with lieutenant of the second class	22

Professors:

Rank	Number
Professors of the first class	25
Professors of the second class	11
Professors of the third class	8
Professors of the fourth class	2
Total number	3,964
Naval police (*gendarmerie maritime*)	18

EXPENDITURES.

The cost of the maintenance of the French navy for the financial year 1877 was:

Personnel of the navy	$6,451,726
All other purposes for the navy proper	20,438,312
Military forces	2,451,462
Colonial service	5,621,804
Total (gold)	34,963,304

According to the *Broad Arrow:*

The enlisted men for the French navy are obtained by what is known as the *inscription maritime.* Every young man who has completed his eighteenth year, and who has made two long voyages, either on board a vessel belonging to the state or a merchant-ship, or who has been leading a seafaring life for eighteen months, or who has been engaged as a fisherman for two years, and who declares his intention of remaining a sailor or fisherman, is inscribed as a sailor, and is liable to be called upon to serve in the French navy. A month before he completes his twentieth year, or immediately after his return, should he at that time be absent from France, he is bound to present himself to the proper authorities. By them he is sent to the chief port of the maritime district to which he belongs, and is there incorporated into one of the divisions of sailors. Should he prefer to do so, he may report himself at any time after he has completed his eighteenth year. The first period of service of the young *inscrit* extends over five years. During this time, he may, if his services are not required, be granted leave, which may be renewed from time to time, receiving no pay, but at liberty to employ himself in seafaring occupations. At the expiration of the first five years he enters upon a second period of service of two years, during which he may also be permitted to remain on leave; Provided that while thus on leave he is engaged only in coasting or fishing vessels, the *inscrit* is allowed to count all the time thus spent as service rendered to the state; after he has completed his second period, he can only be called upon to serve by a special decree in case of an extraordinary emergency. The sailors required for service in the fleet are taken from among the *inscrits* who have not rendered any service to the state, or—should the number of these be insufficient—from among those who count least service, or in case of equal amount of service from among those who have been longest on leave. Certain men are, however, exempted, whenever possible, from actual service. Such are, for instance, the eldest brothers of a family of orphans, men who have a brother already serving, only sons, the eldest son of a blind father, or of a father who has passed his seventieth year, or the eldest grandson of a widow; but it is laid down in the law of 1872 that these exemptions must be specially made in each individual case by the minister of marine upon the recommendation of the chief local naval authorities.

Certain advantages are reserved for these *inscrits maritimes.* They alone have the right of fishing in French waters, or of being employed in French coasting-vessels. They are exempted from all other public service. While employed, and for four months after their return to their homes, troops cannot be billeted in their houses. When serving, they may travel by railway at a fourth of the ordinary fare. If they fall sick during the forty days following their proceeding on leave, they are admitted to the naval hospitals free of charge; and, finally, by paying a small annual premium, which never exceeds three per cent. of his salary, the *inscrit* acquires a right to a pension, termed *demi-solde,* after being at sea for twenty-five years, or on arriving at the age of fifty, whatever may be the amount of service that he has rendered to the state, his widow, should he die, continuing to enjoy a portion of the pension.

Voluntary enlistment is also permitted in the French navy; but, unless he has previously rendered service to the state, the recruit must be between eighteen and twenty-four years of age, the limit being extended to thirty years in the case of musicians, [firemen], carpenters, calkers, and a few other ratings. Time-expired men are also allowed to re-engage for a further period of service of three, four, or five years, provided they can pass a certain examination in their duties, when they receive a somewhat higher rate of pay. The ratings in the navy, beginning from the lowest rank, are those of novice, marine apprentice, sailor of the third, second, and first class—quartermaster, second master, master, and first master. Before a young man can be rated as a sailor of the third class, he must be at least eighteen years of age, must have served in the navy for at least twelve months, or, if he belongs to the *inscription maritime,* must have made two long voyages, or have been employed in coasting voyages for eighteen months or in fishing for two years. Before he can become a sailor of the second or first class he must have served at least six months in the lower rating. Similarly, a man must have served at least six months as a sailor before he can be made a quartermaster. Before he can be promoted to second master he must have served six months as quartermaster on board a ship carrying guns, and he must also be able to read and write; and before he can become master or first master he must have served six months as second master on board a vessel with a complement of 250 men, or have discharged for the same period the duties of acting master on board a ship carrying a crew of at least 150 men.

PART XIV.

THE GERMAN NAVY.

THE DEUTSCHLAND; THE PREUSSEN; THE SACHSEN, AND OTHER VESSELS; TABLE OF DIMENSIONS, ETC., OF ARMORED AND UNARMORED SHIPS OF GERMANY; THE ESTABLISHMENT OF HERR FRIEDRICH KRUPP, AND HIS GUNS.

THE GERMAN NAVY.

The public mind in Europe has for some time been directed to the growing power of Germany at sea, and friendly people have looked, not without some degree of admiration, upon the earnest and business-like efforts of that practical-minded nation to attain a position at sea somewhat commensurate with her position on land.

Few movements are more remarkable than that which has of late years been developing Germany into a maritime power. This navy commenced only about twenty-eight years ago, with one sailing-corvette and two gunboats. Two years afterward it was increased by two steamers and forty gunboats, consequent upon the Danish war and the subsequent blockade of the Baltic ports; and in the two succeeding years the addition was made to it of four more vessels, viz, a brig, two dispatch-boats, and a corvette. From this time up to the year 1860 the German navy was steadily increased by the addition of wooden vessels, until the total number of thirty-one small steamers, eight sailing-vessels, and forty-two gunboats had been reached. And, after the year 1860, when the building of armored ships had commenced, the German navy began to take rank among the navies of Europe. About that time it was determined by the authorities at Berlin to increase the navy by the addition of several armored ships; two, viz, the *König Wilhelm* and *Arminius*, were built in England, others were built in France, at the *Forges et Chantiers*, on the Mediterranean, and one was commenced in Prussia. But it was not until after the close of the Franco-Prussian war of 1870, when Germany had acquired her full territorial and administrative organization, that there seemed to be at length an opportunity for gratifying the naval ambition of the country. Accordingly, after deliberate consideration of what were the purposes for which they wanted a navy, and what number and kind of vessels such purposes would require, a policy was decided upon in 1873 by the present administration, and a plan devised. It was, of course, understood that the plan was liable to be modified from time to time to keep pace with the progress of science, but this consideration has only affected some details of their scheme, and has not prevented it from being systematically carried into effect. They decided on what they needed, and they have ever since been diligently providing for it.

The primary object they had in view was the defense of the German shores from attack and blockade; secondly, the protection of their commerce and colonists abroad; and, thirdly, to be prepared to act on the offensive against hostile fleets.

To secure these ends it was considered indispensable to form a navy of eight armored frigates, six armored corvettes, seven monitors, two floating batteries, twenty unarmored corvettes, six dispatch-boats, eighteen gunboats, and twenty-eight torpedo-boats.

Nothing could be better than the method the Germans seem to have pursued in devising their plan. For the first purpose they took into account the special character of their coasts and harbors, and considered what class of vessels would be requisite at the several points. In the

Baltic a different character of defense from that required for the eastern shores and mouths of the North German rivers was found to be necessary. In the former they determined to utilize defensive torpedoes. For the latter object it was at first intended to provide monitors, but since the development of offensive torpedo-warfare it has been thought wiser to construct a larger number of small but well-armored gunboats. These will carry heavy guns and torpedoes; they are of light draught, fair speed, and are easily handled. Five have been launched, and are in process of equipment, and two are in course of construction.

Of the armored frigates, two, viz, the *Kaiser* and *Deutschland*, were constructed in England by Messrs. Samuda Brothers, from designs of Mr. E. J. Reed, C. B., M. P.; two, viz, the *Grosser Kurfürst* and the *Friedrich der Grosse*, were built at the imperial dock-yards of Wilhelmshafen and Kiel; and a fifth, the *Preussen*, was constructed by the Vulcan Engineering Company at Bredow, near Stettin. The drawings and specifications for the three last-named ships were prepared at the admiralty in Berlin. The two frigates referred to as having been built in England are designed as cruisers. They are broadside-vessels, of the central-battery type, and are in general features similar to the British frigate *Hercules*, but differing from her in numerous points of detail. One of them had just been completed at the date of my visit to the Thames building-yards in the summer of 1875, where I had the advantage of inspecting them. Of the armored corvettes, one, the *Sachsen*, is just completed; one other, the *Baiern*, is in process of construction, and two others have been laid down. The monitors as originally designed have not been ordered, but some of the gunboats and torpedo-boats have been built, and others are in process of building.

Since 1870 much encouragement has been given by the German admiralty to home engineering-works. The *Kaiser* and *Deutschland* were the last ships ordered from foreign countries, and of the ships launched since 1870, fourteen have been supplied with machinery made in German establishments. The result of this patronage has been the development of iron-ship building and marine-engine manufacture to an extent hitherto unknown.

THE DEUTSCHLAND.

This ship is about 280 feet long, with a breadth, extreme, of 62 feet, load draught 24 feet 6 inches, and a displacement of 7,560 tons. The system of framing is unique: both the outer and inner frames are of continuous angle-iron, between which the longitudinal plates and angle-irons are worked. She is built on the bracket-plate system, like the *Invincible* class of ships; and has a depth of 40 inches in the vertical keel, while her longitudinals diminish to 33 inches in breadth, and then increase to 45 inches at the armor-shelf. There are no wing passages, but she is divided into compartments by transverse water-tight bulkheads; and the double-bottom is in 32 water-tight compartments, so that in striking a rock it is expected that not more that four could be filled with water at once, the cubic capacity of the four being about 40 tons.

She has one central battery on the main deck, carrying four guns on each side; of which the forward ones fire 3° within the fore-and-aft line, and thus afford a converging fire ahead, while the embrasures are so arranged that the fire of all the guns on one side may converge at a point 276 feet, or about one ship's length, distant. This battery overhangs the side about 3 feet 6 inches at the forward end, and 1 foot 6 inches at the after end, but it is within the extreme breadth of the ship, and the ports are fitted with heavy forgings on a new plan, so as to protect the gunners as much as possible. The guns are Krupp's 26-centimeter breech-loading steel cannon, having a bore of about 9¾ inches, and weighing about 22 tons each. To complete the all-round fire, a Krupp 22-centimeter gun of 8¼-inch bore, weighing about 18 tons, is placed on the main deck aft, and is capable of being trained to an angle of 15° on each side of the middle line, and protected with armor-plates. The height of the port-sills above the load water-line is 11 feet. The armor at the water-line, in the wake of the engines, boilers, and magazines, is 10 inches thick, and elsewhere on the belt it is 8 inches amidships, tapering to 5 inches forward and aft. In the central battery it is 10 inches at the port-sills, 8 inches on the sides, and 7 inches on the bulkheads. The wood backing is of teak, 10 inches thick, placed upon two thicknesses of ⅝-inch plate, which are supported by 10-inch frames, spaced 2 feet apart. The armor on the belt abaft the battery extends 5 feet 6 inches below the water-line and 6 feet 6 inches above it, while in front of the battery it extends to 2 feet 6 inches above the water-line, up to the lower deck, which is covered with protective plating in two thicknesses, 2 inches thick for 10 feet in front of the battery bulkhead, and 1½ inches thick from this forward. There is an armored bulkhead at each end of the battery, the forward one extending down to the lower deck in order to protect the engines and boilers against a plunging fire. The upper and main decks and part of the lower deck are also protected with plates.

The portion of the upper deck over the battery is covered with ⅜-inch steel plates, and ¼-inch plates before and abaft it; while the entire surface of the main deck is covered with steel plating ½-inch thick abaft the battery, and ¼-inch thick in the battery and forward of it. The lower deck is not plated abaft the forward armor bulkhead. The ship is supplied with a very efficient system of arrangements for pumping out or flooding the compartments, &c. The capstan and fire-engine are worked by an auxiliary steam-engine, which can be supplied with steam either from the main boilers, or, in case steam is not up in them, from a small separate boiler attached to the engine. The hawse-pipes are placed in the upper deck, and therefore special arrangements have been made for the bitts. The steering-wheel, engine-rooms, and battery can be communicated with by means of a telegraph placed on the upper deck, and protected by armor-plates, and close to them a small armor-shield is fitted, behind which two officers could, if necessary, hold a consultation. Deviating from the usual custom, the sick-berth is in this ship placed on the main deck instead of on the lower deck. The bottom of the vessel has been coated with Redman's anti-fouling composition, which proved so successful in the trials in the Medway, and of which mercury is the chief ingredient used to resist fouling.

The machinery was designed and constructed by Messrs. Penn & Sons. The engines are of the old type, horizontal, direct-acting, with trunks; they are of the collective power of 8,000 indicated horses. The diameter of the cylinders is 122 inches, length of stroke 4 feet, and the maximum speed of pistons is 75 revolutions per minute. A speed of 13 knots per hour was attained at sea. The boilers are of the box type, and are eight in number. The *Deutschland* is ship-rigged, has an area of 3,900 square feet of canvas with all sails set, and an area of plain sail of 2,800 square feet. The screw is fixed, and made to revolve when the ship is under sail. It will be seen from the above that the *Deutschland* and sister vessel are powerful armored frigates.

THE PREUSSEN.

For the following particulars of this vessel, I am indebted to the *Zeitschrift des Vereines deutscher Ingenieure:*

This vessel is an armored-turret sea-going ship, similar to the British ship *Monarch.* The length between perpendiculars is 308 feet 6 inches; extreme length, 318 feet; extreme breadth, 53 feet 6 inches; depth from the upper deck to the keel, 34 feet 10 inches; displacement, loaded, 6,748 tons; load draught of water, mean, 23 feet 10 inches.

HULL.

The keel consists of two horizontal plates riveted together, upon which are fastened at the middle, by means of two angle-irons, a vertical plate 3 feet 10 inches high, extending to the two posts, to which all the plates forming the keel are connected by bolts and rivets. Four longitudinal frames stand almost vertically upon the outer skin, and their depth, which up to the fourth longitudinal frame is 31 inches, decreases gradually from the keel, running in the same direction as the latter, but approaching it fore and aft as required by the shape of the vessel. These longitudinal frames are made of plates and angle-irons, and are lightened at intervals by large oval holes. The cross-frames from the keel to the fourth longitudinal frame, and placed at distances apart of 4 feet, are made of short angle-irons extending only from one longitudinal to the other, to which they are connected by full plates, brackets, or angle-irons. The plates have the height of the corresponding longitudinal frames, and those belonging to one cross-frame are connected at the top by means of an angle-iron extending from side to side, through all the longitudinal frames and the keel. The outer skin is riveted to the longitudinal and cross-framing, and to the inner side of the latter is secured the second skin, over a length of 180 feet; the end cross-frames and nine intermediate ones have full plates, so as to form longitudinal water-tight compartments, which are again divided by the vertical keel, and are thus inclosed fore and aft by complete water-tight bulkheads.

Similar water-tight compartments are formed by the frames fore and aft of the double bottom, so that the extreme ends are as far as possible protected against casualty. Above the fourth longitudinal frame the distance between the cross-frames is 2 feet, and the latter extend from this longitudinal frame to the armor-framing, which runs from the front to the back, and which is constructed, like the longitudinal frame, of plates and angle-irons. The cross-framings above the armor-framing are thrown back as much as the thickness of the armor, inclusive of the teak backing, and consist of heavy angle-irons $9\frac{3}{4}$ by $3\frac{1}{2}$ inches by $\frac{1}{2}$ inch, which are fastened to the inner edge of the plates of the cross-frames above the fourth longitudinal, and which are provided at the outside with two angle-irons for the fastening of the outer skin behind the armor. These frames behind the armor extend in the central part of the vessel, that is to say, within the limits of the armored casemate, to the upper deck; but fore and aft of the casemate to the battery-deck only, or the height of the armored belt for the two ends of the vessel. The extension of the frames at both sides of the casemate to the upper deck is formed of light angle-iron; the parts of the vessel above the battery-deck being altogether constructed of light material. A water-tight partition of plates strengthened by angle-irons is erected almost parallel with the outer skin of the vessel at each side, over the length of the double bottom, and extends from the latter to the battery-deck, at a distance from the outer wall of the vessel of about 3 feet; these partitions connected with the double bottom and battery-deck form the bulkheads of the gangway, and they will serve to prevent the entrance of water during action into the other parts of the vessel. The space formed by these partitions and the outer skin is further divided longitudinally by water-tight cross-frames, so that even here only small spaces can be filled with water, which can easily be removed by pumps. The gangway will serve besides, during action, as a free passage to effect provisional repairs of damage done by shells. The outer skin of the vessel in the bottom has the almost uniform thickness of .6 inch, but is doubled at the two extreme ends. The skin behind the armor consists of two .62-inch plates, while that above it at the two ends of the armored casemate has a thickness of .4 inch. Much care was used in riveting up the hull, and rivets with a conical enlargement under the head corresponding with the cone of the holes are used throughout. The decks are carried on girders of T and I irons. The upper deck is covered to a great extent, the battery-deck entirely, and the between-deck at several places, with iron plates from ¼ to ⅜ inch thick, upon which are fastened the deck-planks; the places for the capstan, hawse-holes, timber-heads, &c., are especially strengthened, and the deck-beams are supported by wrought-iron tubes which have under the turrets a diameter of 7½ inches and a thickness of metal of ⅜ inch. The vessel is divided below the battery-deck by eleven water-tight cross-partitions into twelve compartments, which are connected with each other by water-tight doors. Two of these partitions, one at the front and the other at the rear of the casemate, extend to the upper deck, and serve between the latter and the battery-deck for the reception of armor-plates which are to protect the casemate against shells that may penetrate through the two ends of the vessel.

Special care is taken to maintain effective communication between the pumps and the whole of the water-tight compartments. For this purpose an iron pipe, 12½ inches in diameter, is placed close to and parallel with the vertical keel-plate over the length of the double bottom; from this pipe branches extend to the various compartments; the main pipe carries the accumulated water into a reservoir placed under the engine-room, whence it is pumped away by a 12½-inch Downton pump, as well as by all the pumps in connection with the machinery. Four 9¼-inch pumps are besides placed upon the battery-deck, each of which can draw the water from a certain number of compartments; one can also be used for filling the tanks with drinking-water.

ARMOR.

An armored casemate surrounds the two turrets, which project 6 feet 2 inches above the upper deck. This casemate is separated from the fore and aft parts of the vessel by armored transverse bulkheads, while those parts are protected only between wind and water, by an armored belt reaching from about 6 feet 2 inches below water to the battery-deck. The armored casemate is 90 feet 6 inches long; rising through it from the battery-deck are the turrets; also a second steering-wheel, to be used during action. The armor-plates at the water-line are 9¼ inches thick, below the water 7¼ inches, and above water 8¼ inches; these thicknesses decrease toward the ends to 4 inches; behind these plates there is a backing of teak about 10½ inches thick, but varying with the thickness of plates. Angle-irons are used for fastening this layer of teak to the outer skin. The armor-plates are secured by means of bolts 2½ inches in diameter, with conical heads fitting exactly in corresponding holes in the plates. The nuts of the bolts for the armor-plates are provided with double washers, between which a thickness of rubber is placed, in order to prevent as far as possible the tearing off of the bolt-heads when the armor-plates are struck by shot. The armored cross-walls have plates 5 inches thick, with a teak backing 8¼ inches thick.

TURRETS, ETC.

The two turrets are each 26 feet 9 inches in diameter; the shells are made in the usual way. They extend, as already stated, from the battery-deck to 6 feet 2 inches above the upper deck, and are covered with armor only at the parts exposed above the upper deck. The plates of these turrets are 8¼ inches thick, with the exception of those through which the port-holes for the guns are cut, and which have a thickness of 10¼ inches. The teak backing between the shells and armor-plates is 8¼ inches in thickness. These turrets revolve around strong cast-iron center-pins, secured vertically upon the battery-deck, and the outer circumferences of the turrets take their weights, and revolve on conical rollers, running upon rails laid on the deck. Each turret is operated by a high-pressure engine with two cylinders 10¼ by 10¼ inches, besides which, hand-gear for working by manual labor is provided. The ammunition is brought from the battery-deck into the turrets through openings, the tops of which are covered by plates 1 inch thick. The port-sills of the turrets are 13 feet 5½ inches above the water-line.

The *armament* in the turrets consists of four Krupp rifled guns, about 10½ inches in bore and 22 tons in weight, besides a lighter gun forward and one aft.

The upper deck is provided only in the middle above the turrets with a light platform for the reception of the chart-house, and there is a raised forecastle. A light screen for the protection of the crew from the sea is arranged so that it may be laid down, in order to be out of the way of the guns of the turrets, which lie close to the upper deck.

The *Preussen* has three masts, made to be used as ventilating-tubes, and she is a full-rigged frigate. The motive engines are of the three-cylinder type, and the boilers of the box variety. The screw is fixed, and fitted to revolve when the ship is under sail.

MATERIALS.

The following are the weights of materials used for the hull of the vessel, the masts, and the turrets: Plates, 1,375 tons; angle-irons, 600 tons; bar-iron and large forgings, 33 tons; iron for rivets, 115 tons cast iron, 100 tons.

Exceptional strength of material was required. For the plates two qualities were specified, the first to be used for all main parts, such as the keel, longitudinal frames, outer skin, deck-plates, &c., while the other quality was used for the cross-frames, gangways, coal-bunker, and transverse bulkheads, &c. The tests were made as follows: Out of each five tons of plate one sample plate was selected, of which, again, sample pieces were taken, which had to be tested for absolute and relative strengths, with and across the fibers, both cold and hot. In order to test the absolute strength, the sample-pieces were torn asunder by means of a testing-machine; the pressure required for plates of the first quality was, for longitudinal strain, 49,677 pounds per square inch; for cross strain, 40,477 pounds per square inch; for plates of the second quality, in the longitudinal direction, 44,705 pounds per square inch; and for cross strain, 39,000 pounds per square inch. The relative strength of the plates was tested by bending them over the rounded corner of a test-plate up to a certain angle, both cold and hot, by means of slight blows with a hammer. The angles differ for the various thicknesses and qualities of plates, and the iron must show no fracture or cracks. For angle-iron and special sections one quality only is allowed, which must sustain in tension a strain of 49,677 pounds per square inch. Great care was also taken in the forging tests. All the materials were furnished by German manufacturers, except the stern-post and the armor-plates.

THE SACHSEN.

The Sachsen is the first of the second group of German armored ships; she was built by the Vulcan Engineering Works at Bredow, near Stettin, and was launched July 21, 1877, three others of the same class being in process of construction.

The dimensions of the Sachsen are: Length between perpendiculars, 213 feet 3 inches; breadth, extreme, 51 feet 4 inches; depth, 26 feet 3 inches; load draught of water, 19 feet 8 inches, and displacement, 7,135 tons. She is a double-turreted citadel-vessel, similar in many respects to the late English designs, but having the turrets placed on the line with the keel. A German writer describes the vessel in this manner:

The conditions to be satisfied by this class, in their *rôle* of chief defenders of the German coast, are light draught, in order to be able to enter the eastern harbors, and equality in offensive and defensive strength with the armored ships of opposing nations. These demands have been carried out in the calculations for the building and equipment, wherein many new arrangements and departures from earlier contrivances show themselves.

The application of iron-clad frigates to the business of fighting on the high seas demands, also, a strong armor to give sufficient resistance to the ever-increasing caliber of the enemies' guns, without drawing the limits of mobility, maneuvering, and seaworthiness too narrow. After entering into considerations from this point of view, the armoring of the ship the whole length of the water-line was deemed unnecessary; furthermore, it was considered sufficient if this armor existed as a casemate only and in the middle of the ship for the protection of the machinery, &c. The application of side armor was therefore entirely renounced.

In order to limit the depth of the destruction of the unarmored side walls of the ship along the water-line, but chiefly to insure the whole under part of the vessel against destruction by shot, an iron-clad deck is built forward and abaft the armored casemate about 6 feet 6 inches under water, so that the under part of the ship is completely closed from above, and a destruction of the side walls to this deck only is possible. A cork girdle about 3 feet 3 inches wide and of the same thickness is fastened on the inside, forward and abaft the casemate, for the protection of the ship against sinking in case of damage by shot to the unarmored part. For further security to the remainder, the under portion of the ship is divided into a great number of cells. The whole interior is next divided into right and left halves by a water-tight bulkhead, built upon the keel and running the whole length. Each of these halves is again divided by sixteen water-tight cross-walls into just as many closed parts, and each of these parts by the arrangement of a water-tight inner ship-bottom, parallel to the outer, with the intervening space subdivided by the frames, is again split up by vertical and horizontal walls; so that the ship's body under water presents a web of 120 cells. As the joints of each cell are closely made, a leak caused by a ram-thrust or a torpedo can only affect a small part of the ship. A system of pump attachments is provided for the speedy expulsion of water from any leaky cell.

The casemate is protected from above by a 2-inch wrought-iron deck. Upon this are situated two armored turrets in the line of the keel, the after one receiving four 10¼-inch guns, and the forward turret one 12-inch gun. In the after turret is placed a somewhat higher commanding pilot-house. A lance-shaped ram, about 9 feet 10 inches long, fastened to the bow, serves as the second weapon. A contrivance for the launching of torpedoes constitutes the third weapon.

For propelling the vessel, there are two separate pairs of engines, each working up to 2,800 horse-power, and driving each a four-bladed screw-propeller. The requisite steam is generated by eight boilers in four groups of two in each group. They are situated so that the coal can be very conveniently passed into the fire-room.

Forward and abaft the casemate and above the cells are built quarters for the officers and crew, and upon the bow is still another structure, mounted for protection against heavy seas. This last is built narrower than the first, to permit a forward fire from the two after guns.

Special machinery is prepared for ventilation, steering, hoisting anchors, and pumping, and in all remaining relations the contrivances of modern designs are carried out so as to fashion the most complete fighting apparatus possible.

OTHER VESSELS.

Germany has now twenty-four armored ships afloat, five of which are turret-ships, viz, the *Preussen*, the *Friedrich der Grosse*, the *Grosser Kurfürst*, the *Sachsen*, and the *Baiern*. The first of these was built at the private dock-yard of the Vulcan Company, near Stettin, and was launched in 1873; the second was built at Ellerbeck, near Kiel, launched in 1874, and completed at the Kiel dock-yard; the third was constructed at the Wilhelmshafen dock-yard, launched in 1875, and at the date of my visit was sufficiently advanced toward completion to receive the armor-plates and turrets; the last two, of recent construction, the *Sachsen* completed, and the *Baiern* in process of construction, have already been mentioned, and there are two others commenced. The first three vessels are full-rigged ships, constructed after the model of the British ship *Monarch*, and engined by German factories. The boilers of the *Grosser Kurfürst* are of the old box type, with 680 square feet of grate, and 17,736 square feet of heating surface, and the engines are intended to be worked either compound or non-compound, as desired, an arrangement, in view of the low pressure of steam, not to be commended. The last-named two are turreted vessels embodying all the improvements of the present day, as may be seen from the description of the *Sachsen*.

The Germans have also three broadside-frigates, viz, the *König Wilhelm*, built at the Thames Iron Works, London, and at the time of her construction reckoned to be a powerful vessel;* the *Kronprinz*, built by Samuda, on the Thames; and the *Friedrich Karl*, built in France, near Toulon. Besides the two citadel-frigates, the *Kaiser* and *Deutschland*, above described, there are also the armored corvette *Hansa;* the sea-going monitors *Arminius* and *Prinz Adalbert*, the latter built in France, and the former by Samuda, at Poplar, on the Thames; and five armored gunboats, two launched in 1876 and three in 1877, each of 1,000 tons displacement and carrying one 12-inch gun. Besides the above-mentioned vessels, they have eight new unarmored corvettes. The old wooden corvettes, such as the *Hertha* and *Medusa*, and other vessels with auxiliary steam-power, are obsolete, and may be considered useless as fighting-vessels. The Germans also have six dispatch-boats and yachts, and fifteen wooden gunboats.

No originality has yet been exhibited in naval architecture or marine engineering by the German constructors. In commencing to build a navy it has been thought wise, in consideration of the great and varied experience of the English, to repeat the types which they have tried and found successful, and as a consequence of the long time it has taken the Germans to build an armored ship, their earlier vessels, although recently completed, are far below the standard of modern fighting-ships. But those last devised and put afloat in 1877 are represented, as may be seen from the descriptions, to possess the improvements and advantages of recently-constructed armored vessels. And, although Germany will have nothing to match the British mastless armored sea-going ships, or the Italian *Duilio*, *Dandolo*, and other such powerfully-armored craft, their armored fleet will soon possess a degree of strength sufficient, perhaps, to meet the French under any conditions proffered. Besides, the German naval authorities have not overlooked the progress being made in other countries in the construction of rapid unarmored ships, and with a view of keeping pace, in this respect, with other powers, four of

---

* The *König Wilhelm* was designed by Mr. E. J. Reed, C. B., M. P., for the Turkish Government, and was originally called the *Fatikh*. She was sold during her construction to the Prussian Government.—AN ENGLISH NAVAL ARCHITECT.

the recently-constructed corvettes are provided with power sufficient for a speed of 14 knots per hour, the *Sedan* having averaged this speed on the trials. Others of still greater speed are projected, and one fine specimen of a torpedo-boat, the *Ziethen*, hereafter to be described, made 16½ knots per hour on the measured mile; another, also, is designed.

The German ships are armed with the Krupp breech-loading steel guns, the weight and power of which are being steadily increased; their iron-ship-building yards and engineering works are being rapidly developed; their young constructors and engineers are under systematic, practical training, and in all branches of ship construction they are now no longer dependent on foreign countries.

DOCK-YARDS.

The two principal naval dock-yards of Imperial Germany are located at Kiel, on the Baltic, and Wilhelmshafen,* on the North Sea. The latter invited special attention and examination, not only because of its greater importance as a construction-yard, but also in consequence of its being the most modern of all the European dock-yards. It was laid out on clear ground as late as 1871, and the works were commenced with an appropriation for them of about $10,000,000. The engineer selected by the Admiralty at Berlin to make the designs and plans, made visits of observation and study to the leading European yards previous to beginning his work. The plans subsequently produced by him for this important naval establishment, and from which the works and buildings up to this time have been constructed, are most excellent in their general arrangement as well as in the design of the buildings themselves. Those already completed present an outward uniform appearance not found in dock-yards elsewhere. They are built one story high, of stone, brick, iron, and glass; and such of them as are to be used for mechanical operations are grouped in three spans to each building, and well lighted from the sides as well as from the roofs. These are supplied with the best kind of machinery and appliances. Several buildings are yet to be erected on sites properly located on the general plan, and one novel arrangement in the design, but not yet commenced, is a canal to convey fresh water from a distance into the basin in the yard.

To the original harbor, opened in 1870, there have been added in the last five years three stone dry-docks, two building slips, a basin for the equipment of ships, a torpedo-harbor, a boat-harbor, and a mercantile harbor, accessible by special locks. Artillery and torpedo establishments, likewise, have grown in importance.

Similar progress is visible at Kiel. The large naval port at Ellerbeck, with its four dry-docks, will be opened at an early day. Three ship-houses are ready for use, and improvements in the works are progressing.

Dantzig, the third naval center and base of maritime operations, has an iron floating-dock with a shallow basin, and three building-slips are being constructed. To render the Vistula accessible to vessels of large size, the mouth of the river is about to be deepened, which will considerably enhance the naval facilities of the place.

The following tables, kindly furnished for the most part by the liberal German authorities, will show the dimensions, &c., of the armored and unarmored vessels of the empire.

* The plan of the Wilhelmshafen dock-yard has been omitted here, but may be found in the first edition of this report.

Armored ships of Germany.

Name of ship.	Vessel.								Armament.			Machinery.				Cost of ships, in dollars, gold.	Remarks.
	Class.	Length between perpendiculars, in feet and inches.	Extreme breadth, in feet and inches.	Depth of hold, in feet and inches.	Immersed midship section, in square feet.	Displacement, in tons.	Draught forward, in feet and inches.	Draught aft, in feet and inches.	Number of guns and description.	Height of battery above water, in feet and inches.	Thickness of armor at water-line, in inches.	Number of screws.	Indicated horse-power.	Maximum speed in knots per hour.	Makers of engines.		
Kaiser	Casemate-frigate	280 0	62 4	.....	1,336	7,440	24 2	24 8	8 9½-inch and 1 8-inch Krupp gun.	13 7½	10	1	7,890	14.4	Samuda Brothers	1,956,160	Overhanging casemate; bulwarks set in-board; fitted with ram.
Deutschland	....do	280 0	62 4	...	.....	7,560	24 2	24 8	....do	11 0	10	1	7,890	14.5	John Penn & Son	1,768,890	
Sachsen	Two-turret ship	213 3	51 4	26 [illegible]	....	7,135	Mean	19 8	4 10.24-inch and 1 12-inch gun.	.....	8	2	5,600	14	.....	.....	Guns *en barbette.*
Baiern	....do	213 3	51 4	26 [illegible]	.....	7,135	Mean	19 8	....do	.....	8	2	5,600	14	.....	.....	Do.
A and B	....do	213 3	51 4	26 3	.....	7,135	Mean	19 8	....do	.....	8	2	5,600	14	.....	.....	Building.
Preussen	... do	308 6	53 6	34 10	....	6,748	23 0	24 8	4 10.24-inch; 2 6.7-inch; 4 3-inch Krupp guns.	13 6	9.2	1	5,327	14	Vulcan Engineering Company.	1,269,656	Armor-belted; casemate for protection of turret; thickness of turret 10.32 inches; bulwarks arranged for letting down.
Grosser Kurfürst	One-turret ship	298 3	52 0	23 0	1,143	6,558	22 0	23 0	....do	8 5	9.2	1	5,327	14	Egells	903,436	
Friedrich der Grosse	....do	298 0	52 0	.....	2,678	6,558	23 0	24 8	....do	.....	9.2	1	5,327	14	....do	1,177,361	
König Wilhelm	Broadside-frigate	352 8	60 0	25 8	1,299	9,451	24 3	26 6	18 9½-inch and 5 8¼-inch Krupp guns.	10 0	8	1	7,890	14.7	Maudslay	2,424,678	Armor belted.
Prinz Friedrich Karl.	....do	290 0	54 6	....	1,047	5,819	22 0	24 0	16 8¼-inch Krupp guns.	7 6	5	..	3,450	13.5	Forges et Chantiers.	1,548,791	Armor-belted; fited with ram; bark-rigged.
Kronprinz	....do	286 0	...	....	....	5,393	22 8	24 6	....do	.....	5	1	4,735	14.3	Penn	1,511,213	
Hansa	Corvette	235 2	45 0	.....	....	3,497	17 6	21 7	8 8¼-inch Krupp guns.	.....	6.2	..	2,960	12	Vulcan Engineering Comyany.	879,699	Armor-belted; casemated; bulwarks set in-board.
Arminius	Coast-defender	197 4	36 0	.....	.....	1,558	12 5	13 7	4 8¼-inch Krupp guns.	....	4.5	1	1,184	10.5	.....	462,564	Two turrets 3 feet above deck; no rigging; ram; bulwarks removable.
Prinz Adalbert	....do	159 9	....	....	....	1,456	13 2	15 9	2 6.7-inch and 1 8¼-inch Krupp gun.	.....	4.7	2	1,184	9.5	.....	456,484	Two fixed turrets; after bulwarks removable; long ram; brig-rigged.
Wespe	Gunboat	.....	....	...	....	984	.....	...	1 12-inch gun	.....	8	..	690	9	.....	.....	Guns *en barbette;* no rigging.
Viper	....do	.....	.....	....	....	984	.....	....	....do	.....	8	..	690	9	.....	.....	
C, D, and E	....do	.....	...	....	....	984	.....	....	....do	.....	8	..	690	9	.....	.....	Do. Building.

Unarmored ships of Germany.

Name of ship.	Vessel.		Armament.	Machinery.			Remarks.
	Class.	Displacement, in tons.	Number of guns.	Number of screws.	Indicated horse-power.	Maximum speed, in knots, per hour.	
Renown	Line-of-battle ship	5, 432	21	....	2, 960	....	
Leipzig	Corvette	3, 863	12	....	4, 735	14	Iron; covered gun-deck.
Sedan	....do	3, 863	12	....	4, 735	14	Built of iron.
Bismarck	....do	2, 460	16	....	2, 817	15	Iron; covered gun-deck.
Blücher	....do	2, 460	16	....	2, 817	15	Do.
Moltke	....do	2, 460	16	....	2, 817	15	Do.
Stosch	....do	2, 460	16	....	2, 817	15	Do.
Elisabeth	....do	2, 429	18	1	2, 368	....	Covered gun-deck.
Hertha	....do	2, 228	19	1	1, 480	....	Do.
Vineta	....do	2, 228	19	1	1, 480	....	Do.
Freya	....do	1, 954	6	1	2, 368	....	Open gun-deck.
Ariadne	....do	1, 665	6	1	2, 070	....	Do.
Luise	....do	1, 665	6	1	2, 070	....	Do.
Victoria	....do	1, 768	10	1	1, 280	....	Do.
Augusta	....do	1, 768	10	1	1, 280	....	Do.
Medusa	....do	1, 164	9	1	790	....	Do.
Nymphe	....do	1, 164	9	1	790	....	Do.
Hohenzollern	Dispatch-vessel	1, 697	2	....	2, 960	....	New iron vessel.

Two more vessels of the Bismarck class are intended to be added to the fleet. Their length is 244 feet 5 inches; breadth, 45 feet 1 inch; probable draught, 19 feet 8 inches; the calibre of their guns is 5 inches, and their 3-cylinder engines are to be built by Egells.

Besides the above, there are 5 dispatch-vessels, 16 gunboats, 2 transports, 4 torpedo-boats, 14 harbor-service vessels, and 14 sailing-vessels, receiving-ships, hulks, &c.

THE ESTABLISHMENT OF HERR FRIED. KRUPP, AND HIS GUNS.

The most extensive as well as the most important steel-works and gun-manufactory in the world are those of Herr Krupp, located near Essen, in Rhenish Prussia. Established in 1810, it has from small beginnings grown to its present importance by extraordinary skill, business capacity, and energy, chiefly through the life-time of the present proprietor.

The ground occupied for all purposes is about 500 acres, one-half of which is under cover.

The following statistics, taken on the spot, will convey an idea of the magnitude of these works: Number of smelting-furnaces, 250; annealing furnaces, 390; heating-furnaces, 161; welding and puddling furnaces, 115; cupola and reverberatory furnaces, 33; and furnaces of other kinds, 160; coke-ovens, 275; smiths' forges, 264; number of steam-engines, 298, varying in horse-power from 2 to 1,000 each, and supplied with steam by 298 boilers; number of steam hammers, 77, varying in weight from 250 pounds to 50 tons; number of rolling-trains, 18; number of machine-tools, 1,063, viz: 365 turning-lathes; 82 shaping-machines; 199 boring-machines; 107 planing-machines; 42 punching and shearing-machines; 32 pressing-machines; 63 grinding-machines; 31 glazing and polishing machines; and 142 machines of other kinds. In 1876, the number of workmen employed was 10,500, besides about 5,000 more in the mines and at the blast-furnaces.

The articles of steel manufactured are guns, gun-carriages, shot and shell, shafts and other pieces of machinery, and boiler-plates for steamers; axles, tires, wheels, rails, springs, &c., for railways; rolls, tool-steel, &c. Ingots for guns are cast from crucibles in the usual way, but the method of preparing the steel is not made known. In one of the casting-houses accommodation is provided for 1,600 crucibles.

The number of persons who are granted the privilege of visiting the establishment is limited, being chiefly confined to people of distinction known to the proprietor, representatives of foreign governments or officers armed with authority from their governments to acquire information.

This celebrated establishment has furnished guns for armaments on land and sea to every European power, England and France excepted; weapons varying in size from the smallest field-piece used in the Franco-German war to the heaviest gun mounted in the German Empire.

THE GREAT KRUPP STEEL BREECH-LOADER.

For the drawing representing this piece of ordnance, the most powerful breech-loading gun ever constructed, I am indebted to the proprietor of the works. The general dimensions being marked thereon, an idea of its magnitude may be formed. In like manner with all guns manufactured at the Krupp Works, it is made wholly of crucible steel, and the rifling is on the polygroove system, the elongated projectile being rotated by means of a gas-check.

Colonel Wilhelmi, of the imperial royal Austrian marine, in a report on the Krupp breech-loading system, wrote that "Krupp's cylindro-prismatical wedge with Broadwell ring, the best breech apparatus known, requires for its manufacture uncommon mechanical appliances and great technical skill."

Originally the Krupp guns were made from forged solid ingots, but of recent years they have been built up by shrinking hoops over a tube in a manner similar to those of Woolwich, the several parts being made of steel of different qualities, that for the inner barrel of hard and elastic quality, and that for the outer portions becoming more soft and ductile as the exterior is approached. The exact kind of material to be used for each portion of the gun has been ascertained by long experience and at great cost. The design is such that, even after the elastic limit of the material has been reached, much work may be done in permanently stretching the ductile metal before the limit of fracture is arrived at, and thus there is obtained a large margin of safety. It was the smallness of the margin existing between the limits of elasticity and rupture in respect to certain kinds of steel formerly used which rendered a gun so liable to burst explosively. It is believed that Krupp guns in the future will be characterized by success, being the result of patient perseverance, scientific investigation, and a liberal expenditure for experimental inquiry.

The great gun recently manufactured is not only the heaviest breech-loader and heaviest steel piece, but also the heaviest gun of any kind ever made on the continent of Europe. (The largest piece previously made is the one exhibited at the Centennial Exhibition and mentioned already in this report.) It has not yet been tested, but the work expected of it has been closely estimated, and for comparison between the three greatest guns of the present day the following table is presented:*

*For descriptions of the system of constructing these guns, see a *Treatise on Ordnance*, by A. L. Holley, B. P., and a *Report on the Artillery of Europe*, by Captain Simpson, U. S. N.

Krupp's Breech Loader.

40-Centimeter (15·748-In.) Gun.

Dimensions in inches.

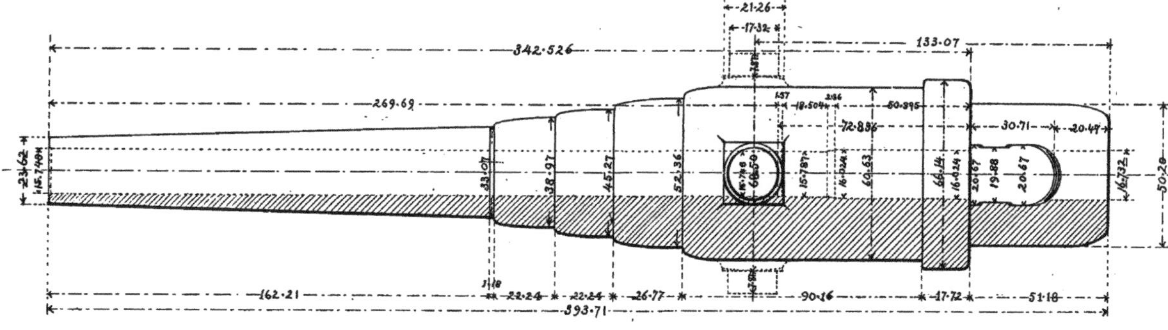

Remarks.

Weight of Gun incl. Breech-block,	158,731. lbs.	Charge, 385·805 lbs. of Prismatic Powder of density, 1·75	
Preponderance of the Breech.	0.	Weight of Chilled, or Steel Shell,	1664·473 lbs.
Number of Grooves,	90.	Initial velocity of Chilled or Steel Shell,	1640·45 ft.
Depth " "	.0787"	Weight of Time fuze Shell,	1375·67 lbs.
Width " "	.3720"	Initial velocity of Time fuze Shell.	1771·7 ft.
" " Lands,	.1772"	Length of twist of Grooves,	708·6744 in.

	Armstrong.	Woolwich.	Krupp.
	100-ton gun.	81-ton gun.	Breech-loader.
Material	Steel barrel and wrought-iron rings.	Steel barrel and wrought-iron rings.	Steel barrel and steel rings.
Actual weight of gun	101.5 tons	81 tons	70.682 tons.
Total length of gun	32 feet 10½ inches	27 feet 4½ inches	32 feet 9¾ inches.
External diameter at breech (greatest).	77 inches	72 inches	60.14 inches.
External diameter at muzzle.	29 inches	24 inches	23.62 inches.
Diameter of bore	17 inches	16 inches	15.75 inches.
Length of bore	30 feet 6 inches	24 feet	28 feet 6½ inches.
Number of grooves in barrel.	27	13	90.
Size of grooves in barrel	1 inch wide by ⅛ inch deep.		0.372 inch wide by 0.078 inch deep.
Weight of projectile	2.000 pounds	1,700 pounds	1,664.47 pounds.
Weight of charge (powder)	375 pounds	370 pounds	385.8 pounds.
Velocity of projectile in feet per second.	1,543.8 feet actual	1,523 feet, actual	1,640.45 feet (estimated).

The Armstrong and the Woolwich guns have been made experimentally in advance of the service weapons for the Italian ships and the *Inflexible.* The velocity in feet per second given for the projectile of the former is far from what is expected to be achieved in the service guns having a bore of 17¾ inches and powder-chamber of 19¾ inches, and the velocity given for the Woolwich gun is that attained before its powder-chamber was enlarged; therefore no correct comparison in this respect can be made until further trial with them, and until the Krupp gun also be put on its trial tests.

It is no less surprising than interesting to look back upon the guns with which ships of war were armed, even up to a comparatively late period, and to compare them with the guns now in existence, to say nothing of those contemplated. At the commencement of the present century the largest naval guns were of cast iron, throwing a projectile of 32 pounds weight. Twenty-pounder carronades constituted a considerable portion of the armament of many of the best ships in the principal European navies. As late as the period of the war carried on by England and France against Russia, 68-pounders were the largest guns in the fleets, these being, like their less formidable predecessors, cast-iron smooth-bores and in most cases mounted upon wooden truck-carriages. Occasionally they were employed for chase purposes, and were mounted on slide-carriages which rested on metal radius plates secured to the deck; these were then considered very heavy ordnance. The revolution in the art of constructing artillery which the past ten or fifteen years have witnessed is chiefly due to improvements in mechanical appliances and in the manufacture of wrought iron and steel.* It would be an impossibility to make satisfactory cast-iron guns of such an enormous size as would be required to discharge projectiles of from 1,700 to 2,500 pounds in weight. No thickness of metal, however great, in a homogeneous cylinder, can withstand a pressure per square inch exceeding the tenacity

*Any one who examines the old guns in the museums of artillery at Berlin, Vienna, Venice, Paris, and in the Tower of London, may see that they are of the same genus as modern smooth-bores, and may even notice some specimens quite as soundly and artistically cast as any of those of the present time. Moreover, from an examination of the wonderful piece of ordnance in the old castle at Edinburgh, Scotland, it may be seen that attempts to produce heavy wrought-iron guns were made nearly two hundred years ago. This piece of artillery was made by forming the barrel of wrought-iron staves and hooping it with rings of the same material. The diameter of the bore is about 12 inches, and the projectiles used were stones, spherical in shape. The want of proper materials and facilities alone prevented success at that early day in the manufacture of ordnance.

per inch of bar of the same metal. Guns manufactured by casting iron or bronze in molds cannot be made of sufficient strength to withstand more than a certain pressure within them; the inner portions receive the brunt of the explosion while the outer portions remain almost quiescent, consequently there is a certain amount of idle metal on the exterior; besides, the expansion of the interior of a thick cylinder of the same metal by explosive heat, while the exterior metal retains a lower temperature, is an element of weakness.

To Sir W. Armstrong is due the credit of first successfully employing the principle of initial tension for all the parts of a gun. He employed wrought-iron coils, shrunk one over another in such a manner that the inner tube was placed in a state of compression, and the outer portions in a state of tension, the amount of tension being so regulated that each coil should perform its maximum amount of useful effect in resisting the pressure from within. The fiber of the coils was arranged in the best position for resisting circumferential strain, and he employed a forged breech-piece in which the fiber ran parallel with the axis of the gun, in order to give longitudinal strength. Steel tubes were gradually introduced, until finally the use of wrought-iron tubes was entirely discarded, except in the case of the cast-iron converted guns. But the Armstrong gun was a costly weapon. The coils were thin and numerous, besides which the large forging for the breech-piece was very expensive. There was also an element of weakness exhibited in the case of a gun which split some of its outer coils while the interior ones remained uninjured, thus proving that the outer coils were unable to take their proper share of the strain. In 1865, Mr. Fraser proposed his important modification of the system. This consists in using larger coils made of thicker bars, and so much stronger longitudinally that the forged breech-piece becomes unnecessary. The greater weight and strength of these coils also allow compression to be given more certainly to the steel barrel and the inner coils. In this construction, moreover, the trunnion-ring, which was merely shrunk on to the Armstrong guns, and occasionally slipped, is welded to the breech coil.

The materials at present used for the Woolwich guns consist of mild steel toughened in oil for the inner tube, and good ductile wrought iron for the remaining portion of the gun. Capt. John F. Owen, R. A., in his recent treatise on *Ordnance in the British Service*, says:

> Thousands of rifled guns have thus been made, from the 9-pounder, weighing but 6 hundred-weight, to the 16-inch of 80 tons weight; and not one has burst explosively on service, nor has the sacrifice of a single life been due to their breaking up under service conditions. No built-up gun of wrought iron and steel of the present manufacture has failed from any defect due to the materials of which it was made. A very different result appears where cast iron has been used, either alone or strengthened by hoops.

As experience proved long before the attack on Fort Fisher, guns made from solid forgings, whether of wrought iron or steel, have a tragic history, the bursting of the Stockton gun on board the United States steamer *Princeton* in the Potomac River in 1843, by which several members of President Tyler's cabinet were killed, being a case in point. It must be added, however, that great improvements in the manufacture of heavy forgings have been made since that day.

The greater cost of steel as compared with wrought iron has of course been a matter of some consideration; besides which, previous to the great advance made in the production of good steel, its uncertainty, especially in the higher and harder grades, weighed greatly against it. This uncertainty having been overcome by long experience and costly experiments, it is admitted that hardness, homogeneity, high elasticity, and great tenacity make steel a most satisfactory gun-metal, by which the same power may be obtained with reduced weight of metal. It is believed that the production of steel will be still further improved, so that cast instead of forged tubes and rings may be used for the manufacture of guns, thereby reducing the cost.

In the construction of the great guns and the experiments with them, which have been carried on at an enormous cost, something of value seems to have been learned at every step; the kind of powder, the form of cartridge, the space to be allowed for the powder, the mode of firing, the cartridge, the system of rifling, and the means of giving rotation to the shot, all are matters which have either undergone change or are undergoing a change at the present time.

The effect of chambering guns is also the subject of experiment, and the value of such an arrangement shows itself both in the 80-ton gun and in smaller weapons. A further change is recognized in respect to the abolition of the studded projectile. The effort to produce an efficient gas-check, so as to prevent erosion, has led to the discovery that gas-checks are capable of replacing the unpopular and objectionable studs.

Captain Owen, R. A., says, again:

> The necessity of progress is recognized, for, in artillery as in other matters, not to advance is to go back. The new guns have shown that they possess admirable accuracy and great power. The results achieved with the 80-ton gun are far in excess of those which were originally demanded. It is calculated that this gun will be made to give its projectile of 1,700 pounds weight a muzzle velocity of 1,620 feet per second, or an energy of 30,935 foot-tons, equal to the penetration of 28½ inches of iron plate at 1,000 yards, or 25 inches at 2,000 yards. These results, however, are to be completely outstripped by the new Armstrong 100-ton guns recently shipped to Italy for the *Duilio*. These guns have a caliber of 17¾ inches and a powder-chamber of 19¾ inches, instead of a uniform bore of 17 inches as in the experimental gun tested at Spezia. The charge of powder is to be about 470 pounds, and the projectiles 2,280 and 2,500 pounds.

As great as the results already achieved are, guns have been designed of still greater magnitude. The drawings made for one intended to be manufactured at Woolwich show that it is to weigh between 160 and 200 tons; it is to be capable of sending a projectile through 36 inches of iron at a range of 1,000 yards.

The great Prussian manufacturer of artillery has a design already prepared for a gun to weigh 124 tons, and to be made on the same plan as the one just described. The larger weapon will have a caliber slightly exceeding 18 inches, and is to throw a steel shell weighing 2,205 pounds, or a chilled-iron shell of 2,271 pounds, and the charge of powder will probably be about 500 pounds.

The designs for other monsters of yet more terrible power have been made at the Elswick Works for the great Italian ships *Italia* and *Lepanto*. It is reported that the length of one of these weapons will be 50 feet, the length of the bore 44 feet, and the diameter of the bore 21 inches. The charge of powder will weigh 950 pounds, and the projectile, 5 feet in length, will weigh 6,000 pounds. This gun is calculated to be able to throw its shot a distance of 12 miles, exceeding in range by one-fourth of the distance the 100-ton gun, which is said to throw a bolt a distance of 9 miles.

Note.—Since the above was written, I have heard from the best authority that it is doubtful whether larger guns than those of 100 tons' weight will be used on these vessels.

PART XV.

THE ITALIAN NAVY.

THE DUILIO; THE 100-TON GUN; THE DANDOLO; THE ITALIA AND LEPANTO; THE CRISTOFORO COLOMBO; CRUISING-VESSELS; DOCK-YARDS; TABLE OF DIMENSIONS, ETC., OF ARMORED SHIPS OF ITALY.

THE ITALIAN NAVY.

The fleet of which Italy could actually dispose at the beginning of 1876 was composed as follows: Fourteen armored ships, of which six were already armed and constituted the permanent squadron of the Mediterranean, four were ready to be armed at short notice, and four required considerable repairs before they could be sent to sea; seven gunboats, of which three were outside the Mediterranean, two armed and in the Mediterranean, and two ready to be armed; nine wooden frigates or corvettes, of which two were beyond the Mediterranean, three armed and in Italian ports, two capable of being armed in a few days, and two others requiring repairs; six dispatch-boats, of which three were armed, one of which could be ready for sea at short notice and two after a few weeks; six transports, of which the largest three were armed; and eighteen smaller vessels, of which one was on foreign service, three armed, and fourteen ready to be armed. These ships carry a total of 490 guns, of which 30 on the armored ships are of heavy caliber. From an English journal we gather the following:

By a decree which has been recently issued [July, 1877], the ships of the Italian navy have been reclassified. Henceforth they are to be divided into three categories, distinguished as A, men-of-war; B, auxiliary ships of the fleet; C, local ships. The first category is again divided into three classes. In the first of these are placed [fourteen] men-of-war, fit to take part in every kind of naval warfare. These vessels are the *Duilio*, the *Dandolo*, the *Italia*, the *Lepanto*, the *Palestro*, the *Principe Amadeo*, the *Venezia*, the *Roma*, the *Ancona*, the *Castelfidardo*, the *Maria Pia*, the *San Martino*, the *Conte Verde*, and *Affondatore*. To the second class belong ten ships, destined for employment on special missions, such as protecting Italian commerce in time of war, defending the coasts, cruising on distant stations, &c. These ten vessels are the armored corvettes *Terribile* and *Formidabile*, the armored gunboat *Varèse*, the screw-frigates *Vittorio Emanuele* and *Maria Adelaide*, the screw-corvettes *Vettor Pisani*, *Caracciolo*, and *Garibaldi*, the cruiser *Cristoforo Colombo*, and the paddle-wheel steamer *Governolo*. To the third class of the same category belong twenty gunboats. The second category, that of auxiliary ships of the fleet, is also divided into three classes—the first consisting of cavalry-transport vessels of over 3,000 tons, the second of transport-ships of between 1,000 and 3,000 tons, and the third of paddle-wheel and screw steamers of between 200 and 1,000 tons. The third category—that, namely, of local ships—comprises all vessels of below 200 tons and tenders employed at the dock-yards and arsenals.

Of the ships enumerated, the *Duilio* will probably not be completed until the beginning of the year 1879, and the *Dandolo* a year later. The *Italia* has only recently been commenced, and the designs for the *Lepanto* are not yet completed. The armored fleet actually afloat and now in service embraces only four really good sea-going ships, viz, the *Principe Amadeo*, *Palestro*, *Roma*, and *Venezia;* neither, however, has more than six inches of armor-plating, and in general efficiency they may be ranked after the British ships of the *Audacious* class. Their armament is considered powerful for ships of their class, especially that of the first-named vessel, and here it may be remarked that the Italian ships generally have been noted for the weight of their ordnance. The *Principe Amadeo* and *Palestro* each carries one 23-ton Armstrong gun and six 18-ton guns. The *Roma* and *Venezia*, although of more recent design, carry six 18-ton and 2 12 ton guns. After these four the most powerful is the *Affondatore*, a turret-ship built at the Millwall Iron Works about eight years ago. Her armor-plating, however, is only five inches thick, and she carries nothing heavier than 12-ton guns. The whole of the completed portion of the armored fleet consists of ships ranging from 2,700 to 5,780 tons displacement, and, except in the cases mentioned, the armor is about four inches thick and the guns mounted in a long tier on each broadside.

At present, therefore, the fighting naval force of Italy may have a value, relative to other navies, of 25.5, England having 100 as assigned by M. Marchal in his work *Les Navires de Guerre les plus récents.*

The present unarmored fleet consists of one corvette of the rapid type, the *Cristoforo Colombo;* three wooden frigates, the *Vittorio Emanuele*, the *Garibaldi*, and the *Maria Adelaide;* two screw-corvettes, the *Caracciolo* and the *Vettor Pisani;* four paddle-wheel corvettes, the *Guiscardo, Governolo, Archimede*, and *Ettore Fieramosca;* six screw-transports, the *Europa, Washington, Città di Genova, Città di Napoli, Conte Cavour*, and the *Dora;* five gunboats, the *Veloce, Sentinella, Guardiano, Ardita*, and *Confidenza;* two screw dispatch-boats, the *Rapido* and *Vedetta;* five paddle-wheel dispatch-boats, the *Sesia, Esploratore, Garigliano, Anthion*, and *Sirena*, with many smaller vessels.

The changes and improvements in all branches of administration and industry during the past quarter of a century, and especially in mechanical works, force themselves on the observation of one acquainted with Italian ports and cities in years past. The spirit of advancement and progress is seen to particular advantage in the reconstruction of the navy. So far has it been pushed here that the late minister of marine at Rome, when he took office, is said to have condemned as obsolete, and recommended the sale of no less than seven armored ships out of the total of fifteen, although some of these condemned vessels were quite new, because they did not reach the standard set up for modern fighting-vessels. The plan has not yet been carried out in its entirety, and probably will not be until after the more formidable ships now under construction shall have been completed.

THE DUILIO AND DANDOLO.

These two powerful armored sister ships, the outlines of which are here given, are building, the former at Castellamare,* and the latter at Spezia, in the Mediterranean. They have been designed to outdo the British *Inflexible* and every other fighting-ship afloat.

The *Duilio* was commenced at Castellamare in 1872, and launched in May, 1876, and the *Dandolo* was commenced a year later. For their general de: ig ı the naval authorities accepted the view of the British committee on designs, and trust for both buoyancy and stability to their unarmored raft; moreover, they placed the turrets as they are located in the *Inflexible.* †

The following principal dimensions and elements of these two ships have been furnished by the kindness of the minister of marine at Rome:

Dimensions, weights, &c.

	Duilio and Dandolo.
Year in which their construction was commenced..........	1872 and 1873.
Year in which they will be entirely completed..............	1878 and 1879.
Navy-yards in which they are building.......................	Castellamare and Spezia.
Navy-yard in which they will be completed................	Spezia.

* The *Duilio* was steamed around the Bay of Naples, November 9, 1877, for a preliminary trial of the machinery, and has probably been taken to Spezia.

† Mr. Reed described the principle upon which the *Duilio* and the *Dandolo* have been built to the committee of naval designs in 1871, the idea then having been quite new. See the printed report of evidence given before the committee.—AN ENGLISH NAVAL ARCHITECT.

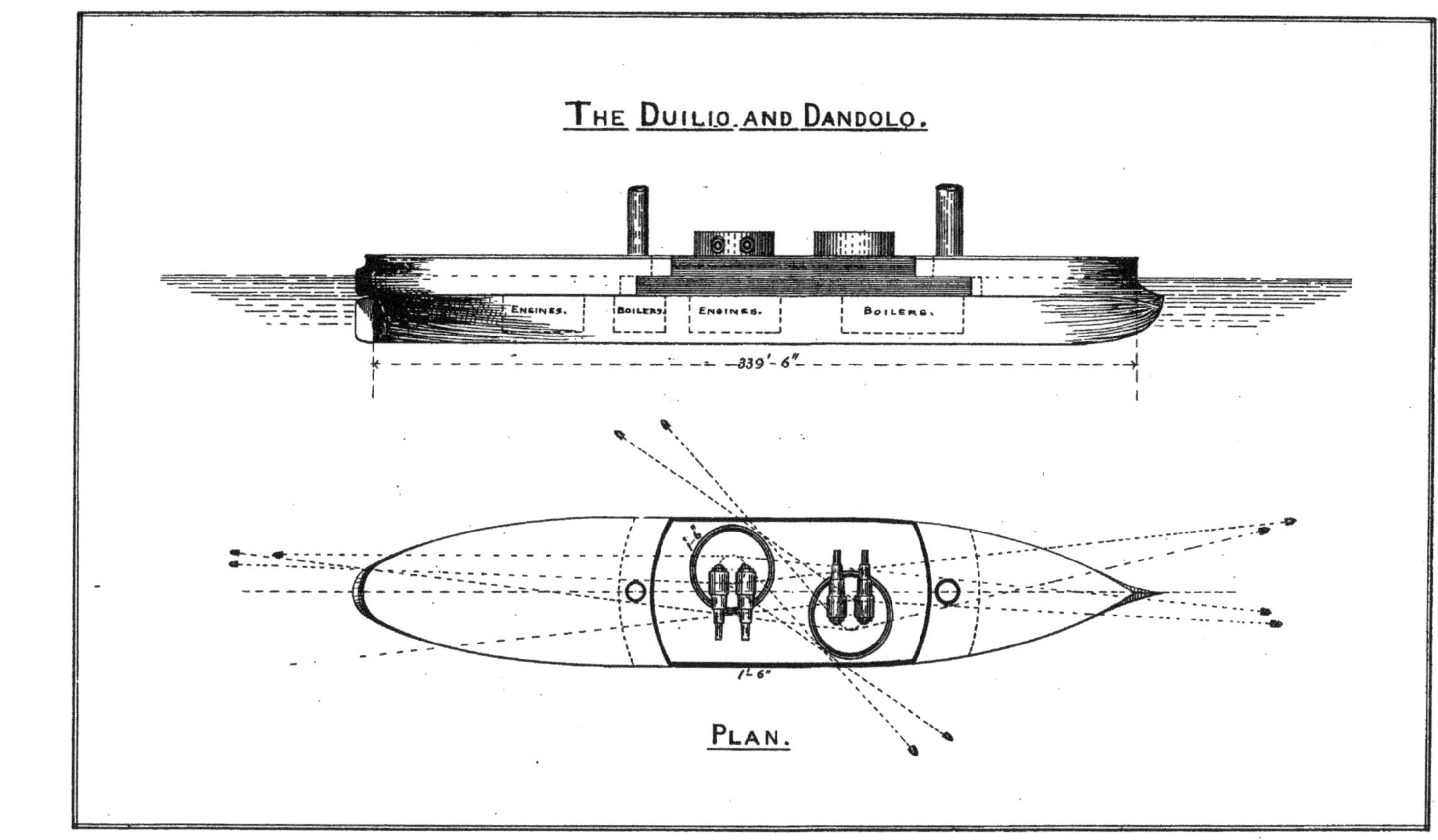
THE DUILIO AND DANDOLO.
ENGINES.
BOILERS.
ENGINES.
BOILERS.
339' - 6"
1' 6"
PLAN.

	Duilio and Dandolo.
Material of which the vessels are built	Iron.
Number of compartments into which each vessel is divided	102
Weight of hull alone, in tons	3,395.5
Length between perpendiculars, in feet and inches	340 11
Breadth of beam, extreme, in feet and inches	64 9
Depth of hold, in feet and inches	21 11
Draught of water, forward, in feet and inches	25 5
Draught of water, aft, in feet and inches	26 5
Displacement at deep load-line, in tons	10,401
Area of immersed midship section, in square feet	1,466.6
Projection of ram forward of perpendicular, in feet	9
Immersion of the same at deep load-line, in feet	14
Number of turrets	2
System of turrets	Revolving.
Diameter of interior of turrets, in feet and inches	25 9
Thickness of armor at water-line, in inches	21.65
Thickness of armor at upper redoubt, in inches	17.71
Thickness of armor of turrets, in inches	17.71
Thickness of backing of wood at water-line, in inches	23.62
Thickness of backing of wood at upper redoubt, in inches	19.68
Thickness of wood backing of turrets, in inches	11.81
Depth of immersion of armor, in feet and inches	5 3
Total weight of armor, in tons	2,559
Total weight of wood backing, in tons	239

Number of guns and their weights, &c.

	Duilio and Dandolo.
Number of guns	4
Weight of each gun, in tons	98.5
Weight of broadside metal, in tons	3.87
Weight of bow-fire metal, in tons	2.9
Weight of stern-fire metal, in tons	1.93
Total weight of guns, their machinery and ammunition, in tons	984.2
Height of main deck above water, in feet and inches	11 6
Height of battery, in feet and inches	15 9
Estimated cost of each ship, in dollars (gold), exclusive of armament and outfit	2,818,800

Motive-machinery.

	Duilio.	Dandolo.
Kind of engines	Ordinary.	Compound.
Name of constructor	Penn.	Maudslay.
Indicated horse-power, contract	7,500	7,900
Number of cylinders	4	4
Diameter of cylinders, in inches	93.5	64 and 120
Stroke of pistons, in inches	39	48
Number of revolutions of engine per minute, estimated	65	80
Number of screw-propellers	2	2
Diameter of screw-propellers, in feet and inches	17 3	16 0
Pitch of screw-propellers, in feet and inches	19 6	19 0
Speed of ships, in knots, estimated	14	14
Number of boilers	10	8
Number of furnaces	40	32
Grate surface, in square feet	899.6	811.6
Heating surface, in square feet	23,775	22,991
Boiler-pressure, in pounds	30	60
Number of smoke-pipes	2	2
Tons of coal to be carried in bunkers	1,279	1,279

The hull is built of iron and steel, on the cellular system. The double bottom extends for upward of 230 feet in length, and is divided both longitudinally and transversely into a great number of water-tight compartments. Each compartment is provided with a branch tube which is connected with one principal tube in communication with powerful steam-pumps. The tubes are, of course, fitted with the necessary valves, so that in the event of damage to the bottom of the vessel, or for any desirable purpose, any one or more of the compartments may be drained or filled with water. There is a central armored citadel or compartment 107 feet in length and 58 feet in breadth, which descends to 5 feet 11 inches below the load water-line. It protects the machinery and boilers, the magazines and shell-rooms, and a portion of the machinery for working the turrets and guns. Forward and aft of this citadel, the decks, which are 4 feet 9 inches under water, are defended by horizontal armor. Over this citadel is built a second central armored compartment which incloses the bases of the turrets and the remaining portion of the mechanism employed in loading and working the guns. Lastly, above this second compartment rise the two turrets. The position of the turrets in the *Duilio* was made the subject of a novel arrangement, and one which was tried for the first time in that vessel. They are placed at each end of the central armored citadel—not in an even line with each other, but diagonally at opposite corners of it, with the centers at the distance of 7 feet 8 inches from the longitudinal center-line of the vessel, so that one turret is on the starboard side and the other on the port side. The effect of this arrangement is to render possible the discharge of three guns simultaneously in a direction parallel with the keel. Only the central portion of the ship and the two turrets will be protected by vertical armor.

As regards the armor of the central portion of the vessel, the thickness of which at the water-line is to be 21.65 inches, it had not been decided, at the date of my visit, whether the plates should be made in one or two thicknesses; that was to depend on the results of comparative experiments made at Spezia by shots discharged by the 100-ton gun, and guns of 10 and 11 inch calibers, against targets constructed on four different systems of steel and iron from three manufacturers.* The decks are protected by horizontal armor of iron and steel, the former being under the latter. The armor of the turrets will be composed of solid plates 17.71 inches in thickness, resting upon teak backing.

ARMAMENT.

The original intention was that the armament should be composed of two 60-ton guns in each turret; subsequently it was decided to employ 100-ton guns, and the opinion prevails that this decision was reached after the fact became known that the *Inflexible* would be armed with 80-ton guns. These 100-ton guns, four for each ship, are being manufactured by Sir William Armstrong, at the Elswick Works, Newcastle-on-Tyne, England; the first one of the number having just been completed ready for shipment to Italy, to undergo the necessary experimental firing tests, at the time of my visit to Elswick.†

The accompanying drawing illustrates the construction of the gun—the heaviest and most powerful piece of ordnance in the world. It is built up according to the well-known Armstrong system, the inner barrel or tube being of steel, rifled with twenty-seven grooves, the spaces between which are nearly equal to the width of the grooves themselves.

* The results of the experiments at Spezia induced the Italian Government to adopt steel for armor in preference to iron. They were led to this determination for the reason that the projectiles fired from the 100-ton gun failed to pass through a target faced with 22 inches of that metal, while similar projectiles from the same gun perforated targets composed of iron plates of the same thickness. They are at the present time engaged in plating the *Duilio* and *Dandolo* with armor of this material.

† According to the latest accounts, two of these guns have been delivered at Spezia for the *Duilio*, and four have been purchased by the British government at the reported price of $80,000 for each.

THE ONE HUNDRED-TON
ARMSTRONG GUN.

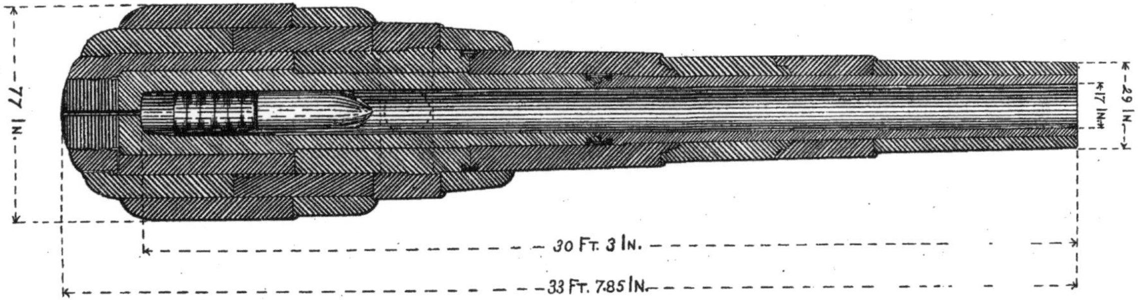

77 IN.
30 FT. 3 IN.
33 FT. 7.85 IN.
29 IN.
17 IN.

The rifling has an increasing pitch, commencing at the chamber with 1 in 150 calibers, and increased to 1 in 50 calibers. The depth of the grooves is $\frac{1}{8}$ inch throughout. The steel barrel is 31 feet 3 inches long, 6 inches thick at the powder-chamber, diminishing to $3\frac{1}{2}$ inches at the muzzle. It is bound by successive layers of wrought-iron coiled cylinders. There are ten of these coils, and they are so arranged as to interlock with and overlap each other at their junctions; six of them are on the rear half of the gun, the remaining four being placed singly end to end around the tube. The first coil at the rear is 13 feet long by 7 inches thick; over this is a coil 8 inches thick and 9 feet 6 inches in length; then the trunnion-coil, which is 11 inches thick and 2 feet long; forward of this is a tapered coil, having an average thickness of $6\frac{1}{2}$ inches and a length of 3 feet 3 inches. Outside of the second long coil is another to the rear, 9 inches thick and 5 feet 6 inches long. Thus the breech portion is bound by three coils, the trunnion portion by two, and the forward portion by one. The trunnions, it may be observed, are not placed on the largest diameter of the gun, but farther forward, where the diameter is considerably reduced. There is consequently a preponderance of weight at the breech end of about 4 tons, which gives stability in working, inasmuch as the weight is always tending in one direction. The exact weight of the gun is $101\frac{1}{2}$ tons, its extreme length is 32 feet $10\frac{1}{2}$ inches, the length of the bore is 30 feet 6 inches, and its diameter is 17 inches. The outside diameter of the gun at the muzzle is 29 inches, and at the breech it is 77 inches.

The weight of the projectile is 2,000 pounds, and that of the proof-shot 2,500 pounds. Rotation is given to the projectile, not by the usual studs fixed in the projectiles to fit the grooves, but by a copper gas-check fixed into the rear end of the shell, and which has projections upon it corresponding with the rifling-grooves of the gun; the shell is so formed that the check on being crushed against it by the pressure of the explosion of the charge presses firmly about it; and the gas-check being caused to rotate by the rifling-grooves, causes the projectile to turn and to take the same rotation. The cartridge measures 52 inches in length by $15\frac{1}{2}$ inches in diameter.

TRIALS OF THE GUN.

The trials decided upon were to settle two points: The one being the test of the gun and machinery for working it under the stipulations of the contract, and the other was the determination of the system of vertical armor to be used on the ships in which the guns are to be mounted. With these objects in view the gun was placed on a pontoon constructed for the purpose, and moored in a firing position in the Gulf of Spezia. It was mounted on its trunnion-blocks, left free to move on the slides, and all the hydraulic apparatus for loading, for regulating the elevation, for running out, and for taking the recoil, was fitted as it will be used when in its place on board ship. As descriptions of similar apparatus for working the guns of the *Inflexible* and *Thunderer* have already been given in this report, it is only necessary to record here the results of the trials, which, with the drawings of the targets shown herewith, have been taken from that excellent paper, the *Engineer*.

If correctly reported, the contract provides that the gun shall discharge 2,000-pound shots with a muzzle velocity of 1,350 feet; that 50 rounds of service charges shall be fired, and 5 of them with projectiles weighing each 2,500 pounds. The experimental firing was commenced

October 20, 1876. The series of rounds fired by the 100-ton gun is given by *The Engineer* of December 29 of that year, in the following table:

Number of rounds.	Date.	Charge, weight and nature, in pounds.	Projectile, weight in pounds.	Velocity, feet per second.	Mean pressure in bore, tons.	Stored-up work, foot-tons.	Recoil-valves set to, pounds.	Recoil, in inches.	Remarks.
1	Oct. 20	200 WA. 1½ inch..	2,000	(*)	......	......	1,050	......	
2	Oct. 21	300 WA. 1½ inch..	2,000	(*)	16.6	......	......	......	
3	Oct. 21	300 WA. 1½ inch..	2,000	1,375	15.9	......	1,150	......	
4	Oct. 23	300 WA. 1½ inch..	2,000	......	......	......	......	......	
5	Oct. 23	300 WA. 1½ inch..	2,000	......	16.0	......	......	35.5	
6	Oct. 23	300 WA. 1½ inch..	2,000	1,374	16.0	......	......	37.5	
7	Oct. 23	330 WA. 1½ inch..	2,000	1,456	20.8	29,400	......	......	
8	Oct. 23	319 WA. 1½ inch..	2,000	1,424	18.0	28,120	......	42.5	
9	Oct. 25	319 WA. 1½ inch..	2,000	......	(*)	......	......	44.75	
10	Oct. 25	336.6 WA. 1½ inch..	2,000	......	19.4	......	......	46.3	
11	Oct. 26	319 WA. 1½ inch..	2,000	1,437	......	28,625	......	42.6	At earth.
12	Oct. 26	341 WA. 1½ inch..	2,000	1,475	19.75	30,150	......	47.1	At Schneider steel plate.
13	Oct. 27	341 WA. 1½ inch..	2,000	1,478	19.75	30,300	......	46.0	At Cammell's iron plate.
14	Oct. 27	341.6 WA. 1½ inch..	2,000	......	......	......	......	......	Shot broken up in bore.
15	Oct. 27	341.6 WA. 1½ inch..	2,000	1,500	20.6	31,200	......	......	At Marrel's iron plates.
16	Oct. 28	341.6 WA. 1½ inch..	2,000	1,493	20.1	30,920	2,180	48.2	At Schneider steel.
17	Oct. 28	341.6 WA. 1½ inch..	......	1,492	19.2	30,880	......	46.0	At Marrel's sandwich t'rg't.
18	Nov. 2	319 WA. 1½ inch..	2,000	......	......	......	......	......	Fired against earth.
19	Nov. 2	319 WA. 1½ inch..	2,500	1,294	19.0	29,027	......	44.75	
20	Nov. 2	319 WA. 1½ inch..	2,500	1,293	19.0	29,000	......	44.5	
21	Nov. 2	319 WA. 1½ inch..	2,500	1,293	18.8	29,000	2,000	44.25	
22	Nov. 3	319 WA. 1½ inch..	2,500	......	......	......	......	42.5	Not taken.
23	Nov. 3	319 WA. 1½ inch..	2,500	......	18.6	......	......	40.5	
24	Nov. 3	276 Fossano......	2,000	1,165	......	......	1,850	23.5	
25	Nov. 3	300 Fossano......	2,000	863	(†)	......	......	17.0	
26	Nov. 3	319 WA. 1½ inch..	2,000	......	......	......	......	......	
27	Nov. 4	319 WA. 1½ inch..	2,000	......	......	......	......	36.0	
28	Nov. 4	319 WA. 1½ inch..	2,000	......	......	......	......	......	
29	Nov. 4	319 WA. 1½ inch..	2,000	......	......	......	......	35.25	
30	......	......	2,000	......	......	......	......	......	
31	......	......	2,000	......	......	......	......	......	
32	......	......	2,000	......	......	......	......	......	
33	Nov. 7	353 WA. 1½ inch..	2,000	1,512	......	31,700	......	42.5	
34	Nov. 7	364 WA. 1½ inch..	2,000	1,514	19.8	31,750	......	42.8	
35	Nov. 7	375 WA. 1½ inch..	2,000	1,542.8	21.4	33,000	......	......	At earth.
36	Nov. 8	319 WA. 1½ inch..	2,000	1,348	......	25,200	......	......	
37	Nov. 8	341 Fossano......	2,000	1,415	......	27,760	......	......	
38	Nov. 8	363 Fossano......	2,000	1,408	......	27,500	......	......	
39	Nov. 8	363 Fossano......	2,000	1,444	13	28,900	......	......	

* Not observed. † Under 10.

The four targets, represented in plates 1, 2, and 3, were faced with three several kinds of plates, viz, wrought-iron plates by Cammell & Co., of Sheffield, steel plates by Messrs. Schneider & Cie., of Creuzot, and wrought-iron plates by M. Marrel, another French maker. The plates were each 11 feet 6 inches long by 4 feet 7 inches deep, and 22 inches thick, mounted on framing representing that of the *Duilio* and *Dandolo*. Referring to the illustrations, the second series shows each target after receiving one blow from a 10-inch projectile. The third series shows each after receiving, in addition to the single round above mentioned, a salvo consisting of one 10-inch and one 11-inch shot. The fourth series deals with the effect produced by the 100-ton gun, either on uninjured or injured targets. The nature of the target and the rounds hitherto fired at it are entered beneath each section.

The targets were built facing the sea. Opposite them was a battery consisting of one 11-inch gun, with two 10-inch guns, one on either side of the 11-inch. Farther to the rear, that is, farther to sea, was the raft on which the 100-ton gun was placed. As this gun is mounted for working in a turret, it has no provision for lateral adjustment or training; the raft, therefore, acted as a turret, the gun was "lined" as to direction by moving the raft, and the latter could be readily turned right round so as to enable the gun to fire for range out to sea. This was the part of the programme that was first

100-Ton Gun Targets. Plate 1.

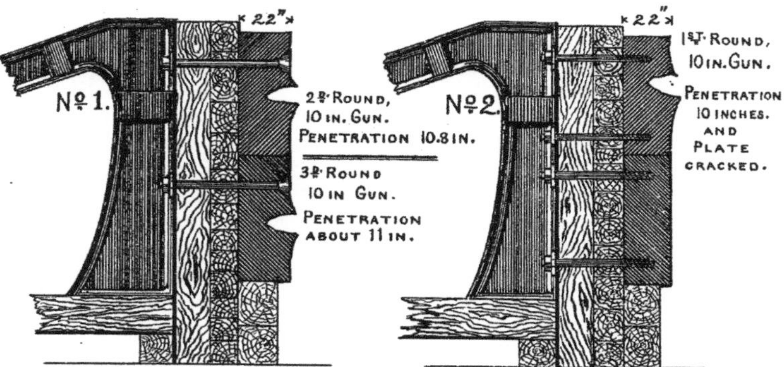

Cammell & Marrel } Iron { Upper Target. Lower Do.
Each Target struck once by 10 in. shot.

Schneider Steel.
Upper Target struck once by 10 in. shot.

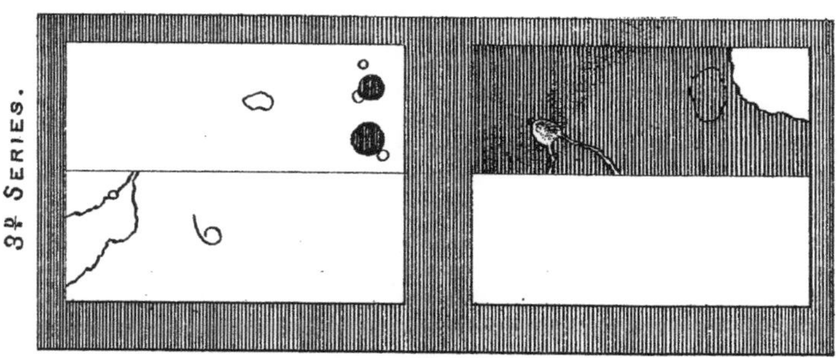

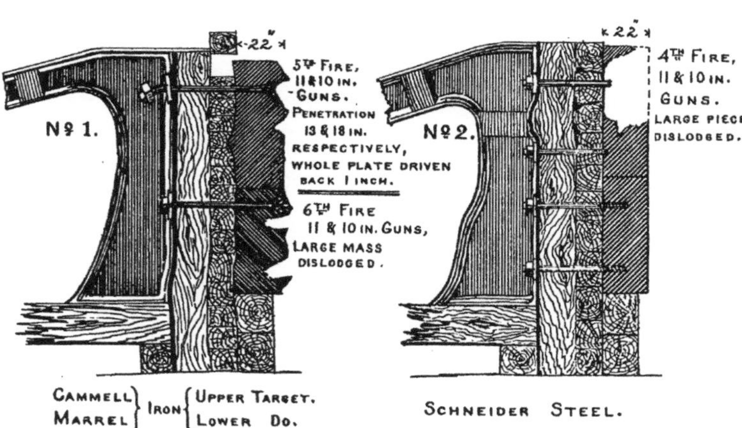

Cammell } Iron { Upper Target.
Marrel } Lower Do.
Each Target after being struck by 1 round, 10 in. shot.

Schneider Steel.
Upper Target after being struck by 1 round, 10 in. shot.

100-Ton Gun Targets.

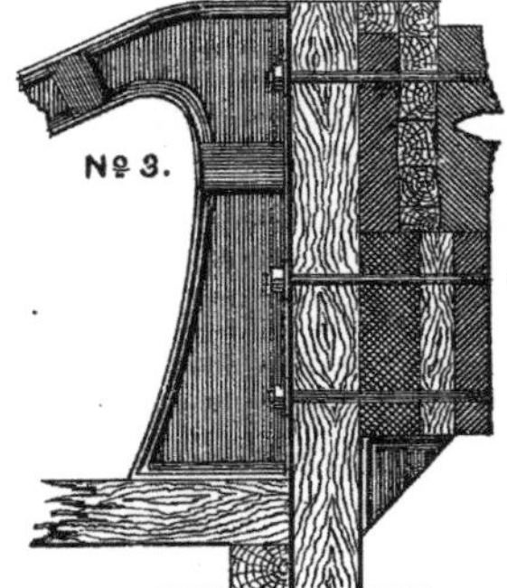

Marrel Iron.
Struck once by 10 in. shot.

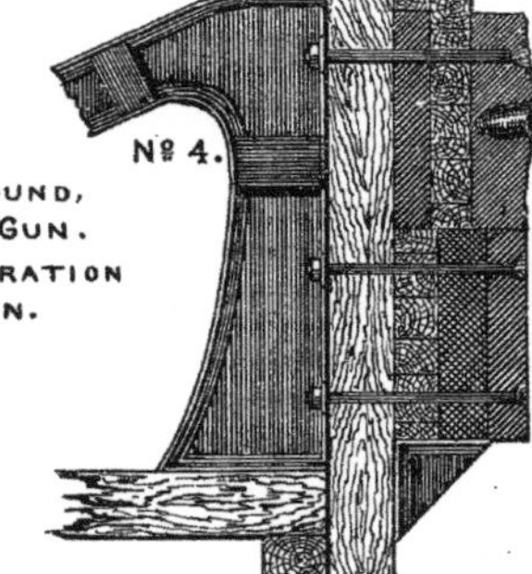

Cammell Iron.
Struck once by 10 in. shot.

3rd Series.

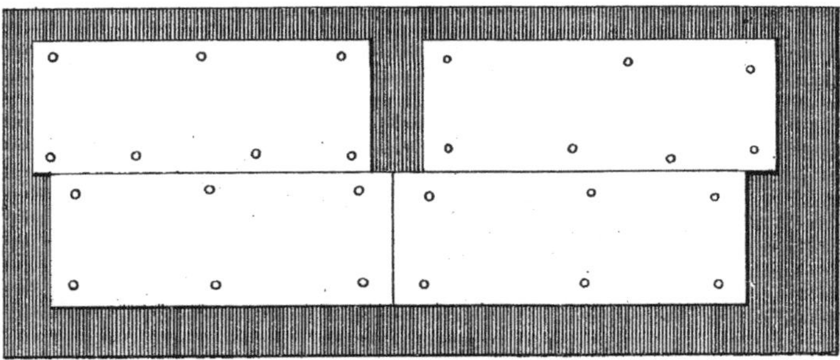

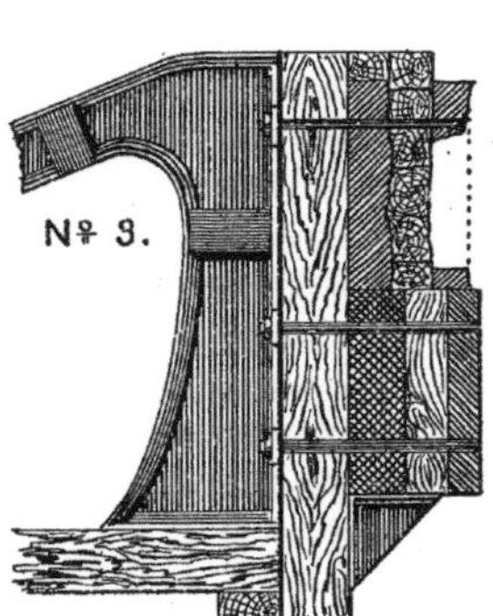

Marrel Iron.

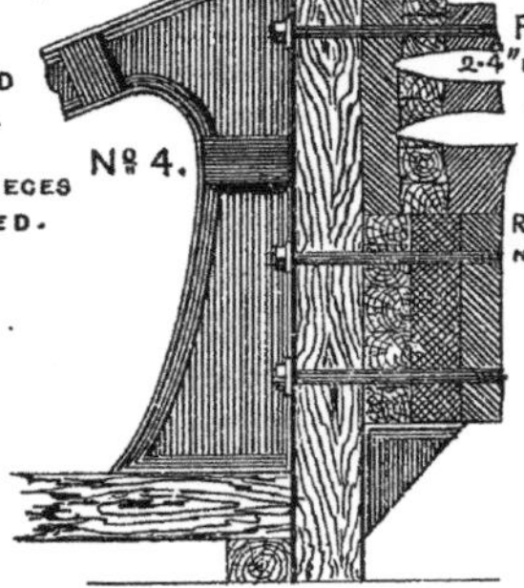

Cammell Iron.

carried out, the rounds from No. 1 to No. 10 being fired to test the performance of the gun as to the velocity it imparted to the shot and the pressure in the bore of the gun, and as to its strength and general behavior on firing, and also to test the action of the carriage. The most important matter in this firing is the velocity that can be obtained with such a pressure as may be deemed allowable for the gun. The highest pressure recorded is 20.8 tons per square inch, and this appears to be rather exceptionally high and irregular. The velocity corresponding to it was 1,456 feet per second. The highest pressure occurring in the 80-ton gun, during the first ten rounds with the full 16-inch bore, was 22 tons, the corresponding velocity being 1,452 feet. This, it is fair to say, was also rather unduly high, being the highest registered at any time with this caliber.

The Italian arrangements as to taking velocities, &c., were not in good working [order] at first; but every portion of the Elswick gear performed its part well, the recoil being checked and the gun working easily.

The plate experiments began on October 25. The construction of the targets is shown in No. 1 series of sections and in Fig. 1.*

No. 1 target consists of wrought-iron plates * * 22 inches thick, on teak backing, with angle-iron and 1½-inch skin. The plate on the upper half of the target is supplied by Messrs. Cammell, that on the lower by Messrs. Marrel.

No. 2 is a similar target, except that 22-inch steel plates supplied by M. Schneider take the place of the wrought-iron plates both in the upper and lower portions of the target.

Nos. 3 and 4 are targets with what has been termed sandwich-plating in the upper halves, that is to say, alternate layers of iron and teak, the front plate in each case being 12 inches thick and the inner plate 10 inches, as shown in Nos. 3 and 4, the only difference between the two being that, in No. 3, Marrel, and in No. 4, Cammell plates, were employed. The lower portions of the two targets had wrought-iron front plates 8 inches thick, No. 3 having a layer of teak next and then chilled cast iron 14 inches thick, and No. 4 having a similar thickness of chilled iron next to the front plates, the teak being all behind. The bolts of all the targets pass through the entire structure except in the case of the Schneider steel plates, into which the bolts were screwed to a depth of not quite half the thickness of the plate. The total thickness of wood in every target was the same, namely, 29 inches; and the iron, exclusive of skin, was 22 inches, the skin being 1½ inches. The sections are given on the plan of the Italian Government sketch.

The object of the programme was as follows: The effect of the lighter guns to be first ascertained; a single round from the 10-inch gun being first fired at each, and then a salvo of the 10-inch and 11-inch guns. The whole three guns were intended to be fired simultaneously, but one 10-inch missing the first round, two only were fired in the subsequent salvos. One steel Schneider half-target was to be reserved untouched for the 100-ton gun; the remainder were to be brought under its fire after the lighter guns had done their worst on them. It was considered that the targets ought to be more than a match for 10-inch and 11-inch guns, and consequently that, after being struck, a very fairly strong target of each kind would thus be fired at by the 100-ton gun. The firing was as follows:

No. 1 round: 10-inch shot against No. 2 upper half, Schneider steel 22-inch plate. Shot penetrated about 10 inches near the center of the plate—*vide* target No. 2, No. 2 series of sections—the plate at the edges of the hole being burred up about 4.4 inches. At first the plate appeared to be but little damaged, but a singing noise was heard in the metal, which shortly after split in two cracks, one running from the hole to the proper right edge of the target, and the other in a downward direction, extending some distance, but not to the plate edge.

No. 2 round: 10-inch gun against upper half of No. 1 target, Cammell wrought-iron 22-inch plate; penetration about 10.8 inches in depth. Two cracks were developed, extending from a left bolt-hole to target edge.

No. 3 round: 10-inch gun at lower half of No. 2 target, Marrel, 22-inch, front plate. Penetration about equal to that in Cammell plate; crack opened from lower left bolt-hole.

No, 4 fire: Salvo from 11-inch and both 10-inch guns at Schneider upper target No. 2. One 10-inch gun missed fire, so that one 11-inch and one 10-inch shot were actually fired at the plates. Both struck from 2 feet to 3 feet of right of target—*vide* third series of sections. A large piece of plate was dislodged from the right-hand top corner. The cracks already made were opened much wider, and fresh cracks were visible, especially in the shot-hole. The entire plate, in fact, had suffered very severely, and was far on the way to destruction. The 100-ton gun was then fired two rounds, Nos. 9 and 10 on table, which we keep distinct from the plate-firing.

On October 26—No. 5 fire at plates: A salvo of one 10-inch and one 11-inch shot was fired at Cammell's 22-inch iron plate— target on third series; the plate was struck rather near to the edge. The 11-inch shot lodged and broke up; the 10-inch drove one bolt some distance inward. The top layer of that was lifted above the target edge. The

* The engravings of the first series are omitted as being unnecessary.

entire plate was driven back 1 inch, and a crack was made from a bolt-hole to the edge of the plate.

No. 6 fire at plates: Salvo of one 10-inch and one 11-inch shot at Marrel iron 22-inch plate, No. 1 target, lower portion, third series. A huge mass of iron was dislodged, and a crack was formed from bolt-hole to toward top edge of plate; penetration not quite so much as in Cammell's plate. The iron of the plate was heard to sing. The general quality of it appeared decidedly more steely than that of Cammell's plate. The 100-ton gun was now fired at an earth-butt 52 feet thick and 27 feet high, with results given in table, No. 11 round.

No. 7 fire at plates: The 100-ton gun was next fired with 340-pound charge at the Schneider steel plate, which had hitherto been left untouched. The effects are shown generally in lower portion of No. 2 target, fourth series. The plate was smashed to pieces, but the shot had broken up, and had not penetrated through the backing. The whole target had been driven 8 inches back, the inner skin was bulged and opened, and the angle-iron torn and twisted. For units of work and pressure, &c., see table, No. 12.

On the 27th—No. 8 fire at plates: The 100-ton gun was fired at the Cammell iron plate, with results indicated generally in the upper half No. 1 section of No. 4 series. For charges, see No. 13 in table. Half the plate was struck away, leaving the wood bare. A part of the shot passed completely through the target, having a velocity measured at 600 feet on the far side. A hole was made nearly 4 feet in diameter, showing *débris* of broken iron and wood in interior. The following round was fired at the steel target, but the shot broke up in the bore of the gun.

No. 9 effective fire at plates: The 100-ton gun was next fired at Marrel's target, on the lower half of the same structure as Cammell's. (See No. 1, series 4.) The whole knocked into ruins. The shot passed completely through, leaving a large hole in the center, as in the case of Cammell's, showing the same heap of *débris*. For charge, velocity, &c., see No. 15 on table. Plate No. 2 shows the state of the targets at this stage of the proceedings.

No. 10 effective fire at plates was from the 100-ton gun at the upper half of Schneider's steel target. For effects, see No. 2 in No. 4 series. Although this plate had suffered severely from the fire of the smaller guns, complete penetration was not obtained. The plate was shivered into fragments. The head of the projectile remained buried in the backing. A considerable part of the body of the shot had been projected outward and bent forward without entirely separating from the head (*vide* figure), a most unusual circumstance with chilled shot.

On October 28—effective fire No. 11 at plate: 10-inch gun at Cammell's sandwich target. Effects, see No. 4 and No. 2 series. Shot-head penetrated 13 inches, the metal remaining in the hole. One bolt started in the rear.

No. 12: 10-inch gun at Marrel's sandwich target. Effects, see No. 3, second series. Shot penetrated 10 inches only, but the plate was split in several directions, some pieces being nearly detached. Cracks through upper bolt-holes.

No. 13: Salvo of one 10-inch and one 11-inch gun at Marrel's sandwich target. Effects, wide fissures opened in plate 5; large pieces of plate and some small ones dislodged altogether and thrown on the ground.

No. 14: Salvo of one 10-inch and one 11-inch at Cammell's. Both shots pierced front plate, and penetrated 2.4 inches into rear plate; right-hand corner of plate dislodged; backing moved; one bolt sheared and broken, and one bolt-head forced into interior. (See section 3 in No. 4 series.)

No. 15: The 100-ton gun at Marrel's sandwich plates as injured above. Effects, the main part of the target carried away, and complete penetration obtained. (See 3 in No. 4 series.)

The remainder of the rounds detailed in the table were then fired from the 100-ton gun.

In reviewing the plate-firing, the principal features to notice are the following:

Series Nos. 2 and 3 show that steel is more liable to be destroyed by the fire of guns not capable of penetrating it than wrought iron,* which under these circumstances suffers but little.

Series No. 4 shows that steel, by transmitting the blow of impact through the plate, is less liable to let the shot through the backing, while more liable to be stripped off and destroyed itself.

It also appears obvious that the *Duilio's* power of offense will be greatly in excess of her powers of defense.

Engineering says that in comparing the results of the thirty-fifth round with those afforded by the 81-ton gun, firing 370 pounds of the same powder and a 1,700-pound projectile, it will be seen that there is a preponderance of energy greatly in favor of the 100-ton gun. The 81-ton

* The steel used was evidently not the best adapted to the purpose.—J. W. K.

100-Ton Gun Targets.

Plate 3.

A.

Effect of
1 Round, 10 in. Gun,
1 Salvo, 10 & 11 in. Guns &
1 Round, 100-Ton Gun.
Half of Plate dislodged
Beams, Knees &c. broken.

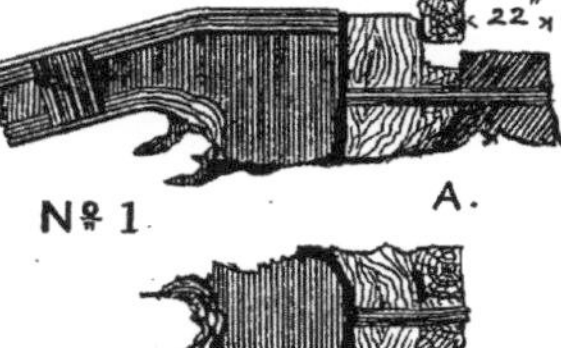

B.

Effect of
1 Round, 10 in. Gun,
1 Salvo, 10 & 11 in. Guns &
1 Round, 100-Ton Gun.
Plate knocked into
Fragments.

Cammell & Marrel } Iron { Upper Target.
Lower Do.

A′.

Effect of
1 Round, 10 in. Gun,
1 Salvo, 10 & 11 in. Guns &
1 Round, 100-Ton Gun.
Plate completely
destroyed.

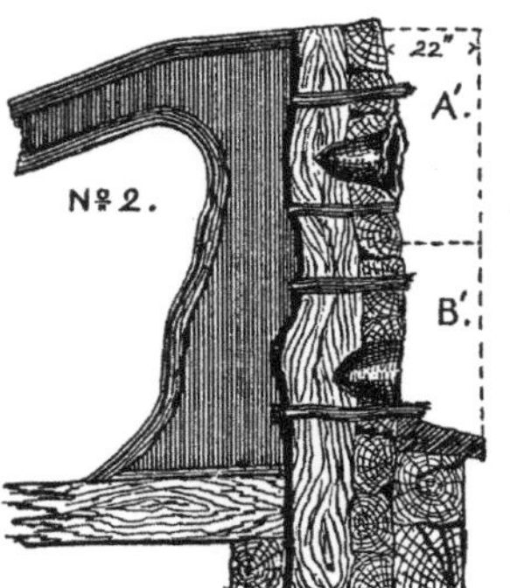

B′.

Effect of
1 Round, 100-Ton Gun.
Plate smashed to pieces,
shot broken up and
the entire Target
driven 8 in. back.

Schneider Steel.

No. 3.

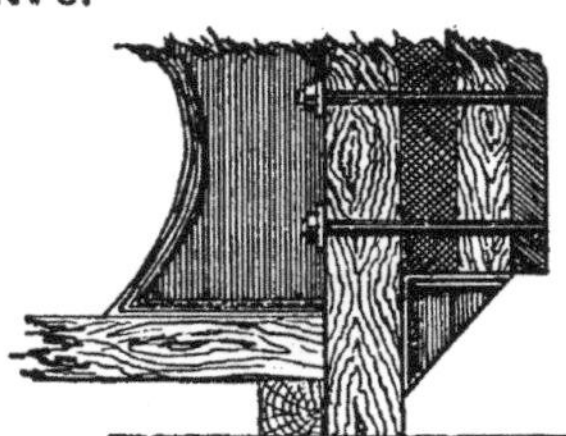

Effect of
1 Round, 10 in. Gun,
1 Salvo, 10 & 11 in. Guns &
1 Round, 100-Ton Gun.
The whole front plate
destroyed.

Marrel Iron.

gun gave a velocity of 1,520 feet to its 1,700-pound shot, with an energy of 27,200 foot-tons, or 540 foot-tons per inch of circumference of the shot. The 100-ton gun gave a velocity of 1,543 feet to a 2,000-pound shot, with an energy of 33,000 foot-tons, or 623 foot-tons per inch of circumference.

This first of the 100-ton guns was in some degree experimental, and important facts have been demonstrated by its use. After the conclusion of the experiments it was reshipped to England for the purpose of being chambered and having its bore enlarged. The other 100-ton guns completed at Elswick and shipped to Italy for the *Duilio* are considered capable of producing much better results than those exhibited by the first of these monster weapons, some important modifications having been introduced in these latter experiments. Instead of the uniform bore of 17 inches which characterized the first piece, these guns have a caliber of 17¾ inches and a powder-chamber of 19¾ inches.

The Italian authorities will probably fire the new guns with a charge of 470 pounds of powder behind a projectile of 2,500 pounds.

These artillery experiments conducted at Spezia by the Italian government are by far the most important ever undertaken. Not only has the greatest gun ever made been tested for velocity and penetration, but armor-plates of different kinds and of almost unparalleled magnitude have been tried for resistance and durability. The interest felt in the experiments by all naval authorities is not due alone to the performance of a single gun, but to the undisputed fact that the Spezia trials must exercise a powerful influence on the warfare of the future. Guns, ships, and armor will all undergo modifications.

The *Duilio* is to be armed with a heavy projecting ram, and she is also to be provided with apparatus for discharging the Whitehead fish-torpedoes. Besides these powerful means of offense, there is to be novel and original arrangement of carrying a rapid torpedo-boat; this boat is to be inclosed at the stern in a tunnel secured by a grated door, and when necessary can be launched and started on its course against an enemy's vessel.

The ship is to be driven by twin screws. The motive machinery is furnished by Messrs. John Penn & Sons, of Greenwich, England, and consists of his old type, a pair of trunk-engines to each screw. The steam will be supplied by the ordinary box-boilers. The two sets of engines are designed to develop the aggregate power of 7,500 horses, and the estimated speed of the vessel on the measured mile is 14 knots per hour. The heavy forgings for the ship were made in Italy. The frames, beams, and plates for the hull, in fact all the iron and steel entering into the construction of the vessel, were being made in France. All the armor-plates were ordered from Cammell & Co., of Sheffield, England, and the guns and machinery for working them were also to be of English manufacture, thus leaving only the construction and labor entering therein to be performed by the Italian engineers. The *Duilio* will be armed with four of the heaviest and most powerful guns ever mounted on any ship, and if completed and successful in all respects, she will be the most formidable fighting-machine, both for offense and defense, ever set afloat in the waters of continental Europe. When launched, the draught of water was found to be just what the designer estimated it would be, and it remains to be seen whether the prediction of the eminent ex-chief constructor of the British navy will be fulfilled, viz, that she will capsize in the first engagement, if seriously injured by shot in the unprotected parts.*

* For the drawings of the *Dandolo*, *Tegethoff*, *Shannon*, and *Nelson*, I am indebted to M. P. Dislere, of the department of the minister of marine, Paris.

THE DANDOLO.

This sister ship to the *Duilio* is under construction at Spezia, but is very far from being in the advanced condition of that vessel, and will not probably be completed until the end of 1879, as the financial resources of the Italian admiralty will doubtless be taxed to the utmost during the coming year to complete the first of the great ships. The *Dandolo* is to be in essential features of construction the same as her sister, but an important distinction is in the motive machinery. This has been designed and furnished by Messrs. Maudslay, Field & Son, of London, and consists of a pair of their inverted vertical compound engines to each of the two screw-propellers. The engines are to be supplied with steam from eight boilers fired from both ends, having in all thirty-two furnaces, four of the boilers being located forward and four aft of the engines.

The pressure of steam is to be sixty pounds per square inch, and the maximum indicated horse-power the same as is intended to be developed in the *Duilio*.

THE ITALIA.

The *Italia*, the construction of which was commenced last year at Castellamare, will, when completed, be a stupendous floating battery; indeed, the largest and most powerful ship of war ever floated in the world. The principal dimensions are approximately as follows: Length, 400 feet 3 inches; beam, extreme, 73 feet 10 inches; draught of water, forward, 25 feet 4 inches; aft, 30 feet 4 inches; height of upper deck above keel. 49 feet 3 inches; displacement, 13,480 tons; weight of unarmored hull, 5,000 tons; area of midship section, 1,848 square feet. The double bottom of the ship is 254 feet 3 inches long, 59 feet wide, and 3 feet 3 inches deep, and is divided into a large number of water-tight cells. Two longitudinal bulkheads extend from stem to stern, and the ship is divided by means of these and transverse bulkheads into fifty-three water-tight compartments, forty of the latter being above the double bottom of the vessel, three in the rear, and ten forward of it. These compartments are again divided horizontally by four water-tight decks; of these, the lowest, which is to be armored with iron 3 inches thick, is about 8 feet 2 inches below the water-line; the second, 5 feet above the line of flotation; the battery-deck, 14 feet 9 inches, and the upper deck 21 feet 4 inches, above the water-line. On the upper deck stands an armored redoubt, of oval form, its longer axis being at an angle of about 20° to the keel, and within it will be the turrets containing the guns. The armor will be mainly disposed in the form of a girdle around the ship, extending from the deck below to the first deck above the water-line. From this latter deck to the upper one the sides of the ship will be consequently unprotected; but the lower smoke-pipes, and also the tubes up which the ammunition is passed from the magazine to the redoubt, will be armored. The ship will be driven by twin screws, each 19 feet 6¼ inches in diameter, worked by four engines capable of developing 18,000 horse-power and propelling the vessel through the water at the estimated speed of 16 knots per hour. It is reported that the guns are to be of the Armstrong manufacture, and each is to weigh 100 tons; whether a greater number is to be carried than on the *Duilio* is, perhaps, not yet fully settled.

The *Lepanto*, a sister ship, is to be built at Leghorn.

There is a positive limit to the dimensions of every construction, a point beyond which it is not wise or safe to go. This point was over-reached in the case of the mercantile ship *Great Eastern*, the cost of which to the owners was very great.

It is reasonable to believe that the Italian naval authorities have gone beyond the judicious limit in the designs of their new ships. The great cost of the guns and the machinery to work them, the still greater cost of the ship, and the expense of maintaining the stupendous and unwieldy bulk, will probably not be the most serious difficulties. The practical question of the effects of firing such ponderous weapons, upon the ship on which they are mounted and upon persons working the guns, is yet to be determined. The Italians seem to have accepted in this respect the results of the firing of the 100-ton gun, mounted on a pontoon in the Gulf of Spezia, in which, as reported, the recoil of the gun, cushioned against air and water, was easy, and the jar or concussion not severe, the oakum in the decks of the pontoon in front of the gun not even being started. It is, however, well known that heavy guns work with more ease when mounted afloat on small vessels than they do on shore, due to the fact that the vessel acts as a secondary slide, taking up much of the recoil by the yielding resistance of the water in which it floats. This experiment is, therefore, inconclusive. Nothing heavier than the 38-ton gun has up to this time been fired from the decks of a ship. What effect the simultaneous firing of the whole of the four guns, of 100 tons each, will have upon the *Duilio*, or the men employed at the guns, or, if that be regarded as not likely to occur in warfare, the firing alternately of the guns in pairs, has yet to be demonstrated. Until it is known what will be the effect of this firing upon the deck-plating and upon whatever is near the guns, to say nothing of other questions, it would seem unwise to proceed with the construction of ships of still greater dimensions and of guns still more stupendous.

THE CRISTOFORO COLOMBO.

This is a wooden corvette, and the first vessel of the rapid type turned out by the Italian Government. She was built at Venice. The beams and knees are of iron, while the hull is composed of seasoned Italian oak; it is put together in the very best and strongest manner in which a wooden vessel can be built. The exact data were not obtainable, but the dimensions, &c., are very nearly as follows:

Length, total	270 feet.
Length on the water-line	250 feet.
Breadth	36 feet.
Draught of water, aft	17 feet.
Draught of water, forward	13 feet.
Area of midship section	476 square feet.
Displacement	2,500 tons.

On the berth or living deck there are twenty-three air-ports on each side, 17 inches by 18 inches, which afford more light and air than are admitted into our vessels of similar class. The quarters for both officers and men are comfortable, and the disposition of the lower-deck arrangements good, but the upper decks seem to be very much hampered by cabins, engine-room sky-light, steering-wheel house, two smoke-pipes, topgallant forecastle, &c., but as the battery is light, and the rig that of a barkentine, less room is needed than in many other vessels. The battery consists of two rifled guns on either broadside, and one bow-chaser of 4¾-inch bore. There are also two mitrailleuses to be used in the event of torpedo attack.

MOTIVE MACHINERY.

This was furnished by Messrs. John Penn & Sons, of Greenwich, England. It consists of one pair of their patent inverted vertical three-cylinder engines applied to a single screw-shaft, with the cylinder so arranged and adjusted by a system of valves as to be worked either on the compound or non-compound principle, as desired. The cylinders are of equal diameters.

Number of cylinders	3
Diameter of cylinders	61⅝ inches.
Stroke of pistons	37½ inches.
Number of air-pumps	2
Number of independent circulating pumps	2
Cooling surface	5,000 square feet.
Screw-propeller, diameter	17 feet.
Pitch of screw	19 feet 6 inches.
Number of blades to screw	4
Number of boilers (cylindrical)	8
Diameter of boilers	12 feet.
Length of boilers	9 feet.
Number of furnaces in each boiler	3
Grate surface	396 square feet
Heating surface	9,000 square feet

The amount of coal stored is 500 tons, and the space occupied by the machinery, boilers, and coal is 127 feet in the length of vessel below the lower deck, besides the space between decks consequent upon the top of the cylinders standing nearly on a level with the spar-deck.

The engines were designed to develop the maximum indicated horse-power of 4,000, and the speed on the measured mile was estimated to be at the rate of 17 knots per hour. I am not aware whether this speed was ever attained, but it has been authoritatively reported that with 86 revolutions the speed of 16.33 knots has been reached, the horse-power being 3,782. The cost of the vessel is reported to have been $514,620.

It was believed by the naval constructors and engineers with whom I conferred in Venice, when the vessel was building, November, 1875, that the hull did not possess sufficient strength to sustain for any lengthened period, the power necessary to produce the speed for which the machinery was designed.

The Italian naval authorities seem to have arrived at definite conclusions regarding their future fleets, and are acting vigorously upon them. They are not concentrating the whole of their attention upon armored ships, for, in addition to the *Cristoforo Colombo*, just described, five other vessels of the rapid type, intended for great speed, are in course of construction, and these later ships are being built with materials to sustain great engine-power, two of them being of steel and three of iron; besides which one torpedo-vessel has been completed, and two others, the *Sebastiano Veniero* and the *Andrea Provana*, are building. Steam-launches built in England, to use both the Whitehead and outrigger torpedo, are also being added to the fleets.

The iron-ship yards of Italy are building ships and training workmen for the government, and the proportion of iron to wooden ships, now building, is greater than in any other country, except England.

In a year hence, if the *Duilio* proves to be a complete success, Italy will possess the most powerful ships in continental Europe, and in 1879 the sister ship *Dandolo* is also to be completed, each having four 100-ton guns; and should the still larger ships, the *Italia* and *Lepanto*, turn out as calculated upon by the designers, Italy will possess a fleet of fighting-ships more than a match for any continental power. These, in addition to her cruisers of the rapid type, will cause her co-operation to be valued and her enmity to be feared even by England, France, or Russia, and certainly by any other European power.

DOCK-YARDS.

The naval dock-yards of the kingdom of Italy are at Venice, on the Adriatic; Castellamare, on the Bay of Naples, and Spezia, between Leghorn and Genoa, on the Mediterranean.

VENICE.

The dock-yard at Venice was in former times, as at present, known as the arsenal. There is no spot in Venice more intimately connected with the period of her powerful grandeur. It was considered one of the most important elements of the power of the republic. Here were constructed the galleys so celebrated for their strength and lightness. Not only were all the stores required in war preserved here, but everything warlike was manufactured within its walls. Before the principal gate, as if to guard it, stand four marble lions, spoils taken from conquered nations. The ancient walls still remain; also many of the former buildings; but they have been internally reconstructed, to suit the many varied and wonderful changes in naval architecture since the days of the Doges. The estimated area within the walls is about one hundred acres. The basin area is very considerable, and one superior stone dry-dock is completed and a second is under construction. The appliances and machinery are as yet rather primitive. The few tools and machinery of value in the buildings of the engineering department are of English manufacture. In fact, none of the buildings contain appliances worth noting.

CASTELLAMARE

is an old ordinary ship-building yard, furnished with improvised sheds, containing the necessary machinery, tools, and appliances for iron-ship building, all purchased in order to be used in the construction of the *Duilio.*

SPEZIA.

The dock-yard at Spezia, on the gulf of this name, is, in common with other Italian navy-yards, known as the arsenal. It has been laid out on a scale of great magnitude, and is intended as the principal naval port. Already about $10,000,000 have been expended on the works, and it may now be ranked among the most important of the naval establishments of Europe. The drawing of it which accompanied the first edition of this report, but which has been omitted here, shows the plan as originally designed. It is intended that there should be nine building-slips and ten dry-docks; but only two of the slips and four of the docks have been completed. The large basin for construction and repairs was designed to be 443 feet long and 80 feet wide at the top, and 56 feet 8 inches at the invert. This basin has only been completed for half of its length. The adjoining fitting-out basin is 360 feet 9 inches long and 72 feet 9 inches and 55 feet 9 inches wide. In addition to this completed basin there are two smaller ones. The basins and docks are emptied by means of two turbines driven by two 150 horse-power engines. A small pump is also kept constantly at work to remove infiltrating water. The buildings completed are of one story, excepting the one containing the offices and mold-lofts. These buildings are as follows: machine, boiler, and smith shops, foundery, forges, steam-hammer buildings, temporary iron-ship-building shop and framing-shop, gun-carriage shops, artillery-shops, store-houses, sail-stores, torpedo-stores, and special stores. The machinery and tools, and system of cranes and appliances, are as yet incomplete.

The following table contains the principal dimensions and other data of the Italian armored fleet now afloat and in process of building:

Armored ships of Italy.

Name of ship.	Vessel.						Armament.			Machinery.		Remarks.
	Class.	Length between perpendiculars, in feet and inches.	Extreme breadth, in feet and inches.	Displacement, in tons.	Draught forward, in feet and inches.	Draught aft, in feet and inches.	Number of guns and description.	Height of battery above water, in feet and inches.	Thickness of armor at water-line, in inches.	Maximum speed, in knots per hour.	Makers of engines.	
Duilio*	Mastless turret-ship.	339 8	64 7	10, 650	Mean	25 11	4 100-ton Armstrong guns	15 9	21½	14	John Penn & Sons	Twin screws; horse-power, 7,500; estimated cost of each, $3,049,600.
Dandolo†	...do	339 8	64 7	10, 650	Mean	25 11	...do	15 9	21½	14	Maudslay, Sons & Field	
Principe Amadeo. (w)	Line-of-battle cruiser.	265 0	57 10	5, 780	24 8	24 8	1 12-inch Armstrong; 6 10-inch Armstrong.	......	6	......	......	Cost, $1,238,900.
Palestro (w)	...do	265 0	57 10	5, 780	24 8	24 8	...do	......	6	10	......	Do.
Roma (w)	...do	250 0	57 5	5, 697	23 6	25 6	6 18-ton Armstrong; 2 9-inch Armstrong.	......	6	......	Maudslay	Cost, $1,155,036.
Venezia (w)	...do	250 0	57 5	5, 697	23 6	25 6	...do	......	6	......	...do	Do.
Castelfidardo (i)	...do	256 5	50 0	4, 194	19 7	22 0	11 guns	6 4	4¾	12. 9	Forges et Chantiers	Immersed midship section 771 square feet; single screws; horse-power, 3,000. The total cost of each ship of this class is stated at $876,800, gold. Total weight of each, 3,789.5 tons.
Messina (w)	...do	......	......	3, 968	......	......	14 guns	......	4¾	......	......	
San Martino (i)	Ordinary station service.	256 5	50 0	4, 194	19 7	22 0	11 guns	6 4	4¾	12. 9	Forges et Chantiers	
Maria Pia (i)	...do	256 5	50 0	4, 194	19 7	22 0	...do	6 4	4¾	12. 9	...do	
Ancona (i)	...do	256 5	50 0	4, 194	19 7	22 0	...do	6 4	4¾	12. 9	...do	
Conte Verde (w)	...do	......	......	3, 932	......	......	7 guns	......	......	......	......	
Varese (i)	......	198 3	42 4	2, 700	13 0	13 0	6 guns	......	5½	10	Forges et Chantiers	Broadside-battery; ram.‖
Terribile (i)	......	198 3	45 0	2, 700	17 8	17 8	30 6½-ton guns	......	4½	10	...do	Cost, $562,270.
Formidabile (i)	......	198 3	45 0	2, 700	17 8	17 8	...do	......	4½	10	...do	Do.
Affondatore (i)	Coast-defender.	290 0	40 0	4, 070	18 6	20 0	2 9-inch Armstrong	......	5	......	Milwall Iron Works	Double turret; ram.¶
Italia*	Turret-ship	400 3	73 10	13, 480	25 4	30 4	4 (?) 100-ton guns	14 9	....	16‡	Penn & Sons	Twin screws; 18,000 h.-p.; est. cost, each, $3,240,200.
Lepanto§	...do	400 3	73 10	13, 480	25 4	30 4	4 (?) 100-ton guns	14 9	....	16‡	......	

* Building at Castellamare, Naples. † At Spezia. § At Leghorn. ‡ Estimated. ‖ Cost, $415,508. ¶ Cost, $762,400. (i) Iron hulls. (w) Wooden hulls.

PART XVI.

THE RUSSIAN NAVY.

CIRCULAR ARMORED SHIPS; PERSONNEL OF THE RUSSIAN NAVY; DOCK-YARDS; TABLE OF DIMENSIONS, ETC., OF ARMORED SHIPS OF RUSSIA.

THE RUSSIAN NAVY.

The Russian navy is now composed of twenty-nine armored ships and one hundred and ninety-six other vessels of all classes, carrying altogether 521 guns. It is divided into two portions, one stationed in the Baltic, and the other in the Black Sea. The former is by far the more formidable, consisting of twenty-seven ships of various descriptions, 911 officers and about 7,000 enlisted men. During severe winters this fleet is securely locked up in harbor, and the majority of the crews are snugly ensconced in barracks, where they are compelled to remain till the ice breaks up. From the official returns received at the beginning of the present war, the Black Sea squadron consisted of nine vessels with 320 officers and about 3,000 enlisted men. In addition to these, Russia has naval vessels in the Caspian Sea, and the Vistula, and the White Sea, also a flotilla in the Sea of Aral, comprising about fifty craft of all classes, with about 250 officers and 3,000 enlisted men.

Except for coast defense, the Russian fleet is numerous rather than powerful. The *Peter the Great* (*Petrr Veliky*) and the *Knaz Minin* are the only two vessels on the list of sea-going armored ships which approach the modern standard of fighting efficiency. The first named was designed after the British ship *Devastation*, and commenced before the sea trials of that vessel; subsequently modifications were made, and, as completed in 1875, she somewhat resembles the *Dreadnought;* though, as may be seen by consulting the list of armored ships, the dimensions are nearly the same as those of the *Devastation*, and the displacement 316 tons more. The indicated horse-power, armor, and battery are equal to those of the *Devastation*. The armor is 14 inches in thickness, with hollow iron stringers in the backing besides, which are alleged to give an additional resistance equivalent to 2 inches of iron. The four guns, two in each of the turrets, are of 12-inch caliber, and the weight of each is 40 tons. In respect, therefore, to armor and guns she is the match of any ship *now in commission* belonging to other nations. But she is not fitted with a spur to utilize the power of the ram, and the estimated speed has not been realized.

The reports from correspondents represent the *Peter the Great* as not having turned out as successful as anticipated. It is reported that during the winter of 1876–'77, when the ship was ice-bound in the harbor of Cronstadt, and exposed to an atmospheric temperature of from 10° to 40° below zero, Fahr., the crews were kept employed in operating and firing the heavy guns, and that when the ice broke up, the ship being ordered out for a few days' cruise, the result of the firing became apparent. The hull was found leaking to a very considerable extent, some of the steam-cylinders were found to be cracked, and other damage done.

A committee of naval experts was immediately convened, and the decision that they arrived at was, that the damage had been caused by the vibrations arising from the firing during the time that the iron composing the hull and machinery was under the influence of very low atmospheric temperatures.

The *Knaz Minin* was constructed as a rigged turret-ship on the Coles system, has a length of 298 feet 3 inches and breadth of 49 feet, with a displacement of 5,800 tons. The armament consisted of six guns of 12½ tons each, and armor of 12-inch plates on 24-inch backing, and the free-board was very low. In consequence of the catastrophe to the British vessel *Captain*, alterations to the *Knaz Minin* were decided upon. As altered she will have a central battery 98 feet long, rising 10 feet above the water-line. The guns are mounted in pairs on two turn-tables, on the main deck, and will fire *en barbette* over the top of the battery. In this shape she is expected to be a formidable ship.

The next ships of the sea-going fleet to be noticed are the broadside belted vessels *Duke of Edinburgh* (*Gertzog Edinburgsky*), originally called the *Alexander Nevsky*, and the *General Admiral*. These ships are of recent construction, and were designed to compete with the fast British unarmored ships *Raleigh* and *Boadicea*. They are built of iron, sheathed with wood, and coppered. The length between perpendiculars is 285 feet 9 inches; breadth 48 feet 2 inches; draught, mean, 21 feet; displacement, 4,438 tons. In weight and dimensions they are therefore intermediate between the two British ships just named. These vessels embody the original conception of the armor-belt on the water-line to protect the vital parts; it is 6 inches thick and 7 feet wide. The battery-deck is similar to that of the British *Invincible* class, open-topped, and arranged so as to give both broadside and right ahead and astern fire from corner ports. It contains four 8-inch rifled guns and two 6-inch chase-guns. An article in the *Revue Maritime et Coloniale*, from which extracts have been taken, represents the lines of these vessels to be fine, the engine-power large, and the speed 13 knots per hour. They are not provided with spurs to be used as rams, and have neither the speed nor the power of battery possessed by the British ships referred to.

Next in the sea-going fleet are the four ships named after admirals, viz, the *Admiral Lazareff* and *Admiral Greig*, carrying each six guns in three turrets, and the *Admiral Tchitchagoff* and *Admiral Spiridoff*, carrying each four guns in two turrets. These vessels are of the Coles type of turret-ships, and differ from each other but little except in number of turrets. The first-named two have a free-board of 4 feet, and the other two of 5 feet. The displacement is about 3,700 tons, and the speed from 9 to 10 knots. The thickness of armor on hulls and turrets is 6 inches, and the caliber of the guns only 9 inches. The sea-going qualities of these four ships, unless it be near home, may be doubted; as coast-defenders, however, they are important additions.

Last among the sea-going fleet are noticed two wooden armored frigates, the *Sevastopol* and *Petropavlovski*, built in 1863 and armored with plates only 4½ inches thick. They have large crews, numbering 609 and 682 men. They displace 6,200 tons, have steamed 11 knots, and carry batteries respectively of sixteen and twenty 8-inch breech-loading guns, and of eight and four 80-pounders. These ships may be regarded as obsolete.

For coast defense Russia has a considerable fleet. The two circular vessels hereafter to be described are the most formidable of the number.

The next in power are ten monitors, of early date, on Ericsson's plans, similar to our harbor and river monitors, drawing nearly 12 feet of water, and armored on the sides with 5-inch plates on a backing of nearly 3 feet. The one turret of each vessel is built up of eleven 1-inch plates without backing. The two guns in the single turret are 10-inch rifles or 15-inch smooth-bores of old pattern.

The *Smertch*, a double-turret vessel, built in England in 1864, is armored both on the sides and turrets with plates only 4½ inches thick, and carries only four 9-inch guns. There are, however, two other monitors of later date and somewhat greater power, built in Russia in 1868.

These are the *Tcharogeika* and *Rousalka*. The side armor is 5 inches thick, and that on the turrets 6 inches. They carry four 11-inch rifles in two turrets.

The speed of all these monitors is given at from 6 to 8½ knots, and they are not provided with spurs for ramming, and must therefore be considered as weak vessels, fit only for operations in shallow water.

UNARMORED CRUISERS.

Of these the following named are the principal, according to their class:

Name.	Tonnage.	Nominal horse-power.
First class:		
The Assletia	2,980	260
Peresvett	3,840	450
Svetlana	3,090	450
Second class:		
Baiane	2,000	300
Four corvettes, Vitiaz type	2,160	360
Third class:		
Vöme	1,820	250
Five cruisers, Almatz type	1,590	350
Two cruisers, Haidamak type	1,100	250
Fourth class:		
Sobol	980	220
Three cruisers, Boïarsine type	900	160
Three cruisers, Zastreb type	800	220

The naval armament of Russia has been for some few years past undergoing the changes necessary to keep up with the progress of the times. Cast-iron smooth-bore guns are gradually being replaced by breech-loading steel rifles. The first of these were supplied by Krupp; now they are manufactured in Russia, some at Perm, and others at Oboukoff, about nine miles from St. Petersburg, on the left bank of the Neva. The Broadwell system of breech-loading mechanism is employed, and the calibers of the guns are 6, 8, 9, and 11 inches. One gun, of 12-inch caliber, was manufactured for the Vienna Exposition, but it is not known that any others of this size have been made. The broadside-ships have not hitherto been armed with guns heavier than 8 inches, with 9 inches for stern or chase guns, but it is intended to substitute 11-inch pieces for these.

RUSSIAN CIRCULAR ARMORED SHIPS NOVGOROD AND VICE-ADMIRAL POPOFF.

During the autumn of 1875, Mr. E. J. Reed, C. B., M. P., ex-chief constructor of the British navy, made a visit to Russia for the purpose, as stated by him, of inspecting two circular armored vessels, one of them being then completed and in commission, and the other under construction. While in Russia he wrote several letters to the London *Times* on the subject of these vessels, in which he eulogized them and attached so much importance to the advantage of the circular form over that of existing types of armored ships, that nearly all the newspapers of London and some scientific papers also contained articles discussing their merits. Subsequently Mr. Gouleaff, a Russian officer, read a paper before the Institution of Naval Architects on the many advantages possessed by vessels constructed on the circular principle. Finally Mr. Reed was invited to deliver a lecture at the United Service Institution in London, and did so February 4, 1876, upon "Circular Iron-clads."

The accompanying illustrations will give a good idea of the outward form and design of these much-talked-of Russian vessels. The noteworthy points in Mr. Reed's lecture are that—

In the first place they are circular only in one sense, *i. e.*, their horizontal sections only are circular, or, in other words, they have circular water-lines. The departure from a circle is a small extension or protuberance at the stern for the purpose of facilitating the arrangement and working of the rudder and steering apparatus. It follows as a consequence from the circular form of water-line, that all the radial sections are alike; * * * * the bottom of the vessel is an extended, plane surface, which is connected with the edge of the deck by a quadrant of a small circle. With this form of section great displacement is obtained on moderate draught of water. The deck of the circular ship is formed in section with such curvature as to give in a ship of 100 feet in diameter a round-up of about 4 feet. There are two Popoffkas already built, named respectively the *Novgorod*, Fig. 1, and the *Admiral Popoff*, Fig. 2, of which the following are the dimensions and other particulars:

	Novgorod.		Admiral Popoff.	
	Ft.	*In.*	*Ft.*	*In.*
Extreme diameter	101	0	121	0
Diameter of flat bottom	76	0	96	0
Depth in hold at center, from under side of beam to top of the frames of the double bottom	13	9	14	0
Draught of water forward	13	2	12	0
Draught of water aft	13	2	14	0
Draught of water, mean	13	2	13	0
Height of barbette tower from load water-line	12	0	13	3
Diameter of barbette tower, outside	30	0	34	0
Height of upper deck at side, from load water-line amidships	1	6	1	6
Height of armor on side above water	1	6	1	6
Depth of armor below load water-line amidships	4	6	4	6
Thickness of armor on sides (including equivalent thickness for the hollow iron girders behind armor)		11	1	6
Thickness of armor on lower strake		9	1	4
Thickness of armor on barbette tower		11	1	6
Thickness of deck-plating		2¼		2¼
Displacement, in tons	2,490		3,550	
Area of midship section, in square feet	1,170		1,416	
Engines, nominal horse-power	480		640	
Coal supply, in tons	200		250	
Propellers, screw, in number	6		6	
Complement of officers and men	110		120	
Armament, breech-loading guns:				
Two in number, each weighing, in tons	28		40	
Smaller guns in unarmored breastwork	....		4	

It is but fair to the distinguished designer of these vessels, carefully to bear in mind that in so far as the *Novgorod* and *Admiral Popoff* are concerned, they have been designed and built purely for service in shallow waters and near the land. * * *

The *Novgorod* and *Admiral Popoff* have extensive unarmored houses erected above the armored decks. The chief of these is a spacious forecastle, which, of course, adds greatly to the buoyancy forward when the sea rises there upon the vessel. I do not think even circular vessels, of very low free-board, could be steamed against a heavy head sea without such a forecastle, more especially when driven at high speed. * * *

The chief characteristic of these circular iron-clads is that they are purely and simply sea-citadels propelled by steam, and without any attempt to make them conform to the shape of an ordinary ship. The question to be determined hereafter is, is this form of vessel thus originated for coast-defense purposes, and proved eminently successful for that purpose, available under proper modifications for sea-going citadels?

I think we may fairly say that, for a sea-going citadel, viewed as a citadel only, apart from other features, the circular form is best, because it requires a minimum amount of armor to protect a given area or volume, or, in other words, a given amount of armor secures the greatest amount of buoyancy. For special purposes some modi-

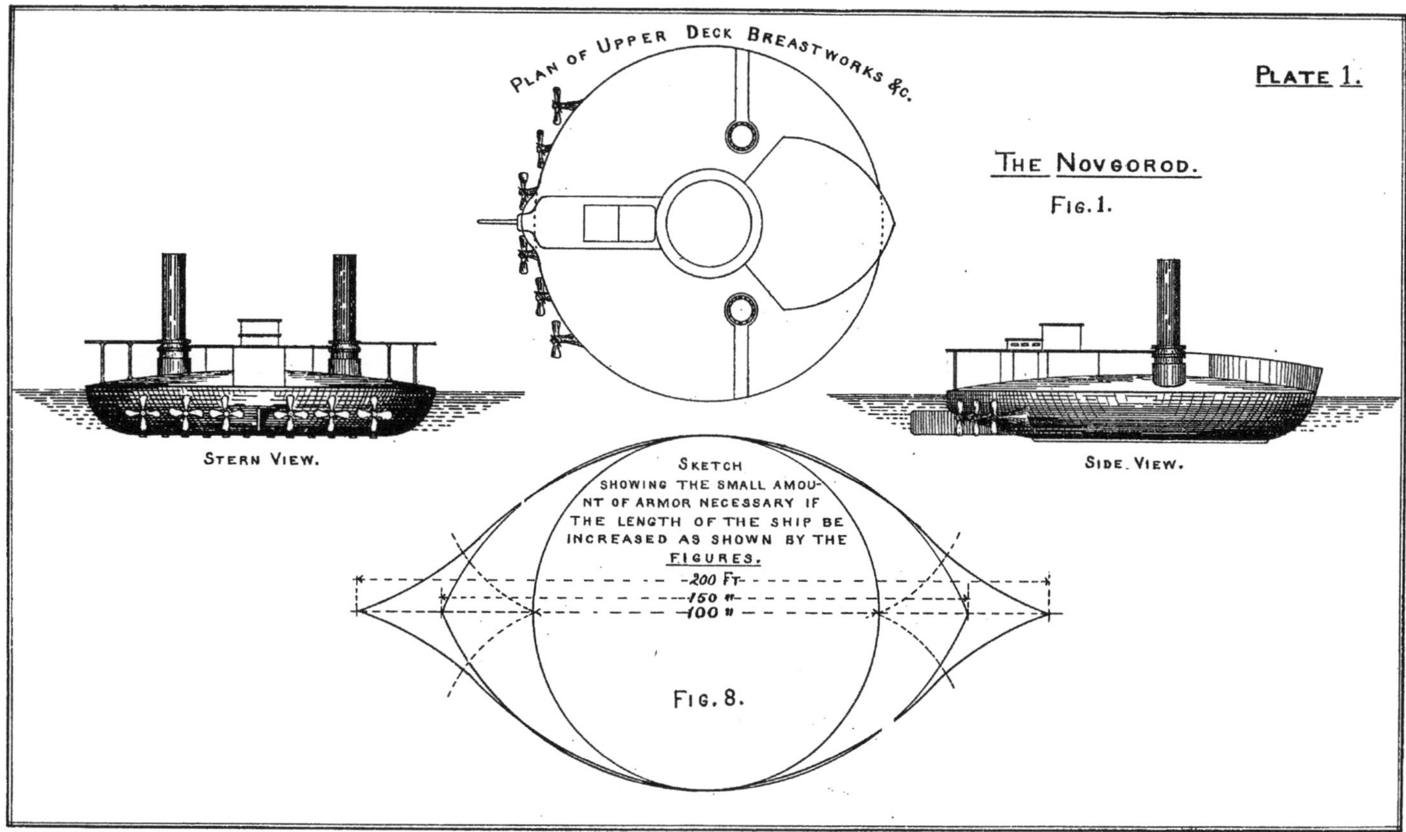
Plate 1.
The Novgorod.
Fig. 1.
Plan of Upper Deck Breastworks &c.
Stern View.
Side. View.
Sketch
showing the small amount of armor necessary if the length of the ship be increased as shown by the figures.
200 Ft
150 "
100 "
Fig. 8.

fied form might be preferable; but speaking generally, the circular form is the best for floating armor to protect an included space, and also for giving that equal all-round cannonade with guns which is so desirable at sea. Starting, then, with this circular armored citadel, and wishing to propel it at a given speed at sea, there are several ways in which we can deal with it:

First. We can put engine-power in it just as it stands without modification; or,

Second. We can build ends to it like those of an ordinary ship, protecting those ends by a belt of armor, as in many other ships; or,

Third. We can build such ends to it and protect the lower parts of them by an under-water deck of armor, as in the *Inflexible;* or,

Fourth. We can build around it an outer circle of thin iron, with a mere narrow belt of armor analogous to the belt of ordinary iron-clads; or,

Fifth. We can build around it such an outer circle of thin iron, with an under-water deck of armor analogous to that of the *Inflexible;* or,

Sixth. We can build short ends to it with either above or under water armored decks, but of greatly reduced length as compared with the ends of ordinary ships of large beam.

The *Novgorod* is the only actual example of the first of these cases that has yet been tried, and we may state roughly that in her, 750 tons of armor and 56 tons of guns are carried on a displacement of 2,500 tons, and driven at 8½ knots, with 2,270 indicated horse-power. This confirms what we already know, viz, that such ships will require great power in proportion to displacement. But taking, not the false standard of displacement, but the better (although not perfect) standard of weight of armor and guns as our guide, we shall find nothing very extraordinary in the power required.

Such are the chief points in Mr. Reed's lecture. I did not see these Russian circular vessels, but from an examination of a completely equipped model of one of them and the drawings, I reached the conclusion that, as floating forts designed for shallow water, they do possess some of the merits stated; but even for this purpose, as at present constructed, there are serious objectionable features, some of the most prominent of which are given below:

1st. As the *Novgorod* is built, there is in the center an open-top fixed turret, or an iron martello tower, having inside it a revolving platform, on which the guns are *en barbette.* This is the system employed in the upper-deck batteries of the French armored ships; but in the French ships the towers are located near the sides of the vessels, and high above water, while in the Russian vessels they are located in the center and on hulls having a free-board of only 18 inches. The barbette principle affords very considerable lateral range, but the disadvantages are, as here applied, that it leaves the guns and men working them fully exposed to the fire of the enemy. On shore, artillery officers rarely, if ever, contemplate mounting guns *en barbette* near the level of the water where serious and close action is expected. They seek for a high and somewhat distant position, where the advantages of an all-round lateral and plunging fire are available, and where the exposure of the men and the guns is reduced to a minimum. With a view to remedy this serious disadvantage in a degree, the second vessel constructed, the *Vice-Admiral Popoff*, has been arranged to work the guns on the disappearing principle of Rendel, in which the gun is loaded in the low position shown in section on Plate 2, and as previously described for the British ship *Téméraire.* This change is only a partial remedy; the disadvantage of the open-top tower slightly above the level of the water still remains.

2d. The second objection is that the side armor-plates do not extend deeper below the water-line than in ordinary vessels, and as the *Vice-Admiral Popoff* is 121 teet in diameter, there is this large target at all times presented for under-water attack by locomobile torpedoes, instead of the bow or stern alone, as would often happen in the attacks on other vessels. The extra defense against torpedoes gained from the cellular construction by means of the circular system would avail nothing.

3d. The third objection consists in the complication of the motive machinery. There are six screw-propellers, operated by three sets of engines. Two of the propellers have diameters greater than the draught of the vessel, the periphery of the blades extending below the keel. These two screws are three-bladed, and are not worked in shallow water. I had considerable experience, during our civil war, on the Mississippi River, in building and operating the machinery of the four wide flat-bottomed gun-vessels of the *Milwaukee* class, provided with four screw-propellers. I therefore have practical knowledge of the complication of the machinery proposed in this case. It is also to be observed that the six screw-propellers are unprotected from attacks of any kind.

4th. The fourth and most serious objection to the circular form of vessels consists in the extraordinary steam-power necessary to drive a vessel, say, of 121 feet diameter, through the water at a speed equal to that of the ordinary vessel of the same carrying capacity. The weight and space occupied by the machinery would be so great as to leave but little room for all the other requirements.

Mr. Reed says that the *Novgorod* made a speed on the measured mile of 8½ knots; that she has steamed a considerable distance at 7½ knots; and when he made a trip in her the speed averaged 6½ knots. This last is probably the real speed when steaming in ordinary weather for a period of twenty-four hours or more. If the weights of the steam-machinery had been given, and other necessary data, the weights, &c., for higher speeds could be readily estimated.

The first objection raised to the circular vessel, viz, the open-top tower, could be remedied in future constructions by substituting the Ericsson revolving turret, and the second objection could be removed by extending the vertical side armor down to the keel. This would entirely protect the vessel from the attacks of moving torpedoes. The third objection may be obviated, but the fourth is insurmountable. Therefore for sea-going vessels, although the form may be changed by adding ends, as shown at Fig. 8, Plate 1, the principle of construction will not be likely to meet with favor from naval architects, much less from naval officers.

HYDRAULIC GUN-CARRIAGES FOR THE POPOFFKAS.

A correspondent gives the following description of the hydraulic gun-carriage which has been manufactured in England for the *Vice-Admiral Popoff:*

In the case of the circular iron-clad, the problem was to accommodate two 12-inch 40-ton guns, 20 feet long over all, and 4 feet 10 inches in diameter, in a turret only 26 feet in diameter and 6 feet 10 inches deep. The task was by no means an easy one, but by adopting many novel expedients and substituting cast steel for cast iron, it has been successfully accomplished. The sides of the carriages are composed of 6-inch wrought-iron plates, connected together by cast-steel frames, so as to form a solid and rigid platform. The guns are placed each side of the center pivot as near as they could be brought together, and are each mounted on a pair of massive wrought-iron levers, connected by a rocking shaft, and elevated or lowered by water-pressure, acting on steel trunks sliding in steel cylinders. From the bottom of each cylinder a 4-inch pipe leads to a valve-chest containing the recoil-valve, which is loaded to the necessary degree of pressure by a series of disk-springs, the tension of which is increased as the gun recoils by means of a pair of chains wrapping round the rocking shaft connecting the levers on which the gun is supported. For raising the gun and lowering it without firing, a special hand-valve is provided and connected to the pressure-pipes. The rotation of the gun-platform is performed by a pair of 40-horse engines, driving a cast-steel annular toothed wheel, weighing seven tons. The wheel is a fair example of the perfection to which steel castings have been brought. The motive power is water under a pressure of between 800 and 1,000 pounds per square inch, supplied by a duplex pump perfectly automatic in its action, stopping as soon as the required pressure is obtained, and starting the moment it falls, working at a rate corresponding to the

Plate 2.

The Admiral Popoff.

Fig. 2.

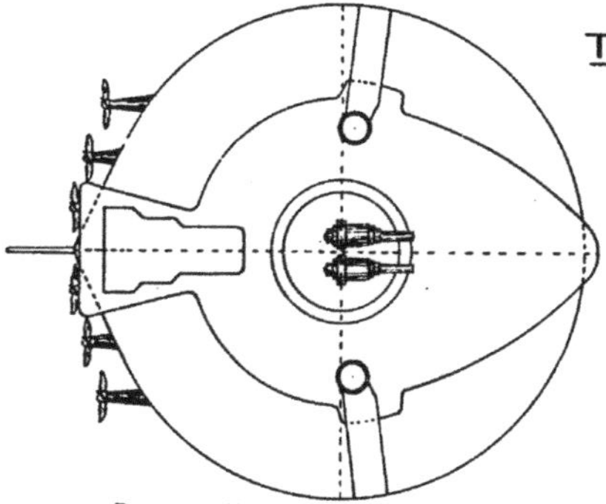

Plan of Upper Deck Breastwork &c.

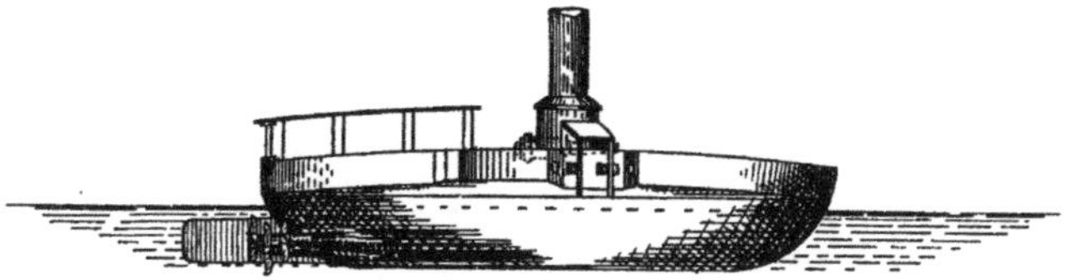

Side View.

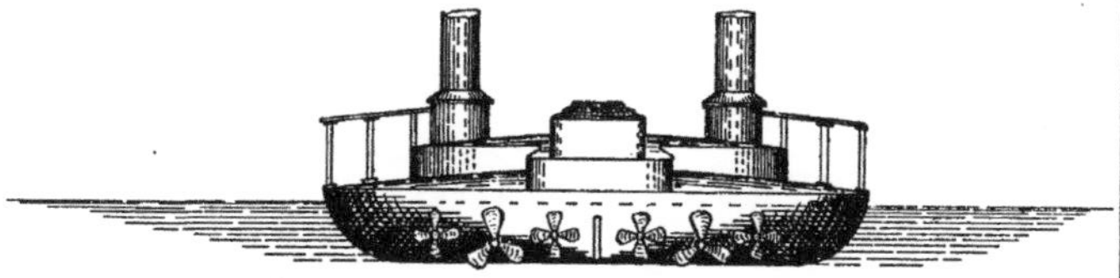

Stern View.

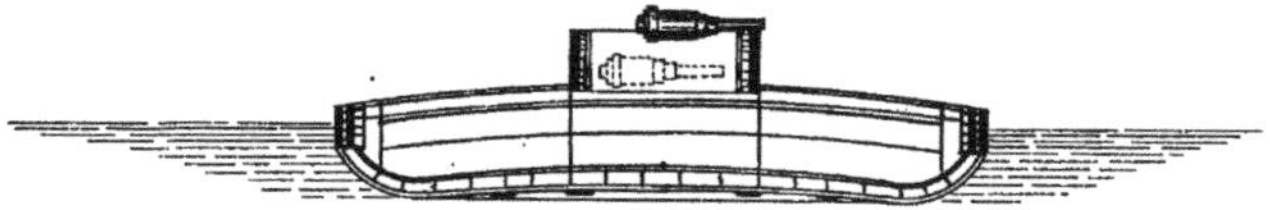

Fig. 6.

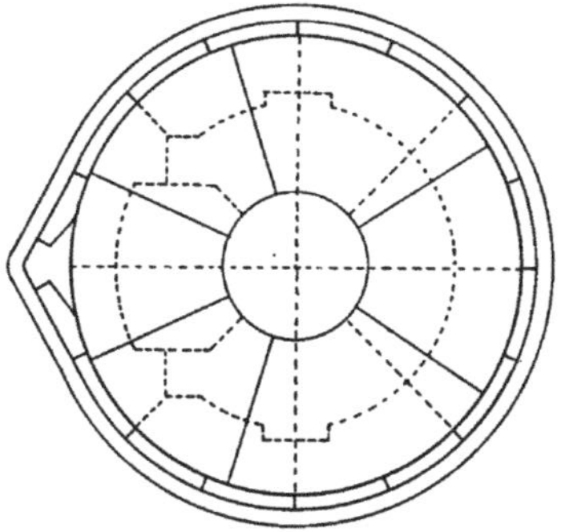

Plan Showing Water Tight Divisions.

Section.

requirement of the guns. These pumps are adequate to work the guns direct, but to keep them of moderate dimensions, and at the same time to retain the power of rapidly elevating the guns, a compressed-air receiver, made of Whitworth's steel, has been provided, together with a number of tubular air-reservoirs. The air is compressed up to a certain point, and remains perpetually ready for use, expanding and driving the water out of the receiver in raising the guns, and being compressed again to its normal state by the water forced out of the receiver by the pumps. Another advantage of the arrangement is that, the reservoir being of sufficient capacity, two shots can be fired before it is necessary to have recourse to pumping. The whole of the machinery necessary for working the guns is arranged on the main deck of the vessel below the water-line, while the parts exposed to some extent in the turret are of such a massive character that they are not likely to be injured by fragments of shells or fire from the enemy's tops. These gun-carriages being of entirely novel and original design, and, moreover, mounting the heaviest guns that have ever yet been worked on this system of protected barbette, it was thought prudent to erect them in the first instance on shore, so that they might be thoroughly tested. They have accordingly been placed at Cronstadt, near Fort Menchikoff, and have been very satisfactorily tried, as we may gather from the following article in the Cronstadt *Vestnik:*

"*The trials of the Moncrieff hydro-pneumatic gun-carriage.*—On Wednesday, the 31st of October, three shots were fired from the two 12-inch 40-ton guns mounted on the Moncrieff disappearing system in the presence of His Imperial Highness the Grand Duke General Admiral and staff. The first charge consisted of 60 pounds of prismatic powder, the second of 81 pounds, and the third of 117 pounds; and in each case cylindro-conical shot weighing 648 pounds. In each of these discharges the machinery worked splendidly, the hydraulic cylinders permitting a recoil of 4 feet, and not a single bolt or nut was in any way disturbed. During the first series of experiments, on Monday, the 29th ultimo, one blank and six shotted charges were fired; on Wednesday, in the presence of the Grand Duke, three shotted charges; and to-day, November 1, again three charges were fired, with 81 pounds, 117 pounds, and 144 pounds of powder, in the presence of the director of the ministry of marine and a number of admirals, who arrived in Cronstadt at eleven o'clock in the steam-yacht *Neva*. We understand that on the 3d instant a systematic course of experiments will be undertaken by a commission which has been appointed for the purpose, and as soon as these have been satisfactorily concluded, the gun-carriage will be taken to pieces and sent to Nicolaeff, to be fixed on the circular iron-clad *Vice-Admiral Popoff*."

PERSONNEL OF THE RUSSIAN NAVY.

The *personnel* of the Russian navy, one year ago, consisted of 81 flag officers of all ranks, 1,224 other officers, 513 mates, 210 artillery officers, 145 engineer officers, 545 mechanicians, 56 constructors, and 260 medical officers. Employed at the admiralty dock-yards, &c., there were 297 officers and 480 civil officials, and the total number of enlisted men was 24,500. No marines are employed, the military as well as the nautical duties of every man-of-war being performed by the officers and sailors.

The expenses of this navy during the financial year 1878 are estimated at $19,970,090. This includes $3,990,578 for the construction of ships, torpedo-boats, and torpedoes. The figures are based upon the assumption that the rouble is equivalent to 79½ cents, silver.

The Russian navy has done little to boast of during the present war with the Turks. Except the attack and destruction of one Turkish gunboat, the *Saiffee*, by torpedoes operated from boats by two gallant lieutenants, and the unsuccessful attacks against two other gunboats, there are no brilliant achievements. The most powerful ship of the fleet, the *Peter the Great*, as seen above, met with peculiar difficulties. Another important armored ship, the *General Admiral*, after being completed last autumn, tried at sea and pronounced successful, came to grief by being driven ashore in a gale at Cronstadt and badly strained and damaged. Besides these mishaps the "cyclads," from which so much was expected, have done absolutely nothing during the present war; the reasons are set forth in the Moscow *Gazette*, a leading Russian journal. It says:

The *Admiral Popoff* and the *Novgorod*, which have long been stationed at Odessa, received orders from St. Petersburg to proceed to the Sulina mouth of the Danube and attack the Turkish monitors there. Having been ordered to put to sea for a few days' trial, the result showed that when the sea is rough the ventilators have to be closed; that the heat of the interior of the vessel was so great as to be unendurable; that they are ill-suited for maneuvering, and that in bad weather they are scarcely seaworthy. It was accordingly decided that the orders could not be carried out.

DOCK-YARDS.

There are two dock-yards in the city of St. Petersburg. The first and larger is employed both as a construction and equipment yard, while the second and smaller is exclusively a building-yard. The batteries of ships of war, however, are not placed on board at either yard, the armaments being supplied at Cronstadt. The principal engineering establishment is at Kolpino, on the river Eshorra ; it was founded by Peter the Great, and in it are now manufactured the steam-machinery, armor-plates, castings, and other articles for the navy. The copper sheathing is also rolled here and chain-cables made. There is also a government manufacture of steel, known as the Aboukoffsky Steel Works; it is located at Alexandrovsky. In addition to the steel made for ordinary use, there is manufactured here cannon-steel from irons drawn from the mountains of Siberia.

The list of Russian armored ships, with their principal dimensions and other data, is given herewith.

*Armored ships of Russia.**

Name of vessel.	Vessel. Class.	Length between perpendiculars, in feet and inches.	Extreme breadth, in feet and inches.	Displacement, in tons.	Draught forward, in feet and inches.	Draught aft, in feet and inches.	Armament. Number of guns and description.	Height of battery above water-line, in feet and inches.	Thickness of armor at water-line, in inches.	Machinery. Number of screws.	Maximum speed, in knots, per hour.	Indicated horse-power.	Makers of engines.	Remarks.
Peter the Great	Two turrets	330 0	64 0	9,510	23 0	22 9	4 12-in. 40-ton guns	13 0	14	2	13	10,000	Baird, St. Petersburg	Cost of machinery, $780,080.
Knaz Minin	Broadside	289 0	49 0	5,800	19 8	23 8	6 12½-ton guns	10 0	7	..	13	6,300		Converted from two-turret.
Duke of Edinburgh	do	285 0	48 0	4,500	19 0	23 0	4 12½-ton guns and		6	..	13	6,300	McPherson	Iron-built ships, carry a large coal supply and a large spread of canvas.
General [illegible]al	do	285 0	48 0	4,500	19 0	23 0	2 7-ton.		6	..	13	6,300	Seminekoff & Politica	
Sevastopol	Frigate	300 0	50 10	6,200	23 11	25 11	16 9½ ton guns		4½	1	14	3,000	Kolpino	Wood-built, and furnished with ram.
Petrop[illegible]vski	do	300 0	56 0	6,200	22 7	25 3	20 9½-ton guns		4½	1	12	3,000	Baird, St. Petersburg	
Knaz Pojarski	Casemate	280 0	49 0	4,500	17 0	21 0	9 9½-ton guns		4½	..	11		do	Iron ship, carries a ram.
[illegible]al Lazareff	Three turrets	260 0	43 6	3,700	19 0	19 0	6 15½-ton guns	4 0	5½–6	..	11	2,000	McPherson	Called the Four Admirals; low free-board, rigged, turret vessels.
Admiral Greig	do	260 0	43 6	3,700	19 0	19 0	6 27-ton guns	4 0	5½–6	..	11	2,000	do	
Admiral [illegible]hagoff	Two turrets	260 0	43 6	3,700	19 2	19 2	4 27-ton guns	5 0	6	..	11	2,000	Seminekoff & Politica	
Admiral Spiridoff	do	260 0	43 6	3,700	19 2	19 2	do	5 0	6	..	11	2,000	do	
Pervenetz	Battery	220 0	53 0	3,300	15 0	17 0	26 19½-ton guns		4½	..	9½		Maudslay, Sons & Field	
Ne-tron-meña	do	220 0	53 0	3,300	15 0	17 0	16 9½-ton guns		4½	..	9			Engine taken from a line-of-battle ship.
Kreml	do	218 6	53 0	3,300			26 9½-ton guns		6	..	9			
Ouragan	One turret	200 0	46 0	1,555	10 6	11 6	2 0-in. rifled guns	6 4	4¾	2	6.7			
Tiphon	do	200 0	46 0	1,555	10 6	11 6	do	6 4	4¾	2	6.7			
Latnik	do	200 0	46 0	1,555	10 6	11 6	do	6 4	4¾	2	6.7			
Lava	do	200 0	46 0	1,555	10 6	11 6	do	6 4	4¾	2	6.7			
Vetchoun	do	200 0	46 0	1,555	10 6	11 6	do	6 4	4¾	2	6.7			
Koldoun	do	200 0	46 0	1,555	10 6	11 6	2 15-in. sm'th bores	6 4	4¾	2	6.7			
Streletz	do	200 0	46 0	1,555	10 11	11 6	do	6 4	4¾	2	6.7			Built at St. Petersburg.
Edinorog	do	200 0	46 0	1,555	10 11	11 6	do	6 4	4¾	2	6.7			
Bronenosetz	do	200 0	46 0	1,555	10 11	11 6	do	6 4	4¾	2	6.7			
Perm	do	200 0	46 0	1,555	10 11	11 6	do	6 4	4¾	2	6.7			
Smertch	Two turrets	172 10	38 0	1,380			4 9-in. rifled guns	4 0	4½	1	11		Maudslay, Son & Field	
Tcharogeika	do	210 0	42 0	1,835	11 9	11 9	4 11-in. guns		5	..	11			Built in England.
Rousalka	do	210 0	42 0	1,835	11 9	11 9	do		5	..	11		Baird	
Novgorod	Circular	101 0	101 0	2,490	13 2	13 2	2 27-ton guns	1 6	11	6	7	2,700	do	Cost of machinery, $215,600; immersed midship section 1,170 and 1,416. Behind the armor-plates is an iron backing of channel-rail of great strength.
Vice-Admiral Popoff	do	121 0	121 0	3,550	14 0	14 0	2 40-ton guns	1 6	18	6	9	3,500	do	

* The spelling of most of the names has been obtained from the secretary of the Russian legation in Washington.

PART XVII.

THE TURKISH NAVY.

TABLES AND DESCRIPTIONS OF ARMORED SHIPS AND PERSONNEL OF THE TURKISH NAVY.

THE AUSTRIAN NAVY.

THE TEGETHOFF; TABLE OF DIMENSIONS, ETC., OF THE ARMORED SHIPS OF AUSTRIA, AND THE PERSONNEL OF THE NAVY.

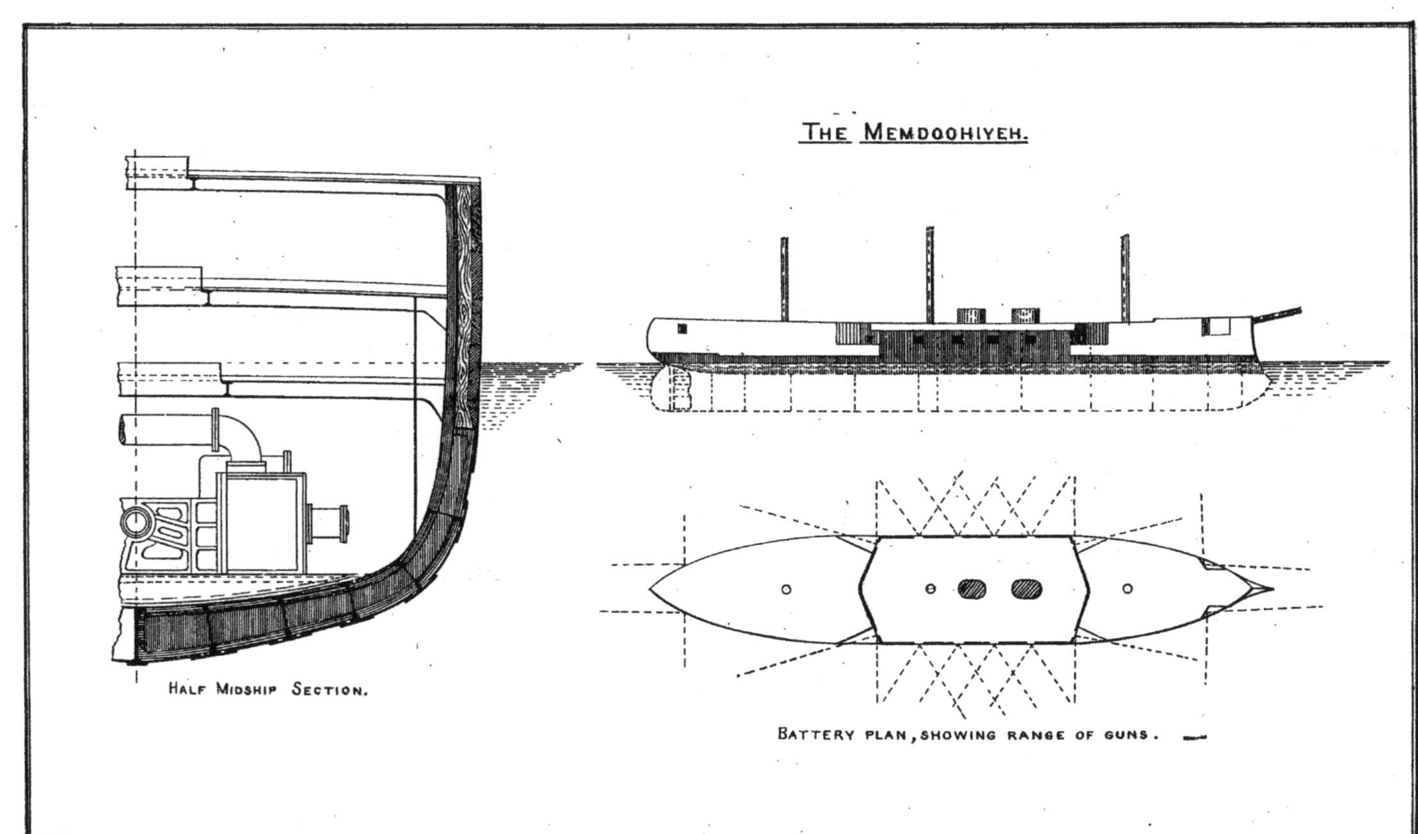
THE MEMDOOHIYEH.
HALF MIDSHIP SECTION.
BATTERY PLAN, SHOWING RANGE OF GUNS.

THE TURKISH NAVY.

The annexed list of armored ships will convey a very fair idea of the strength of this navy in 1877, as relates to its *matériel.*

The ships are for the greater part of English build, of modern type, furnished with Armstrong guns, engineered by Englishmen, and their principal officer, Hobart Pasha, was formerly an officer in the British navy. In addition to the fighting-ships represented on the list, the Turkish navy is possessed of three old wooden ships of the line, mounting an aggregate of 254 old smooth-bore guns, five wooden frigates, and seven corvettes, mounting about 300 guns; also, twenty-one smaller craft, having about 80 guns. All of these vessels are screw-steamers. In addition to the foregoing, there are four paddle-wheel vessels, mounting 4 guns each; three sailing-cruisers, with an aggregate of 8 guns; and twenty-two dispatch-boats, carrying 64 guns in all. Of these last, however, quite a number are very old and doubtless unseaworthy. There are also three royal yachts of high speed, which might be utilized as dispatch-boats, and five old transports, mounting 2 or 3 guns each.

In addition to the naval vessels, Turkey has twenty-nine large steamers belonging to companies or to private individuals, which may be made available for transportation of troops or supplies.

The last vessels built for the Turkish armored fleet are the sister ships *Memdoohiyeh* and *Mesoodiyeh*, built by the Thames Iron Works Company, and the sister ships *Burdj Sheref* and *Peyk Sheref*, constructed by Messrs. Samuda Bros. Favorable opportunities were seized during my visits to the Thames building-yards to inspect these ships, the first named being afloat and well advanced toward completion, the second on the blocks and subsequently launched, and the remaining two in the early stages of construction. The *Mesoodiyeh* was delivered to the Sultan of Turkey in 1876; the other three have been detained in England under existing international law, consequent upon the war between Russia and Turkey, and at least two of them, as has been mentioned elsewhere, have been purchased by the British Government.*

The *Memdoohiyeh* and *Mesoodiyeh* are powerful ships, as may be seen from the data here given, and from a view of the accompanying drawings. The principal dimensions and other data are given in the foregoing table of armored ships of the British navy. They are full-rigged frigates of the broadside central-battery type, with hulls of the usual cellular construction, there being in all 82 water-tight compartments. The battery is 153 feet in length, and the armor-plating on the sides is 12 inches thick, backed by the same thickness of East India teak.

The armaments were furnished by Sir W. G. Armstrong & Co., and consisted originally of twelve 18-ton guns on the gun-deck, two 6½-ton guns on the upper deck forward, and one of the same caliber aft.

---

* Since the above was written, it has been authoritatively announced that all three of these vessels have been purchased by the British Government; they have, therefore, been placed upon the list of British armored ships in this report under the names, respectively, of *Superb*, *Orion*, and *Belleisle*.

The steam-machinery of the *Memdoohiyeh* was constructed by Messrs. Maudslay, Sons & Field. The engines are of that firm's well-known old type, with two piston-rods to each cylinder, connected by an inclined cross-head, one rod passing over and the other below the crank-shaft. The two steam-cylinders are each 116 inches in diameter, and the stroke of pistons 4 feet. The surface-condensers contain 8,800 composition tubes, each 8½ feet in length by ¾-inch diameter inside, giving a total condensing surface of 16,500 square feet. The condensing water is circulated by two 3-foot centrifugal pumps, each pump being capable of circulating 5,000 gallons of water per minute. The diameter of the screw-propeller is 23 feet, and the pitch, as set on the trial-trip, 19½ feet. It is arranged to be disconnected from the screw-shaft, and to revolve when the ship is under sail.

The boilers are eight in number, and of the rectangular type, each containing five furnaces 7½ by 3 feet. The total grate-surface is 900 square feet, and total heating surface, 22,500 square feet.

The results of the trials on the measured mile are given by The *Engineer*. Six runs were made, every alternate one being against the tide—the means of all being, revolutions per minute, 66.3; vacuum, 26.5 inches; boiler-pressure, 28.5 pounds; speed in knots, 13.74.

THE BURDJ SHEREF AND PEYK SHEREF.

These are twin-screw, central-battery, armored corvettes; their general dimensions and particulars are, length, 245 feet; beam, 52 feet; depth of hold, 22 feet; mean draught of water, 19 feet 3 inches; and a displacement of 4,700 tons. The armor on the belt is 12 inches and 8 inches thick, reduced at the ends as usual. On the battery the armor is from 10 to 5 inches thick. The teak backing varies from 12 to 8 inches in thickness. The main-deck plating over the engines and boilers is 3 inches thick, and beyond that 2 inches thick. The armament consisted of four 25-ton Armstrong guns, so arranged as to command an all-around fire, and when firing in broadside to concentrate their fire within 60 yards of the vessel's side. The motive machinery has been furnished by Messrs. Maudslay, Sons & Field, and consists, for each vessel, of two pairs of simple engines the cylinders of which are 65 inches in diameter and have a stroke of 2 feet 6 inches; the engines are intended to run at a speed of 100 revolutions per minute. Steam is supplied to the engines by four boilers having 20 furnaces in all, a grate surface of 466 square feet, and a heating surface of 11,610 square feet. The amount of indicated horse-power contracted for was 3,900; that actually developed on trial was 4,020.

The *Peyk Sheref* was recently put on trial under the *surveillance* of the lords of the British admiralty, to test her speed and sea-going qualities. The trial was made with all the guns, ammunition, coal, and stores on board, and the results obtained were in all respects satisfactory. One important feature developed was the extreme handiness and quickness with which she answered her helm, and in testing this quality the vessel was found to make the entire circle, with engines going at full speed, in 3 minutes and 30 seconds, and in a diameter of 420 yards. The mean of six runs, with and against the tide, gave an average speed of 13 knots, or one knot in excess of the contract speed. The performance of the engines, also, was satisfactory.

*Armored ships of Turkey.**

Name of ship.	Vessel.						Armament.			Machinery.				Remarks.
	Class.	Length between perpendiculars, in feet and inches.	Extreme breadth, in feet and inches.	Displacement, in tons.	Draught forward, in feet and inches.	Draught aft, in feet and inches.	Number of guns and description.	Height of battery above water, in feet and inches.	Thickness of armor at water-line, in inches.	Number of screws.	Indicated horse-power.	Maximum speed, in knots, per hour.	Makers of engines.	
Mesoodiyeh	Broadside-ship	333 0	59 0	8,994	23 0	25 0	12 18-ton guns, 3 6½-ton guns.	10 0	12	1	7,300	13.75	Maudslay, Sons & Field	Rigged and carries canvas.
Noosretiyeh	do	292 3	......	6,900	24 3	25 6	10 12½-ton guns	......	....	..	......	13	do	Do.
Aziziyeh	do	293 0	56 0	6,500	24 0	25 0	1 12½-ton gun, 15 6½-ton guns.	8 8	5½	1	4,500	12	Napier & Sons	Built in England. Rigged vessels. Ten inches of wood backing behind armor.
Osmaniyeh	do	293 0	56 0	6,500	24 0	25 0	do	8 8	....	1	4,500	12	Maudslay, Sons & Field	
Orkaniyeh	do	293 0	56 0	6,500	24 0	25 0	do	8 8	....	1	4,500	12	[illegible]er & Sons	
Mahmoudiyeh	do	293 0	56 0	6,500	24 0	25 0	do	8 8	...	1	4,500	12	Thames Iron-Works	
Assar-i-Tevfik	do	275 0	50 0	3,143	19 0	21 0	8 12½-ton guns	6 4	8	..	3,568	12	Forges et Chantiers	B'lt in France for the Khedive.
Feth-i-Bulend	Ordinary service.	235 0	42 0	2,760	17 6	18 0	4 12½-ton guns	7 0	9	1	3,000	14	Thames Iron-Works	Built in England.
Mookademeï Khaïr	do	235 0	42 0	2,760	17 6	18 0	do	7 0	9	1	3,000	14	......	Built at Constantinople.
Avni [illegible]	do	230 0	36 0	2,320	15 0	17 0	do	6 5	7	1	2,000	12	Thames Iron-Works	
Mooyini Zaffer	do	230 0	36 0	2,320	15 0	17 0	do	6 5	7	1	2,000	12	Samuda Brothers	
Assar-i-Shefket	do	210 0	40 0	2,300	15 0	17 0	1 12½-ton gun, 4 6½-ton guns.	5 11	4½	..	......	12	Forges et [illegible]	
[illegible]	do	210 0	40 0	2,300	15 0	17 0	do	5 11	4½	..	......	12	do	Built at Toulon.
Idjlaliyeh	do	210 0	40 0	2,300	15 0	17 6	do	6 5	4½	..	......	12	......	Built at Trieste.
Hafiz i-Rahman	Coast defender	204 0	43 2	2,300	9 0	9 0	2 6½-ton guns	......	4½	..	......	......	Bordeaux	Monitor for the Danube.
Feth el Islam	Armored gunb't	100 0	24 9	328	6 0	6 0	2 12-ton guns	2 3	3	2	411	9.05	Forges et Chantiers	Each vessel to cost about $87,120. Immersed midship section of each, 124 square feet.†
Bekour Selim	do	100 0	24 9	328	6 0	6 0	do	2 3	3	2	411	9.05	do	
Semendriyeh	do	100 0	24 9	328	6 0	6 0	do	2 3	3	2	411	9.05	do	
Iscodra	do	100 0	24 9	328	6 0	6 0	do	2 3	3	2	411	9.05	do	
Podgoritza	do	100 0	24 9	328	6 0	6 0	do	2 3	3	2	411	9.05	do	

* The authority for the spelling of the names of these ships is the secretary of the Turkish legation at Washington.
† One of these gunboats was sunk on the Danube May 25, 1877, by two Russian spar-torpedo boats.

PERSONNEL OF THE TURKISH NAVY.

The relative strength of European navies has of late been much discussed; the ships composing them have been compared with one another in nearly every conceivable manner; their tonnage, armament, the thickness of their armor, their speed, and their power of maneuvering have each and all been taken as the standard by which to measure and gauge their respective merits. Consequently we are at the present time tolerably well acquainted with the power of every armored vessel afloat, both for offense and defense, so far as this is dependent upon the structural arrangements and weapons of the ship itself. But there is another element, besides the material of which it consists, which must be taken into account, if we wish rightly to estimate the strength of a navy. Its value as a part of the armed forces of a country will depend largely both upon the quality and the quantity of its *personnel*. The most heavily armored and the most heavily armed man-of-war afloat will be of comparatively little use unless it is adequately and efficiently manned; the swiftest cruiser will hardly fulfill her mission properly unless she is skillfully worked. It is undoubtedly much more difficult for a nation to obtain an efficient *personnel* for its fleet than it is to provide the ships themselves.

Turkey has long been in straitened circumstances, and yet she has purchased a fleet which places her, so far as ships are concerned, among the naval powers of the world; but, although a nation may buy ships, it must trust to its own resources to obtain officers and crews to man and work them. Whatever may be the system adopted by the Sultan for the recruiting and maintenance of the *personnel* of his navy, they are notoriously inefficient as sailors; though, if brought to bay, they would fight as desperately now as at Navarino and Sinope.

THE TURKISH FLOTILLA ON THE DANUBE.

Nothing in the present war has been more surprising than the little that has been accomplished by the Turkish fleet, especially on the Danube. When the war broke out the Turks had a flotilla on that river consisting of the following vessels: *Feth el Islam* (Moslem Victory), *Burdj-idelan* (Heart-piercer), *Semendriyeh*, *Iscodra*, and *Podgoritza*, the last three called after names of places. These five vessels were small craft about 150 feet long, fitted with engines of 80 "nominal" horse-power, and carried each of them two 80-pounder Armstrong guns in a battery protected by 2-inch armor. In addition to these armored gunboats, there were two of recent construction and much more formidable qualities, the *Hizber* (Lion) and *Saiffee* (Sword). These vessels each carried two 80-pounder Krupp guns in revolving turrets on the upper deck, protected by 3-inch armor, and a belt of the same thickness was placed around the water-line. Their length was 120 feet, and the horse-power of the engines 100. There were, besides, several wooden steamers armed as gunboats; and two large sea-going monitors, the *Latif-i-djelit* and *Hafiz-i-Rahman*, were sent up the river on the declaration of war. All this formidable force soon disappeared, or was completely checkmated. The *Latif-i-djelit* was the first vessel destroyed—by accident, as the Turks aver; by artillery fire, as the Russians assert. Of the armored gunboats, three have been lost: the *Saiffee*, destroyed by torpedoes, and the *Iscodra* and *Podgoritza*, both of which fell into Russian hands on the taking of Nicopolis. Besides the vessels above named, the Turks have lost four wooden vessels: the *Sulina*, a 60 horse-power gunboat of the old type, designed for the Baltic during the war of 1855; and three river steamers, one destroyed by a torpedo and the others by the fire of Russian batteries. Thus the whole naval power of the Turks in the Danube has been either destroyed, captured, or neutralized. Even in the Black Sea nothing has been accomplished by the Ottoman navy at all in proportion to its immense preponderance of force.

The London *Times* translates from a Russian journal the official report of Lieutenant Dubasoff upon his successful attack against the Turkish armored gunboat *Saiffee*. It is as follows:

My plan of attack was this: Upon entering the Matchin branch of the river I ordered the four cutters under my command to sail in a straight line one after the other. The *Cesarewitch*, which I commanded in person, was to go first; then the *Xenia*, under Lieutenant Shestakoff; then the *Djigit*, under Midshipman Persin; and the last, the *Cesarevna*, under Midshipman Ball. In this order we were to creep along the shore until in sight of the enemy, when speed was to be slackened. Advancing toward the middle of the river, the cutters were then to go two and two, the *Cesarewitch* and *Xenia* in front, and the *Djigit* and *Cesarevna* behind. From the moment of entering the Matchin branch to the moment of attack we were to proceed slowly, to reduce the noise of the engines and the splash of the water to a minimum. As we neared the enemy we were to increase speed. I was to attack, with Shestakoff following close; Persin was to keep ready to assist in case of accident, and Ball to remain in reserve. If the ship attacked first by us was disabled by the explosion, Lieutenant Shestakoff was to attack the second ship, with Persin supporting him, all rendering help, and myself keeping in reserve. Supposing the second explosion to be likewise successful, Persin was ordered to attack the third ship, with Ball supporting him, myself rendering assistance, and Shestakoff remaining in reserve.

The night between the 13th and 14th of May was cloudy, but not quite dark, the moon being mostly visible. There was a light breeze from the northwest, conveying the sound of our approach to the enemy. With the exception of the *Cesarewitch*, however, we went on noiselessly.

Early on the 13th we surveyed the enemy's position from the hills on our side of the river. When we approached their place of anchorage I ordered the steam to be shut off, to prevent any noise, but the steam quickly falling to 32 (I generally kept it at 50), I was four times compelled to stop the engine for a while in sight of the enemy. As this was likely to attract attention, I, after the last stoppage, ordered Shestakoff to readmit the steam, and to follow me rapidly to the nearest monitor, which I intended to attack first. At the moment of giving this order we were 60 sajen (420 feet, English) from the monitor. Notwithstanding, however, the noise with which we were proceeding, we were hailed by the watch only after performing half the distance. I answered what I thought to be the regulation reply, but have since heard that it was not in form, and that my mistake awakened immediate suspicion. The artillerymen, who had laid down for the night on deck, were awoke by the first report of the signal-rifle. I suspect that our excursion the preceding night had attracted attention, and that all the monitors were on the *qui vive*. Indeed the monitors, as we discovered on approaching them, had left their former anchorage and gone nearer Matchin.

The monitor had her steam up, and firing at us from her stern-guns on the upper deck might have inflicted considerable damage. I therefore determined to make for the stern, and thereby escape danger and deprive the vessel of her moving powers. The connecting-wire I ordered to be kept in readiness to be used at any moment. My calculation proved correct. We no sooner neared the ship than the stern-gun opened fire. Three bullets were discharged without effect, and before the fourth could be fired I had passed the stern, and, coming up to the left side of the ship, sprang the mine, which destroyed the stern. It was a torpedo attached to a pole, and hit the ship between stern and midships, a little before the stern-post. The water rushing into the sides of the monitor, the waves washed over the cutter. Many fragments were thrown to a height of about 120 feet. Some bits of furniture falling into the cutter proved the explosion to have taken effect right through the ship up to the deck. The crew of the monitor hastened from stern to prow, the stern sinking considerably into the water. I took measures to save my men, but finding the cutter had righted herself, endeavored to back astern and put the steam-ejector into operation to pump out the water. At this moment the sinking monitor began to fire out of her turret, when I called out to Lieutenant Shestakoff to deal another blow. Quickly coming up, he inflicted the deadly blow a little behind the turret just as the turret-gun was firing a second shot. Lieutenant Shestakoff, it is necessary to observe, actually touching the monitor with his prow, lodged his torpedo under the keel amidships, about twenty feet from the prow-post. As in the first instance, the effect of the explosion was tremendous, as may be inferred from cabin-furniture being hurled into the air and falling afterward into the cutter *Xenia*. After the second explosion the crew of the monitor, finding it impossible to continue their artillery, with remarkable bravery seized their rifles, discharging one salvo after the other. Neither I nor Shestakoff could get away as fast as we wished. The screw of Shestakoff's cutter had got entangled with some of the broken fragments, while my vessel was so full of water that I had to set the whole crew to work to bale it out with pails, the steam-ejector having refused to work. During the whole of this time Shestakoff kept up a raking rifle-fire against the enemy. The two other Turkish vessels (one a steamer, the other a monitor) had kept firing at us ever since the first discharge of the attacked monitor. The steamer evidently was provided with smooth-bore guns, which the crew did not fire with dispatch or precision. Possibly the steamer being 60 sajen nearer to us than the second monitor and having her deck inundated with water, the men found it impossible to handle their guns effectively. The second monitor, being more advantageously placed, could turn her battery upon us without any difficulty. Her shot fell some distance from our stern, and subsequently, when we had got away from the monitor, passed over our heads. The rifle-fire from both ships was kept up incessantly while we were alongside and when we had got away.

The Turkish commander did not avenge his defeat and make reprisals on the crafty foe, but in a dispatch to the Sultan, published in a Constantinople journal, may be read how the admiral who had lost an iron-clad, the best hope of his country against the passage of the Danube by that country's traditional foe, being deeply touched at the disaster, and having determined that it should not be repeated, gave orders to his crews to leave the dangerous neighborhood, and they steamed quietly down the river to a place where there were neither batteries nor torpedoes, there, perchance, to smoke and dream. There has not been a single success, nor even one gallant failure, by the Turkish navy during the war.

THE AUSTRIAN NAVY.

The naval authorities of Austria, in common with those of other European countries, have paid dearly for the error of building wooden armored ships. The *Kaiser Max*, the *Prinz Eugen*, and the *Don Juan d'Austria* constituted the early iron-clad fleet of Austria. The decay of the wooden hulls, insufficient strength, and the advancement made in defense as well as offense of the modern fighting-ships, some time since rendered these vessels useless. As a matter of economy, however, it was decided to reconstruct them, with the view of utilizing as far as practicable the machinery and materials. They were accordingly taken to pieces, and the replacing of the wooden hulls with hulls of iron proceeded with, advantage being taken of the superior lightness of the iron hull in providing for a considerable increase of armor strength in vital places. The original engines and most of the fittings of the wooden ships are being used in the iron ones; and this object is being carried into effect, it is said, with economy. The constructor says:

It is true that the engines of the three old ships cannot give a higher speed to the new ships than 12 knots per hour, but instead of building only one new ship we will possess three rams.

It remains to be seen, after these ships have been completed, what success has attended the making of modern fighting-vessels from the materials of obsolete ships, engined with the old types of machinery.

Of armored vessels the Austrian navy had at the beginning of last year seven casemated ships, two frigates, two corvettes, and two river monitors, an eighth vessel of the first class being on the stocks. Of the seven casemated vessels, the *Custozza* and *Erzherzog Albrecht* are the most heavily armed and armored; the former carries eight 10-inch Krupp guns, and has 9¼ inches of iron armor at the water-line, while the latter has the same armament, but one inch less of armor. The *Custozza*, when loaded, displaces 7,060 tons of water, and can attain a speed of 14 knots when her maximum indicated engine-power of 6,000 horses is developed; she is 302 feet 3 inches long between perpendiculars, has an extreme breadth of 57 feet 10 inches, and a mean draught of 23 feet 8½ inches. The *Erzherzog Albrecht* has a displacement of 5,940 tons, an indicated horse-power of 4,300, and has reached on trial a speed of 13 knots; her length between perpendiculars is 275 feet 7 inches, her breadth 54 feet 5 inches and mean draught 21 feet. Both of these vessels have iron hulls, and were launched in 1872. The *Lissa* and *Kaiser* are wooden vessels of older construction, and are both protected by 6¼-inch plates. The *Lissa* is the larger vessel of the two; her length is 274 feet 6 inches, extreme breadth 55 feet, and mean draught 26 feet 1 inch; she carries twelve 9½-inch guns of Krupp's manufacture, has a displacement of 6,080 tons, and is driven by engines of 3,550 horse-power at a speed of 13 knots; while the *Kaiser*, with a length of 263 feet 8 inches, a breadth of 59 feet 5 inches, and a mean draught of 24 feet 10½ inches, attains a speed of 12 knots with 2,380 horse-power engines, displaces 5,810 tons, and is armed with ten 9-inch guns of Armstrong make. The *Kaiser Max*, *Don Juan d'Austria*, and *Prinz Eugen*, as mentioned above, are all remodeled iron vessels; the first two were launched in 1875, and the last named in 1876; they are sister ships, having the same dimensions throughout and the same armament; each has a length between perpendiculars of 222 feet, a breadth of 44 feet 2 inches, and a mean draught of 20 feet 5 inches, displaces 3,550 tons, and steams at a speed of 8½ knots, with engines having 1,710 horse-power; the armament of each, also, is twelve 7-inch Armstrong guns.

The two armored frigates, *Erzherzog Ferdinand Max* and *Habsburg*, are wooden vessels of much older construction than any of the casemated ships; they are sister vessels, and were launched in 1865; their length between perpendiculars is 253 feet 2 inches, extreme breadth 50 feet 10 inches, and mean draught 22 feet 5 inches; each has a displacement of 5,140 tons, an indicated horse-power of 2,902, and a speed of over 10 knots; the armament consists of fourteen 8¼-inch Krupp guns, and the armor of 5-inch plates.

The *Drache* and *Salamander*, sister vessels of old style, are wooden station-service ships, armored with 4¾-inch plates and carrying each ten 7-inch Armstrong rifles; each vessel has a length of 196 feet 6 inches, a breadth of 43 feet 7 inches, and a mean draught of 19 feet 11 inches; the displacement is 3,110 tons, horse-power 1,418, and speed 7 knots.

There are also two single-turret monitors, sister vessels, which were launched in 1871 for service on the Danube; these vessels, the *Maros* and *Leitha*, have a length of 160 feet 1 inch, with a breadth, extreme, of 26 feet 8 inches, and a draught of 3 feet 6 inches; their displacement is 310 tons, horse-power 314, and speed about 6 knots; each is armed with two 6-inch Krupp guns and protected by 1½ inches of armor.

The unarmored ships of Austria are all wooden vessels; they comprise three frigates, with an aggregate displacement of 9,510 tons and an armament of 63 guns, five spar-decked corvettes, three flush-decked corvettes, five gunboats, five screw-schooners, three paddle-wheel steamers, two dispatch-vessels, three transports, and one torpedo vessel.

The latest design and the most important fighting-ship of the Austrian navy is the armored ship *Tegethoff*, recently completed at the works of the *Stabilimento Technico* at Trieste.

THE TEGETHOFF.

The *Tegethoff* is a broadside central-battery ship, in which the batteries project over the sides of the vessel, as first introduced in the upper batteries of the British *Audacious* class. The following figures give her dimensions, calculated elements, &c.:

Length between perpendiculars	286 feet 11¼ inches.
Length, total	303 feet 1¼ inches.
Breadth on water-line	62 feet 9 inches.
Extreme breadth to the outside of armor	71 feet 1½ inches.
Depth of hold	34 feet 7 inches.
Draught of water aft	26 feet 7½ inches.
Draught of water forward	23 feet 1 inch.
Displacement with one-half of the provisions	7,390 tons.
Area of the midship section	1,301 square feet.
Area at the load water-line	14,308 square feet.
Height of metacenter above center of gravity of displacement	14.623 feet.
Height of metacenter above water	4.770 feet.
Distance of the center of gravity of displacement forward of the midship section	3.356 feet.
Depth of the center of gravity of displacement below water	9.853 feet.
Coefficient of displacement	0.582
Coefficient of water-line	0.782
Coefficient of midship section	0.82
Displacement of an inch immersion at the load water-line	34.47 tons.
Weight of armor and backing	2,160 tons.

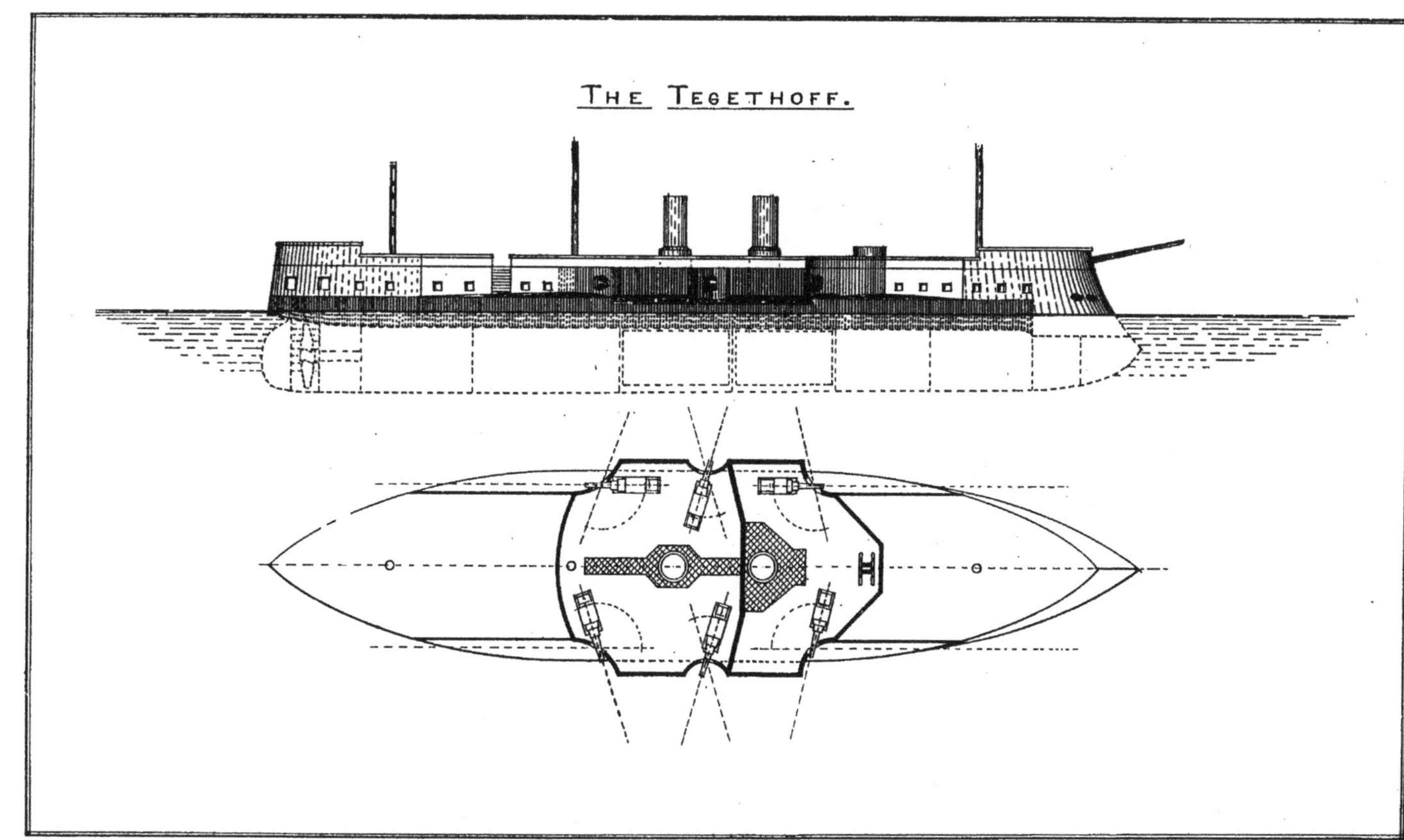

THE TEGETHOFF.

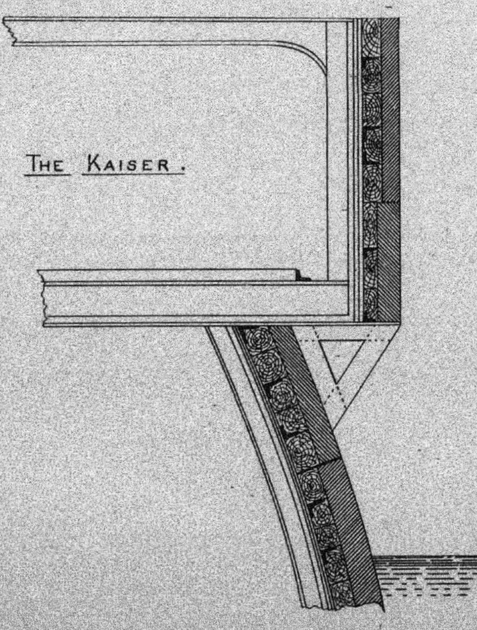
THE KAISER.

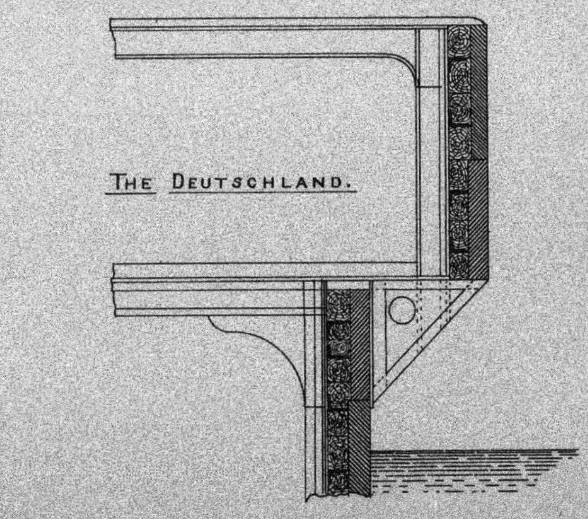
THE DEUTSCHLAND.

The armament consists of six 11-inch Krupp guns.

Area of sails	12,165 square feet.
Cost of hull, estimated	$839,759
Cost of engines and boilers, estimated	$397,135
Nominal horse-power	1,200
Number of cylinders	2
Diameter of cylinder, effective	125 inches.
Length of stroke	4 feet 3 inches.
Griffith's propeller, diameter	23 feet 6 inches.
Pitch	24 feet.
Number of blades	2
Revolutions per minute	70
Number of boilers	4
Area of fire-grate	850 square feet.
Heating surface	25,500 square feet.
Superheating surface	1,800 square feet.
Pressure of steam	30 pounds.
Number of furnaces	36
Mean indicated horse-power	8,000
Speed, estimated	14 knots.

Mr. E. J. Reed, C. B., M. P., before the Institution of Naval Architects in London, in April, 1876, spoke as follows of this ship:

From these figures it will be seen that although we are not dealing with a ship of the *Inflexible* type, in which armor of excessive thickness is placed over a central citadel of extremely limited extent, we nevertheless have a very powerful ship indeed, with armor of apparently about 13 to 14 inches thick, and with a concentrated battery of six 11-inch Krupp guns, each weighing, I presume, about 27 tons. The ship has a belt of armor extending from the stern to within about 30 feet of the foremost perpendicular, where it terminates in a transverse armored bulkhead, and a stout iron deck going forward to the stem at about 7 feet below water. It would appear from this that the Austrian authorities consider that a strong iron stem, supported by a stout deck near the point of the ram, is sufficient for ramming purposes, whereas in our navy we have thought it better, beginning, if I remember rightly, with the *Rupert* and *Hotspur*, to keep the bow armor and carry it down at the stem to considerably below the ram point. We take it for granted that the latter, or English, arrangement would at least have the advantage of protecting the ram bow from much local damage in ramming iron vessels, and this is no doubt very desirable where a ship is designed primarily as a ram, as were the *Rupert* and *Hotspur;* while, on the other hand, where the ram is a subordinate feature, as in the *Tegethoff*, it may be unnecessary to burden the bow with so much armor protection.

It is worth while to observe in this connection that the Austrians, who have had practical experience of the effects of ramming in actual warfare, have in this, their largest and most powerful ship, preserved a very great length of underrunning or sheer projection. * * * The projection is 9 feet from the stern at the load water-line and 19 feet from the stem head.

I observe next in this ship, the Austrian admiralty have adopted an improvement in armor to which I have for a long time past attached great importance. I refer to the getting rid for the most part of great curvature. Of course armor-plates, if carried round the ends of a ship, must be bent to the curvature of the water-lines; and when imbedded, so to speak, in the sides of a ship of ordinary form and curvature, as has been usual in sea-going ships, they must also be bent crosswise. Now, this double curvature of the plates is not only an expensive process, but it is also injurious to the armor-plates in some degree.

* * * * * * *

By so designing the ships that the armor-plates have only to be curved in one direction, all these disadvantages and difficulties are practically got rid of. The next feature to be remarked in this ship is that the battery is of the projecting type, which so greatly facilitates the obtainment of direct fire ahead and astern from a midship-battery without excessive recession of the unarmored parts of the ship before and abaft the battery.

The admiralty constructors and myself introduced this arrangement in many cases of upper-deck batteries in ships designed while I was at the admiralty (most notably in the case of the *Audacious* class), but I do not think the same thing has been done at the admiralty in the case of the main-deck battery. I have, however, done it myself in several ships for foreign governments since I left the admiralty, and I give examples in Figs. 1 and 2, which represent outline sections of the *Kaiser* and *Deutschland* (German), and the *Almirante, Cochrane*, and *Valparaiso* (Chilian), armor-clad ships.

In the *Tegethoff*, the overhang is very low and considerable in amount, the battery projecting between 4 feet and 5 feet, the spread commencing at 18 inches above the water, and terminating at a height of 6 feet.

The armored citadel of the *Tegethoff* is to be furnished with a transverse armored bulkhead abaft the two foremost guns, an arrangement which will prevent the battery from being raked in chasing. This improvement exists in the *Alexandra*, where it was, I believe, introduced for the first time by the present admiralty constructors. The foremost bulkhead of the battery is inclined forward at a considerable angle to within about 4 feet of the middle line, where it becomes transverse, as shown. Immediately over this foremost portion of the battery, at the middle, is a very strong pilot-tower, standing well above both the gunwale and the forecastle. This shows that the Austrian officers who have been in an action with iron-clads do not consider such towers unnecessary. The above appear to me to be the principal features of the *Tegethoff*. It may be interesting to add that, while the outer skin and angle-irons of the hull are of iron, all the remainder is of Bessemer steel, varying in tensile strength from 30 to 33 tons per square inch of section, and possessing this, as I am informed, in combination with 25 per cent. of ductility. This Bessemer steel is produced very successfully in Styria and Carinthia, from which districts of Austria the chief supplies of the *Tegethoff* are derived. I may further add that, in designing this ship, much consideration has been given to securing both strength and subdivision by means of water-tight bulkheads between the coal-spaces and the boilers and elsewhere.

PERSONNEL OF THE AUSTRIAN NAVY.

The *personnel* of the Austrian navy consists, when on a peace footing, of 483 officers and 5,836 men; the list of officers comprises the following: one admiral, two vice-admirals, sixteen captains of line-of-battle ships, seventeen captains of frigates, eighteen captains of corvettes, eighty lieutenants of the first class, forty lieutenants of the second class, and, finally, one hundred and fifty-four cadets. On a war footing the list contains one admiral, six vice-admirals, eighteen captains of line-of-battle ships, nineteen captains of frigates, twenty captains of corvettes, ninety lieutenants of the first class, forty-five lieutenants of the second class, and one hundred and eighty-five cadets.

The men of the Austrian navy are divided into twelve companies, which are quartered at two depots, and, according to their qualifications, are trained either as sailors and gunners, or as stokers and engine-artificers. Those belonging to the first two categories are instructed on board hulks and school-ships, the gunners being further taught the details of their profession on board the artillery school-ship *Adria*, which, at present, has a complement of 500 men.

DOCK-YARD.

The great naval port and dock-yard of Austria is at Pola, in the province of Istria, on the east coast of the Adriatic, about sixty miles south of Trieste. The whole naval resources of the country, belonging to the government, are concentrated here. The location is excellent, the bay and harbor are capacious, and in all respects well adapted for the purpose, besides which it is susceptible of being strongly fortified. The dock-yard, which has existed for a long period, was regarded as of very small importance until 1856, at which time its improvement was commenced, and gradually continued since then until it has been brought to the present very complete state as an extensive repairing and outfitting yard for the fleet. Although this dock-yard is equipped with ship-houses, buildings, and the necessary appliances for ship construction, nearly all vessels for the navy have hitherto been built, by contract, at the private building-yards of San Marco and San Rocco, near Trieste.

Armored ships of Austria.

Name of ship.	Vessel.							Armament.			Machinery.			Cost in dollars, gold.	
	Length between perpendiculars, in feet and inches.	Extreme breadth, in feet and inches.	Depth of hold, in feet and inches.	Immersed midship section, in square feet.	Displacement, in tons.	Draught forward, in feet and inches.	Draught aft, in feet and inches.	Number of guns and description.	Height of battery above water, in feet and inches.	Thickness of armor at water-line, in inches.	Number of screws.	Indicated horse-power on trial.	Maximum speed, in knots per hour.	Machinery.	Ships completely equipped.
Line-of-battle cruisers.															
[illegible]ff *	287 0	71 1	34 7	1,301	7,390	23 1	26 7½	6 11-inch Krupp guns	.....	13.5?	1	8,000	14	397,135	1,236,84
Custozza † (i)	302 3	57 10	37 1	1,092	7,060	21 7	25 10	8 10-inch Krupp guns	9 9	9¼	1	6,000	14	387,100	2,030,60
Lissa (w)	274 6	55 0	22 3	1,047	6,080	24 0	28 2	12 9.45-inch Krupp guns	6 6	6¼	.....	3,550	13	323,419	2,022,20
Erzherzog [illegible]t † (i)	275 7	54 4.8	35 0	969	5,940	18 9½	23 3½	8 10-inch Krupp guns	10 2	8, 3	1	4,300	13	362,600	1,752,20
Kaiser † (w)	263 8	59 5	.....	960	5,810	24 1	25 8	10 9-inch Armstrong guns	8 0	6¼	1	2,380	12	282,240	1,652,80
Don Juan [illegible]ia ‡ (w)	222 0	44 2	.....	706	3,550	18 4	22 6	12 7-inch Armstrong guns	6 0	4½	1	1,710	8.5	204,441	.....
Kaiser Max ‡ (w)	222 0	44 2	.....	706	3,550	18 4	22 6	do	6 0	4½	1	1,710	8.5	204,441	1,028,376
Prinz Eugen ‡ (w)	222 0	44 2	.....	706	3,550	18 4	22 6	do	6 0	4½	1	1,710	8.5	204,441	.....
Erzherzog Ferdinand Max § (w)	253 2	50 9.8	20 3	831	5,140	20 0	24 10	14 8.27-inch Krupp guns	6 11	5	1	2,902	10.3	239,120	1,403,80
Habsburg § (w)	253 2	50 9.8	20 3	831	5,140	20 0	24 10	do	6 11	5	1	2,902	10.3	239,120	1,403,80
Ordinary station service.															
Drache † (w)	196 6	43 7	.....	693	3,110	18 7	21 3	10 7-inch Armstrongs, rifled	6 3	4¾	1	1,418	7	174,240	800,928
Salamander † (w)	196 6	43 7	.....	693	3,110	18 7	21 3	do	6 3	4¾	1	1,418	7	174,240	.....
River monitors.															
Maros ‖	160 1	26 8	.....	95	310	3 6	3 6	2 5.9-inch Krupp guns	3 9	1.5	1	314	5.6	19,600	98,000
Leitha ‖	160 1	26 8	.....	95	310	3 6	3 6	do	3 9	1.5	1	314	5.6	19,600	98,000

* In process of construction. † Central battery. ‡ In process of reconstruction; rams. § Broadside-battery. ‖ Single turret. (i) Iron hulls. (w) Wooden hulls.

PART XVIII.

THE NAVIES OF HOLLAND, SPAIN, DENMARK, SWEDEN, NORWAY, AND PORTUGAL.

TABLES OF DIMENSIONS OF ARMORED SHIPS; PERSONNEL OF THESE NAVIES.

BRAZILIAN AND JAPANESE WAR-VESSELS.

RÉSUMÉ OF EUROPEAN ARMORED SHIPS, AND ACTIONS IN WHICH THEY HAVE BEEN ENGAGED.

THE DUTCH NAVY.

Holland possesses a navy of considerable strength, as may be seen from the appended lists of armored and unarmored ships. The armored vessels were built some years ago, and cannot, therefore, now be classed with the types recently constructed. There are, however, two sea-going ships on the list, viz, the *Prins Hendrik der Nederlanden* and the *Koning der Nederlanden*, both of them turret-ships with ram bows; the first named was built by Laird Bros., of England, and is of the type of the British ship *Monarch*, but fitted with tripod masts. Her length between perpendiculars is 230 feet 2 inches; extreme breadth, 44 feet; depth of hold, 26 feet 6 inches; mean draught of water, 18 feet 4 inches; and the thickness of armor on the water-line is 5¾ inches. She is fitted with two turrets, constructed after the English system, in each of which are mounted two 9-inch Armstrong rifled guns. The height of the battery above water is 7 feet 2 inches. The motive power is of the old type, the indicated horse-power is 2,426, and the speed on the measured mile, as reported, was 12 knots.

For coast defense Holland is provided with nineteen monitors, the *Buffel* and *Guinea* being the most powerful. These two sister vessels have each a displacement of 2,190 tons; length between perpendiculars, 205 feet; breadth, extreme, 40 feet; mean draught of water, 15 feet 6 inches; and depth of hold, 24 feet.

The hull is constructed on the bracket-plate system, and provision is made for admitting water between the outer and inner skins, to bring the ships lower in the water when necessary.

The battery is in a single turret, and consists of two 9-inch 12-ton Armstrong guns, mounted on the Armstrong carriages in their completeness; also, four 30-pounders on deck.

The vessels next of importance are the *Schorpioen* and *Stier*, built at the *Forges et Chantiers*, France. These two sister vessels have single turrets, ram bows, and a reported speed, on the measured mile, of 12½ knots. The length between perpendiculars is 193 feet 6 inches; mean draught of water, 15 feet 6 inches; displacement, 2,113 tons; indicated horse-power, 2,238. The armament consists of two 9-inch Armstrong guns.

The next ten vessels are known as the *Cerberus* class, and, as may be seen from the following list, the principal data of each are: length between perpendiculars, 187 feet; breadth, extreme, 44 feet; draught, mean, 8 feet 6 inches; armor on the water-line, 5½ inches; and the speed is 7 knots. The armament consists of two 12-ton Armstrong guns.

The naval authorities of Holland intend to keep up with the progress of the times in constructing unarmored ships. One iron corvette of the first class, having the bottom sheathed with wood, has already been completed, and two others of the same class are building. These vessels are designed for fast cruisers, and are fitted with all modern appliances.

DOCK-YARDS.

The principal dock-yard of Holland is at Amsterdam, but there is also a naval station at Nieuwe Diep. The first-named yard is chiefly devoted to construction and repair, and in it several of the smaller monitors were built. At the latter place, the ordnance stores, including powder and shells, are put on board vessels preparing for service. No articles of this description, torpedoes excepted, are kept at Amsterdam. At Nieuwe Diep is also located the naval academy.

At the Hague there is a government foundery for the manufacture of bronze cannon; the gun-carriage shop, pyrotechnic school, and small-arm factory are at Delft; the powder-mills at Mindow, and the practice firing ground is at Scheveningen.

PERSONNEL.

The *line officers* of the Dutch navy include two vice-admirals, four rear-admirals, nineteen captains, forty-three lieutenant-captains, one hundred and twenty-three lieutenants of the first class, one hundred and eighty-five lieutenants of the second class, and fifty-two midshipmen.

Engineer officers.—These consist of one chief engineer, director of marine architecture; four chief engineers; four engineers of the first class, and two of the second class; also, eight engineer officers for service in ships, and thirty-five artificers with fixed appointments.

Medical officers.—One inspector-general, six directing medical officers, thirty-two medical officers of the first class, twenty-three of the second class, and three surgeon-apothecaries. Also five medical officers of the army, detailed for duty in the naval service.

The permanent petty officers consist of two hundred and twenty-one persons, divided into three classes. At the Royal Naval Institute there are one hundred and eight cadets, divided into three classes. The marine corps is composed in the aggregate of forty-five officers of all grades. The naval administration has attached to it three inspectors of administration, eighteen officers of the first class, thirty of the second class, and thirty-seven of the third class.

His Majesty the King is the commander-in-chief, with a staff consisting of four princes of the royal house, who hold rank as lieutenant-admiral, admiral of the fleet, lieutenant-admiral commander-in-chief of the fleet, and captain of the staff, respectively. Also, two naval officers act as adjutants to His Majesty in ordinary service, or six in extraordinary service, two adjutants to the commander-in-chief of the fleet, and one to the captain of the staff.

Armored ships of Holland.

Vessel.						Armament.				
Name of ship.	Class.	Length between perpendiculars, in feet and inches.	Extreme breadth, in feet and inches.	Draught forward, in feet and inches.	Draught aft, in feet and inches.	Number of guns and description.	Thickness of armor at water-line, in inches.	Maximum speed, in knots, per hour.	Engines, where built.	Remarks.
For general service.										
Prins Hendrik der Nederlanden.*	Double-turret and ram	23 02	44 0	17 10	18 10	4 9-inch Armstrong guns	5.75	12	Birkenhead	Depth of hold, 26′ 6″; indicated horse-power, 2,426.
Koning der Nederlanden.	Turret-ship and ram	......	......	19 8	19 8	4 11-inch Krupp guns 4 4½-inch Krupp guns	......	......	Amsterdam	Indicated horse-power, 3,500.
For coast, harbor, and river defense.										
Schorpioen†	Single-turret with ram bow	193 6	36 0	15 6	15 6	2 9-inch guns, rifles	6	12.82	Forges et Chantiers	Indicated horse-power, 2,238.
Stier†	do	193 6	36 0	15 6	15 6	do	......	12.82	do	Total cost of each, $448,840.
Buffel‡	do	205 0	40 0	15 6	15 6	2 9-inch rifles; 4 30-pounders	......	......	Glasgow	Displacement, 2,190 [illegible]
Guinea‡	do	205 0	40 0	15 6	15 6	2 12-ton guns; 4 ( )-pounders	8¼	13	Amsterdam	Depth of hold, 24′ 0″.
Bloedhond	Single-turret	187 0	44 0	8 6	8 6	2 12-ton Armstrong guns	5½	7	do	Twin-screws 650 ind. h.-p.
Cerberus	do	187 0	44 0	8 6	8 6	do	5½	7	do	Do.
Heiligerlee	do	187 0	44 0	8 6	8 6	do	5½	7	Birkenhead	Do.
Krokodil	do	187 0	44 0	8 6	8 6	do	5½	7	do	Do.
Tijger	do	187 0	44 0	8 6	8 6	do	5½	7	Glasgow	Do.
Adder	Single-turret with ram bow	187 0	44 0	8 6	8 6	do	5½	7	Amsterdam	Do.
Haai	do	187 0	44 0	8 6	8 6	do	5½	7	do	Do.
Hyena	do	187 0	44 0	8 6	8 6	do	5½	7	do	Do.
Panter	do	187 0	44 0	8 6	8 6	do	5½	7	do	Do.
Wesp	do	187 0	44 0	8 6	8 6	do	5½	7	do	Do.
[illegible]k	Turret and ram bow	......	......	10 10	10 10	2 11-inch rifles	......	......	do	Unfinished. 800 horse-power.
Matador	do	......	......	9 10	9 10	do	......	......	Fijenoord	Unfinished. 650 horse-power.
Luipaard	do	......	......	8 8	8 8	1 11-inch rifle	......	......	do	650 horse-power.
Rhenus	For river service	......	......	4 3	4 3	2 4½-[illegible] guns	......	......	Amsterdam	Unfinished. 50 nom. h.-p.
[illegible]la ‖	do	150 11	18 4	4 3	4 3	do	......	......	do	50 [illegible] horse-power.
Vahalis	do	......	......	4 11	4 11	......	......	......	Fijenoord	60 [illegible] horse-power.
No. 1	Gunboat	......	......	6 11	6 11	2 60-pounders	......	......	do	450 horse-power.

NOTE.—All these vessels have iron hulls. * Height of battery above water 7 feet 2 inches. † Midship section 542 square feet; displacement, 2,113 tons; height of battery above water 7 feet 3 inches. ‡ 2,400 horse-power. ‖ Displacement, 370 tons; steel breech-loading guns.

UNARMORED SHIPS OF HOLLAND.

In addition to the *armored fleet*, Holland possesses a fleet of unarmored *cruising-vessels*, which are classed as follows:

1 steam-frigate, wooden, armed with one 60-pounder, forty-two 30-pounders, eight 6-inch rifles.
3 screw-corvettes, first class (iron hull sheathed with wood), armed with six 6½-inch rifles (2 unfinished).
5 screw-corvettes, first class, wooden, carrying from twelve to sixteen guns, 30-pounders and 6-inch rifles.
3 screw-corvettes, second class, wooden, carrying six 5-inch and 7-inch rifles.
1 screw-corvette, third class, composite, carrying two 6-inch and one 7-inch rifle.
2 screw-corvettes, third class, wooden, carrying two 6-inch rifles and four 4½-inch guns.
3 screw-corvettes, fourth class, composite, carrying from three to four guns, 4½, 6½, and 7-inch rifles (2 unfinished).
1 side-wheeler, first class, wooden, carrying six 4½-inch guns.
18 small unarmored gunboats, carrying one 9-inch rifle or one 11-inch gun, for coast, harbor, and river defense.
3 torpedo-boats, for coast, harbor, and river defense.

A number of old vessels of all classes are used as guard, school, and practice ships.

In addition to the above there is the naval force of the East India service, consisting of 13 side-wheelers of the second, third, and fourth classes; 15 screw-corvettes of the fourth class, and 5 sailing-vessels.

THE SPANISH NAVY.

The ships belonging to the navy of Spain, as registered on the official list published by order of the *ministro de marina* at Madrid, 1876, consist of eleven armored vessels, nine wooden screw-frigates, six screw-corvettes, fifteen screw-sloops, and sixty-five gunboats. One of the frigates, the *Asturias*, is armed with fifty-one guns, four others mount each forty-eight guns, and four have from thirty-three to twenty-eight guns each; two of the corvettes mount each eighteen guns, and the others from five to three guns; the sloops from two to three guns. Some of the gunboats are provided with two guns and others with one. The frigates, corvettes, and sloops constitute the cruising-fleet. They are mostly, if not all, of obsolete types, having neither guns of sufficient power to fight nor speed sufficient to run from the enemy.

It will be seen from the following meager list of fighting-ships that none of them have either the armor, speed, or armaments possessed by lately-constructed armored vessels.

The *Numancia* and *Vittoria* are the most powerful of the number. The former was built at the *Forges et Chantiers*, France; she is rated as a line-of-battle ship, and has a broadside-battery and ram bow. The thickness of armor on the water-line is 5 inches; the length is 313 feet 7 inches between perpendiculars; breadth, extreme, 56 feet 11 inches; depth of hold, 28 feet 11 inches; draught of water, mean, 25 feet; and displacement, 7,053 tons. The armament consists of six 18-ton guns, three of 9 tons each, and sixteen of 7 tons each, all Armstrong rifles. The height of battery above water is 7 feet 6½ inches. The single-screw propeller is operated by machinery of the non-compound type, and the maximum horse-power is 3,708. The total cost of the ship was $1,546,440, gold.

The *Vittoria* is of the same general dimensions, but is protected by 5½ inches of armor instead of 5 inches, and carries four 12-ton, three 9-ton, and twelve 7-ton Armstrong rifled guns. The smaller vessels are protected by only 4½-inch armor, and two of them mount each two 18 ton rifled guns besides others of less caliber.

There is but one turret-vessel on the list, the *Puigcerda*. This vessel is a monitor of small dimensions, carrying three guns, and is engined with only 60 "nominal" horse-power.

If all accounts of the *personnel* of the Spanish navy be true, it is in no better condition to work the ships than the ships are to meet modern fighting-vessels in combat.

The following list contains all the obtainable data of the armored ships of Spain:

Armored ships of Spain.

Name of ship.	Vessel.								Armament.	Thickness of armor at water-line, in inches.	Remarks.
	Class.	Length between perpendiculars, in feet and inches.	Extreme breadth, in feet and inches.	Depth of hold, in feet and inches.	Midship section, in square feet.	Displacement, in tons.	Draught forward, in feet and inches.	Draught aft, in feet and inches.	Number of guns and description.		
Numancia	Line-of-battle ship	313 7	56 11	28 11	1,151	7,053	23 4½	26 8	6 18-ton Armstrongs; 3 9-ton Armstrongs; 16 7-ton.	5	Broadside-battery; ram bow; built at the *Forges et Chantiers*. Indicated horse-power, 3,708; speed, 13 knots; single screw; total cost, $1,546,440; height of battery above water, 7 feet 6½ inches.
Vittoria	do	316 0	57 00	28 11	1,151	7,053	23 4½	27 0	4 12-ton Armstrongs; 3 9-ton Armstrongs; 12 7-ton.	5½	
Saragossa	do								4 11-inch; 2 8¾-inch; 14 7¾-inch	4½	
[illegible]es	do								2 18-ton; 5 9-ton; 10 7-ton	4½	
Mez Nunez	Ordinary station-service ship.								4 9-inch Armstrongs; 2 8-inch		Late the Resolucion.
Sagunto	do								2 18-ton; 5 9-ton; 10 7-ton	4½	
Aragon	do	No data							3 guns		In course of construction.
[illegible]a	do								do		Do.
Navarra	do								do		Do.
Duque de Tetuan	Light draught battery										
Puigcerda	do										

NOTE.—It has been proposed to add to this fleet during the year 1878, three sea-going iron-clads of moderate power, and three iron-clads of small power; one of the latter being a monitor and another a floating-battery.

The unarmored fleet is also to be probably increased by six steamers of moderate tonnage, eight of small size, and several of still smaller capacity.

THE DANISH NAVY

The naval authorities of Denmark have been wise in avoiding the errors of other continental countries as to the materials of which their ships are composed. Not a wooden armored ship has been built since 1864; and since 1868 vessels of all the other classes have been built of iron and steel.

As may be seen from the following list, kindly furnished this year by Mr. Ravn, the assistant minister of marine, the Danish fleet comprises seven armored ships, all built of iron except the original one. The *Helgoland*, broadside-ship now building, will be the most powerful. Her length is 257 feet 5 inches, extreme breadth 59 feet 2 inches, displacement 5,265 tons, armor 12 inches thick on the water-line, and she is designed to carry one 12-inch, four 10-inch, and five 5-inch rifle-guns.

In addition to the armored fleet, the Danish navy is possessed of three wooden screw-frigates of the old type, viz, the *Jutland*, *Sjoelland*, and *Niels Juel;* the first named is 201 feet and the others are 195 feet 8 inches between perpendiculars, with a breadth of 44 feet 3 inches, and displacement of 2,263 tons. Each mounts two 8-inch and twenty-four 6-inch guns. There are also two screw-sloops and three screw brigs, wood-built. Besides these, there are fifteen iron screw gunboats, varying in displacement from 547 tons to 140 tons. Two of the latter, the *Absalon* and *Esbern Snare*, are plated with iron 2½ inches in thickness; some of these mount one 10-inch and some two 5½-inch guns.

There are also two transports, two yachts, one Thornycroft torpedo-boat, and one other boat converted for torpedo purposes.

Armored ships of Denmark.

Name of ship.	Vessel.					Armament.	Thickness of armor at water-line.	Machinery.	
	Class.	Length between perpendiculars, in feet and inches.	Extreme breadth, in feet and inches.	Displacement, in tons.	Mean draught of water, in feet and inches.	Number of guns and description.		Indicated horse-power.	Maximum speed, in knots, per hour.
Helgoland* ..	Broadside...	257 5	59 2	5, 265	18 6	1 12-inch, 4 10-inch, and 5 5 inch guns.	12	....	......
Danmark	Frigate	269 3	49 5	4, 664. 5	19 0	12 8-inch and 12 6-inch guns.	4½	1, 007	8. 1
Peder Skram	....do	225 5	49 5	3, 321	21 7	6 8-inch and 12 6-inch guns.	4½	1, 680	11.
Odin	....do	236 10	48 5	3, 0[illegible]6. 5	15 11	4 10-inch guns ...	8	2, 260	12. 4
Gorm	Corvette ...	231 8	40 0	2, 308	14 5	2 10-inch guns	8	1, 670	12. 2
Lindormen ..	....	216 3	39 4	2, 044	14 1	2 9-inch guns	5½	1, 560	12
Rolf Krake..	....	184 10	38 1	1, 323	10 3	3 8-inch guns	4½	750	7. 8

* Building.

THE SWEDISH NAVY.

The naval force of Sweden has been for a number of years divided into two branches, one comprising the navy proper, and the other known as the "coast artillery."

The fleet consists of thirty-eight vessels, armed with about 325 guns. One ship, the *Stockholm*, of 2,850 tons and carrying sixty-six guns; one frigate, the *Vanadis*, of 2,130 tons and sixteen guns; four corvettes namely, the *Balder*, of 1,880 tons and six guns; the *Gefle*, of 1,280 tons, and eight guns; the *Saga*, of 1,530 tons and seven guns; and the *Thor*, of 1,070 tons and five guns; the *Vanadis* and *Saga* are under construction. Besides these, there are eighteen gunboats, carrying twenty-six guns. All the above named are screw-propeller vessels of moderate power and armament, the guns being rifled breech and muzzle loaders of 9, 5½, and 4 inch caliber, and more powerful guns are devised. In addition to the above, twelve sailing-vessels belong to the fleet, viz, one ship, the *Skandinavien*, of 2,380 tons and sixty-two guns; five corvettes and six brigs, carrying in all one hundred and twenty-nine guns.

The coast artillery is considered the most important arm of defense; its total force is one hundred and twenty vessels of all kinds. This includes the armored vessels, the four largest being counterparts of our *Passaic* class. In the above are included, also, the small turreted boats built from the designs of Captain Ericsson, and provided with machinery to be worked by hand, independently of steam; there are, also, forty-four sloop-rigged galleys, six mortar-launches, and fifty-three yawls.

The headquarters of this force is at Stockholm, the island of Skeppskolm being occupied as the repairing and outfitting yard. There is also another naval establishment at Carlscrona.

The number of vessels employed on foreign service is limited to an extremely small force, generally to three or four. The *personnel* of the navy, as authorized by the law of 1870, was: Navy proper, officers, 88; petty officers, 181; seamen and others, 4,900. Coast artillery, officers, 55; petty officers, 70; seamen and others, 2,500; total, 7,794.

THE NORWEGIAN NAVY.

This small fleet is composed of four monitors, one frigate, two corvettes, one sloop, and twenty-two gun boats; also, four sailing-vessels and two torpedo-boats. In addition to these, there are eighty-three small boats of various kinds, carrying from one to six guns each.

Of the monitors, one, the *Thor*, has a displacement of 2,003 tons and an indicated horse-power of 600; a second, the *Thrudnang*, is of 1,575 tons and 500 indicated horse-power; a third, the *Mjalner*, is of 1,515 tons and 450 indicated horse-power; and the fourth, the *Skarpianen*, is of 1,447 tons and 350 horse-power. Each of these monitors has an armament of three guns.

The frigate, the *Kang Suerre*, has a displacement of 3,472 tons, an indicated horse-power of 1,500, and mounts fifty guns. Of the corvettes, one, the *St. Olaf*, has a displacement of 2,182 tons, an indicated horse-power of 1,100, and mounts thirty-eight guns; and the other, the *Nardstjernen*, 1,609 tons, 720 horse-power, and nineteen guns. The sloop is of 958 tons displacement, 250 horse-power, and sixteen guns. Of the gunboats, the *Sleipner* has a displacement of 580 tons, 800 indicated horse-power, and mounts two guns. The remainder are very small vessels, varying in displacement from 230 to 59 tons; four of them carry two guns and the rest one gun.

THE PORTUGUESE NAVY.

The navy of Portugal is composed of one armored ship, nine screw-corvettes, seven gunboats, one sailing-frigate, one sailing-corvette, three transports, two tugs, and one yacht. The *Estephania*, the largest of the corvettes, is under repairs at Lisbon, and of the other eight corvettes, three only are in commission.

The armored ship *Vasco da Gama* was built at the Thames Iron-Shipbuilding Yard, London, and launched in 1876. Her length, according to *Iron*, is 215 feet 11¾ inches, breadth 43 feet 3¾ inches, depth of hold 25 feet, and displacement 2,479 tons. The octagonal turret is protected by armor 10 inches thick on teak backing of the same thickness, and the sides by armor 9 inches thick. She carries two Krupp guns 10¼ inches in caliber, one 6 inches in caliber, and two 40-pounders, and is furnished with a ram. Her cost is said to have been $519,400.

The hull is constructed on the cellular system, with a double bottom, iron decks, water-tight bulkheads, and 47 compartments in all. She has three masts.

The vertical engines by Messrs. Humphrys & Tennant operate twin screws; on the trials they developed 3,625 horse-power, and the ship realized a speed of 13¼ knots per hour. In four minutes she made a complete circle 433 feet in diameter.

The *personnel* of the navy consists of 579 officers on the active list, as follows: 1 vice-admiral, 6 rear-admirals, 16 captains, 25 commanders, 80 lieutenants, 74 second lieutenants, 73 midshipmen, 7 engineer-constructors, 61 engineer officers, 38 pay officers, 24 officers of the medical department, 6 chaplains, and 168 officers of inferior rank. The enlisted men and boys number 3,189.

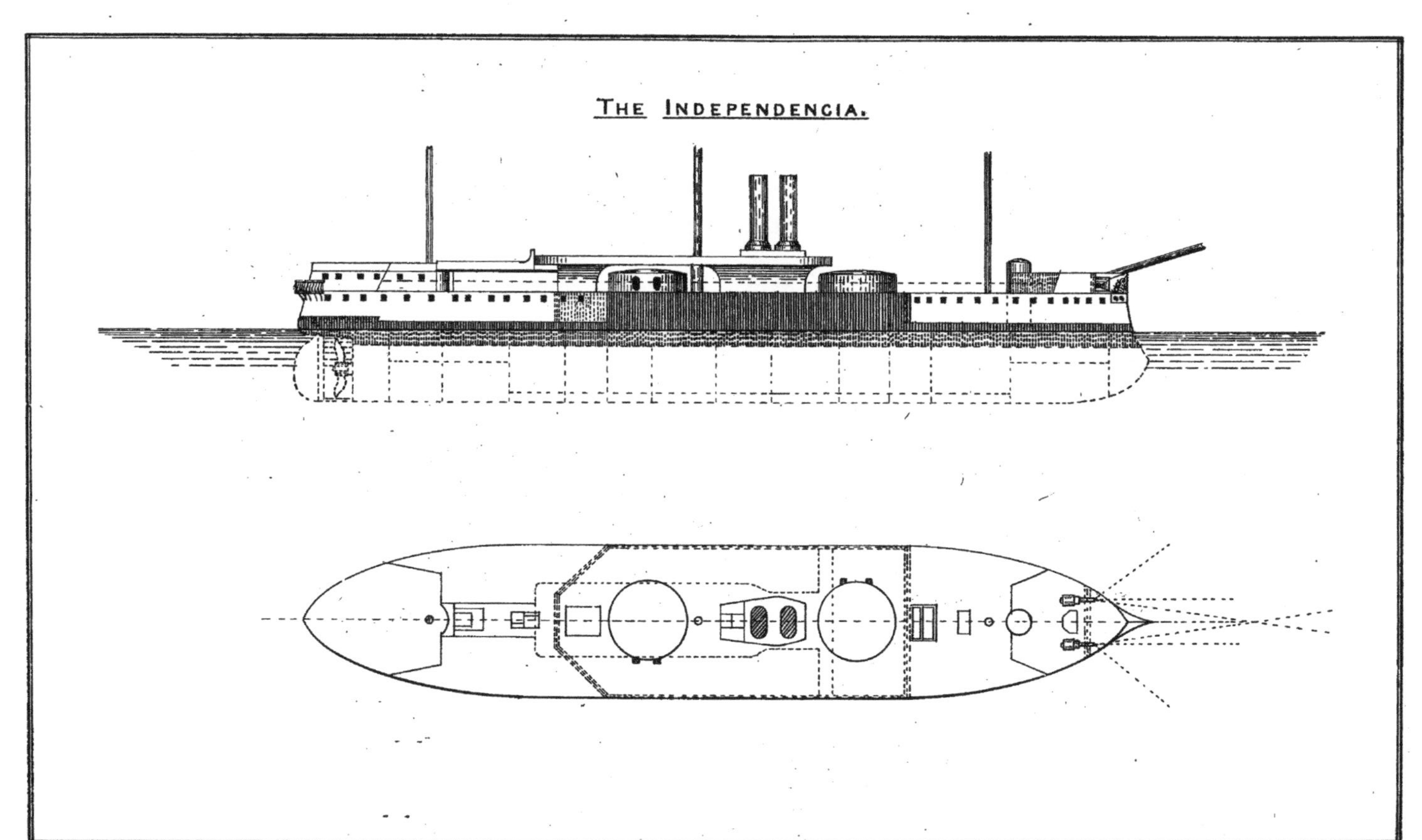
THE INDEPENDENCIA.

THE BRAZILIAN ARMORED SHIP INDEPENDENCIA.

This report, as its title indicates, was intended to be confined to European ships, therefore the Brazilian fleet, South American and Asiatic ships, have not been described in whole; but as the above-named vessel came under my notice during its construction, and as it is the most formidable ship belonging to any power not of Europe, and is a unique specimen of naval construction, containing many novelties, a brief notice is believed to be advisable.

The *Independencia* was designed in 1872, by Mr. E. J. Reed, C. B., M. P., in accordance with conditions prescribed by a commission of Brazilian officers. The dimensions, draught of water, thickness of armor, size and number of guns, speed, sail-power, and other primary qualities were arranged by Mr. Reed in concert with this commission. At that time the British ship *Devastation* was being advanced toward completion in the dock-yard at Portsmouth, and was exciting great attention in other countries as well as in England, by reason of the enormous powers of offense and defense which were being developed. Her 12 and 14-inch armor on the hull and turrets, 2½ and 3-inch armor on the decks, and her 35-ton guns, constituted such an advance upon all the fighting-ships then built or building as caused her to be looked upon as the type to which first-class fighting-ships would, in the future, have to approximate. The Brazilian officers desired a ship equally powerful, with the addition of sails. The *Independencia*, as completed in 1877, is different in many respects from any other ship, but her typical features may be best described by calling her a rigged *Devastation*. She bears, however, a superficial resemblance to the unfortunate *Captain*, chiefly due to the upper works formed by the forecastle, hurricane-deck, and poop, and by her being full-rigged.

The *Independencia* is a two-turreted, breastwork ship of 9,000 tons displacement. The principal dimensions, &c., are: Length between perpendiculars, 300 feet; breadth, extreme, 63 feet; depth of hold, 16 feet 6 inches; mean draught of water, loaded, 24 feet 9 inches. The armor on the water-line is 12 inches thick, tapering to 10 and 9 inches below and above, constituting a belt 8 feet 6 inches broad, which extends forward and aft so as to completely surround the vessel. The central breastwork is 130 feet in length at the top of the belt, and extends to the upper deck, 11 feet above the water-line. This breastwork incloses the boiler and engine hatches, the scuttles to magazines and shell-rooms, the principal openings for ventilation, and the two turrets. There is one turret at each end of the breastwork. Over the breastwork and between the turrets is an erection somewhat similar to the hurricane-deck of the *Devastation*. It consists of a deck about one-half the breadth of the ship, extending from the fore turret to some distance abaft the after turret, this deck being supported by the casings of the boiler and engine hatches. Upon this deck is a rifle-proof house, containing the steering apparatus and appliances for navigating the ship, the boats, hammocks, steam-winch, ventilating-shafts, &c. There is also a poop and a forecastle, the hurricane-deck amidships being narrowed abaft the breastwork and continued aft to the poop. Upon this continuation of the hurricane-deck are placed the standard compass and steering-wheel. The poop is fitted for mounting mitrailleuses, and ports are cut in the after corners of the hurricane-deck for fighting a 9-pounder gun on each side. Under the poop are spacious apartments for the admiral and his staff, and cabins for other officers. The forecastle is fitted for working the anchors, and has an armored bulkhead across the fore part, behind which are placed two 7-inch guns. At the after end of the forecastle is an armored pilot-tower containing telegraphs and voice-pipes to the engine-room, steering-wheels, battery, &c., and from which the captain will work the ship in action.

ARMAMENT.

The armament, exclusive of the 9-pounder guns and mitrailleuses referred to, consists of four 35-ton Whitworth guns, two in each turret, and two 7-inch guns forward. The guns are all made of the Whitworth compressed-when-fluid steel, and are rifled upon the hexagonal principle, so that the guns of this ship will possess the advantages of being able to fire very long shells containing large bursting-charges of powder, and also of penetrating the enemy's ship below water with flat-fronted projectiles. The presence of the poop and forecastle prevents a complete all-around fire, as in the *Devastation*, but the obstruction thus caused is limited to a few degrees from the fore-and-aft line, supposing the guns to be laid right ahead or right astern. These obstructions are necessarily caused by the determination to make the *Independencia* a full-rigged sailing-ship. The manner in which this has been done, and the difficulties in the way of it removed or minimized, is one of the most notable features of the design. The foremast is just abaft the forecastle, and is worked from the breastwork-deck; all fittings in connection with it, and all bitts and the lead of ropes being so arranged that in clearing for action they will all be out of the line of fire. The shrouds to the foremast and also to the mainmast will all be cleared away for action except two shrouds on each side of the mast, which are made larger than the rest and will remain fixed and take their chance of being shot away. The mainmast is between the boiler-hatch and after turret, and all ropes connected with it will ordinarily be worked on the breastwork-deck; but in clearing for action they will be raised upon the hurricane-deck, and can be worked there if required. The rigging of the mizzen mast is worked entirely from the poop.

The ship is also fitted with hydraulic machinery for working and loading the guns, similar to what will be employed in the Italian ships *Duilio* and *Dandolo*.

The motive machinery has been furnished by Messrs. John Penn & Sons. It was contracted for before the compound system of working steam had been fully established. The engines are of that firm's usual type of low-pressure trunk-engines, similar to those fitted in the British ship *Sultan*. There is a pair of cylinders 127 inches in diameter, with trunks 47 inches in diameter. The stroke of piston is 4 feet 6 inches. The guaranteed indicated horse-power is 8,500. There is also the usual number of engines as fitted to recently-constructed armored ships, exclusive of the motive engines.

The only full-rigged turret-ships built for the British navy are the *Monarch* and the *Captain*; the former has proved to be a thoroughly safe and sea-going ship. The *Independencia* is 30 feet shorter than the *Mon-*

arch and has 3 feet less free-board, but she has $5\frac{1}{2}$ feet more beam. In comparison with the unfortunate *Captain*, she is in many respects quite different; her armor is 3 and 4 inches thicker, and her guns are 35-ton against the *Captain's* 25-ton; besides, she has 3 feet more free-board than the *Captain* was intended to have, and nearly twice as much as she actually did have; in addition, she is 20 feet shorter and 10 feet broader; this extra breadth, combined with the free-board, is doubtless what is relied upon for giving that ample safety against capsizing which the *Captain*, unfortunately, did not have; at any rate, these differences must give the *Independencia* very great stability and power to carry sail, as compared with the *Captain*.

The long time the *Independencia* has been under construction is owing to an error in launching by which serious damage was done to the hull, requiring a great amount of work to be refitted. In consideration of this and the time that has elapsed since the ship was designed, during which interval such rapid advances have been made in the powers of offense and defense in ships of war, the *Independencia* is a very powerful vessel. Her side and turret armor is only one inch thinner than the *Devastation's*, and her guns, although of the same weight, are worked by hydraulic power instead of hand, and they throw heavier projectiles and have greater penetrating power.

The official trials took place February 2 of this year. Six runs were made over the measured mile, with the following results: revolutions per minute, 70; indicated horse-power, 9,000; speed, 14 knots.

NOTE.—Since the foregoing was written, this vessel has been purchased by the British Government; she has therefore been placed upon the list of armored ships of Great Britain in this report under her new name, *Neptune*.

THE JAPANESE ARMORED SHIP FOO-SOO.

Having described the most powerful ship of the Brazilian fleet, the last productions for the navy of Japan may also be noticed, since they embody some new features.

The *Foo-Soo* is the first armored ship built in England for His Imperial Majesty the Mikado of Japan. She was built at Poplar, on the Thames, by Messrs. Samuda Bros., from designs furnished by Mr. E. J. Reed, C. B., M. P., and was launched April 14, 1877.

She is a broadside vessel with a central battery of somewhat similar type to those built in England nearly three years ago for the Chilian Government. She is bark-rigged, spreading 17,000 square feet of canvas, is handsome in appearance externally, and will, no doubt, be easily handled, the length being little more than four and a half times the beam, and the speed being high in porportion to her size.

The principal dimensions and particulars are:

Length between perpendiculars	220 feet.
Breadth, extreme	48 feet.
Depth in hold	20 feet 4½ inches.
Draught of water { forward	17 feet 9 inches.
Draught of water { aft	18 feet 3 inches.
Height of port from load water-line	7 feet 6 inches.
Displacement	3,718 tons.
Indicated horse-power	3,500
Speed, estimated	13 knots.
Complement of men and officers	250

Armament:

Main-deck battery, four 9½-inch guns.

Upper-deck battery, two 6.7-inch guns to fire forward and aft.

The annexed drawings of the midship section, main and upper decks, will convey a correct idea of the general design.

The system of framing introduced in this vessel is said to be new. The frames behind the armor-plating and below it, from the main deck down to the keel, were made continuous. By reference to the midship section, it will be noticed that the method of attaching the armor to the hull is by brackets outside the frames of the ship, thus allowing a clear line from floor to deck for the frame, and avoiding all the expensive and often unsatisfactory work of the armor-shelf as hitherto constructed. The ship has an inner bottom, divided by bulkheads into water-tight compartments in the usual way, and a central fore-and-aft bulkhead extending the length of the engine and boiler rooms and magazine. The armor is confined to a belt on the water-line 9 inches thick, and to the main-deck batteries, where it is 8 inches thick. The total weight of armor is 776 tons.

The armament consists of Krupp guns of the dimensions given above; it will be seen that they are disposed on the central-battery system. The guns on the main deck, four in number, have a weight of about 15½ tons each, and, being fitted in embrasures, command the broadside and within 30 degrees of the fore and aft line. The guns on the upper deck weigh about 5½ tons each, and, commanding the horizon as they do, must be valuable as chasers.

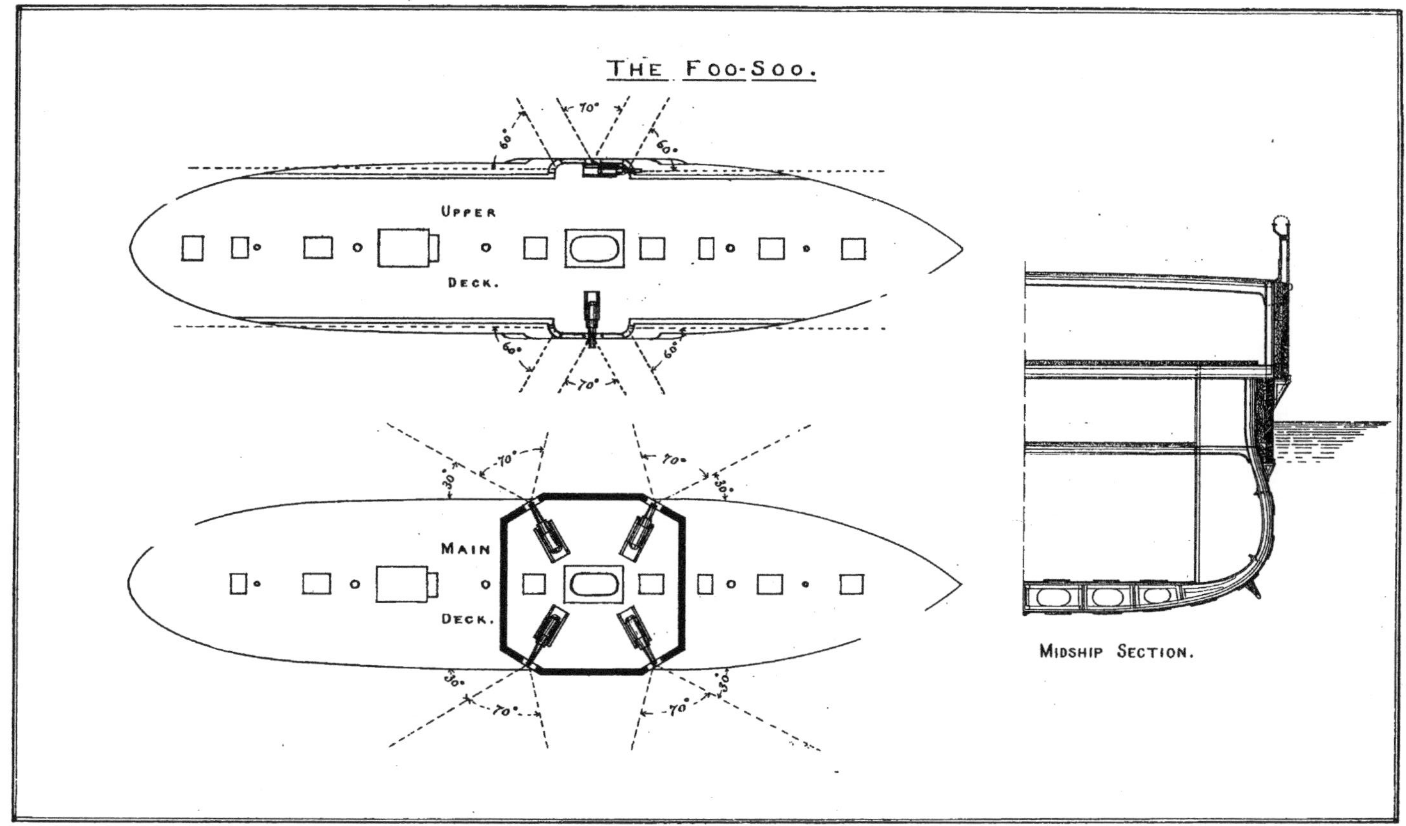
The Foo-Soo.
70°
60°
60°
Upper
Deck.
60°
60°
70°
70°
70°
30°
30°
Main
Deck.
30°
30°
70°
70°
Midship Section.

The motive machinery consists of two pairs of compound, horizontal, surface-condensing trunk-engines, designed and constructed by Messrs. Penn & Sons. The cylinders have diameters of 58 inches and 88 inches, with trunks 30 inches in diameter; the stroke of piston is 30 inches. The engines operate twin screw-propellers, 15 feet 6 inches in diameter and 16 feet in pitch, and the contract indicated horse-power is 3,500. The boilers, of the cylindrical type, are eight in number, each having a diameter of 11 feet 3 inches, and containing three furnaces 2 feet 11 inches in diameter; the pressure of steam is 60 pounds per square inch.

On the trial, with a draught of 17 feet 1 inch forward and 18 feet 1 inch aft, and with an immersed midship section of 788 square feet and a displacement of 3,639 tons, the vessel steamed at full speed for three hours; six runs made upon the measured mile resulted in a mean speed of 13 knots per hour, with an indicated horse-power of 3,824, the number of revolutions being from 93 to 94 per minute.

THE JAPANESE ARMOR-BELTED CORVETTES KON-GO AND HI-YEI.

These two vessels were built last year (1877) in England, and were also designed by Mr. E. J. Reed. The former, which is composite, was built at Hull by Earle's Engineering Works; the latter, which has an iron hull, at Pembroke, Wales, by the Milford Haven Company. The machinery for both vessels was supplied by the first-named firm. Some of the principal data relating to the *Kon-Go* are as follows: Length, 231 feet; extreme breadth, 40 feet 9 inches; mean draught, 17 feet 6 inches.

The leading feature of the design is that the armor-belt, 4½ inches in thickness on the water-line, has been worked between the two wooden skins of the ship in the wake of the engines and boilers, and behind this armor-plate, and extending from it both to the bow and stern, a broader strake of thin iron plating was also worked and riveted to the iron frames of the ship, thus adding greatly to the strength both structurally and defensively. The armament consists of six Krupp guns on the broadside, having a caliber of 6 inches and weighing about 4 tons; and two guns at the bow, 6.7 inches in the bore, weighing about 5½ tons, and capable of firing either right ahead or 32 degrees abaft the beam, and one 6.7-inch gun at the stern, capable of a range from right astern to 35 degrees forward of the beam. The motive machinery consists of a pair of horizontal, compound, return-connecting-rod engines of 2,500 indicated horse-power. The cylinders have diameters of 60 inches and 99 inches, and a stroke of 33 inches. The mean speed attained in six runs on the measured mile, December 7, 1877, was at the rate of 13¾ knots per hour, with 60.5 pounds pressure of steam per square inch, 82 to 87 revolutions per minute, and a mean indicated horse-power of 2,450. The screw-propeller is fixed, not lifting, has a diameter of 16 feet and pitch of 17 feet 6 inches, and the free revolution of the screw when the vessel is under sail alone is provided for. The *Kon-Go* is regarded as a very beautiful vessel, being in appearance more like a yacht than a man-of-war.

The *Hi-Yei* is built entirely of iron, and is in some respects a precise counterpart of the *Kon-Go*. A few of the most important data are: Length, 220 feet; extreme breadth, 48 feet; mean draught of water, 18 feet; load displacement, 3,718 tons. The machinery and armament are the same for the two vessels. The *Hi-Yei* made an official trial December 26, 1877. Four runs over the measured mile gave a mean of 2,490 horse-power and 14 knots per hour.

RÉSUMÉ OF ARMORED SHIPS.

As a summary of interesting facts relating to armored fleets, it may be stated that the grand total of all the European armored ships built from the commencement amounts in the aggregate to more than 1,060,500 tons; that the grand total of all the armored ships built and building for the British navy up to January, 1878, amounts in the aggregate to 340,000 tons, and the cost thereof, in round numbers, is given at considerably over eighteen million pounds sterling, or eighty-seven million four hundred thousand dollars in gold.

According to numbers, and excluding the small craft, such as the *batteries démontables* of the French and the few floating batteries still remaining on the English and French lists from the Crimean war period, it appears, by reckoning in all the other classes, rigged and mastless, sea-going and coast-defenders, that twenty years' continuous effort and the expenditure of enormous sums of money have produced the following results: England heads the list with 64 vessels. France comes next with 53 vessels, having an aggregate tonnage of 184,000, including those now building. Russia is possessed of 29 ships, with a tonnage of 89,500, among which are four sea-going ships. Italy can muster 18 vessels, of the aggregate tonnage of 94,000, and when the *Duilio* and *Dandolo* are completed will possess the two most powerful fighting-ships in the continental waters of Europe. Turkey, if possessed of little else, can show a fleet of 20 fighting-ships, with an aggregate tonnage of 40,000. The German Empire has 17 vessels, eight of them sea-going, and the aggregate tonnage of all is 73,000. Holland possesses 23, Austria 14, Spain 11, Denmark 7, Sweden 14, Norway 4, Portugal 1, and Greece 2.

No European power claiming to have a navy is entirely without armored ships. The grand total of European armored ships, including those building, may at the present time be taken as 277; of which England possesses about one-fourth, France about one-fifth, Russia about one-eighth, and the remaining powers still smaller proportions. Crossing the ocean, it is found that Brazil has 17 armored vessels, Peru 6, Chili and the Argentine Confederation each 2; while in Asia, Japan boasts of 4 armored vessels (one of them the *Stonewall Jackson* of well-known fame), and, lastly, the Chinese have entered the field with armored gunboats.

It is proper to add that neither the number of vessels, the number of guns carried, nor the total tonnage will convey correct ideas of the power and efficiency of the fleets.

With a single exception, leaving out of account the skirmishes on the coast of Spain during the late civil war in that country, none of the numerous European armored ships have ever been engaged in mutual warfare. The exception is the short and ill-contested fight of one hour's duration at Lissa, between the Austrians and Italians. The account of this battle published by Admiral Persano has shown that the Austrian fleet consisted of seven armored vessels and fifteen wooden frigates, and that the

Italian fleet consisted of twelve armored vessels and eight wooden vessels; that the Italian flag-ship *Rè d'Italia*, a wooden iron-clad, was sunk by the ram of the Austrian flag-ship *Ferdinand Max;* that the Italian corvette *Palestro* was struck abaft in the unarmored part by a shell, which set the vessel on fire and she blew up, all hands on board being lost. The sea was somewhat rough, and the Italians managed their ships and guns so unskillfully that but little damage was done to the Austrian ships; and although one of the Italian ships delivered two concentrated broadsides at one of the enemy's ships as she passed within 40 yards, not one shot struck. All, on both sides, were broadside-ships. This battle cannot be quoted as decisive testimony of the value or weakness of armored ships, nor does it prove that future naval engagements will be mainly decided by ramming, for if any one will carefully read the account of the *Fatti di Lissa* he will perceive the extreme difficulty which was experienced by both Austrian and Italian captains in delivering the fatal and decisive blow against a ship in motion. It was not until the rudder of the *Rè d'Italia* was disabled that the Austrian iron-clad was able to ram her. No lesson of value can therefore be drawn from this contest.

An engagement of more recent date, but of quite a different character, which occurred in the Pacific off the Bay of Ylo, May 29, 1877, between the British unarmored vessel *Shah*, assisted by the corvette *Amethyst*, and the Peruvian armored ship *Huascar*, may be mentioned as evidence of the indecisive action of the guns of unarmored ships against armor of even such a limited thickness as 3½ inches.

This action, if worthy to be called by that name, was brought about by very strange procedures on the part of the officers of the *Huascar*. The gist of the reports is as follows: During a revolution common to South American governments, the adherents of an insurgent leader—Nicolas de Pierola—persuaded the officers of the *Huascar* to rebel against the Peruvian Government; and with their consent a number of these men seized the vessel in the harbor of Callao, under the cover of darkness, and put to sea, sailing to the southward. At Cobija, in Bolivia, Pierola embarked on the *Huascar*, which then steamed to the north with a view to effect a landing. Very shortly after this, the *Shah*, with Admiral de Horsey on board, arrived at Callao, and being informed of the above facts, also that depredations had been committed by the *Huascar* on British property and against British subjects, Admiral de Horsey made complaint to the Peruvian Government, and receiving in response a decree declaring the *Huascar* a pirate, offering a reward for her capture, and repudiating all responsibility on the part of Peru for acts committed by her, he determined to proceed against the *Huascar* with his flag-ship, the *Shah*, and the *Amethyst*. Having put to sea for this purpose, he sighted the *Huascar* off the town of Ylo on the afternoon of May 29, and summoned her to surrender. This summons the commanding officer refused to entertain; the *Shah* then fired, first a blank cartridge, and then a shotted charge, but the *Huascar* still refusing to surrender, a steady and well-sustained fire from both the *Shah* and *Amethyst* was directed against her.

The *Shah* is an iron, unarmored, single-screw frigate, sheathed with wood, 334 feet in length between perpendiculars, has a beam of 52 feet, a mean draught of 23 feet (21 feet forward and 25 feet aft), a displacement of 6,040 tons, and an indicated horse-power of 7,477. The armament at the date of my visit on board, April, 1876, consisted of two 18-ton guns, sixteen 6½-ton guns, and four 64-pounders, all rifles. The *Amethyst* is a wooden corvette, built in 1873, is of 1,934 tons displacement, and 2,144 indicated horse-power, carrying fourteen small guns.

The *Huascar* is an armor-belted, single-screw, sea-going ram, built in England by Laird; her length is 200 feet, mean draught 14 feet, free-board 5 feet. She is built of iron, and is divided into compartments by three transverse bulkheads. The armor on the hull is 4½ inches thick, tapering to 2½ at the bow and stern, and backed by 14 inches of teak. The single revolving turret is armored with plates 5½ inches thick on teak 14 inches thick. The armament consists of two 300-pounder Armstrong rifles in the turret, also two 44-pounders and one 12-pounder outside the turret.

The fight was partly in chase and partly circular, the distance between the combatants being for the greater part of the time from 1,500 to 2,500 yards. The time employed in the engagement was about three hours, the fight being terminated by darkness coming on and the *Huascar* running close inshore, where the *Shah* could not follow consequent upon her greater draught. Of the projectiles thrown from the English ships, it is reported that some seventy or eighty struck the iron-clad, principally about the upper deck, bridge, masts, and boats; one projectile from the heavy gun pierced the side on the port quarter, two feet above water, where the armor was 2½ or 3 inches thick, and brought up against the opposite side, killing one man and wounding another; two other heavy projectiles dented in the side armor to the extent of 3 inches. The turret was struck once by a projectile from the heavy guns of the *Shah;* it was a direct blow, but penetrated 3 inches only. The hull showed that several 64-pound shot had struck it, only leaving marks. When at close quarters, which the *Huascar* sought for the purpose of ramming, the Gatling gun in the *Shah's* foretop drove the men from the deck-guns of the former. On one of these occasions a Whitehead torpedo was launched at the iron-clad, but as she altered her course about the same instant, the torpedo failed to strike its mark.

Excepting the damage done to the boats, smoke-pipe, casing, and wood-work, the *Huascar* was unharmed by three hours' cannonading from the heaviest guns carried by the most powerful unarmored ship in the British navy. The fact that a single shot entered the armored hull, only 3 inches thick, was doubtless owing to the accident of the full broadside being presented to the enemy at the instant when the gun was fired.

The officers of the Peruvian ship, consequent upon her being manned by "a heterogeneous crowd of insurgents," managed their guns and worked their ship with so little skill that not a single shot from the 300-pounder guns struck either of their antagonists, neither did any projectile from the smaller deck-guns, except one which passed through the rigging of the Shah, carrying away some of the ropes. The *Huascar* frequently tried to ram the *Shah*, but in this attempt, too, failure was the result. It is true the *Shah* had the important advantages of much greater speed and two powerful long-range rifled guns. On these advantages Admiral de Horsey doubtless counted when he decided on the attack, and after the action commenced he soon ascertained that the guns of the *Huascar* were being wildly handled.

If the iron-clad had been manned with a properly-drilled crew and skillful officers, and had been in good working order, the contest at from 1,500 to 2,500 yards ought to have resulted, apart from questions of right or law, in sinking both the English ships.

The reasons given by Admiral de Horsey for his determination and subsequent action in this case are shown in his official report to the lords of the admiralty, as follows:

1. The *Huascar*, in boarding and detaining the *John Elder* at sea, in boarding and demanding dispatches from the *Santa Rosa,* in forcibly taking coal from the *Imnuncina*

in forcibly taking a Peruvian officer out of the *Colombia*, and in forcibly compelling the engineer, a British subject, to serve against his will, committed acts which could not be tolerated.

2. The *Huascar* having no lawful commission as a ship of war, and owing no allegiance to any state, and the Peruvian Government having disclaimed all responsibility for her acts, no reclamation or satisfaction could be obtained [except] from that ship herself.

3. That the status of the *Huascar*, previous to action with the *Shah* and *Amethyst*, was, if not that of a pirate, at least that of a rebel ship having committed piratical acts.

4. That the status of the *Huascar*, after refusing to yield to my lawful authority, and after engaging Her Majesty's ships, was that of a pirate.

5. That had the *Huascar* not been destroyed or captured, there would have remained no safety to British ships or property on this coast, not even to Her Majesty's ships, as the *Huascar* might have destroyed the *Shah* or the *Amethyst*, by ramming, any night at any port they were found.

6. That I trust the lesson that has been taught to offenders against international law will prove beneficial to British interests for many years to come.

7. That I have carefully abstained from any interference with the interests of the Peruvian Government, or those of the persons in armed rebellion against that government, my action in respect to the *Huascar* having been entirely for British interests.

It may be seen from this engagement, as well as from previous ones, that the chances of a shot penetrating thick armor at sea are extremely remote, owing to the difficulty of striking it fairly, or at right angles. Too much stress is laid upon the results of artillery experiments as affecting actual warfare. Persons read of a projectile penetrating so many inches of solid iron, but do not bear in mind that this result is only obtained by such a combination of the most favorable conditions for the gun as could never occur in a battle at sea. Nor is it generally understood how very uncertain must be the practice from ships' guns in action, for the extreme precision of modern weapons depends altogether upon accuracy of range and steadiness of platform, and with ships in rapid movement, as would be the case under present conditions of motive power, the distance and position would be rapidly changing every minute, while it rarely happens that ships are entirely free from wave motion. The engagements by ships against forts, in which the writer has participated, corroborate these views.

PART XIX.

TORPEDO-WARFARE.

THE WHITEHEAD, HARVEY, AND SPAR TORPEDOES; THORNY-CROFT'S AND YARROW & CO.'S TORPEDO-BOATS; THE ZIETHEN; THE UHLAN; THE UZREEF; THE PIETRO MICCA; THE OBERON EXPERIMENT, AND DEFENSE AGAINST TORPEDOES.

TORPEDO-WARFARE.

Torpedo-warfare has now attained to a recognized and important place in maritime contests. There is, however, in this, as is generally the case in striking inventions, a tendency to exaggerate its importance. Recent experiments have somewhat tended to diminish the estimate previously set upon the power of submarine mines, and it has been shown in practice that care and precaution can do much to render them innocuous. There are so many items of detail, without close attention to which it is impossible to operate with them successfully, that, even in the quietly-conducted experiments of peace, failures occur without number. Still the effect that can be depended upon is so considerable, that we must accept as proved the necessity for a more thoroughly scientific investigation of the problems relating to the subject.

Two elements have contributed to make torpedo-warfare what it is—electricity and the new explosive compounds. It is true that in the Whitehead or fish torpedo recourse is had only to the latter of these, but it is the sole material exception, and all the work effected by this branch of marine warfare has been, so far, the result of electric torpedoes.

The torpedoes used in the naval and military services of different countries are of various kinds. They may be classed generally as stationary or defensive, and locomotive or offensive. The former are employed for the protection of harbors, channels, and roadsteads, being moored in selected positions at the bottom or below the surface of the water, and exploded by contact with vessels passing over them, or from the shore by means of an electric current through a wire. They vary infinitely in minor details, but a general similarity is to be noticed in all. The latter, in the shape of movable torpedoes, carried to the attack by swift-moving steam-launches, or by self-contained power within, are employed for offensive work.

In England they use, as a rule, compressed gun-cotton in their machines, while on the continent they seem to entertain a predilection for nitro-glycerine, or rather dynamite. Both substances are what chemists term nitro-compounds, in contradistinction to gunpowder, which comes under the class of nitrate-compounds, and they appear to exercise an explosive force of almost similar violence, measuring the substances weight for weight. Compressed gun-cotton, it need scarcely be said, is cotton yarn acted upon by nitric and sulphuric acids and then pulped and washed, so that the result is a finely divided mass which may be made to assume any shape.* The material is generally pressed into cakes of disk-like form, which weigh from a few ounces to a pound or more, and, while still moist, the slabs are stored away in magazines. In this moist condition the compressed pulp is not only non-explosive, but actually non-inflammable, unless one possesses the key to its detonation.

* For a description of the manufacture of gun-cotton at the factory of Waltham Abbey, which has a capacity to turn out 4,000 pounds per day, see the report of Colonel Laidley, U. S. A., ordnance document, 1877.

This is nothing more than a dry cake of the same material, or, as it is termed in military parlance, a "primer," which, on being detonated by a few grains of fulminate, brings about the explosion of any gun-cotton in its immediate neighborhood even though it be moist. Thus, if simply a net is filled with gun-cotton slabs and thrown into the water, the whole charge may be ignited by a primer contained in a water-proof bag, having an electric fuse and wire attached and placed in the interior of the mass.

Dynamite, as is well known, is a mixture of nitro-glycerine and sand, or it may be described as siliceous earth impregnated with explosive fluid, and owing to the chemical action continually going on within it, it needs to be reworked about every three years; if this is not done the parts become chemically separated. The Swedish engineer, Mr. Nobel, who was also the first to manufacture true nitro-glycerine on a large scale, manufactures dynamite by mixing together nitro-glycerine and silica; it fuses at a temperature of about 150° Fahrenheit. This description of dynamite is largely used in the Swedish service, where it is looked upon with more favor than even gun-cotton.

Gun-cotton and dynamite explode with much greater force than gunpowder, and for this reason a very destructive charge may be confined within a comparatively small space; moreover, they are particularly adapted to submarine mines, since nitro-glycerine is no more affected by water than is gun-cotton.

TORPEDOES FOR OFFENSIVE OPERATIONS.

Three principal varieties of these are employed in European navies—the Whitehead, the Harvey, and the spar torpedo.*

The principal features of construction of the Whitehead-Luppis fish-torpedo are unknown to the public, for its mechanism has been successfully kept a secret since its first introduction to notice. Each of the several European governments which have purchased the secret appointed a small number of officers to be put in possession of a complete set of drawings furnished with the weapons, and they are bound on honor not to reveal their workings.

Many notices have appeared from time to time, purporting to be descriptions of this remarkable invention, the most widely known and generally adopted of any of the instruments in the field of movable torpedoes. None of them can be regarded as entirely accurate; still, for general information, it may be said that the Whitehead torpedo is an explosive submarine boat. It is made of different sizes, from 14 to 19 feet in length and from 13 to 16 inches in diameter in the center. It is constructed of the very best thin steel plate, and of the cigar shape common to many of these implements of destruction. It is divided into compartments internally. The anterior portion, or nose, contains a bursting-charge of gun-cotton or dynamite, together with the fuse and detonating apparatus, which is arranged to explode on contact, by forcing a pin into a cap of fulminate; this compartment is made separately and secured to the main body when desired. The tail or posterior portion of the torpedo, as most authorities state, contains a set of small three-cylinder engines of the Brotherhood type, and they drive the screw-propeller which forms the organ of propulsion. Such deli-

* Two lectures, the one on movable torpedoes in general, and the other on the Whitehead torpedo, delivered at Newport, R. I., December, 1874, by Lieut. F. M. Barber, U. S. N., are remarkable for the information contained therein, and as showing careful research of past records on the subject of submarine warfare.

cacy in materials and workmanship has been attained in the manufacture of this machinery, which is operated by compressed air, that the three little working cylinders, exerting a force of 40 indicated horses, do not weigh, according to Mr. Donaldson, Mr. Thornycroft's manager, more than thirty-five pounds. The central portion or chamber contains the air for actuating the engines. This chamber constitutes nearly the whole rear half of the implement, and its strength must be very great to withstand the pressure of the air, which is compressed by pumps worked by steam-power to a tension, it is reported, of about 1,000 pounds to the square inch, or upward of sixty and a half atmospheres.

There is a horizontal rudder or external mechanism capable of regulating and maintaining the depth under water at which the torpedo is intended to travel, and also for keeping it in a straight line or sending it on any curve which may be desired, taking into account currents and eddies.

There is also an apparatus intended to throw the detonating arrangement out of gear, in case of the failure of the torpedo to strike the object at which it is aimed, so that the instrument shall either sink to the bottom or float, so as, in the latter case, to be regained without danger.

It is obvious that the speed of the torpedo will diminish as the pressure of the air in the reservoir diminishes by the working of the engine. Thus it will travel more rapidly for a short distance than for a longer one. It is stated by Mr. Donaldson that for 220 yards the velocity attainable is at the great rate of 24 miles per hour, and that 1,000 yards can be accomplished at the rate of 16 miles, or 4,000 yards at 5 miles, per hour, the speed being least at the end of the journey. At the speed of 16 knots the transit through 1,000 yards would occupy about two minutes, or, more exactly, one minute and fifty-two seconds. As to the effect of the blow there can be little question, if the fuse is perfectly instantaneous in action, so as to insure explosion at the moment of contact. Minute portions of time are here of the utmost importance, as, if the weapon recoils before exploding, the effect is prodigiously diminished. This is the main objection to the use of chemical acid fuses, the action of which is slow enough to allow of sensible recoil between impact and explosion.

The only instances of the employment of the fish-torpedo in war, which have occurred up to the time of writing, have been failures. A Whitehead torpedo was discharged by the *Shah* at the *Huascar*, when the latter vessel was passing the former at no great distance, but failed to strike the mark. The second failure, reported by the naval correspondent of the London *Times*, was the attempt by the Russians, on the night of the 20th of December last, to destroy by means of Whitehead torpedoes one or more of the ships of the eastern division of the Turkish Black Sea squadron, then lying in the harbor of Batoum, and consisting of three armored and three wooden vessels. The *Avni Illah*, an armored corvette, was lying moored to a buoy in the center of the harbor, unprotected by spars or guard-boats, and from this circumstance was chosen as the object of attack. The torpedoes having been towed to a convenient position by boats, and pointed in the direction of the Turkish ship, the air-valves were opened and the machines started upon their mission, but without success. The escape of the ship, however, was due not to the precautions adopted for her safety, nor to the vigilance exercised by the officers on board, but to the precipitancy with which the attack was made by persons not thoroughly acquainted with their work. In the morning two torpedoes were picked up, one floating on the water with the nose containing the explosive material gone, and the other lying stranded on shore. The latter was intact, the nose still loaded with the gun-cotton charge, some thirty pounds. This torpedo was carefully unloaded by unscrewing the magazine section, and, as it was in all respects perfect, Mr. Whitehead's secret is now in possession of the Turks without their having paid for it. Although the Russians have been in possession of this machine during the war, we have no account of its successful use by them.*

The German Government have not realized the results in their experiments which they anticipated, and the practical test made by the English from the armored ship *Téméraire* have been reported as follows:

November 14 was devoted to a practical test of the torpedo-fittings, and to a series of experiments with the Whitehead torpedo. Two shots were fired from the starboard and two from the port broadside, while the ship was steaming at the rates of 5½ and 8½ knots per hour, the mark aimed at being a man-of-war's cutter, stationed at a distance of about 320 yards. The projectiles in both instances passed within a few feet of the object, and the result was considered satisfactory. But the shots subsequently fired from the bow tubes were far otherwise, notwithstanding that the sea was smooth and the ship going only 5½ or 6 knots. As soon as the torpedoes entered the water and felt the impulse of their own engines they deliberately crossed the ship's bow, and went off on the port tack at a speed of 12 knots.

This seems to be the natural consequence of the action of the water as it is pushed forward by the vessel, and confirms previous experiments, that these projectiles, when fired from the bow of the ship in motion, are exceedingly erratic, and that when so ejected little dependence can be placed on them. If it could be rendered practicable so to control the path of a locomotive torpedo as to allow of a modification of its course, under the direction of an observer on shore, or at a safe distance from the point of attack, the torpedo would be raised to the rank of an irresistible weapon of offense.

England, France, Germany, Austria, Italy, and Russia have purchased the right to use this torpedo. The first-named government has paid, it is reported, about $72,300 for the privilege of its use, with examples and drawings. Besides which, about two hundred of these weapons are said to have been purchased when the Turko-Russian war commenced, at the cost of $2,430 each, and a large number subsequently. The improvement and development of the instrument and system of working it are objects of continued study and experiment by the English at the royal arsenal, Woolwich, and in the river Medway; also at the special school for torpedo instruction on board the *Vernon* at Portsmouth; by the French at the mouth of the Charente, and at Boyardville, where officers are instructed in the science of electricity and explosives; by the Italians at Venice; by the Austrians at Fiume, where Mr. Whitehead has his factory; by the Germans at Wilhelmshafen and Kiel; and, finally, a distinct torpedo service has been organized by the Russians at Kertch and Cronstadt, where torpedo appliances are to be used for the defense of the Black and the Baltic Seas †

---

* Since writing the above paragraph the following has been received:

"AN OTTOMAN STEAMER BLOWN UP.

"ST. PETERSBURG, *Wednesday, January* 30, 1878.

"The commander of the Russian steamer *Constantine* reports that he left Sebastopol for a cruise on the 22d instant. He approached Batoum on the 26th, where there were seven Turkish vessels. The *Constantine* sent a Whitehead torpedo against a screw-steamer which was on guard outside, and sank her immediately. The crew were all drowned. The *Constantine* has returned to Sebastopol."

† The most successful automatic or self-propelling torpedo tested in the United States is the one invented by Mr. John L. Lay, late an engineer officer, United States Navy. A specimen of this torpedo was shown at the Centennial Exhibition. It was made of

THE HARVEY SEA-TORPEDO.

The principle of Captain Harvey's torpedo is by no means new, it having been designed from a wooden float, called an "otter," a contrivance extensively used by poachers in Scotland for the purpose of conveying their lines out into mid-stream, thereby enabling them to dispense with the use of long fishing-rods. This torpedo has been adopted in the British service, also into a number of the navies of continental Europe. It is intended to be used at sea by ships engaged with other ships, and it is so shaped and so slung that when towed from a vessel in motion it diverges from her path and thus enables her to pass by an opponent at a certain distance from her and yet near enough to bring the instrument into sharp contact with some point of her submerged hull. Two kinds of these torpedoes are served out to each ship, though they differ only as regards the position of their respective planes, so that they may diverge, the one to port, the other to starboard.

Each torpedo consists of an external case of well-seasoned elm, about 1¼ inches in thickness, screwed together, with water-tight packing between the joints and bound with iron, the interior being usually cemented with pitch. * * * The inner case is constructed of stout sheet-copper carefully soldered at the joints, and provided with a cylindrical tube running through the center, into which is fitted the exploding-bolt and priming-charge; it has also two circular ports about 3½ inches in diameter, one on either side of the bolt, for charging the torpedo. These ports are rendered water-tight by means of screw-caps forced firmly down on to suitable water-tight packing. The priming-case is made of stout sheet-copper, and contains a large bursting-charge of rifle-grained powder on gun-cotton disks.

In the center of the priming chamber or cylinder is a brass tube in which the exploding-bolt works, and at the bottom of this tube is a steel-pointed anvil which, when the bolt is forced down, pierces the capsule and, striking the muzzle, ignites the detonating compound, and, through the bursting-charge, the main charge itself. At the side of the brass tube, and near the base of the pin, is a small hole covered with thin brass foil which will allow of an escape of water into the priming-case should any have collected at the bottom of the tube. The priming-case is charged from the bottom on the same principle as the loading-ports for the main charge, a box-spanner being employed for screwing in the caps. The priming-chamber may be charged separate from this if preferred, but this precaution is quite unnecessary, unless the explosive used be a dangerous one, in which case it would, of course, be impossible to be too careful. * * * *

Descriptions of the Harvey torpedo have been widely published, perhaps the most complete, with illustrations, being in *Engineering*, November 23, 1877.

Various modes of attack with this instrument are proposed, but it would seem that the only case in which it is likely to be used with success is in a night attack on a vessel at anchor by a fast steam-launch, running across the bow or by the quarter, launching the torpedo between the moorings in passing, and leaving the tow-rope slack until the launch is a sufficient distance off, checking the passing out of the tow-rope by

---

boiler-iron; was shaped like a spindle, 28 feet 3 inches in length, and was divided into four compartments. The front end, technically called the nose, contains the explosive material, viz, about 300 pounds of powder and 75 pounds of dynamite. The motive power, which consists of carbonic-acid gas generated in the usual way, is conducted from the generating apparatus, through iron tubes, to the engine, which is in the fourth section, and which operates the screw-propeller. The initial pressure is about forty atmospheres. The third or middle section contains a roll of two or more miles of insulated wire which is paid out from the torpedo itself, and serves to keep up electrical connection with the firing station. The torpedo is submerged to about four-fifths of its volume. Its maximum speed claimed is 10½ knots, but it seems from the Newport experiment that 7 knots only were attained. The instrument is entirely under the control of the electrician, who, by means of a series of contacts, opens or closes the communicating valves, and thus increases or diminishes the speed, and stops or starts the torpedo as required. It may be fired either by contact or by closing the electric circuit. Its position is always kept in view by a rod projecting from its upper surface.

means of the brakes provided for the purpose, causing thereby the torpedo to diverge into contact with the vessel attacked, and by so doing explode the mine and destroy the ship.

When the torpedo is launched on its mission, very much of the probable success of that mission depends on the man in charge of the brake; indeed, skill and precision are necessary for its management, and although Captain Harvey has used it successfully in many and varied experiments, the results, when intrusted to hands less practiced in its manipulation, would probably be different, especially in actual warfare, when contingencies would arise requiring skill, courage, and good judgment.

The spar-torpedo is simply a copper case, in which are contained explosives, carried at the end of a spar or pole projecting from a boat, the latter being rowed or steamed to the object, and the torpedo discharged either by concussion or electricity. This type of torpedo will be noticed presently.

TORPEDO-BOATS.

The question of how best to use torpedoes in offensive marine warfare has received considerable attention and study in Europe during the last few years. Spar-torpedo launches adapted to the purpose have been extensively introduced into the service of all European navies.

Messrs. Thornycroft & Co., whose building-yard is at Chiswick, on the Thames, above London, have attained distinction as the builders of these miniature war-vessels. The famous river-launch *Miranda*, built by this firm in 1871, and which in the spring of 1872 attained the astonishing speed, on the measured mile, of nearly 16¼ knots per hour, though less than 50 feet in length, was apparently the prototype of the modern fast torpedo-launch. She was built of thin steel plates, and fitted as lightly as possible, engined to the utmost, and was conspicuous in every detail for that perfection in design and workmanship which is never absent from the work of those who achieve, in new fields, rapid and complete success.

The speed of the *Miranda* seemed to offer just what was needed, and continental governments were not slow to recognize such apparent advantages. In 1873 the Norwegian naval authorities gave Messrs. Thornycroft & Co. their first order.*

THE NORWEGIAN TORPEDO-LAUNCH.

This boat was 57 feet in length, 7 feet 6 inches in the beam, drew 3 feet of water, and the stipulated speed was 16 English statute miles, or nearly 14 knots, per hour, which speed was not to be ascertained by a mere measured-mile trial, but was to be 16 miles through the water in a run of one hour's duration. The hull of the vessel was constructed entirely of steel plates and angle-bars, and was divided into six watertight compartments. The compartments in the extreme stem and stern were for stores, those next adjoining were fitted with seats for the crew, and were provided with movable steel covers, so that, on going into action or during rough weather, they might be completely covered.

The compartments amidships were for the steersman and the machinery, and were covered completely by steel plating $\frac{3}{16}$ inch in thickness. * * *

The compartment for the steersman was furnished with a hood, having slits ¼-inch wide all round, through which he could see with sufficient distinctness to direct his course easily. Motion was communicated from the wheel to the tiller by means of steel-wire ropes, which it was originally intended should be incased in wrought-iron tubes. The possibility, however, of these tubes being bent by a shot and so jamming the wire ropes led to this arrangement being abandoned, and the ropes were simply run through eyes at intervals along the side.

* The descriptions of Thornycroft's boats are from a paper by Mr. Donaldson, read before the Royal United Service Institution.

The engines were compound, of the usual inverted double-cylinder direct-acting type, capable of developing about 90 indicated horse-power, and were fitted with a surface-condenser, so that the vessel could run in salt water without danger of injuring her boiler. A small tank contained a supply of fresh water, to make good deficiencies arising through leakage, and from steam escaping at the safety-valve, &c. The circulating, air, and feed pumps were driven by a separate engine. The boiler was of the locomotive type, the shell being made of Bessemer steel, the fire-box and its stays of copper, and the tubes of solid drawn brass.

The armament consisted of a cylindro-conical shaped torpedo towed from the top of the funnel, round which a ring was fitted with two pulleys for the towing-ropes, the strain being taken off by means of two stays attached forward. The length of this torpedo was 13 feet, and the diameter 9 inches, and with a speed of 11 knots it has diverged to about 40 degrees from the direction of the boat's motion when running in smooth water. The torpedo is worked by means of a small winch and brake fixed on the after part of the engine-room skylight; davits are provided for dropping the torpedo overboard.

On the official trial, which took place on the Thames on the 17th of October, 1873, the number of revolutions done in the hour was found to be 27,177, and the number required to do a mile in still water was 1,578. The distance run in the hour was then $\frac{27177}{1578} = 17.22$, or very nearly $17\frac{1}{4}$ miles. The steam-pressure during the trial averaged 85 pounds per square inch, and the vacuum $25\frac{1}{2}$ inches.

The Government of Norway in some eight years has expended for torpedoes and experiments with them more than a million of marks, or, on an average, nearly $28,000 annually. In the yearly experiments at Drobak, near Christiania, all officers and subofficers of the navy, as well as some of the army, take part; and besides these, the Norwegian authorities have made other experiments at Carlscrona, in conjunction with the Governments of Sweden and Denmark.

Boats which I inspected at Mr. Thornycroft's yard in 1876, of the same size and similar in all particulars, excepting the engines, to the one described above, were made for the Swedish and Danish Governments. In these there was an increase of speed to 17.27 miles in the case of the Swedish boat, and to 18.06 miles, or $15\frac{5}{8}$ knots, in the case of the Danish vessel.

There is no information regarding the armament of the Swedish boat, but the Danish boat was armed with two spindle-shaped torpedoes 12 feet long and $11\frac{1}{2}$ inches in diameter, somewhat like the Whitehead torpedo externally. They were placed on deck longitudinally, near the funnel, so as to facilitate launching, and were arranged to be towed from an upright pole 8 feet high, placed about 6 feet from the stem. A small winch was fixed on either side aft, to pay out the towing-line and to bring back the torpedo. By these arrangements the torpedo could be projected at a large angle from the direction of the boat's motion and at considerable velocity. The speed of the boat when towing one of these torpedoes is about 10 knots.

The Norwegian launch proved to have very fair sea-going qualities, as was proved by her voyage from Götheburg to Horten, Norway, a distance of 150 nautical miles.

The cost of two Norwegian boats was $8,748 and $16,524, gold.

AUSTRIAN TORPEDO-LAUNCH.

The boat built for the Austrian Government was of the same general design, but of larger dimensions, the length being 67 feet, breadth 8 feet 6 inches, and draught of water 4 feet 3 inches. This boat was built of somewhat thicker plating than the 57-foot type, and the guaranteed speed was 15 knots in a run of one hour's duration. The speed trials took place on the 11th of September, 1875, when 24,700 revolutions in one hour's run on the Thames were attained. The number of revolutions required to the knot in still water was found to be 1,357, making the distance run in an hour 18.202 knots, or 3.202 knots over the contract speed. During the run the steam-pressure averaged 105 pounds per square inch and the vacuum 25½ inches.*

The boat was divided into six water-tight compartments, and it differed from the Scandinavian boats in having the spaces forward and aft of the machinery permanently decked, instead of being covered with movable steel covers only. The following is from an English journal:

The machinery was somewhat similar to that in the Scandinavian boats, excepting that the engines were capable of developing 200 indicated horse-power, and that the air was supplied to the furnace by being forced into an air-tight stoke-hole instead of being forced directly under the fire-grate.

The armament of these vessels consisted of two torpedoes attached to the end of wooden poles, 4½ inches in diameter and about 43 feet long, connected to the battery by insulated wires, and arranged to be fired either by coming in contact with the enemy's vessel or at any distance from it, at the will of the operator.

The torpedoes themselves were simply copper cases, * * * and of sufficient size, in the case of the Austrian boat, to contain 11,000 cubic centimeters of explosive, and in the case of the French boats, to contain 25 kilograms of dynamite. At one end is the socket for the pole, and at the other the contact arrangement, which consists of a metallic plate capable of being pressed against the ends of the studs to which the wires are attached. This plate and its connections are covered by an India-rubber cap, so as to render the cases water-tight.

In the middle of the case is the aperture for charging the torpedo. This is a hole 3½ inches in diameter, into which, when the torpedo is filled, is screwed the cap F. The wires are introduced by the aperture C, fitted with a screw-gland, so as to prevent the ingress of water.

The battery is a modification of Smee's well-known single-acid battery, and consists of six cells, fitted with platinized silver and zinc plates, which, in order to prevent unnecessary oxidation, may be lifted and kept clear of the acid by means of the roller N.

The fuse * * * consists of two strong copper wires, kept apart by means of a non-conducting composition and connected by a very fine platinum wire, imbedded in fulminate of mercury, which is protected by a tinfoil casing. These fuses are used with a detonator, a long copper cap half-filled with fulminate of mercury. The connecting-wires are arranged in the neat and effective way patented by Captain McEvoy, of the London Ordnance Works, by means of which, with only three wires, the torpedo may be made to explode either on contact with the enemy's vessel, or by means of a firing-key, at the will of the operator. This arrangement is shown in the accompanying diagram. D^1 and D^2 are the wires leading from the poles of the battery to the torpedo. The fuse is inserted in the wire D^2 at a point within the torpedo-case, so that, when the case is charged, the fuse is entirely surrounded by the explosive. The connecting-wire D^1 is attached to the wire D near the battery, and to the wire D^2 at a point between the fuse and the stud to which that wire is attached in the torpedo-case. A firing-key is inserted in the wire D^1 at E^1, and a contact-breaker in the wire D at E^2.

The firing-key is simply an apparatus for connecting the two ends of the wire quickly. It consists of two pieces of vulcanite, through each of which the wire is led and fastened over the end. These pieces are kept together by means of a vulcanite nut, and a spring keeps the ends of the wire apart until pressure is applied.

The contact-breaker is similar to the firing-key, but there is no spring in it, and the two parts may be screwed backward and forward, so as to separate or connect the wires when required.

The object of having the contact-breaker in the circuit is to prevent the torpedo from being exploded by contact with the enemy's vessel, and so to place the control of the explosive entirely in the hands of the operator. If it is in use, it will be seen that no current can pass through the wire D, and that it is only possible to fire the torpedo by pressing the firing-key and sending a current through the wires D_1 D_2. Should it be desired to fire by contact, the contact-breaker is screwed up, so that the wire D may

* The boats built by Mr. Thornycroft are driven by screw-propellers of his own invention. They are generally three-bladed, with the driving-faces of the blades slightly concave and the opposite faces convex. In his letters patent granted in the United States 24th of August, 1875, he says: "I claim as my invention: 1st. A propeller, the blades of which are projected rearward from the hub, said rearward projection being most prominent nearest the hub and decreasing toward the tips of the blades; 2d. A propeller, the blades of which have an increasing pitch and are projected rearward from the hub, said rearward projection being most prominent nearest the hub and decreasing toward the tips of the blades."

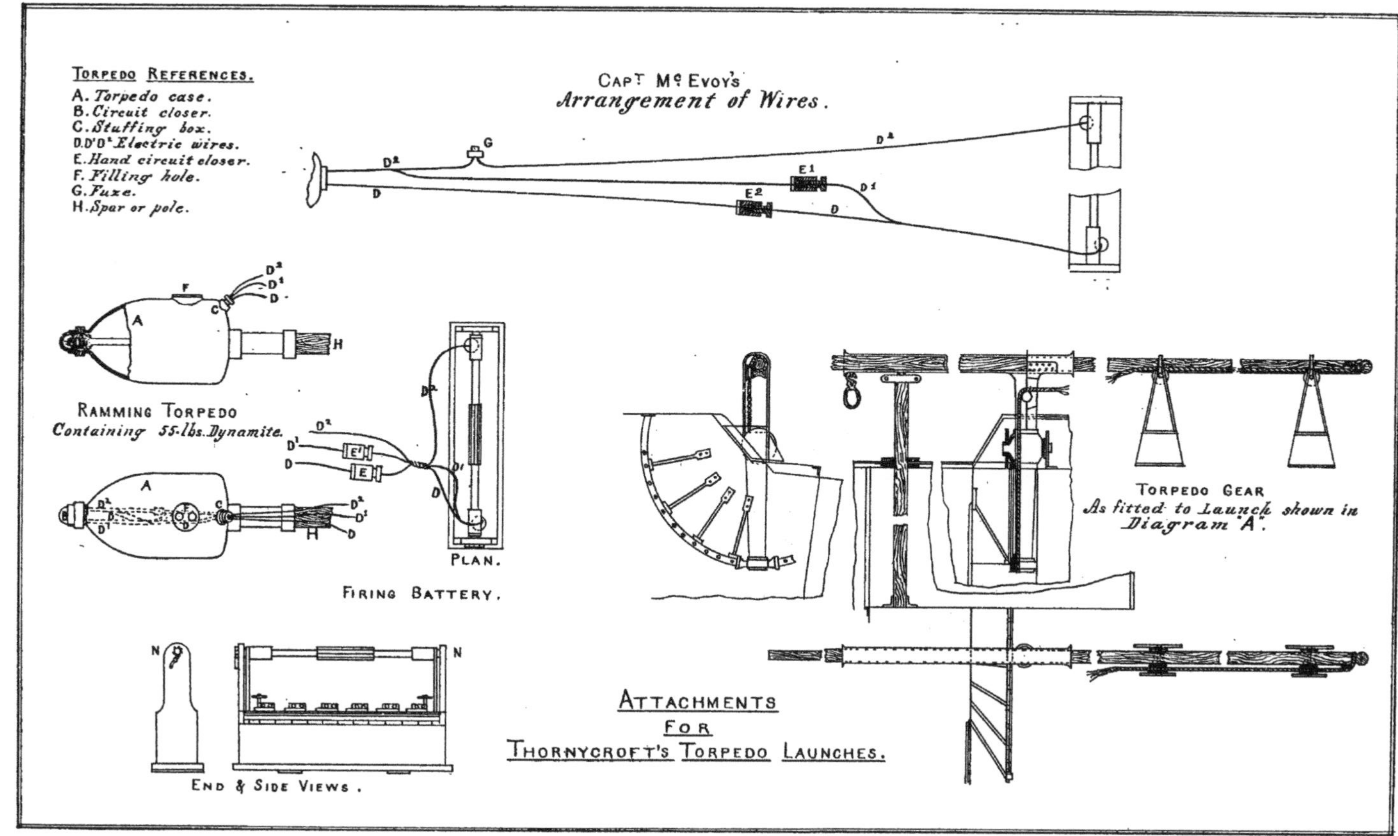
Torpedo References.
A. Torpedo case.
B. Circuit closer.
C. Stuffing box.
D.D'D² Electric wires.
E. Hand circuit closer.
F. Filling hole.
G. Fuze.
H. Spar or pole.
Capt. McEvoy's
Arrangement of Wires.
Ramming Torpedo
Containing 55 lbs. Dynamite.
Plan.
Firing Battery.
End & Side Views.
Torpedo Gear
As fitted to Launch shown in Diagram "A".
Attachments
for
Thornycroft's Torpedo Launches.

be put in circuit; a current is then possible through the wires D and D_2 as soon as the circuit is completed by the contact-plate being pressed against the studs.

However effected, whether by the firing-key, or by contact with the enemy's vessel, as soon as a strong current passes through the fuse the small platinum wire is heated to redness, and the fulminate of mercury exploded; this explodes the detonator, and with it the charge of the torpedo, the force of the explosion finding its way along the nearest path to the air, which path, if the torpedo is sufficiently close, is through the enemy's ship.

The arrangement for working the torpedo-poles is shown in the diagram, and consists of two tubes riveted together at right angles so as to form something like the letter T. The torpedo-pole is put through the horizontal tube, which is free to move round the center of the vertical tube, and the vertical tube is free to move through a quarter-circle at right angles to the center line of the vessel.

In attacking in front, the vertical tube is laid over till it is parallel with the water-surface, and the horizontal tube is allowed to incline sufficiently far to allow of the end of the pole, when run out, to be depressed from 8 to 10 feet below the water-line. It is held in this position by a pair of blocks attached to the top of the short mast.

In attacking on the broadside, the vertical tube is laid over till it assumes a position such as to allow of the pole when swung round to touch an enemy's vessel at about 8 or 10 feet below the water-line.

FRENCH TORPEDO-LAUNCHES.

Eight torpedo-boats have been built by Messrs. Thornycroft & Co. for the French Government. The first two were of the same dimensions as the one last described, but more powerful, the indicated horse-power being 200, and the speed 18 knots. The armament as at first fitted was the same, except that the copper torpedo-cases were charged with 55 pounds of dynamite instead of the 670 cubic inches of explosive used in the Austrian boat; but after the vessels reached Cherbourg they were altered so as to attack in front only, as the French officers found that these small vessels were better adapted for resisting the effects of an explosive at the bow than at any other part.

The arrangement consisted of a steel pole, about 40 feet in length, having one end about 6 inches in diameter and solid, and the other about 1½ inches in diameter and hollow; this pole was mounted at its solid end on small pulleys, which ran upon two ropes stretched fore and aft of the vessel; the other end, to which the torpedo was attached, was led over a pulley fixed on the bow. Ropes passing over pulleys to a windlass in the after compartment were attached to the inboard end, and by turning the windlass the pole was drawn backward or forward as required. When the pole was drawn forward, the inboard end being constrained to move in a line parallel to the deck, the outer end was depressed in the water, and was so adjusted that when the pole was run out to its full extremity the torpedo was depressed to about 8½ feet below the water-level. The arrangements for firing were similar to those described.

The speed trials were made in the roadstead off Cherbourg. The total revolutions in two hours were 49,818, and the number required to the knot in still water was found to be 1,382, so that the distance run in two hours was 36.05 knots. During the run the pressure of steam was kept at 108 pounds and the vacuum 25 inches.

These two boats made the passage boldly from the Thames to Cherbourg without hugging the coast, and thus proved their efficiency as sea-boats.

In February and March, 1877, some experiments were made off Cherbourg with these boats, which have attracted a great deal of attention. The *Bayonnaise*, an old wooden frigate, which had been damaged by one of the earlier experiments, and on this occasion kept afloat by empty casks, was towed by another vessel at a speed of about six knots per hour, and was attacked by one of the torpedo-boats. No attempts at defense being tried, a hole was made in the *Bayonnaise* large enough to admit a whale; a result not at all surprising, nor calculated to teach anything not known many years previously, but from its sensational character it attracted considerable attention from the public press. The torpedo used contained 33 pounds of damp gun-cotton, and was fired 8½ feet below the surface of the water, at the end of a steel pole 40 feet long. The real interest of the experiment and the point it was intended to throw light upon was the effect of the explosion upon the attacking boat. The accounts published at the time were absurdly exaggerated.

Six more vessels of a different size and type were built for the French in 1877. These vessels are 87 feet long, 10 feet 6 inches in beam; the plating is heavier than in any boats previously built, and the contract speed is 18 knots per hour in a run of three consecutive hours.

The propeller is placed forward of the rudder instead of aft as in other Thornycroft boats, so as to give increased readiness in steering. With the view to prevent oxidation of the hulls, the plates and frames below the water-line are galvanized.

The cost of these larger boats was $26,730. Their armament may consist of an outrigger arrangement similar to that described, or they may be fitted for the Whitehead torpedo, as desired, being equally well adapted to either weapon.

BRITISH TORPEDO-BOAT LIGHTNING.

This vessel is illustrated in diagram B. The length over all is 84 feet breadth 10 feet 10 inches, draught of water 5 feet, and guaranteed speed 18 knots on the measured mile. The hull of the *Lightning* is made of heavier plating than was formerly used, and her lines are fuller, as she is intended for use in a tolerably rough sea if necessary; and in order that she may be able to remain at sea for some time, cabin accommodations on a scale larger than in any of the other boats have been provided for the officers and crew. The steering-gear is arranged so that the vessel may be steered from the deck or from the conning tower. The top of the conning tower is supported on three screws, so arranged that it may be raised or lowered and the space for sight adjusted according to the range of vision required or the risk to be run from the enemy's missiles.

The motive machinery is similar to that already described, and is capable of indicating 350 horse-power. The armament is to consist of Whitehead torpedoes, and the discharging apparatus is fitted to operate from the deck forward.

The *Lightning* on her preliminary runs in the Thames obtained a speed on the measured mile of 19.4 knots per hour, a speed which will be considerably reduced when all the weights are on board. The cost of this boat is $25,515.

A large number of torpedo-boats are in process of construction for the British admiralty, by private firms and at the dock-yards.

In diagram A will be seen another size of boat built for the Italian and Dutch Governments. These vessels are 76 feet long, 10 feet broad, and the guaranteed speed is 18 knots on the mile run. They are similar to the smaller sized French boats, but have more free-board forward, and the indicated horse-power is 250.

The Italian boats are armed with the Whitehead torpedo, and the Dutch boats are fitted with the outrigging apparatus. The cost of the latter vessel was $23,085.

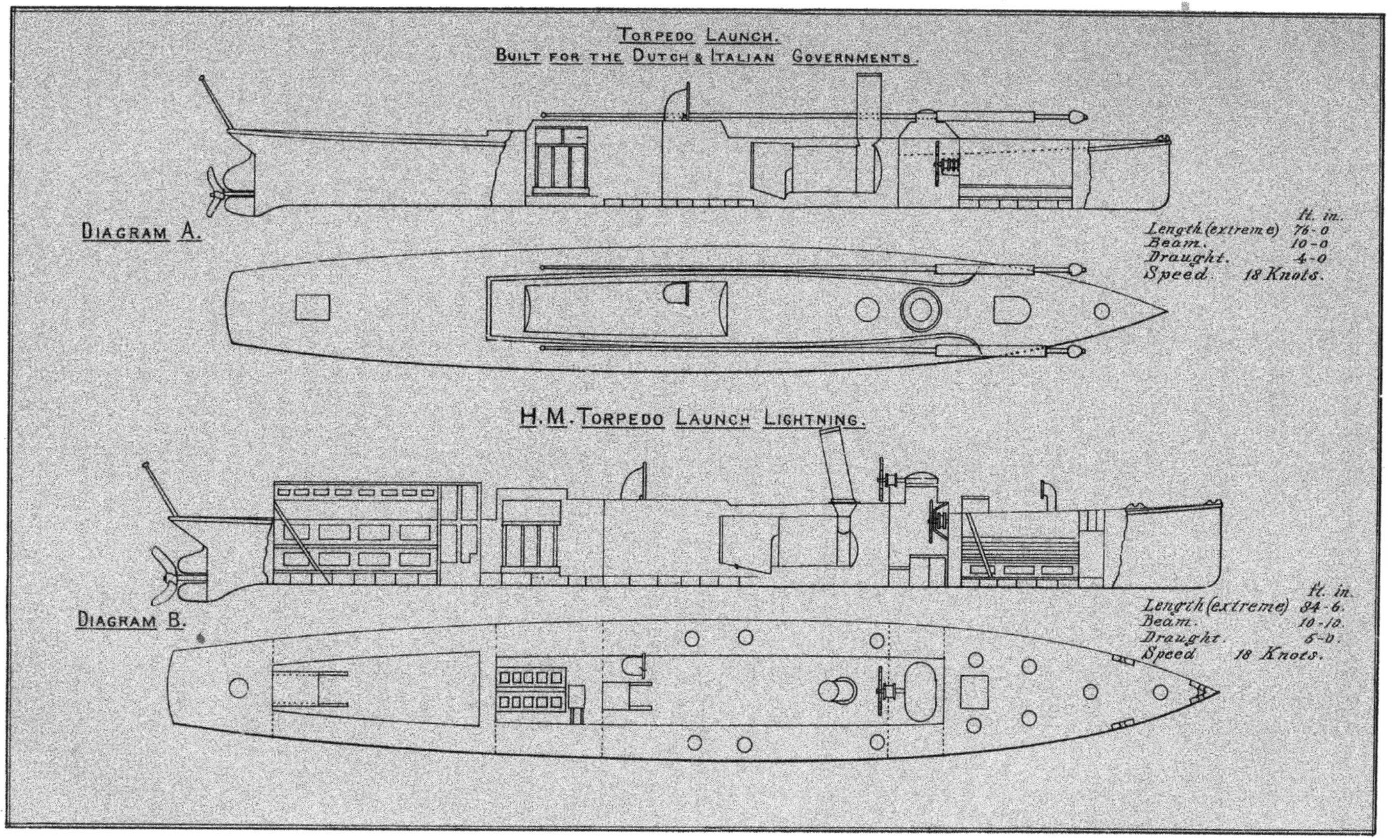

Torpedo Launch.
Built for the Dutch & Italian Governments.
Diagram A.
ft. in.
Length (extreme) 76-0
Beam. 10-0
Draught. 4-0
Speed. 18 Knots.
H.M. Torpedo Launch Lightning.
Diagram B.
ft. in.
Length (extreme) 84-6.
Beam. 10-10.
Draught. 5-0.
Speed 18 Knots.

All the boats described are without sails, and carry coal for a few hours only, at their maximum speed.

TORPEDO-BOATS BUILT BY YARROW & CO., ISLE OF DOGS, LONDON.

This firm has also attained distinction as the builders of these recent and improved engines of naval warfare which science has given to the world.

Nothing short of actual inspection can give an adequate idea of the amount of ingenuity expended in overcoming difficulties and arriving at the required results in these craft. During 1875 I had the opportunity of inspecting a torpedo-boat for ocean purposes, built by Messrs. Yarrow & Co., for the navy of Holland. It is 66 feet long, 10 broad, and 5½ feet deep, and is driven by a pair of inverted direct-acting engines, of 11 inches diameter of cylinders and 14 inches stroke of pistons. The boiler is of the locomotive type, with a total heating surface of 450 square feet, and a working pressure of 140 pounds per square inch; the estimated maximum horse-power is 200. This firm has constructed one of their fast torpedo-steamers for the Russian service in the Black Sea. It is 85 feet in length, and the guaranteed speed on the measured mile is 20 knots per hour.

A number of boats have been built by the same firm since then for continental navies.*

The following outline-drawings will convey a good idea of the general nternal arrangement of two sizes of a type of torpedo-boat of which a considerable number have been built by Messrs. Yarrow & Co. The dimensions of the one showing the section at A are: Length, 75 feet; beam, 10½ feet; draught of water, 3 feet. The length, therefore, is little more than seven times the beam, a condition which renders the extraordinary speed which has been attained the more remarkable. The boat just referred to is, like all other torpedo-launches, built of steel of the best quality, no other metal possessing the requisite strength and stiffness for scantling and plates of such lightness. It is divided into eight compartments by seven transverse bulkheads, the forward and after compartments being used for stores, the two central compartments being occupied by the machinery, while the steersman and the officer managing the torpedoes are placed in the compartment immediately abaft the engines. The steersman's head projects above the deck, and is protected by an iron truncated cone, the top part of which is movable like the visor of a helmet; when it is lowered he sees his way through a series of slots made all around the circumference of the cone. The hull is decked over from end to end with a curved shield, the plates of which are intended to resist rifle-shots even at comparatively close quarters; it also adds strength to the structure and prevents the sea from washing inboard. Strictly speaking, there is no deck on which men can stand; two wire life-lines are rigged at each side, and between these a space about 4 feet wide forms a precarious kind of floor.

The motive machinery consists of a pair of vertical, condensing, compound engines, the cylinders being respectively 10 inches and 18 inches in diameter, with a stroke of 12 inches. Steel is introduced into the stationary and working parts to secure lightness. The revolutions per minute, when running at full speed, are about 470, and the indicated

* The London *Times* of January 22, 1878, reports that one hundred torpedo-boats have been ordered by the Russian Government to be built in St. Petersburg, to be copies of those built by Messrs. Yarrow & Co. Fifty of them are to be completed in six months and then to be transported by rail to Odessa.

horse-power 275 to 280. The piston speed is thus 940 feet per minute, a velocity not often exceeded.

It need scarcely be said that great care is required in designing and building machinery of this kind; not only is it necessary that the moving parts should be properly balanced, but also that the materials and workmanship should be exceptionally good.

The steam is condensed by a surface-condenser. The air-pump, circulating and feed pumps, are worked by a separate vertical engine, running at a comparatively slow speed. The propeller is of steel. The steam is supplied by a boiler of the locomotive type with a very large grate surface, and the pressure of steam is 120 pounds per square inch. The smoke-pipe is fixed at one side of the line of the keel, to be out of the way of the torpedo-pole. The fires are forced by a fan driven at 1,100 revolutions per minute by a small separate engine, so that the weight of coal consumed per square foot of grate surface is very large. To secure the absence of smoke, Welsh coal is used; and the condition requiring that the machinery should be noiseless is satisfied by the employment of condensing engines.

During a run of two hours, a speed of 17 knots per hour is reported to have been attained.

Boats of larger dimensions, viz, 84 feet in length and of 11 feet beam, are in process of construction; the estimated speed of these is 18½ knots. The torpedo-gear has been described as follows:

The mechanism of attack consists of three torpedo-spars, which are made of steel in order that they may be strong enough to pierce torpedo-nets put down for the protection of the ships likely to be attacked. The bow-pole consists of a hollow steel tube 5 inches in diameter and 40 feet long. Under ordinary conditions this rests snugly on the top of the curved shield, but when going into action it is forced out and lowered by a small steam-engine provided for the purpose, which hauls on it with ropes, and is under the control of the steersman; the engine also raises the pole up and hauls it in. When the pole is out to its fullest extent it projects 25 feet from the bow, and the torpedo itself is 10 feet below the surface of the water.

The bow-pole is only applicable for attacking ships at rest, as it is found from actual experience that, in spite of its strength, it is quite unable to withstand the strain which would be brought on it by the resistance of the water when the launch is moving at seventeen or eighteen knots an hour. Besides, the presence of the torpedo in the water would materially diminish the speed of the boat, and interfere with her steering qualities, as may well be imagined. For what we may term a running fight, the torpedo-boat is provided, as we have said, with two other poles, which are so fitted that they swing out from the side, as oars will turn in pin-rowlocks. When using these, the boat endeavors to pass alongside the vessel which she is attacking, at a distance of 15 or 20 feet. The end of a pole is then disengaged by the steersman. It drops overboard, and is immediately swung round by the resistance of the water, and, if all goes well, comes into contact with the ship's side and explodes. Should the distance be too great it falls astern, the pole lying alongside like an oar. It can then be recovered and used for another attempt. The torpedo-boat throughout the operation moves at full speed. It will be seen that the use of these torpedoes is extremely hazardous, unless the distance between the boat and the ship is carefully calculated. But considerations of this kind do not militate against the employment of a very ingenious expedient. So long as there is a chance of destroying an enemy's ship, brave men will not be lacking to try it at any personal risk.

The torpedo itself is a steel or copper case, holding 40 pounds of dynamite, and is arranged to be exploded by an electric current, the current being closed either by the torpedo coming in contact with an obstruction or at the will of the operator.

SEA-GOING TORPEDO VESSELS.

Several sea-going vessels of small dimensions have been built and fitted especially for ejecting Whitehead fish-torpedoes. The first one of this variety constructed by the English was the *Vesuvius*, put afloat in 1874. This vessel is of iron, of 260 tons displacement and 379 indicated horse-power. She is fitted with a launching-tube in the forward end,

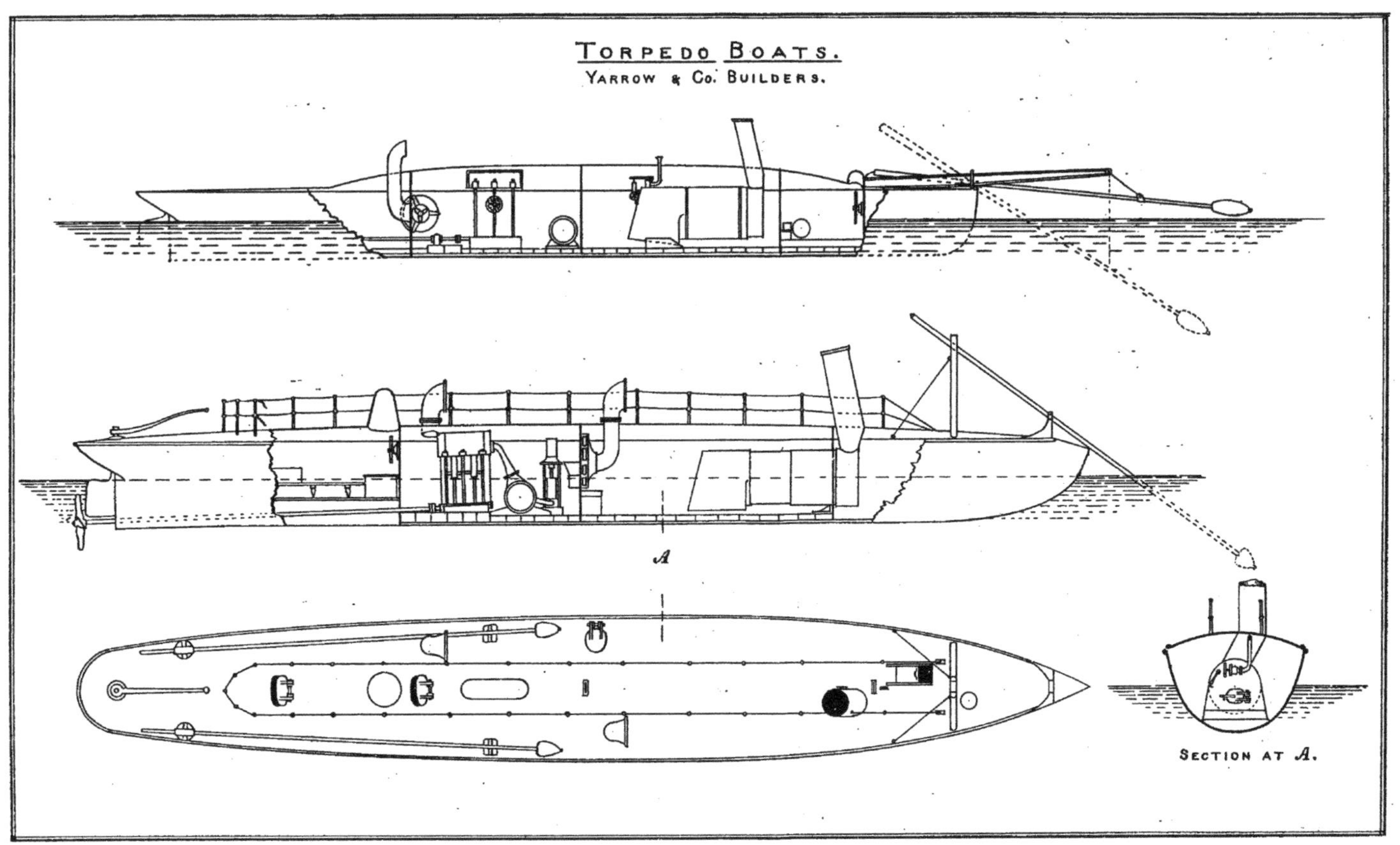
TORPEDO BOATS.
YARROW & Co. BUILDERS.
A
SECTION AT A.

and has been used generally in Porchester Lake, near Portsmouth, as a school of instruction in the use of the Whitehead torpedo for both executive and engineer officers. Quite a number of other vessels, for torpedo service exclusively, have been ordered by the English admiralty, and some are in process of construction, in addition to which it has been decided to apply the apparatus for manipulating the Whitehead torpedo to all ships, both armored and unarmored, fitted for service, and all of those equipped for sea within the last two years have been so fitted, and are provided with the torpedoes as a part of their armaments.

THE TORPEDO-VESSEL ZIETHEN.

The *Ziethen* is an unarmored iron-built German vessel, 226 feet long, 28 feet broad, 18 feet 6 inches deep, and has a load-draught of 11 feet 8 inches. This vessel was built and completed in June, 1876, by the Thames Iron Works at Blackwall, London, for the torpedo service of the German Imperial Government, and she was intended to carry out a series of experiments at sea with the Whitehead or fish torpedo. She has twin screws and two pairs of non-compound engines, by Messrs. John Penn & Sons, designed to indicate 2,500 horse-power. There are six cylindrical boilers, each containing two furnaces, set in the vessel back to back, occupying the whole width and about 30 feet fore and aft of the vessel, besides the fire-rooms, one of which is forward and the other aft, next to the engines. The maximum speed is 16 knots per hour at sea, and as the vessel is not to be used as a cruiser, economy of fuel was not a consideration in the design.

The *Ziethen* is built with two tubes, not much unlike screw-propeller-shaft tubes, placed in a line with the keel, one forward and the other aft, having valves at either end. They are 6 feet below the water-line; the after tube projects just over and beyond the rudder, the outlet of the forward tube is about 16 feet back in the fore body of the vessel, and a triangular portion of the hull is made to hinge in the stem and lift into a water-tight well, so that there is no projection to interfere with the speed of the vessel. From the tubes Whitehead torpedoes are expelled by means of compressed air forced in by pumps worked by a steam-engine. A small pipe connects the tube with a Kingston valve in the bilge of the ship. The torpedo used in this vessel is represented to be a great improvement over the weapon which the British government purchased the right to use. It is cigar-shaped, is about 17 feet 6 inches long, by 15 inches in diameter at the center, and the pressure of air contained in its chamber, when launched from the tube, is stated to be 1,000 pounds per square inch. To use the torpedo it is first charged with air to the desired pressure by means of an air-compressing pump, and then set to the depth and distance intended to be run, after which the fore end containing the explosive charge is secured, and the percussion arrangement properly adjusted. The torpedo being then ready for action, it is pushed into the launching-tube, the valve behind is closed, and the water from the Kingston valve admitted into the tube. A pump in the engine-room supplies air to a reservoir under a high pressure, and when it is wished to discharge the torpedo, the exit-valve of the tube being open, communication is opened between the reservoir and an apparatus in the rear end of the launching-tube, which forces the torpedo out with great velocity, at the same instant tripping its air-valve, which sets in motion the engine that works the screw-propeller by the self-contained power accumulated in the air-chamber. The speed at which the torpedo leaves the ship is to be 20 miles per hour, and it is said that it will maintain this rate for a short distance with a given immersion, after which the speed, consequent upon lessened pressure in the air-chamber as the engine works it off, will gradually diminish until the distance of about 2,500 feet from the starting-point is reached, when it will have run its course. The slower the velocity at which it is started the greater the distance it will run, depending on the capacity of the air-chamber and the pressure of air contained therein.

The German Government intends to determine, by careful experiment at sea, whether it can be delivered with certainty from a ship more or less in motion riding on waves, against another vessel similarly circumstanced, and perhaps moving under sail; in fact, under all conditions likely to happen in war. For this purpose, together with the cost of the *Ziethen*, the patent-fees and torpedoes, about $350,000 have been appropriated, and it is intended to expend $150,000 additional if the experiments justify it.

The *Naval and Military Gazette* reports the following, under date of October 10, 1877:

THE WHITEHEAD TORPEDO.

Experiments have once more been made in Germany—for the third time this year—with the Whitehead fish-torpedo, and, as in the two previous trials, the torpedo has again issued triumphant. It has, according to the testimony of the judges, almost surpassed their expectations, and military and naval authorities alike predict for it a great future. The first experiments were made in June last on a small scale, but sufficient to establish the character of the weapon, then as good as new to Germany. In consequence of this satisfactory result, the naval minister (General von Stosch) ordered the experiment to be repeated on a larger scale, in his presence, on the 18th of last month. Blank torpedoes were fired from the torpedo-steamer *Ziethen* at a submarine target 2,300 feet distant. The target was not missed a single time. After this the experiment was varied with a view to ascertaining if the torpedoes may be used without the protection of coast batteries, as an independent means of defense for harbors. For this purpose the services of the gunboat *Scorpion*, which is provided with a novel apparatus for discharging torpedoes, were put into requisition. Again the torpedo gave complete satisfaction. The experiment was tried a third time. The steamer *Ziethen* was ordered to aim torpedoes as if in battle, while sailing at full speed, at a target representing the broadside of a corvette. Of four shots fired, two from the bow and two from the stern, two struck the target right in the middle, which was as satisfactory a result as could at all have been looked for. In conclusion, the new apparatus for launching torpedoes from the upper part of a vessel was tried, and notwithstanding the apparatus had only just arrived and the crew were not yet perfectly familiar with its machinery, satisfactory results were obtained. So satisfactory, indeed, are the results altogether as to induce the minister of war (General von Kameke) to order a fresh trial in his presence for the purpose of ascertaining if the new weapon may be used by land batteries for coast defense. These experiments were proceeded with on the 28th of last month, in the presence of a considerable number of distinguished officers of the artillery and the engineers, and though no details are given in the published reports, the results are stated to have been highly creditable to the torpedo, and to have satisfied the military authorities that it may be used with advantage in the way suggested. It is reported to be likely that before long torpedoes will be served out to coast batteries for regulation practice.

GERMAN TORPEDO-BOAT UHLAN.

This boat was built in Germany by the Stettin Engine Company, and launched early in the summer of 1876. I did not see her, but the German papers announced the leading features of the boat, and, as they are peculiar and unusual, I copy the description:

This vessel will receive a torpedo charged with dynamite, to be carried on a 10-foot ram, lying deeply under the water-line; which torpedo is to explode on contact with the hostile ship. To protect the torpedo-boat from the results of the discharge of its own torpedo, the vessel is built with two complete fore parts, sliding one within the other, and having a considerable extent of intermediate space between them. This space is filled with a tough and elastic material (cork and marine glue), and thus if even the bows were carried off, there would be a second line of resistance. The object of the filling is to act like a buffer, deadening the blow and protecting the stem.

Another striking feature is the great power of the engines. The *Uhlan* carries an engine of 1,000 indicated horse-power. The steam is supplied by Belleville's tubular generator. The vessel, in fact, is all engine, only a very small space being left for coal and crew. The great power of the engines is necessitated by two circumstances. In the first place, the steamer has to be propelled at a high speed, and it has a very great draught, so as to offer but little scope to projectiles. In the next place, the greatest facility of steering or maneuvering had to be attained; hence the proportion of width to length, 25 to 70 feet. In order to save the crew at the worst, a raft has been constructed, which is filled with the above mixture of cork and marine glue, and is placed near the helm. When the *Uhlan* enters into action the dynamite cartridge is to be fixed by divers at the point of the ram. The rudder is then to be fixed; and the crew are to open a wide port on the ship's side, and with their raft jump into the water. The steamer is then allowed to rush forward and burst its cartridge on the enemy's armor. The crew, however, are to hold on to the torpedo-boat by a line while they are awaiting the result of the explosion; and in case their boat is not hurt they are to board it again, in order, if necessary, to repeat the maneuver. The price of this torpedo-boat is about 200,000 thalers.

RUSSIAN TORPEDO-BOAT UZREEF, (EXPLOSION).

This vessel, built at St. Petersburg, is constructed solely for the purpose of ejecting Whitehead torpedoes. She was launched August 13, 1877. Some of the data are, length about 115 feet, breadth 16 feet, draught at bow 7½ feet, and at stern 10 feet. The engines, which have two low-pressure cylinders and one high-pressure cylinder, are designed for an indicated horse-power of 800, and the estimated speed is 17 knots per hour, at which rate she will carry coal for twenty-four hours. Steel is the material used in portions of the hull.

THE ITALIAN TORPEDO-BOAT PIETRO MICCA.

The *Pietro Micca*, a vessel intended to discharge the Whitehead torpedo, was launched in Venice in the month of August [1876]. In the matter of construction this vessel is truly a novelty. According to *L'Année Maritime*, a few of her dimensions and other particulars are as follows:

Length between perpendiculars	202 feet 11 inches.
Extreme width	19 feet 7 inches.
Mean draught of water	11 feet 10½ inches.
Displacement	526.545 tons.
Immersed midship section	109.9672 square feet.

The bottom, which is entirely flat in the central part, is joined by arcs of circles with the sides, which are vertical above the bilge; these vertical sides again are connected by very pronounced inverse curves with the upper works, so that the transverse section at the widest part has the form of a mallet of which the handle would be very thick and very short.

The object sought in the construction of this vessel was great speed, in order that she might surprise an enemy and escape after accomplishing her object or upon being seriously threatened. Almost the entire hold is occupied by machinery, only about one-fourth of the forward portion containing the mechanism for ejecting torpedoes.

The machinery, built by the Ansaldo Company at Sampierdarena, is of 1,400 effective horse-power. It consists of vertical trip-hammer engines, and some of the other data are as follows:

Number of cylinders	2
Diameter of cylinders	30 inches.
Stroke of pistons	16 inches.
Number of boilers	4

Number of furnaces in each boiler	2
Pressure of steam	90 pounds.
Grate surface	136.5 square feet.
Heating surface	5,511 square feet.

There is a surface-condenser; the circulating-pumps may draw from the bilge. There are two chimneys and a superheater. In each boiler-compartment is a special 8 horse-power blower to be driven 1,200 revolutions per minute, and to furnish the boilers with air. The rudder and capstan are actuated by steam, and the former may be operated from three different stations.

The hull is of iron, and not armored excepting as to the lower deck, which is horizontal for a width of 7 feet 1 inch, and inclines slightly beyond that point toward the sides of the ship. The central horizontal part consists of three sheets, one of steel .6 inch thick, and two of iron each .8 inch in thickness. The inclined portions are .4 inch and .8 inch thick. The armament consists of ten Whitehead torpedoes and two mitrailleuses.

The vessel was launched with all the machinery on board, so that a week after, they were able to make the preliminary trial of the machinery, which worked with great regularity. As, however, the feed-pump of the forward group [of boilers] broke down, the trials for speed were delayed. The estimated speed is 18 knots, and cost $171,540.

In the *Pietro Micca* great speed is almost exclusively the point in view, the proportions of length to breadth being 10.36 to 1, and 12.73 horse-power being employed per square foot of immersed midship section; and this is imperative, for vessels of this type, naturally vulnerable, must remain for the shortest possible time exposed to danger from artillery. They must fall upon the enemy unawares, and escape if they are seriously threatened, or avoid an adversary similarly armed.

Another vessel like the *Pietro Micca* was to be built by the Italians.

SWEDISH TORPEDO-VESSEL RAN.

The *Ran*, which was built near Stockholm, and launched in July, 1877, is an unarmored vessel, rigged with two masts and fitted for ejecting Whitehead torpedoes. Her dimensions and other data are: Length on the water-line, 165 feet 6 inches; beam, 25 feet 4 inches; draught, 9 feet 6 inches, and displacement, 625 tons. She is provided with twin screw-propellers which are operated by engines capable of indicating 960 horse-power, and the speed is estimated at 13 knots per hour. In addition to eight torpedoes, which may be increased to twelve, if necessary, there is an armament of one 4¾-inch rifle-gun, and four Palmkrantz mitrailleuses. Her crew is said to consist of 65 men, and her estimated cost is reported to be $138,950.

THE OBERON TORPEDO EXPERIMENTS.

The *Oberon* is an iron vessel, of 649 tons B. M., which has for severa years past been used for carrying into effect a series of extensive and costly experiments, by exploding torpedoes near to and against her bottom, the chief object in view being to ascertain the effect of torpedo explosions on the double bottoms of armored ships, under different forms of construction and various conditions of attack. For this purpose the vessel has had an outer bottom built to the inner skin, which is intended to represent, as nearly as possible, the bottom of the broadside armored ship *Hercules*, and to be of about the same strength.

The *Oberon* has an original ½-inch plate bottom, single-riveted, with support afforded by angle-iron transverse frames 21 inches apart. These frames consist of two pieces of angle-iron riveted together. The second or outer bottom, ¾ inch thick, is fixed at a distance of 3 feet 6 inches from the inner one at the keel, and 2 feet 3 inches above the water-line. This bottom has every alternate plate with both its edges double-riveted outside those of the adjacent plates. Along the center of each of these alternate or outside plates, which are those that first come in contact with any external object, runs a longitudinal frame attached also to the inner bottom. Transverse frames run around at intervals of 4 feet, and are riveted to both bottoms by means of an angle-iron. Thus the space between the two bottoms of the ship is divided up by the longitudinal and transverse frames into spaces something like cubes or boxes, the top and bottom of such boxes being furnished by the inner and outer bottoms of the ship, and the four sides by the frames crossing one another. To lessen the weight of the metal, circular holes are cut in the middle of these box-sides in most cases. Every fourth transverse frame, however, is thoroughly closed and makes a water-tight bulkhead. It will be seen, therefore, that this is an exceedingly strong bottom.

The several experiments made with this vessel have been published in detail. The only one carried out during my sojourn in England was in June, 1876, and as it was considered the most important of all, I reproduce here, from *Engineering*, the results for the benefit of those interested in this special branch of naval warfare:

The *Oberon* torpedo-hulk was placed in No. 10 dock at Portsmouth on Wednesday afternoon, and the water having been let out, on Thursday morning it was possible for the first time to observe clearly the injuries she sustained from the torpedo experiments of Monday. The ship is divided into seven water-tight compartments, of which the two in the immediate neighborhood of the discharges were destroyed and filled with water. The bulkheads of four of the others remained intact, but permitted the water to leak through, but not beyond the capacities of the ordinary ship's pumps to keep down. The center compartment amidships remained perfectly dry; and as this was the largest in the vessel, it sufficed, with the artificial flotation which was afforded by upward of 300 casks which were packed away in the fore and aft compartments, to float the *Oberon* at high tide, and enable her to be taken in tow with little difficulty. In consequence of the buoyancy thus imparted to her, she settled with great deliberation, and it was the general impression at the time that she had not been severely hit, and least of all by the Harvey torpedo, which had been suspended from the starboard bow. This impression was effectually dispelled by the melancholy spectacle which presented itself on Thursday morning when the ship was fully exposed in dock. Notwithstanding the lightness with which she lay in the water—she only drew 11 feet, and consequently bore only a distant comparison with an iron-clad with its machinery and weights on board—every charge seems to have told with terrible effect, any one of the holes being sufficient of itself to have sunk the best of iron-clads, in spite of the Makaroff mat or any other leak-stopping devices that could have been applied. The Harvey torpedo, which contained 66 pounds of gunpowder, has split and bulged in an area of plating of the outer bottom about 16 feet square, extending downward through two longitudinals to the garboard plates, and laterally to the water-tight frame on each side of No. 4, utterly destroying the intermediate brackets. The injury here is very clearly defined, the longitudinals and frames having apparently acted as knives, so cleanly have the plates forced in upon them been cut through in the direction of the fiber of the iron. Had the longitudinal girders been placed closer together, the resistance would have been greater, and the damage to the inner bottom would at least have been less. The bracket-frames, which are only kept in position by angle-irons, seemed to have been snapped and doubled up with alarming ease by the force of the concussion. The inner bottom has been extensively damaged and bulged in, but not so much as might have been supposed from the appearance of the outer skin, the straightness of the bows having allowed much of the explosion to spend itself vertically. As might have been expected, the greatest damage is exhibited under the bilge on each side of No. 30½ frame, against which two charges, respectively of 33 pounds of slab gun-cotton and 33 pounds of granulated gun-cotton, were fired. Here frightful wounds were visible—wounds which are plainly past redemption. The holes are about 18 feet square each, and extend from the third strake below the armor-shelf well-nigh to the keel-plating. The greatest force appears to have been exerted on the starboard side by the granulated preparation. The iron skin has been torn from the rivets, the girders and bracket-frames shot away, and the upper plating wrenched completely off from their supports and blown away. The port side of the same frame presents a similarly ruinous aspect. The only difference is—and practically it is one without a distinction—that the plates, instead of being broken off, are lacerated in all directions and forced upon the inner bottom, which here as also on the opposite side is torn and forced inward. With the exception that the taffrail is blown away and the galley dismantled, the explosive forces seem to have been confined for the most part within well-defined limits. The wounds left by the previous experiments have not been reopened, and though the ship must have been lifted fore and aft, the fissures amidships do not appear to have extended. It is probable that, after a careful survey has been made, the *Oberon* will be filled with coal and submitted to a series of shell experiments. She can be of no further use for torpedo purposes.

It is impossible not to be struck by the suggestiveness of the above-mentioned experiments with reference to the future not only of naval warfare, but, in a still more impressive degree, of naval architecture. The terrible injury which the *Oberon* has sustained proves beyond all cavil that no iron-clad could withstand the bursting of a torpedo in contact; a torpedo would prove destructive almost wherever it struck, and a ship could hardly be saved by any turn of the helm. A projectile from the 81-ton gun would probably not prove utterly destructive, unless it hit at right angles with the keel; but a torpedo, so long as it hits, no matter where, would dislocate the integrity of the ship within an area large enough to prove fatal. The larger the vessel the more likely it is to fall a prey to the torpedo or the water-rocket. Its size would render it more susceptible to attack, and its slower speed would make it more difficult to escape from an active enemy. Smaller ships, therefore, would seem a necessity of the time, leaving details of construction for further consideration.

The question of stopping leaks from the outside, which was raised in the case of the *Vanguard*, again suggests itself here. It would appear that this was the most hopeful way of dealing with a leak of this character, where there is so much bent plate, which, while it is very difficult to repair, affords support to the sail or other material let down over it from the outside. Some experiments as to the possibility of closing such leaks would be valuable.

Contact charges and the action of the movable torpedo lead naturally to the contemplation of the probable effect of the Whitehead fish-torpedo. On every ground this has become a grave question for the British, especially since it has been found that it can be dispatched from the decks of armor-clad or other ships. As long as a special and subordinate class of vessels was devoted entirely to this species of attack, and had to carry tubes below the water-line, it was argued that approach would be excessively difficult; but when the heaviest armor-clads can, without sacrificing anything, avail themselves of this additional weapon, the question wears a different aspect.

A very important point in the investigations on the *Oberon* was the total absence of damage sustained by certain steam-launches, moored at the distance of 22 feet from each torpedo; the object being to ascertain the effect likely to be produced upon a boat supposed to carry the torpedo at the end of a spar; the result showed that the effect upon the supposed aggressor was almost *nil*, and this seemed to be corroborated by the experiments mentioned elsewhere with the *Bayonnaise*. It had already been proved that the explosion of 120 pounds of gunpowder in the open air, at the extremity of the spar of a torpedo-launch, worked no injury whatever to the boat.

DEFENSE AGAINST TORPEDOES.

The various methods by which torpedoes may be detected, warded off, exploded by counter-torpedoes, or otherwise rendered harmless to the party assailed, are now the subject of patient and exhaustive study in the British service.

In the case of moored torpedoes depending for their ignition upon electricity, many points of scientific interest have recently been brought to light. Some experiments undertaken in Denmark about three years ago showed most conclusively that dynamite torpedoes cannot be placed close together without incurring the danger of one charge bringing about the explosion of others. A dynamite torpedo of 150 pounds, ignited in 10 feet of water, was found capable of exploding other charges at a distance of 300 feet by the mere vibration imparted to the water, so that in supplementing coast defenses with dynamite torpedoes it is absolutely necessary to keep them far apart from one another. A second point noted was that a mere current of electricity, if it emanates from a powerful frictional machine, traversing one of a bundle of wires, will induce a current in the other wires, and thus bring about the explosion of torpedoes other than that which the operator on shore desires to ignite. It is these facts, particularly, which have led to the development of a system of counter-attacks, and have enabled seamen to devise a means of defending themselves from the insidious weapons.

Both dynamite and gun-cotton are peculiarly sensitive to vibration; indeed, their detonation is brought about by no other cause; hence, by exploding counter-mines in the channel which it is desired a ship of war shall enter, any lurking torpedoes may soon be disposed of—that is, if they contain a nitro-glycerine compound—and so a way for the ship be speedily cleared.

The successes of both Russian and Turkish divers in the present Eastern war, in searching for and severing the connecting cables and removing moored torpedoes, has shown a second way of clearing the road.

As a protection against the fish, Harvey, and spar torpedoes, many engineers and men of science have devoted their energies to determine some satisfactory means of disclosing the maneuvers of the attack and preventing its being effective. Hobart Pasha fitted his flag-ship, the *Assar-i-Tevfik*, with a large net of half-inch rope, the top of which was made fast to the bowsprit, while the lower part stretched along a strong iron bar, and was boomed out in advance of the hull by spars. When the net was not required it was triced up to the bowsprit and drawn inboard. It has also been suggested to employ a flexible wire-rope netting, surrounding the submerged portion of the vessel, as a shield to ward off the attack by recoil. Such contrivances, however, can only be regarded as a cumbersome appendage, and, at best, only a partial defense; moreover, increased velocity in the movable torpedo may render such precaution unavailing. The better plan would be to surround the vessel with a number of boats to repel the attack.

The Turks succeeded in defending one of their monitors in June last, near the mouth of the Aluta, from a most daring and persistent attack of four Russian torpedo-boats. This occurred in broad daylight. The boats lay in wait behind an island, and when the monitor was steaming past, suddenly darted out from their hiding-place and bore down on her. This vessel, it soon became evident, was handled in a different fashion from others with which the Russians had to deal. With wonderful quickness and skill she was prepared for action, and made defense against the four little enemies by thrusting out torpedoes on the ends of long spars, thus threatening the boats with destruction, at the same time opening fire on them with mitrailleuses and small-arms. The torpedo-boats escaped with the loss of only four or five men wounded, and considerable damage to the little vessels. The attack lasted about an hour, and that the boats should have suffered so little loss shows how difficult it is to hit these launches when in motion. It is reported that they were fitted out in the same manner as those which blew up the monitor at Braila, but this attempt, as well as the one at Giurgevo, was made in broad daylight, and neither of them succeeded.

The most valuable aid to defense, and the one to which special attention seems to be directed, is that of illumination—light of sufficient power to disclose any object attempting to enter a zone of illumination around a ship. If a sufficient and continuous illumination can be maintained at a given distance from a ship, no torpedo-launch or boat would venture to approach it. The launch would be doomed to destruction by the mitrailleuses, Gatling guns, or other small weapons now carried by ships of war.

The important adaptation of an old invention to a new need, that of the electric light to the discovery of an approaching boat, is called by its introducer, Mr. Wilde, the "torpedo-detector." It was applied to the *Comet*, which vessel was sent to sea one night from the Isle of Wight in order to receive the attack of two torpedo-boats, of whose whereabouts and direction of approach she was completely ignorant. They were discovered while more than a mile distant, and in different directions, and when they sought to escape from the cone of luminous rays enveloping them, and to renew the attack from other points, the apparatus followed them so relentlessly that they were soon convinced of the uselessness of attempting concealment. The light was sufficiently brilliant to enable the *Times* to be read on board vessels at a distance of a mile and a quarter. The whole apparatus weighs about 1,100 pounds and occupies a bulk of less than 5 cubic feet. The rotatory power was derived on the *Comet* from the fly-wheel of a hoisting-engine. The expense of producing the light, apart from first cost and steam-power, is about four cents per hour.

Several ships have already been fitted with electric lights, and recently some advance has been made toward solving this problem of illumination at sea by a trial of what is known as the "Holmes distress signal," in the form of a projectile for illuminating purposes, to be fired from mortars at a range varying from 500 to 2,500 yards. These signals possess the property of emitting a very powerful white light the moment they come into contact with the water, and when once ignited are absolutely inextinguishable by either wind or water, and burn with a persistency that is almost incredible, thirty or forty minutes being an average duration. The missile containing this light is made so as to be buoyant upon water, and at the same time with sufficient rigidity of form to withstand the concussion of powder. It is thus further described:

> Upon striking the water at the required range, the shot, floating up to the surface, immediately bursts into a brilliant flame, with great illuminating power. Some half-dozen of these shots fired from an iron clad or gunboat would effectually surround her with an impassable cordon of light at any required range, and by such a device, while the vessel herself would remain in darkness, the enemy's movements of attack would become plainly discernible, and any attempt to break through the illuminated zone of light be at once detected, however dark the night.

Mr. A. M. F. Silas, of Vienna, is also reported to have invented a light of much the same description.

In fact, a whole system of tactics is in course of being elaborated with a view to defense against the new weapon, thereby signally illustrating the changing and progressive condition of the art of naval war.

PART XX.

SEA VALVES AND COCKS; STEERING-GEAR.

19

SEA VALVES AND COCKS.

It is well known that all the water from the sea required for the purpose of working the engines and boilers of the vessel is admitted through several holes cut through the bottom or bilge, and that these holes are covered by what are known as sea valves or cocks, made either of cast iron or brass; and that they are secured to the skin of the vessel by bolts, and connected to the engines and boilers by pipes of copper, brass, or iron. Heretofore, these valves or cocks have generally been placed low down on the floor of the ship, so as to be out of the way; the engine-room as well as the fire-room floor-plates being above them, they are out of sight and very difficult of access, so that if a leak or accident occurs, it is not always discovered until the water rises to the floor, when it may be too late for remedy, because the engineer cannot ascertain the cause in consequence of the valves and pipes being covered with a considerable depth of water. Many serious accidents have happened in consequence of this arrangement. Among them may be mentioned the loss of the *Knight Templar;* also, the case of the French trans-Atlantic steamer *Europe*, of 4,000 tons, abandoned in the Atlantic April 2, 1874; the crew and passengers, in all about four hundred, being taken off the sinking vessel by the English steamer *Greece*. When abandoned there were only about 6 feet of water in the engine and fire rooms, and 1 foot in the adjoining holds. It is evident that the leak was in the engine department from sea-connections, as was proved by a very singular coincidence, for only a few days afterward her sister ship, the *Amérique*, of 4,000 tons, belonging to the same owners, was abandoned in the English Channel, afterward picked up by an English steamer and towed into Plymouth Harbor, with the engine and boiler rooms full of water. When it was pumped out, nothing was found wrong with the ship, and the difficulty with the sea connections was never made public. These two ships were valued at nearly $3,888,000, exclusive of cargoes, and their abandonment at sea, under such circumstances, was altogether owing to the want of practical engineering skill and courage, added to the captain's lack of presence of mind in emergencies. The *Ormesby*, of 930 tons, was abandoned in the Bay of Biscay in January, 1874; after having encountered a very heavy gale, water was suddenly discovered coming into the engine and fire rooms, and in a short time it put out the fires; as no leaks could be discovered in any other parts of the vessel, it was supposed to have come from the exposed and dangerous condition of the sea-cocks. There are many known cases of narrow escape, and, among naval vessels, notably that of the *Iron Duke;* besides which it is believed that unreported losses have occurred from the same causes. In consequence of these accidents, Lloyd's Register of Shipping now requires the sea valves and cocks of all descriptions to be placed above the bilge in all ships constructed or remodeled.

There is, however, no British law regulating the number of bulkheads in iron vessels, or the height to which they should extend. Neither Lloyd's nor the Board of Trade makes other provisions than that there must be a collision-bulkhead at each end of a steamer and at one end of a sailing-vessel. A vessel may be 500 feet long, and have 400 feet of that length in the center practically in one compartment. In some ships the bulkheads are sufficiently numerous, but do not extend above a deck which is nearly level with the load water-line. In that case, if one compartment fills, the vessel sinks enough to bring the tops of all bulkheads under water. In other ships where there are proper bulkheads, doors are cut in them, which destroy their integrity. The pumps afford but feeble refuge against the dangers thus incurred. That greater precautions are consistent with commercial success is proved by the recent practice of some English owners.

STEERING GEAR.

The most essential features in any arrangement of steering-gear are simplicity, certainty of action, immunity from derangement, and general efficiency. Several varieties of gear have been patented to accomplish these objects. It may be profitable to consider those most successfully employed, of which there are three systems of recent date, viz, the screw-gear, the steam-gear, and the hydraulic gear.

The introduction of steam-power and screw-propellers brought into greater prominence the demand for an increase of power in the steering-gear, owing to the increased speed, and different conditions under which the rudder acts in a steam and sailing vessel, besides which it became important to relieve the steersman from the shocks transmitted by the rudder when the vessel is being driven through heavy seas. To meet these demands the screw-gear (the well-known right and left handed screw) was introduced, which in some form or other is now generally employed in steamers either as a primary gear or as a stand-by in heavy weather. A large number of steamers of the mercantile marine are fitted with two sets, one placed aft, immediately in connection with the rudder, and the other in such a position that the steersman who works it commands a view ahead and is immediately under the eye of the officer on watch. That placed aft is as a rule a screw-gear, and the forward one is of the old-fashioned variety, either a combination of tiller with blocks, with wire-rope or chains carried along the deck to the steering-wheel, or else a quadrant is keyed on to the rudder head, to each extremity of which are attached chains which pass through snatch blocks on either quarter, and the communication is carried along the decks by rods or rope as before to the hand-gear. But in order to obtain sufficient power by the hand-wheels, geared wheels are used, and a pitch-chain which connects the rods or wire rope on either passes over a sprocket-wheel or chain-pulley, which is keyed on to the same shaft as the spur-wheel. By the arrangement of the screw-gear the danger consequent on the rudder taking charge is partially avoided, but the whole of the gearing and rudder-stock require to be of increased strength to meet the additional strain.

These appliances or modifications of them, for ships of moderate dimensions answer the purpose, and do not require under ordinary circumstances a very great amount of manual power, but in the largest steamers the best devised of any of these hand-gears meet their wants inadequately, and for heavy ships of war the efficiency of the steering apparatus becomes a most important question. Here, again, mechanical skill has enabled the problem to be dealt with, for both steam and hydraulic power have been successfully applied, and one or the other is now generally employed in such vessels, so that they can be readily handled in narrow channels or crowded harbors and maneuvered in action.

The steam-gear which up to the present time has met with the largest measure of success, and has been adopted in many ships of the British navy, also in a large number of commercial ocean steamers, is that designed by Mr. McFarlane Gray, and now made by Messrs. Forrester, of Liverpool.

As generally applied, it consists of a pair of engines with barrel and pitch-wheel, which are placed aft, so as to be in close communication with the tiller or quadrant. The valve which controls and reverses the engines is in connection with a light steering-wheel placed forward in the ship in some convenient position, the connection being made by means of shafting. The rudder is also placed in communication with this valve, in order that when the position of the rudder required by the steersman is reached the engines may come to rest, and on further motion being given to the steering-wheel the rudder can be moved either to port or starboard. The engines can be fitted amidships, as is sometimes the case, and various modifications made. This steering apparatus possesses simplicity, is rapid in its action, the engines are not liable to disarrangement, it is as efficient as a simple steam-winch, and the rudder can be easily controlled by *one man in the heaviest weather.* The first cost of this system of working the rudder is considerably more than that of the old plan worked by manual power, but the saving in the number of men at the wheel is a large item; added to which (and of more importance) is the greater security of the vessel by steering more easily in narrow or crowded places, and the less risk of collision owing to rapid handling.

The other steam steering-gears patented, and in some cases fitted, vary from the above principally in the manner in which the engines are brought to rest whenever the desired movement of the rudder has been effected, and in the arrangement of the slide-valves.

The only objection raised to steam as the power to move the rudder is the condensation in the long pipes when the engines are placed aft; but with a proper adjustment of the valves this need not interfere with the quick and sensitive action of the engines.

The system of steering-gear that has for some time been competing with steam as an agency is the hydraulic. Several types of this have been patented. The most successful is that designed by Mr. Brown, of Edinburgh, Scotland. It is by far the most complete of its kind, possessing many of the essentials required for a steering apparatus. The large ferry-steamers running between Liverpool and Birkenhead, double-enders, vessels which can be steered from either end, are successfully handled by this gear. It has also been applied to some large passenger-steamers nearly 400 feet in length, and is reported upon favorably. In these vessels a small hand-tiller is used for steering, which requires but a comparatively small amount of pressure to move from amidships to port or starboard, and the rudder quickly follows the movement of the hand-tiller, remaining in a similar position until this is again moved.

A clear description of the apparatus, by Mr. W. J. Pratten, I. N. A., is here given:

In this plan the tiller is acted upon by rams placed athwartships, or in any other convenient position; the water being conveyed to them by small pipes from an accumulator in the engine-room. This accumulator is kept filled with water by a pair of small pumping-engines; it consists of a long cylinder fitted with piston and ram. The space on the top side of the piston is in communication with the boiler, thereby having the same pressure per square inch as the boiler, the space under the ram being, as usual, filled with water. It is so arranged that the pumping-engines take their steam from the top of the cylinder, so that when the piston rises to the top, owing to the water-chamber being filled, the steam is shut off and the engines come to rest. As the water is used and the piston falls, the engines at once start, and fresh water is pumped into the accumulator. The relative areas of piston and ram are such that the water is under a pressure of about 700 pounds per square inch at the full boiler-pressure of steam. Iron pipes of one inch bore convey the water from the accumulator to a small slide-valve, which is in close proximity to the hand-wheel or tiller, for the use of steersman, and other pipes lead from this valve to the driving-rams placed aft. This valve, which resembles a small locomotive one, has a double connection, viz, with the tiller used by the steersman, by means of a vertical shaft and levers, and with the rudder by wire-ropes. The action is very simple. On moving the hand-tiller far enough to port or starboard, as may be required, water passes from the accumulator through the open valve-port to one of the driving-rams aft; the water at the same time escaping from the other ram through the exhaust-port of slide to a tank in the engine-room, from which the engines take their supply of water. The rudder is also connected to this valve, which is closed by the former getting into the position indicated by the hand-tiller. The valve is thus entirely self-acting in closing, after any movement of the hand-tiller. The device by which this valve is worked is very ingenious, and other contrivances in the design are well worthy of notice.

As the water is continually returned from the rams to the accumulator, there is only the loss by leakage to be made good. This leakage will be principally from the driving-rams, and, when these are packed, is reduced to a very small quantity. One great advantage arising from always using the same water, is in case of frosty weather, when it is necessary to add some non-freezing mixture to protect the pipes from injury. The patentee of this design advocates in such cases the use of about 20 per cent. of methylated spirits as a cheap and most effective preventive against frost.

The liability of water to freeze in the pipes has been urged as a main objection to the use of water-power in any form for a steering apparatus; but, except in cases of extreme frost, there is little risk of injury if even a smaller proportion of spirits than the above be used. The pipes can be carried along the main or lower deck, and the driving-rams can also be placed below instead of on the exposed upper deck.

It is highly necessary that the rams should have sufficient power to place the rudder hard over when the ship is running full speed; in fixing the diameter of the rams and leverage on tiller, it must be borne in mind that the boiler-pressure of steam is not always constant. If the rams have only been designed to force the rudder hard over when using the highest steam-pressure, it is obvious that it will partially fail to work effectually when the pressure of steam falls to one-half or two-thirds; but this is a minor point.

In order to relieve the pipes from a sudden increase of pressure, such as would take place when a sea struck the rudder, spring escape-valves are fitted on the barrels of the rams, allowing the water to pass from behind one ram to the other, thus preventing any loss of water and giving relief to the pipes.

Of the other hydraulic gears patented, the one by Mr. Esplin has been fitted to several ships; also one by Admiral Inglefield has been fitted to a vessel, but it was not free from objections, and after trial was removed. When hydraulic gear of this nature is fitted, the water in the accumulator may be usefully employed for other purposes, such as working independent engines for the starting-gear or anchor-hoister.

In war-ships or large merchant-ships entire trust is not placed in the steam or hydraulic gears, where they are fitted; for they are supplemented by a screw-gear or other suitable arrangement. In some cases the engines are so fitted that they can be readily disconnected from the remaining gearing, which can then be worked by hand in the ordinary manner.

Besides the steering-gear successfully applied as above described, there has been recently produced by Messrs. P. Brotherhood & Hardingham,* engineers, of London, a new system of steam steering-gear by means of which the rudder of a ship can be laid over at any desired angle, and, when released, always returns to its normal position. The power is transmitted through the Brotherhood three-cylinder type of engines. The apparatus is illustrated in the *Revue Industrielle*, and described fully in *Engineering* of October 13, 1876. It has been approved by the British admiralty, and has been fitted on board the armored ship *Téméraire*.

The advantages to be derived from the application of power other than manual for steering, as well as for hoisting anchors and for receiving and discharging weighty articles on board vessels, are so well known that enumeration of them is unnecessary.

---

* A set of Brotherhood engines was exhibited at the Vienna Exposition, and has been described by Professor Thurston as follows: "The cylinders are arranged with their axes making angles with each other of 120°. They are single-acting, and the pistons are coupled to a single crank. The valve is a revolving one, its axis is coincident with that of the crank-shaft, and it is turned by a prolongation of the crank-pin. The central chamber is the steam-chest, and the valves permit steam to pass to the rear of the pistons successively, thus disturbing the existing equilibrium and producing rotation."

PART XXI.

COMPOUND ENGINES.

NAVAL COMPOUND ENGINES; COMPARATIVE MERITS OF THE SIMPLE AND THE COMPOUND ENGINE; STATISTICS OF THE PERFORMANCE OF ENGINES.

COMPOUND ENGINES.

The efforts of the engineering profession have long been directed to the reduction of the cost of the production of power, and for a lengthened period the question of the relative merits of the simple and compound engine occupied the attention of the engineering world. Many experiments have been made, and much ability and ingenuity displayed, to prove that the compound system possessed no advantages over the ordinary simple engine working high-pressure steam; the most important of these experiments will be noticed presently.

My annual report to the department, as Chief of the Bureau of Steam-Engineering, dated October 30, 1871, just after returning from a short tour of investigation of the subject, gives the facts at that time; and as they hold good to-day, I cannot do better than copy that part of the report which bears on the subject. It is as follows:

The tour proved very instructive and interesting in many respects, especially in Great Britain, where immense fleets of iron ships are constructed and put afloat yearly, where ships are built for nearly all European nations, and where the British navy is systematically improved and strengthened by additions of nearly twenty thousand tons of iron-clads annually, besides transports and unarmored vessels.

The vast and varied experience of their constructing engineers and the sharp competition between rival building firms have given rise to rapid improvements, and compelled the abandonment of many designs only a few years ago regarded as the best productions of engineering skill.

The pressure of steam carried in marine boilers has gradually risen with corresponding increase in the extent to which expansion is carried, until boilers have been introduced, and are coming into universal use on board ship, in which the steam is kept from sixty to seventy-five pou ds per square inch, and expanded in the cylinders from ten to fourteen times.

The form of boiler used for generating the steam differs from the variety so long employed in all European steamers. It is built with a cylindrical shell of from 9 to 15 feet diameter, and of thickness from $\frac{3}{4}$ inch to $1\frac{1}{4}$ inches, according to the pressure to be carried. It is sometimes from 9 to 10 feet long, and is placed in the vessel athwartships, with fire-rooms fore and aft in the ordinary way, and sometimes about 18 feet long, placed fore and aft and fired from either end. There are two or three cylindrical flues, according to diameter, in which are the grates, and above them a set of horizontal fire-tubes, through which the gases return to the front, thence to the funnel. This boiler is strong, of moderate cost, and generates steam freely and economically.

The type of engine employed for working the steam is known as double-expansion or compound, the steam being admitted from the boilers first into a small high-pressure cylinder, there expanded to a third or fourth of its original pressure, then passed through a receiver into a large or low-pressure cylinder, where, after propelling the piston to the end of the stroke, it is exhausted into the condenser. This system of working steam expansively was introduced by Woolf seventy years ago; but at that date steam was used at a very low pressure, and the project met with no success. For many succeeding years the subject was discussed and experimented upon by engineers, but no advantages over the ordinary method of working steam expansively in a single cylinder were reached until a talented engineer of Scotland, Mr. John Elder, proprietor of the Fairfield Works, near Glasgow, several years ago took up the subject, and, in the face of immense opposition, zealously pursued it, designing and constructing every year several sets, each being more and more improved in design and detail, until they have been brought to the present state of perfection.

Of the thirty-seven engineering works and iron-ship building-yards on the Clyde, the Fairfield Works is now the most extensive and important. At the time of my visit this firm had already completed one hundred and thirty pairs of marine compound engines, and accompanying boilers, and had then twenty-two pairs under construction—all for ocean steamers; besides, in their ship-yard, ten large iron vessels were on the blocks, four at the wharf being engined and completed, and orders were on hand for others as soon as room could be made for them. This firm is justly regarded as the pioneer of the compound system, and their productions are accepted as the best types.

The largest establishment on the river Tyne, "and, as a combination of engineering and manufacturing," the most extensive in Great Britain, the Palmer Company, had under construction about the same number of compound engines, two iron-clads for the British navy, and several other vessels. In short, all engineering works in Great Britain are now building their marine steam-machinery on the new system. These engines are also manufactured by Messrs. Schneider & Cie., at Creuzot, and M. Mazeline, at Havre, France. They are also built in Germany and in Belgium. Indeed, the success of the system is so certainly established that no European owner will now contract for an ocean steamer unless it is stipulated that she shall be propelled by compound machinery.

Many commercial vessels containing very good machinery of the ordinary kind have been laid up, the machinery removed, and other machinery, on the new system, substituted. An example of this came under my notice in the case of one of the West India mail screw-steamers—the *Tasmania.* This vessel, twelve years old, had the best machinery, with jet-condenser, at the date of construction; with it fifty round trips had been made. All the machinery was then removed, and compound machinery substituted. At the date of my visit one round trip had been made with it, and an inspection of the logs showed the average consumption of coal for the round trip with the old machinery to be three thousand tons, while the round trip with new machinery had been made with fourteen hundred and forty-five tons, the speed of vessel remaining the same. In some cases, such as that of the *Ville de Paris* and *Pereire,* of the French Atlantic line, both comparatively new ships and having excellent machinery, orders have been issued for its conversion by the addition of a third steam-cylinder, to be used for the admission of the high-pressure steam, the removal of the present boilers, and in lieu thereof high-pressure ones placed on board.

The Cunard Company, the most conservative of all steamship owners, who were the last of the companies to substitute iron vessels for wooden vessels, who were the last to give up the paddle-wheels for the screw, are now the last to accept the compound engine and high-pressure steam in the vessels of their line in lieu of the old system. But in this, as in former cases, sharp competition of several lines has caused this company to convert the machinery of one of their vessels, and to place the new type in two small vessels, the *Batavia* and *Parthia;* besides, at the time of my visit, the models were being prepared for two vessels of dimensions larger than any now on the line, to be constructed and supplied with the compound type of machinery.*

The British admiralty, charged with the administration of by far the largest and most powerful navy in the world, always cautious in the application of new inventions, rarely adopting any untried plans, but surely accepting the most successful in practical operation, have in this particular made the test of the new system by its application to two ordinary wooden vessels, the *Tenedos* and the *Briton,* of 1,331 and 1,261 tons register, respectively. The success attending the trials of these two vessels justified them in ordering the third set for the *Thetis;* also two other sets for the twin-screw iron-clads *Hydra* and *Cyclops,* all now under construction; and although several iron-clads and unarmored vessels are being completed with the ordinary type of screw-machinery, ordered two or more years ago, all future ships constructed for the British navy will doubtless be engined on the new system. It is not probable, however, that any of their six hundred registered vessels will be ordered to undergo the process of alteration, because their cruising steamers are intended to be essentially sailing-vessels, and their steam-power is only used when the sails cannot be depended upon to accomplish the purpose required; the cost, therefore, of the alteration from the old to the new machinery would be too great for the results desired to be obtained.

The original cost of steam-machinery on the compound system, its total weight, and the space occupied by it in a vessel, does not materially differ from that of the ordinary type. The advantages are chiefly in the reduced consumption of fuel, less space for it on board, and smaller number of men required to handle it. These advantages have been found very great in commercial vessels making long passages, so great that the new type of machinery is as surely displacing the old on the ocean as the screw displaced the paddle, and surface-condensation displaced the old jet-condenser. In the two steamers of the British royal navy to which the compound engines have been applied and tested, and in the Atlantic steamers plying between New York and Liverpool, having this kind of machinery on board, the power developed is obtained with less than two pounds of good English coal per horse-power per hour, or half the fuel for the same power developed under the system of non-expansion and jet-condensation. In other words, a steamer is propelled the same distance and at the same speed by compound machinery with five hundred tons of coal, that required one thousand tons with the old type, and from thirty to forty per centum less than with the latest and best kind, having surface-condensation and superheaters.

The progress made in this direction, of economy in fuel, may be briefly noted in the

* These two ships, since completed, are the *Bothnia* and *Scythia.*

following facts: The paddle-wheel steamer *Scotia*, of the Cunard line, put afloat in 1862, and at that date regarded as the best and latest type of engineering skill, a vessel having a midship section of 841 square feet, consumed 160 tons of coal per day, or 1,600 tons on the ten days' passage between New York and Liverpool. The *City of Brussels*, a screw-steamer of the Inman line, put afloat in 1869, and having a midship section of 709 square feet, consumed 95 tons per day, or 950 tons on the ten days' passage. The *Spain*, a screw-steamer of the National line, put afloat in 1871 with compound machinery, and the largest ship on the Atlantic, having a length of 425 feet 6 inches on the load-line, beam molded 43 feet, draught, loaded, 24 feet 9 inches, made the passage in September with 53 tons of coal per day, or 530 tons on the ten days' run; all three vessels having the same average speed; a small percentage only of the gain being due to the finer lines and proportions of the last-constructed vessels.

The field of observations in this direction has been thoroughly gone over, the subject carefully investigated, and the bureau is informed of the extent of European operations and results, of which a brief outline has been given. The information is positive, and there can be no hesitation in recommending that all cruising steamers for the Navy hereafter put afloat be engined on the compound system, and that all the steam-machinery stored in the navy-yards that cannot be used to advantage in the old vessels, or converted into compound, be disposed of by public sale, or broken up and used as old material; and in view of the necessity of timely preparation, the bureau has taken steps for the execution of complete sets of working-drawings for machinery of vessels on the stocks, and that will suit classes of vessels that may be designed as cruisers. It is not, however, recommended to remove the machinery from any of the old vessels for the purpose of replacing it by the new, except in cases where the repairs to the old shall be found so costly as to condemn it altogether; for the appropriations are desired to be as small as compatible with the public interest, and the cost of alteration would be too great, considering the advantages desired to be gained under the regulations, of using sails as the motive power, except in cases of emergencies.

Subsequent experience has still more firmly established the compound system. In European nations simple expansion engines for all commercial and naval vessels are now as obsolete as the paddle-wheel on the ocean. The records of the surveyor's office in London show that of the several hundred commercial sea-going vessels built in the United Kingdom since 1871, including one hundred and twenty-three under construction when the report of 1876 was received, all, excepting a very few for special purposes, have been engined on the compound system. Besides this, a large number of steamers, set afloat previous to the successful application of the compound engine on the ocean, have, since my last report, been docked, the engines and boilers removed to the scrap-heap, and in lieu thereof compound machinery erected on board; a number of the vessels having the machinery so altered not being more than eight years old, and some only five years old. The steamers of the commercial fleet of Great Britain still retaining the old types of engines and boilers, retain them because the vessels cannot be spared from employment sufficiently long to remove the old machinery and substitute the new, or because the vessels are not worth the expense of the change. So far has this system of compounding the machinery of old vessels been carried, or, more properly, of substituting the compound for the simple type, that for vessels of 1,000 tons and under, the proprietors of several engineering works have built in anticipation, and retained on hand to a considerable extent, compound engines and accompanying boilers suitable for vessels probably requiring them, and in many cases they have contracted for and accomplished the feat of removing the old machinery and erecting on board the new in ten days' time. Many of the old channel, coast, and river steamers of England still retain their old machinery, for the reasons above stated. Some, however, have been altered; one of them a paddle-steamer, the *Wolf*, in which I chanced to make a passage from Havre to Southampton. This vessel has a length of 250 feet, breadth 27 feet, and draws 9 feet of water. She was furnished originally with the old type of inclined engines, jet-condensers and box-boilers, carrying 30 pounds pressure of steam. For eight years she made the passage regularly between Havre and Southampton, of eighteen hours, on 48 tons of coal for the round trip. Two years ago these engines were compounded, and cylindrical boilers, carrying 60 pounds pressure of steam, substituted for the old ones. The results of this change were increased speed and an average consumption of coal for each round trip of 28 tons. Many other cases could be cited, but it is believed to be sufficient to state that Lloyd's inspecting engineer's books show the saving of fuel resulting from compounding engines of British steamers to be from 30 to 40 per cent.

One of the steamship companies visited, having steamers that make long passages, was the Royal Mail. This company owned in 1876 twenty-four steamers, eighteen of which are engined on the compound system, three still retain the old type of engines, and three have been altered from the old to the new system, with an average gain in fuel of 40 per cent. The *Moselle*, a vessel of this line, to which my attention was called as one economical in fuel, has a length of 376 feet, breadth of 40 feet, and a draught of water of 21 feet. She was built by Elder & Co. in 1873. The engines are of the inverted vertical compound type, with a high-pressure cylinder of 60½ inches and low-pressure of 112 inches in diameter, having a stroke of 4 feet. The boilers are cylindrical, and the pressure of steam is 60 pounds to the square inch. This vessel is on the West India line; the distance steamed for each round voyage is 10,000 miles, the average speed 12 knots, and the average consumption of coal 40 tons per day of 24 hours.

NAVAL COMPOUND ENGINES.

It has been seen that as early as 1871 the British admiralty, though always cautious in the application of new inventions, had set afloat two corvettes engined on the compound system, and that the same type of engines were under construction for three other vessels. About this time a special committee on the designs of ships of war, consisting of sixteen of the most distinguished naval officers, naval architects, and engineers in the kingdom, were holding their sessions in London. In 1872 their voluminous report was presented to Parliament, and the following is an extract from all that part of the report relating to compound engines:

> * * * The carrying-power of ships may certainly be to some extent increased by the adoption of compound engines into Her Majesty's service. We are aware that this modification of the ordinary marine engine has not escaped the notice of the constructive department of the navy, and that some few of Her Majesty's ships have been so fitted; but its use has recently become very general in the mercantile marine, and the weight of evidence in favor of the large economy of fuel thereby gained is, to our minds, overwhelming and conclusive. It is unnecessary for us to say that in designing a ship, economy of fuel may mean either thicker armor, greater speed, a smaller and cheaper ship, or the power of moving under steam alone for an increased period, according to the service which the ship is intended to perform. We beg leave, therefore, earnestly to recommend that the use of compound engines may be generally adopted in ships of war hereafter to be constructed; and applied, whenever it can be done with due regard to economy and to the convenience of the service, to those already built.

Since the date of this recommendation, viz, 1872, and up to 1877, there have been designed and ordered to be constructed nine armored ships of large size; twenty-three unarmored corvettes; twelve unarmored sloops; two armed dispatch-vessels; one troop-ship, and twenty-eight gunboats, besides several vessels designed but not ordered to be built. All of these vessels, excepting a very few of the gunboats, have been or are being engined on the compound system. The excepted gunboats referred to have been supplied with simple expansive engines, carrying the same pressures of steam on the boilers, viz, from 60 to 70 pounds per square inch.

The British admiralty have not converted the engines of any war-vessels, for the reason previously stated, but some of the troop-ships employed in making long passages to India, and as a consequence using coal in large quantities, have been converted—one, the *Euphrates*, built and engined by the Laird Bros. ten years ago, at a cost for the machinery of $364,500. Though the machinery was in all respects in good order, it was removed to the scrap-heap in 1876, and compound engines and accompanying boilers substituted by the same firm at a cost of $267,300 and the old machinery. The *Euphrates* is of 6,211 tons displacement and 5,000 indicated horse-power.

The compound system was introduced and adopted in the national navy of France after its success was established in the British navy; and I was informed at the engineering department of the *ministre de la marine* in Paris that no ships are now built or ordered to be built for the French navy with other than compound machinery.

In the Italian navy one ship, the *Duilio*, building at Castellamare, is being engined with Penn's old type of machinery. This machinery was ordered before the Italian naval authorities felt sure of the success of the compound system, but all other new vessels are being fitted with compound engines; and it is quite certain that all vessels for this navy hereafter ordered to be built will be engined on the new system.

In the German imperial navy the compound system is in use, and, as far as I was able to learn, it is also in use in all other continental navies, as well as in the vessels of the commercial marine of every country.

If our naval vessels, built since the successful introduction of the compound engine in foreign navies, had been engined with the old types of machinery, we should have been regarded, both at home and abroad, as being far behind the times in steam-engineering; and the very officers who have raised objections to the compound engine for naval vessels would doubtless have been the first to denounce the Engineering Department as being uninformed and incompetent for the responsible duties of construction.

If objection to the compound system is based on the necessarily higher pressure of steam used than formerly, it may be stated in reply that the working pressure of steam has gradually risen from about five pounds per square inch, as at first introduced on board ship, to 80 pounds per square inch. Also, that in all the gunboats employed on the Western rivers during the civil war, about 140 pounds pressure per square inch was used.

The objection to high-pressure steam on board fighting-ships is said to be the increased danger to life in the event of a projectile piercing a boiler during an engagement. Now, all experience in boiler explosions, that of the *Thunderer* included, has proved that no persons can breathe steam, at whatever pressure it may escape from the boiler in which it was contained, and live. If, therefore, a boiler having a pressure of 80 pounds per square inch should be penetrated, the result would be the same as if the pressure had been only 20 pounds, viz, probable death to the engineers and men employed in and near the fire-rooms, but to none others necessarily.

To meet the objections raised by some British naval officers against going into battle with steam in the boilers of their ships at high pressure, the engines of the *Iris*, and other vessels recently built, have been provided with valves so adjusted as to let the steam on direct to all the pistons, and to allow the exhaust steam from the high-pressure cylinders to pass directly to the condensers, and by this means reduce the pressures in the boilers to 4 or 5 pounds above the atmosphere, when so desired. But in this case what becomes of the power required for maneuvering, for ramming, or for choosing the position from which to fight?

COMPARATIVE MERITS OF THE SIMPLE AND COMPOUND ENGINE.

I have previously stated that many experiments had been made to test the relative merits of the compound engine, and the simple engine working high-pressure steam. It is quite unnecessary to go into any lengthened statement or detail of these experiments. Many of them have been published in full and are familiar to the engineering profession. I shall therefore refer only to those made by the British admiralty, and to the more important and expensive experiment of the Allan line of steamers.

In 1874 two gunboats for the royal navy were built. One, the *Swinger*, was supplied with simple engines, and the other, the *Goshawk* with compound engines. The same kind of boilers and the same pressure of steam was used in each case, and the indicated horse-power was to be the same in both, viz, 360. The limited trials made with these boats were at the time published, and as different opinions existed as to the merits of the two systems for small vessels with low power, it was decided to supply three others ordered to be built, after the same fashion. Accordingly the *Sheldrake* and the *Moorhen* were fitted with simple expansion engines, and the *Mallard* with compound engines. These vessels are composite gunboats. The two former were engined by Messrs. Napier & Sons, and the latter by Earle's Engineering Company at Hull. The boilers in all are practically identical. They are cylindrical, each containing two cylindrical furnaces 2 feet 6 inches in diameter and 4 feet 6 inches long. These two furnaces terminate in a common combustion-chamber 3 feet 2 inches deep. The tubes do not return over the furnaces, but are continued from the combustion-chamber to the smoke-box at the forward end of the boiler, the boilers being placed in a fore-and-aft direction. On the top of the combustion-chamber, between the ends of the furnaces and the tubes, is suspended a bridge, which deflects the flame and heated gases before they enter the tubes, and the use of which has been attended with satisfactory results. The engines of the *Sheldrake* and *Moorhen* are simple expansion engines, direct-acting, the connecting-rods working between the cylinders and the cranks; expansion-valves, capable of cutting off the steam as early as one-fifteenth of the stroke, are fitted on the backs of the main slide.

The experiments above named were with small vessels, in which either system of working the steam will answer the purpose. I am not aware of any large sea-going vessel with screw-propeller, and as a consequence short-stroke engines, in which the simple engine using high-pressure steam has been successful. The most expensive and notable attempt to realize the benefits of the compound system by the simple engine at sea was made three years ago by the proprietors of the Allan line of steamers. This company made the comparative test on a large scale. Two ships were built, the one fitted with compound engines and the other with simple expansive engines. The boilers were identically alike, made from the same drawings, having the same grate and heating surface, and the same pressure of steam was used in each vessel.

I have not at command all of the details of these ships, nor all of the results of the performances at sea, but I received from the designer of the machinery the following particulars:

Name of ship, *Polynesia.*	Name of ship, *Circassian.*
Length, 400 feet.	Length, 360 feet.
Breadth, 42 feet.	Breadth, 40 feet.
Draught of water, 25 feet 6 inches.	Draught of water, mean, 23 feet.
Kind of engines, compound.	Kind of engines, non-compound.
Number of cylinders, 4 (two high and two low pressure).	Number of cylinders, 2.
Diameter, high-pressure, 43 inches.	Diameter of cylinders, 62 inches.
Diameter, low-pressure, 80¾ inches.	Stroke, 4 feet.
Stroke, 4 feet.	Number of boilers, 10.
Number of boilers, 10.	Number of furnaces, 20.
Number of furnaces, 20.	Pressure of steam, 60 pounds.
Pressure of steam, 60 pounds.	

The expansion-valves were fitted in the simple engines to work the steam, when desired, to the same degree of expansion as in the compound engines. The workmanship and materials were equally good, and the parts equally strong, in each set of engines. The two ships were put on the line between Liverpool and Quebec, Canada; and, as was anticipated, the results as to economy of fuel were not materially different, about two pounds of good Welsh coal per indicated horse-power per hour being expended in each ship. This satisfactory result, however, soon found an offset in the shape of unexpected difficulties with the simple engine, consequent upon the serious shocks resulting from the rapidly-varying pressures on the crank-pins. So serious were these, that not only the crank-shaft, but also the stationary parts of the engines, began at an early day to show signs of weakness, and in a short time gave out altogether. The superintending engineer of the company was the designer of the machinery, and it was only after his skill and efforts failed to keep the ship running, that he reluctantly decided to remove the engines and to substitute compound engines in their stead. The engines substituted had a pair of vertical inverted cylinders, with a diameter for the high pressure of 55 inches, and for the low pressure of 92 inches.

The performance of the *Polynesia* was satisfactory from the first, the voyages never having been interrupted; and the performance of the *Circassian* has also been satisfactory since the substitution in her of the compound engines for the simple ones.

In so far as simple expansion is concerned, it is not of consequence whether it takes place in one cylinder or in several; or, as Professor Rankine says:

> The energy exerted by a given portion of a fluid during a given series of changes of pressure and volume depends on that series of changes, and not on the number and arrangement of the cylinders in which those changes are undergone. The advantages of employing the compound engine are connected with those causes which make the actual indicated work of the steam fall short of its theoretical amount, and also with the strength of the engine and its framing, the steadiness of action, and the friction of its mechanism.

The economy of working steam many times expanded may be readily seen when it is stated that a decrease of one-half pound of coal per horse-power per hour may give, on a large trans-Atlantic steamer, a saving of about four hundred tons of coal for a single round voyage.

This economy is obtainable only by using steam of high pressures, whether it be worked in a compound engine with its divided expansion, or in a single-cylinder engine with its irregular strains. As yet, by no satisfactory device except compounding have great expansion and consequent economy of fuel been obtained at sea.

In the face of these facts, further discussion on the subject of adopting the compound engine for the vessels of our own Navy is as useless as would be a discussion of the relative merits of the screw-propeller and paddle-wheel for ships of war.*

PERFORMANCE OF COMPOUND ENGINES.

The following paper, read at the Institution of Naval Architects, in London, last year (1877), by the well-known mechanical engineer, Mr. J. R. Ravenhill, gives some interesting statistics relating to the performance of compound engines of British construction:

* * * * Since 1871 the introduction of the compound engine into our commercial fleet has been very great and very rapid; the Liverpool Red Book shows on the figures being abstracted that the total number of screw-steamers fitted with compound engines is as follows:

TABLE III.

	No. of vessels.	Total amount of nominal horse-power.
For the year 1871-'72	849	133,547
For the year 1876-'77	2,765	436,041
Giving as the increase in the horse-power during the five years		302,494

In looking through the register one could not fail to remark the large application of the compound engines to small vessels engaged in the coasting trade or in short sea voyages, some of them with less horse-power than 90 horses; but between 90 horse-

* With regard to this and the rest of Mr. King's remarks upon the comparative merits of simple and compound engines, it is worthy of note that the engines of the *Northampton*, which have recently been tried at their full power of 6,000 horses as simple expansive engines, have given most satisfactory results, and have developed none of the extraordinary phenomena exhibited by the engines of the *Circassian*. The knocks and shocks in those engines, attributed by Mr. King to the varying pressure, were due, as is pretty well known, to the inefficient action of the Corliss valve-gear as applied in this case. Putting together the remarkably smooth and regular working of the *Northampton's* engines and the acknowledged economy of the engines of the *Circassian* over long sea voyages, we cannot at all concur in Mr. King's sweeping condemnation of simple expansive engines for war and for commerce. Besides the engines of the *Ajax* and the *Agamemnon* for the British navy, each intended to be worked as simple engines at their full power of 6,000 horses, Messrs. Penn have recently received orders for engines of the same type for the Italian Government, which will develop a power never before even approached on shipboard. Again, in a paper read during the present session of the Institute of Civil Engineers, Mr. Alfred Holt, the well known ship-owner and engineer, himself the originator of one of the most successful types of compound marine engine, expressed the view that, notwithstanding the universal adoption at present of the compound engine at sea, the reintroduction of improved simple expansive engines for commercial ships may yet be found desirable. The full discussion the subject has received in this country has recently led to the adoption in the British navy of means for enabling the compound engine to be worked as a simple engine when occasion requires; and the more glaring defects of the commercial type of compound engine for purposes of war have thus been to some extent neutralized.—AN ENGLISH NAVAL ARCHITECT.

The engines of the *Northampton* and other naval vessels mentioned, are designed to work as compound engines primarily and under normal conditions—as simple ones only exceptionally and for a short time; it is intended further, in the latter case, to work steam of very low pressure in all the cylinders—a concession to the danger of shot piercing the boilers in action. That the engines of the *Northampton* have given satisfactory results at a short trial, proves nothing for their capability of endurance for such lengths of time as the *Circassian* steamed, and with the same high initial pressure. By the way, the effect of a particular valve-gear upon the shock in the main shaft-bearings of an engine is not apparent. The testimony of present facts is a thousand-fold in favor of the compound engine; what the future may bring forth is matter of speculation.—J. W. K.

power and 99 horse-power I found the increase worthy of special notice, as the following table shows:

TABLE IV.

Number of vessels having compound engines of—

	1871-'72.	1876-'77.
90 horse-power	61	188
91 horse-power		3
92 horse-power	1	1
93 horse-power		1
94 horse-power		2
95 horse-power	26	56
96 horse-power	16	24
97 horse-power	1	1
98 horse-power	62	139
99 horse-power	57	124
Total	224	539

Increase, 315.

It is not my intention to do other than thus make passing allusion to what I have heard happily spoken of as the "Mystic Nines," but the number of vessels with 100 horse-power stood only in these respective years at twelve and fifty-three. The most general form of compound engine is the one having two cylinders of the ordinary inverted type, one small and one large, side by side, the cranks standing at an angle of ninety degrees to each other, and the air-pumps being worked by beams, notwithstanding that those fitted with three cylinders—one high-pressure cylinder and two low-pressure—standing in a similar way, and those with four cylinders, the small ones being placed above the large ones, as well as those with two cylinders only of the same construction, are each reported on favorably by their respective partisans; but in the first-named class, which possesses some advantages possibly over the other three, the large diameter of 120 inches has been reached in the low-pressure cylinder, with a length of stroke of 5 feet; and although the reintroduction of loose liners is opportune in reducing the risk of casting them, and being in themselves easy of renewal in the event of undue wear taking place, the action of the weight of such large pistons in vessels rolling at sea must at times prove detrimental to their full efficiency, in consequence of the liability of the piston packing-springs to yield, and thus allow the pistons to leak. I do not, therefore, anticipate seeing any material increase in diameter to those now at work being proposed. I need scarcely add that all classes are fitted with expansion-valves. The present practice, as regards the multiple of the cylinders, appears to be generally to make them three to one, but the Liverpool and Great Western Steamship Company, Guion Line, adopted, in the case of the *Dakota* and *Montana*, a proportion of 6.21 to 1. Their original intention was to work at a pressure of 115 pounds to the square inch, and I make this passing allusion to them, not only because they proposed both a higher pressure and a higher rate of expansion than anything before attempted, but were the first owners who introduced on a large scale the water-tube boiler as a marine boiler; and while we may regret the failure of their endeavors, we cannot but admire the boldness which led to the introduction of such machinery. I gather from the Registers of Shipping of 1876, periodically corrected up to a recent date, that the Pacific Steam Navigation Company has thirty-four vessels fitted with compound engines; the Peninsular and Oriental Company, thirty; the Royal Mail Steam Packet Company, fifteen; the National Company, nine; the Oceanic, or "White Star," nine; the Guion Line, Great Western, four; the Inman Steamship Company, four; working at pressures varying from 40 pounds to 80 pounds on the square inch. In Table No. V. may be seen the results obtained from vessels A, B, C, D, E, steaming over many thousand miles on their several stations, each of them masted and rigged on the most approved plan for making the fullest effective use of its sail-power whenever it can be beneficially employed, and all burning, as far as it lies in their power, the description of coal that experience has shown to be the most advantageous; and here let me incidentally mention that many owners are directing their attention most closely to having an accurate entry kept in the log-book of the number of hours their vessels are under canvas and the number and description of the sails set.

TABLE V.—*Dimensions of screw-steamships and their relative performances at sea.*

	A.		B.		C.		D.		E.	
	Ft. in.		*Ft. in.*		*Ft. in.*		*Ft. in.*		*Ft. in.*	
Length between perpendiculars	200 3		319 4		345 0		346 0		348 1	
Breadth	26 4		40 2		42 0		39 2		40 4	
Depth	16 2		32 7		32 6		35 3		33 2	
Gross registered tonnage	611		2, 833		{3, 128 before alteration.} {3, 429 after alteration.}		2, 956		2, 999	
	Ordinary engines.	Compound engines.	Ordinary engines.	Compound engines.	Ordinary engines.	Compound engines.	Ordinary engines.	Compound engines.	Ordinary engines.	Compound engines.
Total distance run at sea, in knots	31, 990	58, 040	284, 279	125, 562	76, 658	75, 406	49, 028	56, 265	25, 335	25, 201
Total hours under weigh	4, 478	9, 300	30, 229	12, 213	7, 868	7, 161	4, 595	5, 115	2, 250	2, 201
Average speed per hour, in knots	7. 14	6. 24	9. 40	10. 28	9. 74	10. 53	10. 67	11. 00	11. 26	11. 45
Total coals consumed, in tons	2, 007	2, 007	70, 368	15, 481	15, 141	11, 276	13, 875	9, 083	5, 072	3, 485
Consumption of coals per day	10. 75	5. 18	54. 66	30. 42	46. 18	37. 7	74	42	54	37. 5
Average mean draught of water	14 ft. 1 in.	14 ft. 1 in.	19 ft. 3 in.	20 ft.	19 ft. 1¾ in.	20 ft. 9 in.	22 ft. 8 in.	24 ft. 2 in.	18 ft. 11 in.	19 ft. 7 in.
Area of midship section, in square feet	342	342	644. 3	674	677. 6	745. 6	787. 56	850. 7	612	645
Displacement, in tons	1, 320	1, 320	3, 727	3, 925	4, 444	4, 946	4, 600	5, 068	3, 990	4, 200
Coefficient of merit, $\frac{D^{\frac{2}{3}} \times V^{2.3}}{C}$*	950. 6	1, 448	608. 8	1, 377. 7	1, 099. 8	1, 729. 7	865. 7	1, 745. 2	1, 221. 2	1, 891. 1
Improvement in percentage in favor of compound engines	....	52. 3	....	126. 2	....	57. 3	....	101. 5	....	54. 8

* Where D = displacement of vessels in tons; V = speed of ditto in knots per hour; C, consumption of coals in tons per day.

In Table V. the *Deccan* is included as the letter C. Her performances we have had before us on a previous occasion, but I have thought it as well to introduce her for purposes of comparison. I have also adopted the formula introduced by Mr. De Russett and his paper on the lengthening of the Peninsular and Oriental Company's ship *Poonah*, read at our last annual session, without offering any opinion as to its correctness or otherwise. The percentage of merit thus shown as obtained in cases where no alteration has taken place in the form of the vessels I take to mean a percentage of commercial merit, that is, the comparative commercial advantages obtained on every ton of coal expended in the use of compound engines on the present system over ordinary jet-condenser engines working with 20-pound or 30-pound pressure on their boilers. The first vessel, A, comparatively of small dimensions, is one of the few instances in which owners, on making the change to compound engines, have reduced the speed of their ships; and, although this performance shows an improved co-efficient of commercial merit in favor of the compound engine of 52.3 per cent., some deduction would have to be made for the extra expenses of the ship's crew, &c., in consideration of such a reduction. All the other examples show an increase of speed with an improved co-efficient of commercial merit, the difference in the value of the co-efficients in the case of B and D being the most marked of any that have come under my notice; but, passing on, I am desirous of drawing your attention very particularly to the performance of the vessels F and G at sea in the following:

TABLE VI.

	F.	G.
Length between perpendiculars	358 ft. 2 in.	413 ft.
Breadth	41 ft.	41 ft. 7 in.
Depth	33 ft. 6 in.	27 ft. 4 in.
Gross registered tonnage	3, 252	3, 130
Total distance run at sea	55, 963	34, 252
Total hours under weigh	4, 967	3, 131
Average speed in knots per hour	11. 11	10. 94
Total coals consumed in tons	7, 604	4, 775
Consumption of coals per day in tons	36	36. 6
Average mean draught of water	21 ft. 9 in.	20 ft. 4 in.
Area of midship section in square feet	760. 28	639
Displacement in tons	4, 925	4, 832
Coefficient of merit per Mr. De Russett's formula	2, 042. 6	1, 917. 6

The above performances were stated by the owners to represent their best type of ship at the present time; and, in addition, Table VII. shows their relative performances at the measured-mile trials. The results are as follows:

TABLE VII.

	F.	G.
Average speed in knots per hour	14. 81	13. 953
Mean draught	19 ft. 8 in.	17 ft. 9.5 in.
Area in midship section in square feet	674. 06	524. 7
Displacement in tons	4, 310	3, 970
Indicated horse-power	3, 245. 3	2, 590
Diameter of screw	18 ft.	17 ft. 6 in.
Pitch	28 ft.	25
Revolutions per minute	61. 75	63
Pressure on boilers	55	65
Diameter of cylinder (small)	60. 5	56
Diameter of cylinder (large)	112	97
Proportion of cylinders	3. 26 to 1	3 to 1
Length of stroke	4 ft. 0 in.	4 ft. 6 in.
Coefficient of performance as per Admiralty displacement formula, $\frac{\text{speed}^3 \times \text{displacement}^{\frac{2}{3}}}{\text{I. H. P.}}$	265	262. 7

The best Welsh coal is used on board both at sea, but the services on which they are employed are very different, as is also their rig. It is admitted that F can make the better use of her canvas. Doubtless some of the $6\frac{1}{2}$ per cent. is to be found here. In these cases I have no percentage of improvement to show, but their coefficients of commercial merit are very satisfactory, and are higher, you will notice, than any obtained in the compounded vessels in Table V. It is a striking example of how closely results approximate in vessels designed by our leading naval architects, having their machinery supplied by our best engineering firms. And in obtaining these results the naval architect deserves his share of credit equally with the marine engineer, for while the latter is strenuously endeavoring to improve his machinery, the former spares no effort on his part in bringing science to bear so to apportion the use of the material in the hull as to obtain the maximum of strength with the minimum of weight, with the object of improving the design of his vessel. But while these performances show very conclusively the advantages obtained by the introduction of the compound engine, I do not find any reduction in the coal consumed per indicated horse-power per hour, reliably recorded, below the figures we heard of as far back as 1871, viz, 1.75 pound. The relative consumption of coal is now so closely criticised by engineers, and the difference in some cases so small, that the condition of the indicator used and the aptitude of the engineer who takes the diagram become matters of serious consideration, and I believe the total consumption of coal in tons will continue to be the test most preferred by owners. The great question with them is to ascertain how much cargo in cubic contents or dead weight can be carried over a given distance at a certain speed for a certain weight of coal. This weight to them is a known money value, seriously affecting their pockets, the indicated effects on which they thoroughly understand. It is their commercial diagram, and most of them prefer the study of this rather than the indicator diagram forwarded to them from the engine-room. While the older companies and steamship-owners are thus deriving benefit from the great reduction in the consumption of coal above recorded, new companies have been able to commence other service which a few years back would have never been attempted.* At the commencement of the present Holyhead mail service in 1860, I heard many well-known officers prognosticate that the service must be accompanied by considerable risks and dangers through collisions, but no such accident has occurred during the seventeen years the service has been regularly performed day and night in all weather, and some of them have lived to see the Atlantic crossed to and fro at an average speed of over 15 knots an hour during a period of twelve months. Never has there been a stronger proof as regards the saying that "circumstances make men," and great credit accrues to all parties engaged in such performances. The Inman Steamship Company have two large steamers running between Liverpool and New York. But taking them as a fleet, there is nothing that will compare with the Oceanic or White Star Line. This fleet consists of nine vessels with machinery all on the compound principle. Their large vessels, from 600 horse-power upward, have the four-cylinder engine of the vertical type, with a working pressure of steam from 65 pounds to 70 pounds, the stroke being 5 feet. These vessels deservedly hold the pride of place for their speed maintained as ocean-going steamers between Liverpool and New York. The *Britannic* is recorded as having made the shortest homeward passage yet known, in December, 1876, having run the distance, 2,882 knots, from Sandy Hook to Queenstown in seven days twelve hours forty-one minutes, at an average speed of 15.95 knots per hour, while the shortest recorded outward voyage has been made by the *Germanic* in April last, when she performed the passage in seven days eleven hours thirty-seven minutes, running over a distance of 2,830 knots, at a speed of 15.75 knots per hour; the mean speed of these two voyages is 18.459 statute miles per hour. The difference in the distances run in the outward and homeward voyages is 52 knots, a distance something shorter than the length of the passage between Holyhead and Kingstown. The service of the company between Liverpool and New York is performed by six vessels, and Table VIII. shows the performances of the *Britannic* on eleven voyages, from June, 1876, to June, 1877. You will observe that the average indicated horse-power is recorded at 4,900, with 52.3 revolutions per minute, and the copies of indicator diagrams * * * * * * have been handed to me as a fair average performance of her machinery. The *Britannic* is of the following dimensions: Length between perpendiculars, 455 feet; breadth, extreme, 45 feet 2 inches; depth, 33 feet 7 inches; gross register tonnage, 5,004. She is fitted with four masts, was built by Messrs. Harland & Wolf, of Belfast, and has engines by Messrs. Maudslay, Sons & Field, of London. Four cylinders, two large and two small. Diameter of small cylinders, 48 inches; diameter of large cylinders, 83 inches; length of stroke, 5 feet.

---

*Since this paper was in type the arrival of the *Lusitania* has been announced from Melbourne in thirty-nine days from London.—[NOTE BY MR. RAVENHILL.]

TABLE VIII.—*Performances of steamship Britannic, of the White Star Line, United States mail steamers, from Liverpool to New York, on eleven voyages, from June, 1876, to June, 1877, inclusive.*

Total distance	64,230 knots.
Total number of hours under weigh	4,269 hours.
Total coals consumed	18,214 tons.
Average speed per hour	15.045 knots.
Average coal consumption per day of twenty-four hours	101.932 tons.
Average mean draught of water	23 feet 7 inches.
Area of midship section at 23.7 feet draught	926 square feet.
Displacement at 23.7 feet draught	8,500 tons.
Diameter of propeller	23 feet 6 inches.
Pitch of propeller	31 feet 6 inches.
Surface of propeller	128 square feet.
Average number of revolutions	52.3 per minute.
Average working pressure in boilers	65 pounds.
Average indicated horse-power	4,900.
Average consumption per indicated horse-power	1.94 pound.
Average per knot	5 cwt.
Average per hour	4.25 tons.
Average per furnace per hour	2.67 cwt.

S. GORDON HORSBURGH,
Supt. Engineer.

The figures require no comment from me further than that I have been requested to draw your particular attention to the small amount of indicated horse-power exerted in proportion to the displacement of tonnage of the ship at this high velocity. By way of comparison, I introduced the following particulars of the Great Western steamship the first vessel that earned a great name in trans-Atlantic voyages:

TABLE IX.—*Great Western steamship.*

Name of builder	Patterson.
Where built	Bristol.
Date of launch	1837.
Wood or iron	Wood.
Length	212 feet between perpendiculars.
Beam	35 feet 4 inches.
Depth	23 feet 2 inches.
Tonnage	1,340 tons builders' measurement.
Name of engineers	Messrs. Maudslay, Sons & Field.
Horse-power (nominal)	420.
Type of engine	Side lever.
Size of cylinders	74 inches diameter.
Stroke	7 feet.
Paddle-wheels	28 feet diameter.
Paddle-wheels	28 common floats 1.10 wide.
Number of revolutions per minute	From 10 to 18.
Boilers	Four, iron, return-flue.
Steam-pressure	5 pounds per square inch.
Average time of voyage between Liverpool and New York	About 15 days.
Date of ship's decease	Broken up in 1856.

The average length of voyage of the Great Western appears to have been about fifteen days from Liverpool to New York. In forty years we have thus seen the time between these two ports practically reduced one-half—another proof, if one is required, of the progress made in steam navigation.

The books of Lloyd's Register in London showed that in July last (1877), there were building in the United Kingdom two hundred and twenty-five pairs of marine engines, all on the compound system, to carry pressures of steam in the boilers from 60 to 90 pounds per square inch.

PART XXII.

MARINE BOILERS.

CORROSION OF MARINE BOILERS; PRESERVATION OF BOILERS IN HER BRITANNIC MAJESTY'S VESSELS; WATER-TUBE BOILERS; EXPLOSION OF THE BOILER OF THE THUNDERER; BOILERS OF THE MERCANTILE MARINE; HIGH PRESSURE SAFETY-VALVE; LLOYD'S RULES AND FORMULÆ FOR BOILERS; RULES OF THE BOARD OF TRADE.

CORROSION OF MARINE BOILERS.

This subject has attracted the attention of engineers, ship-owners, and chemists in England from the time of the introduction of the surface-condenser up to the present day. Much diversity of opinion has always existed as to the exact nature of the action which, accompanying surface condensation, destroys the boiler-plates so rapidly and so curiously. Most of the various explanations which have been given attribute the process of deterioration to galvanic action arising from passing the repeatedly distilled water over large surfaces of copper or brass tubes, but the tinning of condenser-tubes and the substitution of iron boiler-tubes and pipes in place of brass or copper has not arrested it. It has been attributed to the corrosive action of certain acids arising from the decomposition of the lubricants; but such action is too limited to do the mischief, and even where little, if any, lubricant has been used in the cylinders, the results have been the same. Many panaceas have been proposed, from time to time, to cure the malady, such as the injection of solutions of soap and soda, of muriatic acid, &c. Much was at one time expected as the result of the simple plan of placing plates of zinc at different points in each boiler. The corrosion was, it was stated, confined to the zinc, while the iron of the boilers escaped. The zinc-cure was reported in some cases to have been an advantage, while in other cases it appears to have been entirely inoperative. The use of strainers or filters, by which grease was taken out of the feed-water, was also held to be a remedy, but this scheme, like many others, died a natural death. In fact, many cases are known where boilers working with engines into the cylinders of which no lubricant whatever was admitted, decayed just as rapidly as though oil or tallow had been admitted, and others in which all the remedies referred to were applied have also met the same fate. The profession now seem inclined to account for the corrosion by the action of the redistilled sea-water itself; and this view seems to be substantially confirmed by the Perkins system, where it is seen that a boiler thirteen years in constant daily working, using pure rain-water over and over again, without change for all of that time, is found to be in a good state of preservation. This case, with many other well-authenticated instances which came under my notice, seems to establish the fact that the evil must be attributed to the use over and over again of distilled sea-water, and that no remedy, except that of changing the water in the boilers at frequent intervals, has been found effective. It has been believed by marine engineers that the formation of scale on the surfaces of the iron protects it from corrosion, but more recent evidence has been produced that the protection is not due to the scale, but to the condition of the water. The circumstance that scale is found on the iron is simply evidence that the water has been kept in such a condition that it would not cause corrosion, and that is all the scale has to do with the matter. Very many illustrations of the proof as to changing the water being the sure remedy were found; among them were noted, as a comparison, two ships of the same size in the royal navy serving on the same station at the same time. They were paid off and recommissioned at the same period. In one ship the water had been used in the boilers over and over again; in the other the feed-water had been mixed largely with sea-water. The boilers of the former had been found free from scale, but so completely deteriorated as to be useless. Those of the latter had a scale formed on the plates, but the boilers were found in excellent preservation. The experience of many intelligent engineers has been presented on the subject, among the most important of which is that of Mr. Milan, read before the Cleveland Iron-Trade Association. The following is an extract:

I may begin this paper by stating that what data I shall have the pleasure of laying before you, as to the corrosion of boiler-plates when generating steam from distilled or redistilled water, will be generally such as has resulted from my personal observation. Much of it will doubtless be familiar enough to those whose professional experience has included the management of boilers while working in conjunction with surface-condensers. In the earlier days of the practical application of these condensers (I refer to fifteen or sixteen years ago), you must be aware that engineers were often driven to their "wit's end" in striving to account for or to arrest the extraordinary and insidious decay which was steadily and rapidly consuming their boilers before their eyes, while yet there was no recognized means of preventing it. And, indeed, up to the present time, although we have, by a primitive and withal a make-shift expedient, been able in great measure to mitigate the evil, I am not aware that the light of scientific research has yet been brought to bear upon the subject with a view to determine the nature or cause of the action itself, or to devise any means of counteracting it.

Fully a decade of years has passed since the writer first endeavored to ventilate this question by introducing it to the notice of gentlemen whose social and professional positions and interests warranted him in assuming that they would have influenced scientific inquiry in respect to the question. More recently I have tried to set forth the necessity for immediate and decisive action in the matter.

The decay of boilers in Her Majesty's ships has been so extraordinary since the introduction of surface-condensers that the officers have been greatly exercised in regard to it. In spite of all their care the decay has gone on silently and certainly; boilers have become disabled after a few years' service, and it became evident that the causes baffled discovery and remedy in the ordinary course of duty. So impressed were the admiralty with the vital importance of the subject, and with the futility of trusting to experience alone to find a way out of the difficulty, that about two years ago they appointed a committee to inquire into the boiler management of ships in commission and in the reserve, the nature and extent of the decay of boilers, and to ascertain, if possible, what are the causes and what the remedies. This committee consists of two executive officers, two engineers, and one chemist. Their labors have been directed to the examination of boilers in ships arriving from different seas of the world and in different stages of condition or decay; also in experimenting at the dock-yards.

The report of this committee is looked for by the naval and commercial services with unusual interest, but has not yet been received; and the voluminous evidence, comprising more than eight hundred pages, is conflicting, as might have been expected.

The following is an extract from an order issued by the admiralty for the care and management of the boilers in the vessels of the royal navy; the entire circular has been furnished the Bureau of Steam-Engineering:

In order to protect the plates and stays from corrosion it is essential that the surfaces should be coated with some impervious substance. A thin layer of hard scale, deposited by working the boilers with sea-water, has been found to be the most effectual preservative, and therefore all boilers when new, or at any time when any of the plates or stays are bare, are to be worked for a short time with the water at a density of about three times that of sea-water, until a slight protective scale has been deposited. * * * After this, in the ordinary working of the boilers, the engineer officers in charge of machinery are to use their discretion as to the most suitable density at which the water in the boilers should be kept for the service on which the ship is employed. This density, which is in no case to exceed three times or to be less than one and a half times that of sea-water, will probably vary to some extent on different stations and under different conditions of working.

No tallow or oil of animal or vegetable origin is to be put into the boilers to prevent priming, nor for any purpose. Boilers should, if possible, be kept empty and dry when they are not at work. * * * The engineer officer must bear in mind that unless the boilers can be made thoroughly dry, it will be better to keep them entirely full of water, as a simply damp state of the boilers and stays is a most fruitful cause of decay.

Boilers are not to be used as tanks to contain fresh water for the use of the ship's company, and they are to be used as little as possible for the purpose of trimming the ship when under sail, as these practices tend to keep the boilers in a damp state, and conduce to their decay. When boilers are used for distilling water, care is to be taken that the stop-valves be quite tight, so that no steam can leak into the boilers not in use.

A late circular on the care and management of machinery and boilers, directs that Crane's mineral oil shall be used in cylinder and valve-chests, as it is not readily decomposed and possesses no acid properties; but as an additional safeguard, and to neutralize any possible acid properties in the water of the boilers, about one pound of carbonate of soda for each ton of coal used, is to be pumped into the boilers with the feed water once or twice during a watch; also that zinc slabs are to be suspended in convenient parts of the boilers as additional protectors for the iron.*

Particular directions are also given as to regular inspection of the interior of the boilers, and if signs of decay are detected, to reports of the facts with sketches showing the parts attacked, &c.

In the mercantile marine also, the system of protecting the interior surfaces of iron boilers by a thin coating of scale is practiced. Much importance is also attached to the care and management of the boilers, and greater care taken than was formerly the case in the selection of competent engineers to operate the machinery. In some lines of steamers the plan has been adopted of *not* discharging (blowing off) any water from the boilers at sea, but steaming from port to port with the water taken in before starting, and making up the deficiencies caused by leaks, &c., with sea-water, even though the density should reach $\frac{4}{32}$. The average life of usefulness of boilers in the commercial vessels of England is nine years.

PRESERVATION OF BOILERS IN HER BRITANNIC MAJESTY'S VESSELS.

To preserve the boilers in vessels laid up at the dock-yards from deterioration, two different systems have been tested. The one is known as the wet and the other as the dry system. The former consists in keeping the boilers filled completely with sea-water mixed with carbonate of soda, 25 pounds if soda ash, or 50 pounds if ordinary crystal soda be used, for every 100 cubic feet of sea-water in the boiler; this is to be dissolved and placed in the bottom before running up the boiler. The quantity must be tested by placing a piece of clean new iron in a bottle for a night with some of the mixture; if the iron rusts, more soda must be added. To secure the perfect filling of the boiler, an air-hole may be provided at the topmost point, and closed with a cock, and a small pipe for maintaining a column of water above the boiler will insure this. The dry process consists in removing all water from the boilers, to dryness, by stoves if necessary; after which depositing pans containing altogether two or three hundred-weight of dry caustic (quick) lime, lime newly burned, in the bottoms, over the furnaces, and above the tubes, the quantity depending on the size of the boiler; in addition, a sheet-iron tray of burning coal is to be placed in the ash-pit or furnace till the coal is coked, then introduced into the boiler and the latter immediately closed; this consumes much of the oxygen of the air in the boiler, and increases the efficiency of the dry lime. At least every six months, inspection

* The latest indications seem to be that the zinc should be in actual contact with the shell of the boiler, and not merely suspended from some of the braces.

is to be made, and if the lime is found to be much slaked, the pans at the bottom, which can be removed without considerably changing the air, are to be taken out and refilled with fresh lime. In addition to this, when the atmospheric dampness is extreme, light fires in the ash-pits are sometimes found to be necessary. The boilers of ships in commission are to have one or other of the above methods applied for their preservation, according as the conditions of service of each ship admit; or a pan of burning coked coal may be introduced with benefit, or, if the boilers be kept filled, soda may be added without interfering with getting up steam at any time. The wet system, although successful to a considerable degree, has the objectionable feature of keeping the boilers and surroundings damp and disagreeable, and as a consequence it has not met with favor among the engineer officers, while the treatment as applied by the dry system has received universal favor, and has been so successful as to be now generally adopted. New boilers stored in the dock-yards waiting to be placed on board vessels are, except in two locations, left out in the open weather, but care is taken to preserve them by paint, and to keep all openings and doors closed.

WATER-TUBE BOILERS.

Ever since the successful introduction of the compound engine on the Atlantic Ocean, the desire has been to extend the expansion of steam beyond the limits at which it is now worked; and in order to meet this demand, the attention of many engineers has been directed toward producing a safe and reliable steam-generator, capable of carrying steam of 100 pounds and upward per square inch. Many plans have been projected and quite a number of them tested practically at sea, some of which have been on a scale of magnitude involving large expenditures of money. Every one of these new types of boilers, practically tested, has, up to this time, proved a total failure. In England considerable professional as well as public interest has been drawn to these failures, and many discussions took place and numerous papers appeared on the subject during my stay abroad. Perhaps the most able of the papers referred to, and one which gives a clear and full account of the noted failures at sea, was produced in April, 1876, by J. F. Flannery. This paper so completely covers the ground, that I reproduce it here. It is as follows:*

The almost universal adoption of high-pressure steam at sea has brought with it much inconvenience, notwithstanding its great economy of fuel. The anxiety attending upon the construction and the working of the marine boiler is much greater now than formerly, its first cost is largely increased, the expense of repair is much more, and the duration of its life is much less. I think I will have the concurrence of the members in saying that one of the gravest, if not the very gravest and most important, of the questions now arising in connection with the machinery of steamships is the improvement of the marine boiler. And not only does the present practice leave much to be desired, but the feeling in favor of still higher pressures is checked chiefly by the difficulty of designing a boiler which will sustain them safely and efficiently. It is an axiom laid down by theory, and confirmed by practice, that the higher the boiler-pressure and the greater the ratio of expansion, the greater is the economy of fuel. Theoretically this axiom is sustained by the fact that to generate steam of the pressure of 30 pounds per square inch about 1,190° of heat are required, while to generate steam of 120 pounds per square inch about 1,218° of heat are required; that is to say, to produce steam of four times the boiler-pressure, less than $1\frac{24}{1000}$ times the heat is required. Again, a boiler-pressure of 240 pounds per square inch requires about 1,235° of heat, or $1\frac{4}{100}$ times the heat of steam at 30 pounds, and contains eight times the initial force. Practically these differences are so small, that the same consumption of coal may be taken to evaporate the same quantity of water, whatever the pressure. Theoretically, therefore, we have only to increase our boiler-pressure to reduce our consumption of

* Paper read before the Institution of Naval Architects.

fuel, were it not that two well-ascertained considerations intervene. First, we have not hitherto been practically able to construct a boiler which, under usual conditions of steamship requirements, will stand a pressure greater than is now in use; and, second, the heat of very highly pressed steam burns up the lubricant in the engine-cylinder, quickly destroying the working parts. With the second consideration, however, we have not now to deal; and if we could overcome the difficulty of making a boiler to bear even 120 pounds with safety, much improvement would be effected.

To start from the beginning, let us remember that the thickness of the shell of a cylindrical vessel containing high-pressed steam must, in order to maintain the necessary strength, be increased directly in proportion to increase of pressure and diameter, and that the thickness of the shell may, consistently with the strength necessary for any given pressure, be reduced, if its diameter be also reduced. A very usual size for modern high-pressure cylindrical boilers is, say, 12 feet in diameter, and the thickness of the shell of a boiler this diameter, double-riveted, to bear a pressure of 120 pounds per square inch, would be fully $1\frac{1}{2}$ inches. I am not aware of any examples of boiler construction where the thickness of the shell-plates has exceeded $1\frac{1}{4}$ inches or $1\frac{3}{8}$ inches, and it is almost doubtful if, considering the practical difficulties of manufacture, a perfectly good job is made with shell-plates of this thickness; at all events, it is generally accepted that to make sound workmanship with a thickness greater than this is practically impossible. On the other hand, a cylindrical vessel or tube, say 12 inches in diameter, may, for this same pressure of 120 pounds per square inch, be made, if welded, only $\frac{3}{16}$ inch thick, and still possess the same proportionate strength, or, for a pressure of 150 pounds per square inch, it would be made practically about $\frac{1}{4}$ inch thick, and this would give a factor of safety nearly three-fourths times greater than that possessed by the cylindrical boiler-shell 12 feet in diameter, $1\frac{1}{2}$ inches thick, and pressure of 120 pounds. It is practically certain, therefore, that the limit of pressure with the present type of boiler has been reached; and when we consider how much economy of fuel has been effected by increasing the pressure to its present height, and how much more economy may be effected by still further increasing the pressure, it is seen that the adaptation of an efficient high-pressure marine boiler would meet a great and increasing want.

Many look for the solution of this problem to the type known as the water-tube or tubulous boiler; that is, the system by which the cylindrical portions of the boiler subject to internal pressure are reduced in diameter and increased in number, the result being that for a given weight of material a much higher pressure of steam is carried, and, indeed, by the reduction of the necessary thickness of the shells a higher pressure of steam is carried, and more economy of coal insured than is possible with any other type of boiler. The other advantages claimed for boilers of this type are, that the thinness of the metal interposed between the fire and the water enables the heat to be conveyed more efficiently and with less waste, and also enables the steam to be raised more rapidly; that the danger of explosion is limited; that a damaged tube can be easily replaced; that greater strength proportionate to the pressure being possible in first manufacture, the wear of the metal in the boilers need not, at a later date, enforce reduced pressure, and therefore reduced speed of ships, as is the case with existing types of boilers. It is pointed out also that where boilers of the ordinary type require renewal the decks must be broken away, whereas the separate portions of the tubulous boiler may be passed down the hatchway; and, further, that a war-ship on a foreign station must either return home or to the nearest dock-yard in the event of any serious wear upon ordinary boilers, but so many spare tubes of the water-tube system might be carried as to greatly extend the period of service upon a foreign station.

*   *   *   *   *   *

The direction of the flame in the ordinary boiler is always along the faces of the heating surfaces, whereas in the water-tube boiler it is in a direction at right angles to them; the heating efficiency of flame striking the plate directly, as in this latter case, must be much greater than when sliding along the surface, and we see, therefore, that this system not only enables thinner plates conveying the heat more perfectly to the water to be used, but also directly increases the action of the flame itself. Again, the escape of the steam-bubbles from the metal surface upon which they are generated is more rapid and direct when that surface is disposed horizontally. The condition most essential to safety is beyond doubt good circulation. The heat to which a tube immediately over a fire is exposed amounts to upward of 2,000°; the specific heat of water, or its power to absorb heat, is twice that of steam, and, in addition, the evaporating water absorbs the latent heat of steam, making its receptive power many times that of steam; there is, therefore, no difficulty in arriving at the conclusion that a tube of, say, $\frac{3}{8}$ inch thickness, which is not relieved by freely-circulated water, will become hot and dangerous in a very few seconds, and it may be shown that the absence of provision for this good circulation has been the primary cause of the difficulties the water-tube boiler has yet encountered. It is noticed, on reference to the diagrams, how large a body of water is contained by the ordinary boiler, as compared with the volume of water held by the water-tube boiler, in proportion to their heating and grate surfaces.

The effect of this is that the ordinary boiler has a large reserve of water at nearly boiling-point, and is not therefore exposed to such sudden fluctuations of pressure, with their attendant evils, as the water-tube boiler. If the strength of the fires be suddenly increased in a water-tube boiler, a spasmodic rise of pressure, leading to accumulation of steam in the water-chambers, priming, loss of water, and overheated tubes, takes place, and this defect is quite absent from the ordinary boilers; and not only so, but the large reserve of water in the ordinary boiler, the whole of which above the fire-bar level must be heated to nearly boiling-point, is the cause of a much more gradual, and, therefore, more easily effected circulation of the water; indeed, the small volume of water is still further reduced by the isolated subdivision of the parts, each tube having to supply, circulate, and evaporate in a very independent manner, and with little reference to the volume of water contained by its neighbors. Clearly the smallness of the contents of a tubulous boiler is in some respects a serious drawback, although it is in other respects an advantage. The difficulty of obtaining a good circulation is from the nature of the case greater in the tubulous boiler, and the increased friction of the water due to high pressure adds to this difficulty, while at the same time it increases the necessity, and it is a matter for the earnest consideration of inventors whether artificial circulation, produced by subdivided feed-inlets or otherwise, may not after all be resorted to. This idea has been expressed to me by one of our most eminent marine engineers, and it does seem that the certainty of good circulation thus obtained would compensate for the complication of working parts. One element of danger absolutely possessed by the water-tube boiler, and not shared by the ordinary boiler, lies in the fact that the metal exposed to the direct action of the fire is at the same time subjected to the tensile strain of internal pressure. In the ordinary boiler the parts subjected to tensile strain are not acted upon by the fire, and those parts acted upon by the fire have the water surrounding them with a compressing strain. Although the crushing strain of wrought iron is less than its tensile strain, the case in point of the fire playing upon iron under tension is more dangerous, because any flaws will be stretched out, the flame will penetrate them, and promote their increase to bursting-point.

The water-tube boiler has, under several designs differing in their details, been extensively used for the generation of steam in factories, and for other stationary purposes, both in England and America; and, although some important defects exist and some unfortunate failures have occurred, still it cannot be denied that, as a rule, its results have been fairly satisfactory. Why, then, has it not come into more extensive use at sea? It may be answered, generally, that the conditions to be fulfilled on board ship are more stringent, the space allotted is more circumscribed, and the boiler must conform to its limits. A less leisurely and methodical action is necessary than in a machine used for stationary purposes, and defects of little importance on land develop into serious evils at sea. We may best elucidate this question, however, by considering some of the cases in which the water-tube boiler has been tested for marine purposes, and by analyzing the cause of failure in these cases. Several practical trials at sea have been made, and the enterprising and truly admirable spirit displayed by two steamship-owners especially cannot be too highly commended. The gentlemen alluded to are Mr. S. B. Guion, of Liverpool, the managing owner of the steamship *Montana*, and Mr. W. H. Dixon, of Liverpool, the owner of the steamship *Propontis*. The examples in these vessels being the most recent and by far the most important trials of the water-tube system, I propose to ask your attention to them first.

The *Montana* is a steamship of more than 4,000 tons, and was built for service on the Guion line between Liverpool and New York. Among other innovations, she was fitted by Messrs. Palmer, of Jarrow-on-Tyne, with boilers intended to work at 100 pounds pressure, and it was anticipated that the economy of fuel would be very great. Figs. 1 and 2 * * * will convey a good idea of the nature of the design. Each boiler was composed of tubes 15 inches in diameter and 15 feet long, sealed at the ends, but communicating with each other by vertical necks about 6½ inches in diameter. There were five horizontal, or nearly horizontal, rows of these tubes in each boiler, and each row contained seven tubes. Noticing the profile, it will be seen that to carry off the steam generated and to promote the circulation of the water each horizontal tube was supplied with two necks. As the steam must all rise to the top, the necks between the bottom tube and the one immediately above must take up the steam generated in the bottom tube; the neck between the second row from the bottom and the one immediately above it must take not only the steam generated in the second, but also that in the first row, and so on, the neck nearest the top having to convey the steam generated in the whole of the tiers of tubes below it. In addition to these tubes, there are in the uptake three steam reservoirs or superheaters 3 feet in diameter. In designing the boiler, the two vertical rows of tubes nearest the fire-door were set apart for the purpose of heating the feed-water, and the six other vertical rows of tubes farthest from the fire-door were set apart to generate steam from the feed-water so heated. In accordance with the duty thus assigned to them, the two rows of tubes vertically nearest the fire-door had the feed-pipes connected to them at the top and had small pipes

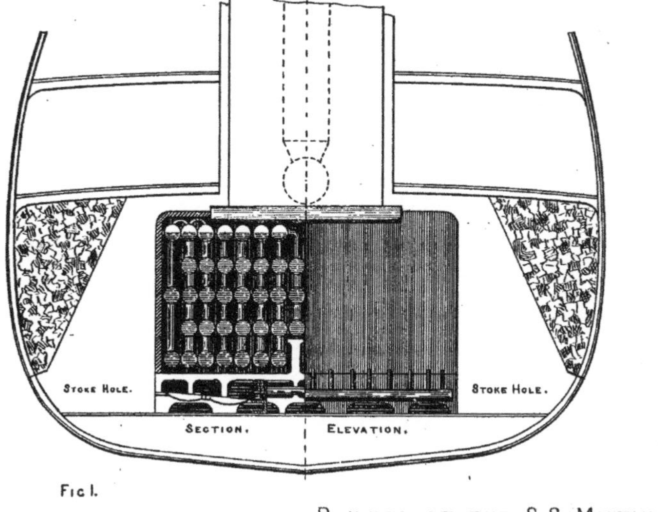

Fig 1.

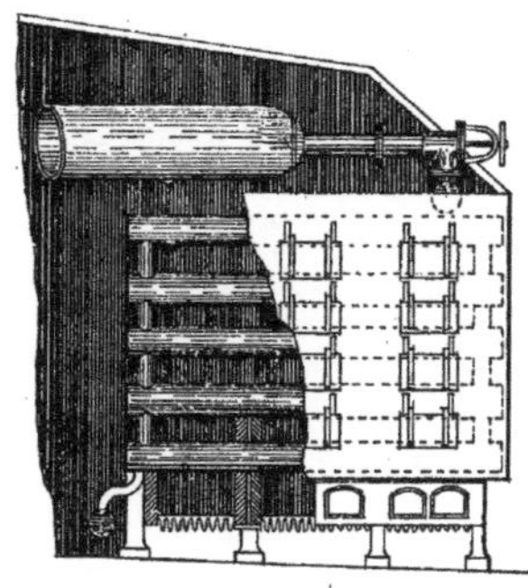
Sectional Elevation.
Fig 2.

Boilers of the S.S. Montana.

only $\frac{7}{8}$ inch in diameter fitted also at the top, to carry away any steam which might happen to be generated. It was, of course, expected that the current of the feed-water passing through these two feed-heating sections would be so rapid and that the fire under them would be so much less vivid nearest to the dead-plate that the water could not, while in them, absorb sufficient heat to evaporate. The water was delivered from these feed-heating sections into a large feed-water chamber, which supplied the other sections. The two sections designed as feed-heaters are so called, though as a matter of fact they were not feed-heaters only, but calculated to generate steam nearly as much as any of the other tubes in the boiler, and the steam had really no outlet, because the $\frac{7}{8}$-inch pipe provided for the purpose was so small as to be practically useless. The natural action of the steam generated in these sections was, therefore, to rise to the top tubes, to accumulate there, and to depress the level of the water, driving it out through the bottom pipes from all that vertical section of tubes, and finally exposing to the direct action of the fire a bottom tube filled only with steam, or steam and a small quantity of water. This is exactly what did take place, and on the first trial passage of the vessel from the Tyne toward the Mersey, five of the tubes burst. These tubes were all situated in corresponding places in the different boilers; that is, the lowest tube in the inner feed-heating section. After this experience, the connections were altered so as to join this inner feed section to the rest of the steam-generating part of the boiler, leaving the outer vertical row only to act as a feed-heater. For some few hours this arrangement gave promise of better success; but, after a day and a half's working, two more tubes, one of them the fourth from the front in the bottom row, gave way; and the owners, completely discouraged by these failures, and taking into consideration also the fact that the boilers were heavier and more bulky than boilers of the same power of the ordinary kind, decided to condemn them and replace them by ordinary cylindrical boilers. The cause of the failure of the sections devoted to feed-water heating appears sufficiently obvious, and as the separation of these sections was a special feature in this example, and is not likely to be repeated in other water-tube boilers, we may dismiss it from further consideration. But the bursting of the other tubes is more serious, and appears to more closely affect the general principle of water-tube boilers. A very able article, from which some of the particulars given above have been drawn, appeared in the Nautical Magazine on the subject of the *Montana's* boilers, and it is there suggested that the cause of the explosion of the inner tube on the bottom row was the absence of a sufficiently large steam connection between the top or steam-holding chambers to counteract the inequality of pressure in them. It will be noted that the feed-pipes for all the sections are common to one feed-water chamber, and supposing any inequality of steam-pressure to exist, the water would, of course, be forced from the chamber containing the highest pressure to those containing a lower pressure. The steam-pipes connecting the upper chambers were 2 inches in diameter, and the area of the fire-grate under each section was about 7$\frac{3}{4}$ square feet. A very simple calculation shows that these 2-inch pipes were very small, if any large difference in pressure was to be equalized. Another explanation, and one which appears quite reasonable, was advanced by Mr. John Watt. He calculates that the grate surface, 7.75 feet, acting upon the water in each bottom tube, is equal to the evaporation of 12.83 pounds of water, or 78 cubic feet of steam at 50 pounds pressure, per minute. Now, the only orifices through which this quantity of steam may escape are the before-mentioned pair of 6$\frac{1}{2}$-inch vertical necks, which have each an area of 33.18 square inches; one of these would probably be a downcast for the water, and the other, at the higher level, being an upcast for the steam, the steam would therefore rush through this neck at the rate of nearly 340 feet per minute, a speed so great that the assumption that the steam would drag the water upward from the bottom tubes, leaving them exposed to the fire, does not seem exaggerated. Whichever of these explanations be correct, it is evident, in the light of the experience now gained, that the failure of these tubes in the *Montana's* boilers proceeded from easily-preventable causes, because in a future design the feed-water arrangement could be omitted, and the steam connections to equalize the pressure in the different sections could be enlarged, the vertical necks for the escape of the steam could be much increased in area, and it is possible that if these alterations had been effected upon the *Montana*, a result very nearly approaching success would have been obtained. Indeed, it was most unfortunate, in the interests of science, that no further experiments were made with this vessel; she was condemned after a series of trials extending altogether over not more than five or six days.

The *Propontis* is a steamer of about 2,000 tons, belonging to Mr. W. H. Dixon, of Liverpool. In 1874 she was fitted by Messrs. John Elder & Co. with boilers on the water-tube principle, under Rowan & Horton's patent. * * * They are four in number, having one chimney, and each boiler consists of seven horizontal cylindrical vessels, connected together by vertical tubes 12 inches in diameter, supplemented by numerous pipes 2$\frac{1}{2}$ inches in diameter; the space for steam in the upper vessels being increased by four diagonal domes. To further guard against priming, tubes are carried directly from the steam-dome down to the lowest vessels, so that any water carried up by the steam will be broken from it in the upper chamber and fall down again to the lowest level; the feed is also delivered at the bottom chambers. Baffle-plates are fitted to circulate the flame among the tubes. * * * The lower or water chambers were connected, but no connection was made between the upper or steam chambers; the importance of this omission is obvious, and was discovered in the working of the boilers, as will be seen further on. The working-pressure intended was 150 pounds. While the defects in the boilers of the *Montana* exhibited themselves after a very few hours' work, the boilers of the *Propontis* continued their promise of success for a much longer period, and, indeed, were exhaustively tested at sea for a time extending over several months. The first voyage of the *Propontis* was from Liverpool to the Black Sea and back, and very careful observations of the working of the boilers were made. The consumption of coal, taken during a run of 10 hours, was 1.54 pounds per indicated horse-power per hour; the pressure of steam being easily maintained at 130 pounds to 140 pounds. It is by no means certain that the full advantage of the *Propontis's* boilers in point of economy is represented by the above figures, because the baffle-plates for the deflection of the flame were not arranged just in the way that experience afterward showed they should be arranged, and consequently much unconsumed gas escaped up the chimney. So far the career of the tubulous boiler in this ship was fairly satisfactory and sufficiently promising; but before long, evils which had been gradually accumulating, developed themselves in an unpleasantly conspicuous manner. * * * Some of the smaller sized vertical tubes terminate in S-bends at the points where they join the horizontal vessels. The peculiarity of this bend rendered it impossible to clean the tubes from scale, so that salt water could not be used; indeed, precautions of an unusually perfect character were taken to admit pure distilled water only to the boiler, and the action of this distilled water, unmixed with any salt, was just what might have been expected: a corrosive action was set up in the tubes; some parts of the internal surface were quite untouched, while other parts close to them were perforated by circular blotches of corrosion of ½ inch to 1½ inches diameter. These were continually giving out, allowing the steam and water to escape upon the fires. The defects were temporarily remedied by drawing the fires and binding a ligature round the injured tube over the hole; and these little explosions were of such frequent occurrence that one of the four boilers was almost constantly disconnected.

In the early part of 1875 the boilers underwent a thorough repair, being supplied with some 300 new tubes, and were tested by water-pressure to 275 pounds. After this a small quantity of sea-water was allowed to enter the boilers to make up the waste, and after completing another voyage to and from the Levant, her boilers were opened up and found to be coated with a slight scale. In September of last year she commenced another voyage, but on the third or fourth day the wing-chamber of the forward starboard boiler burst, injuring two men, the pressure at the time being 150 pounds. The rent was nearly two feet long, and was in the solid plate near the lower side of the tube, and about its middle length. At Lisbon this was repaired by riveting a patch ⅜ inch thick. Leaving Lisbon, she had been but a very few days out before another explosion took place, the pressure at the time being only 105 pounds. The fracture was in the wing-chamber of the starboard after boiler, and was again in the solid plate, at the lower side of the chamber; the rent measured 12 inches by 7 inches. The diameters of the tubes which gave way were 21 inches, the plate ⅜ inch thick, and the bursting pressure of this would be 1,200 pounds, provided the plate were cool and uninjured. Putting into Algiers, the boilers underwent an inspection by French engineers, and in accordance with their advice the furnaces were partially bricked up and the pressure reduced to 60 pounds. It may be here remarked that the opinion of the French engineers that a reduction of pressure meant, in the case of these boilers, greater safety, was very questionable. It may be a paradox to say that the lower the pressure the greater the danger, but in the case of water-tube boilers, under certain conditions, that is the fact. Let us remember that a cubic inch of water will make a cubic foot of steam of the same tension as the atmosphere, and that the greater the pressure the less is the bulk of any given quantity, say a pound weight of steam. Now, as the power of a boiler may be measured by the number of pounds weight of steam that it evaporates, the *Propontis's* boiler would evaporate no really greater quantity of steam at 60 pounds pressure than it cou'd evaporate at 150 pounds pressure; but whatever steam it did evaporate would at 60 pounds pressure occupy more than twice the space it would occupy if at 150 pounds pressure. Again, evaporation cannot take place without circulation, and circulation is the mutual changing places of the steam as it is generated and the water about to be evaporated; if, therefore, the steam occupies twice the volume, then there will be twice the space for it to travel through to circulate or change places with the new water; and as one dangerous defect in the *Propontis* was insufficient circulation, it is seen that by lowering the pressure the French consulting engineers may not at all have increased the safety of the boiler. It is a most interest ng investigation, but one which I fear can only be conducted practically, to earn how far the increased friction, varying directly as the pressure, compensates for the decreased volume of the evaporating steam. The vessel was towed from Algiers to Gibraltar, and then the extra fire-bricks were taken out, a baffle-plate between the chambers and the fires was fitted, and test-cocks were placed on each water-chamber. Under the protection of the baffle-plates, and the careful watching of the water-level made possible by the test-cocks, the vessel steamed home with three boilers.

Upon final examination it was found that the vertical tubes were bulged and distorted in several places, and all the horizontal tubes exposed to the direct action of the fire were more or less bulged upon their lower sides, thus affording clear evidence that in most of the water-chambers of the boiler steam was sometimes collected. These bulgings or partial injuries to the boiler evidently arose from defective circulation, and the bursting or more serious injuries were the result of the unequal pressure of steam in different sections. There was no pipe connecting the steam-chambers of the two boilers, and although the steam-chambers were not connected, the water-chambers were, so that any increase of steam-pressure in one boiler simply had the effect of driving the water out of its own chamber into the other boiler. Not only was the unprotected plate exposed to the fire, but the abnormal increase in the water-level of one chamber held out every inducement to priming. This serious structural omission explains the bursting due to overheating, and at the same time the priming which occurred. It is very remarkable that the partial absence of a similar pipe contributed to the failure of the boilers of the *Montana*. It has been suggested that the bursting of the tubes in the *Propontis* arose not so much from defective circulation of the water as from corrosion and consequent weakening of the plates; this is the fact as regards the small circular perforations, but the same explanation does not apply to the longitudinal rents, and is not borne out by an examination into the character of the larger fractures. The longitudinal rents which caused the loss of life occurred in portions of the tubes practically uninjured by corrosion, and it was found that the fracture showed only a slightly reduced thickness of metal, and that at a distance of about 1 inch from the lip of the rent the plates resumed their original thickness. The bursting pressure of the rent tube was, as already stated, 1,200 pounds per square inch, and the uncorroded appearance of the fractured portions points to the conclusion that defective circulation and overheating caused the explosion. The plates, if hot, would stretch a little before giving way, and the confined space over which the thinness extends seems to show that it should be traced to stretching only.

The *Birkenhead* is one of the large passenger ferry-steamers running between Liverpool and Birkenhead. She was built in 1872, and fitted by the Patent Steam-Boiler Company, Birmingham, with tubulous boilers on Root's patent. The boilers * * * were each divided into fifteen sections, each section consisting of eight horizontally-inclined tubes 11 feet long by 6 inches in diameter ; these tubes were at the ends connected to small cellular vessels. Communication between these vessels in their respective horizontal tiers was effected by bent pipes bolted to them * * * * * The working pressure was 40 pounds, although in designing the boiler the pressure was intended to be 60 pounds. * * * It is only fair to the manufacturers to say that they labored under the difficulty of very confined head-room; these vessels are of limited draught, and as a flush deck is necessary for the passenger traffic, the vertical space at the disposal of the designer was quite insufficient to give boilers of this type a fair chance of success. The boilers were for the first few months of their existence worked with impure water, and considerable trouble from overheated tubes was the consequence. A fresh-water tank was afterward fitted, and with an improved result, but still it was found that the tubes in the bottom row were continually bulging and giving out from overheating, and the vessel was frequently off duty for repair. The exit of the steam was no doubt retarded by the turnings and counter-turnings it underwent before reaching the surface. The greater vertical distance through which the steam rises from the surface of the water before rushing to the engine, the less is the tendency to carry water away with it, not only because there is more time given for the water to fall back from the ascending steam, but because the pull caused by each stroke of the engine is less severely felt at the bottom tubes if they are some distance below the steam outlet. * * *

In this design a very small height was between the stop-valve and the bottom tubes, and the engines being paddle-engines with large cylinders and slow stroke, there is no doubt that the tendency of the water to forsake the bottom tubes was greatly enhanced from this cause, although primarily due to imperfect circulation. This defect would also be increased by the stoppages and the lying-to of the vessel in her work as a ferry-steamer allowing accumulations of the steam-pressure. Not only so, but the deficient room rendered it impossible to get in so much boiler-power as was afterward thought necessary, and the manufacturers allege that consequent forcing of the fire contributed in no small degree to the failure of the tubes. The sister ships to the *Birkenhead*, working on the same station and with very similar engines, but boilers of the ordinary type, offered an excellent standard of the average consumption of coal, and the saving in the *Birkenhead* by comparison with one of the boats amounted to 17 per cent., and by comparison with the other to 25 per cent. This result is the more remarkable when it is remembered that the pressure was only 40 pounds, and that the expansion in the engines of the *Birkenhead* could not be greater than in the sister ships; the comparative economy was therefore entirely due to the more effective arrangement of the parts, the thinness of the metal in the tubes, and the direct action of the flame upon them favoring the evaporation; had the pressure been higher and the expansion greater, the economy would no doubt have been largely increased. Notwithstanding the excellent result in point of economy, the ferry management condemned the boilers after nearly three years' work, on account of the continual breakdowns, and the vessel is now fitted with boilers of the ordinary type made by Messrs. James Taylor & Co. The cause of the comparative failure of these boilers has already been indicated; the twisted passages for the exit of the steam and the low vertical height, added to heavy firing, caused the bottom tubes to be occasionally full of steam, and being exposed to the full action of the fire they were quickly destroyed.

Another vessel fitted with boilers on a Root's patent, and of very similar design, but still more contracted passages for circulation, was the steamship *Malta*, of 2,000 tons, belonging to the Merchants' Trading Company, Liverpool. They were in this case found to give off steam with great rapidity, and steam could be raised in them in about half the time necessary in boilers of the ordinary type. The working pressure was 50 pounds. The boilers went on about twelve months, during which time the vessel visited the Mediterranean and the Baltic. Very little priming took place, but the same defect as in the other cases developed itself, the circulation of the water was insufficient, some of the bottom tubes nearest the fire bulged and gave out, and the boilers were finally taken out to convert the vessel into a sailing-ship. The engines were very old before the Root boiler was fitted, and their worn-out condition contributed to the owners' decision to alter the vessel. I am informed that many of these Root boilers, of the same manufacture, are working successfully on land, and where large space may be given and leisurely delivery of the steam is possible, they have been found highly successful.

* * * * * * *

Boilers were fitted to the steamships *Amalia* and *Palm* under Ramsden's patent. These boilers worked very satisfactorily for about four years, but at the end of that time it was found that the construction of the boilers not fully providing means for the removal of the incrustation, the plates had seriously deteriorated, and the boilers were consequently removed.

Trials were made of the Howard patent safety-boiler in the steamships *Fairy Dell*, *Meredith*, and *Marc Antony*. These boilers were all constructed about the same date, 1870. Their working on board ship in all these cases was quite unsatisfactory; the lower tubes burst, others leaked, and very great priming took place. These results were traced to the same general causes that led to the difficulties experienced with the *Birkenhead* and *Malta*, and it is unnecessary to discuss them in detail.

After discussing so many failures of the water-tube system, I cannot close this paper without describing the most recent working example of the tubulous boiler for marine purposes. One of the earliest advocates of the tubulous boiler is Mr. John Watt, of Birkenhead, and his system is shown on the diagrams. The diagram represents the boiler of the steam-flat *Gertrude*. "The boiler consists of a series of inclined tubes connected at their ends to rectangular water-chambers. To the top of the chamber is connected a steam-receiver, from whence the steam is taken to the engines. The rectangular chambers are stayed, like the ordinary locomotive fire-box, by means of stays through the door and tube-plates, one end of each stay being left sufficiently long to enable the tube doors to be secured. The tubes are in diagonal rows, and the course of the flame is in zigzag direction among them." A very brief consideration of the natural action of the steam and water in this boiler will show that it possesses features superior to some of the boilers already described. As the steam is generated in the tubes, the angle at which they are inclined will facilitate its ascent to the upper rectangular water-space B, and once there, its ascent to the steam-receiver is easy and certain, its upper passage being unobstructed; at the same time the water to take the place of the steam generated contained in the lower rectangular chamber A, will, by a natural action, ascend from it, and so a good circulation is secured. The necessary feature in a boiler of this arrangement is a separate outlet for the steam and a separate inlet for the circulating water; any contention between the ascending steam and the descending water must be fatal.

The shorter and wider the tubes are made the greater will be their safety, because the amount of evaporation will vary as the heating surface of the tube. Now, the surface of the tube increases directly as the diameter, but the volume contained increases as the square of the diameter, and this fact goes far to explain the danger of long narrow tubes, the length still further increasing the difficulty of the exit of the steam. In the case of the *Propontis*, some of the smaller vertical pipes were so far attenuated as to be 8 feet long and 2½ inches in diameter. I am informed that the *Gertrude's* boiler has worked for some months, and with so perfect a circulation that no deposit or scale has been formed, although partially salt water is used, and the engine is not fitted with a surface-condenser. Another design, strongly advocated by the inventor, is that of

WATT'S BOILER.

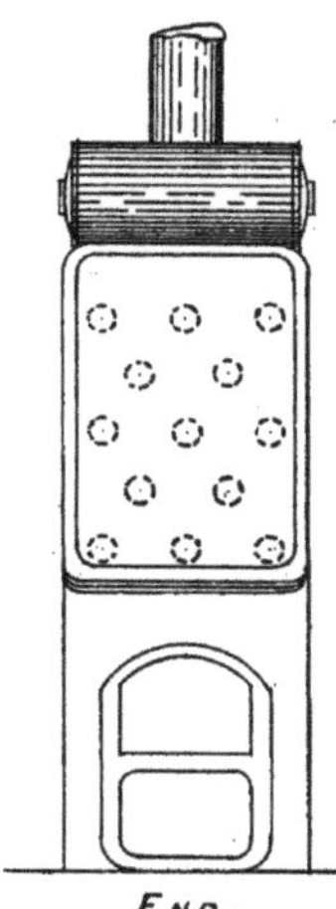
END.

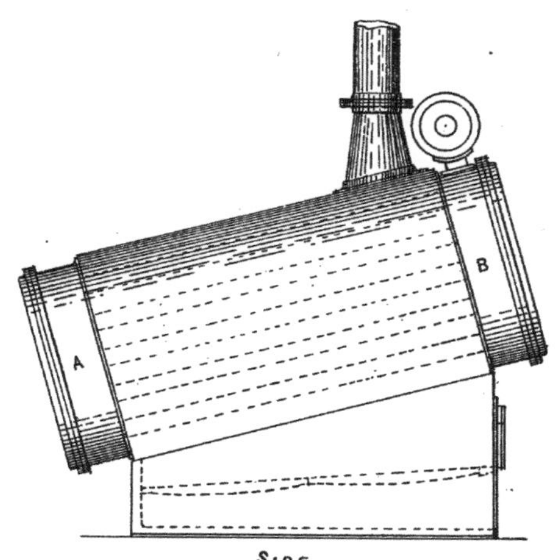

SIDE.

S.S. GERTRUDE.

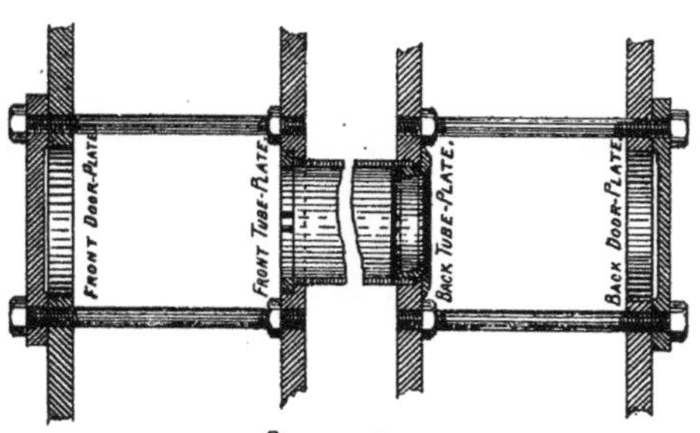

DETAILS.

Mr. Wigzell. * * * This is a compromise between the tubulous boiler and the boiler of ordinary type; the flame will surround the inclined cylinders, and the other cylinders being perforated with tubes, it will pass through these on its way to the chimney, its exit through them being made compulsory by blocking up the lower spaces between the cylinders by diaphragm plates. No boilers of this design have yet been tried at sea, though the inventor assures me that most satisfactory results are obtained by its use for land purposes.

We set out with the statement that, so far as present lights enable us to judge, greater economy in the use of coal at sea can only be obtained by increased boiler-pressure; that cylindrical vessels of small diameter are the only practical generators of high-pressure steam; that up to the present time no water-tube boiler on a large scale has given perfectly satisfactory results at sea; and that the practical trials already made are in themselves sufficient to teach useful lessons, and perhaps to indicate to some extent the practice of the future. If there is some specific feature in the design of the boilers of the above-mentioned ships which is common to them all and has contributed to their failure, then a valuable lesson will have been drawn from them. Insufficient circulation of the water has been more or less a characteristic of the above cases, and to get at the root of the matter it is necessary to discover why the circulation has been insufficient. In the ordinary boiler the reservoir for steam is vertically above the surface upon which it is generated, and no obstruction to its ascent is offered except the friction through the superincumbent water. Looking at the diagrams, it will be seen that the steam generated at certain portions of the boiler surface, in all the examples of water-tube boilers, must travel a greater or less distance horizontally, or nearly horizontally, before it can make its movement in a vertical direction, and in some of the cases the steam encounters repeated obstructions to the horizontal deviations from its continued ascent through the water. No difficulty of this kind is to be found in the ordinary marine boiler; the steam as it is generated ascends by a free and natural action through an unobstructed body of water, and it is only when steam can move in a vertical, or nearly vertical, direction that it has any tendency to circulation at all. To design a boiler in such a manner that the steam shall move in a horizontal direction for ever so short a distance, is to lay down a wrong principle, because the horizontal movement can only be borrowed from the force developed in those particles of steam which are ascending, and which jostle, so to speak, their neighbors along the horizontal portion of their journey. Now, the greatest obstruction to circulation at high pressures, apart from the arrangement of the tubes, is the great friction. An arrangement which at a lower pressure would give fairly good results will quite refuse to produce proper circulation when the friction is seriously increased by high pressure, and it is evident that a high-pressure water-tube boiler must, if it does not depend on artificial circulation, have no portion of its interior so arranged that a horizontal flow of the steam through the water is necessary. The features which the successful marine high-pressure boiler must possess are, a clear vertical lead for the steam as it is generated, separate downcasts for the water and upcasts for the steam, and accessibility for the removal of incrustation. Whether the designs introduced, but not yet fully tested, may be found to give better results than some of those we have named is a problem yet to be solved.

* * * * * * *

So far as we may now judge, it does appear that the water-tube boiler, designed scientifically and in accordance with the experience the past failures have taught, holds out prospects of much more satisfactory and profitable work than has yet been obtained by its use at sea.

The most extensive and costly of the boiler experiments represented in the foregoing paper was the one made by the Guion line of steamers. In addition to the boilers fitted on board the steamship *Montana* and tested at sea, the *Dakota*, a sister ship, was also fitted with the same kind of boilers, and although those of this ship were not quite completed when the others in the *Montana* were condemned, they too were at the same time removed to the scrap-heap. The cost of the boilers for each of these ships was $175,000, and it is believed that the total cost to the company for this experiment has been not less than $486,000. It has been estimated that during the seven years ending in 1876, the total cost in Great Britain resulting from the failures of boilers designed to be worked at sea with pressures of upward of 100 pounds per square inch has nearly reached the enormous sum of $1,000,000.*

* For a description of sectional boilers used in Europe, see report, volume III., on the International Exposition held at Vienna, 1873, by Prof. R. H. Thurston, A. M., C. E., late engineer officer, United States Navy.

BOILER EXPLOSIONS.

It is well known that in several countries there are associations which insure boiler-owners from damage arising from boiler explosions, by a thorough system of periodical examination of their boilers, and directing necessary repairs to be made. These bodies publish reports from time to time of all accidents to boilers, detailing the causes which led to them, and also giving accounts of experiments made by themselves and others to test various practical points relative to the strength of boilers. The following is a synopsis of the proceedings at a meeting of one of these useful and important associations:

At the annual meeting of the Manchester Steam-Users Association, held at the Manchester town-hall [1876], there were exhibited some large boiler-plates showing the rents that had been formed by hydraulic pressure in a full-sized mill-boiler of the *Lancashire* type, constructed entirely for experimental purposes. The plates shown contain the fractures developed. The experiments on this boiler, which have just been brought to a conclusion, have extended over about eighteen months. The boiler has been burst eleven times, being substantially repaired after each bursting, the entire outer shell at one time being remade. These experiments have led to interesting and important results, of which the following is an official summary. They have shown the weakening effect that steam-domes have upon cylindrical boilers, and the importance at high pressures of having manhole mouth-pieces, as well as all the fitting blocks, of wrought iron instead of cast. They have also proved that the furnace-tubes when strengthened at the ring-seams of rivets with encircling hoops or flanged joints, and the flat ends, when suitably stayed, are stronger than the cylindrical portion of the shell, which rends at the longitudinal seams of rivets. In the recent fatal explosion at Rochester on board the steam-tug *Prince of Wales*, the flat end failed before the cylindrical portion of the shell, showing that the boiler was malconstructed. The experimental boiler, which was 7 feet in diameter, and made of plates 7-16ths of an inch in thickness, rent in the outer casing at a cast-iron manhole mouth-piece at a pressure of 200 pounds on the inch; at a machine-made, single-riveted longitudinal seam of rivets, at a pressure of 275 pounds on the inch; at a double-riveted longitudinal seam of rivets at a pressure of 300 pounds on the inch when hand-made, and at a pressure of 310 pounds on the inch when machine-made. The experiments showed the superior tightness as well as strength of double-riveted to single-riveted seams, and of machine-work to hand-work. The factor of safety adopted by the association, and which they find efficient, is 4 to 1. It had been calculated that the bursting pressure of the boiler would be 300 pounds, which has been verified by the experiments. Boilers of the dimensions just given are guaranteed by the association at a working pressure of 75 pounds on the square inch. The plates of which the boiler has been made are about to be tested in an accurate machine as regards cohesion and elasticity, so as to check the results obtained by water-pressure. When these are completed the entire results will be given to the public.

The labors of the association have conclusively shown that steam-boiler explosions are not accidental nor mysterious, but that they arise in the great majority of cases from the use of boilers unfit for work.

REPORT ON EXPLODED BOILER OF THE THUNDERER BY THE CHIEF ENGINEER SURVEYOR TO LLOYD'S.

In consequence of a request made by Messrs. Humphrys, Tennant & Co. to the committee of Lloyd's Register, that I should form one of the committee of experts to investigate the cause of the explosion of one of the boilers of this vessel, I received instructions from the secretary to accompany the various gentlemen deputed by the coroner, the admiralty, and the manufacturers, to inquire into the cause of this accident.

The investigation commenced on the 26th July [1876]. * * *

The *Thunderer* is an armor-clad turret-ship, built at Pembroke about 1870. The engines, which are constructed to indicate between 5,000 and 6,000 horse-power, were made about the same time. They receive their steam from nine rectangular wet-bottom boilers, of the form at that time adopted in Her Majesty's navy, the boilers having 33 furnaces in all. They are arranged in a fore-and-aft direction on each side of the vessel, and are situated in two stoke-holes, there being four boilers in the after stoke-hole and five in the forward one. The boilers are all connected to the main steam-pipes by screw-down stop-valves, situated on the tops of the boilers, and so arranged that each can be shut off from the others at will. Each boiler is also fitted with a smaller stop-valve connecting it to a system of steam-pipes, so that each or any of them can be used for supplying steam to the numerous auxiliary engines fitted for various services in all parts of the vessel.

Properly-constructed steam and water gauges, feed-cocks, blow-off cocks, and other necessary appliances are fitted to each boiler, together with a sufficient number of safety-valves. These safety-valves are inclosed in cast-iron chests bolted on the fronts of the boilers, and have internal pipes or pockets fitted so that the steam is taken from the highest part of the boiler.

The valves themselves are made on the ordinary direct-weighted principle, with solid spindles, and are guided at the bottom by three feathers working in the valve-seats. Each valve is loaded to a working pressure of 30 pounds per square inch. They are such as are generally fitted in Her Majesty's navy, and have nothing novel about their construction.

The steam-gauges are on the ordinary Bourdon principle, graduated to 35 pounds per square inch, and are fixed in conspicuous places on the fronts of the boilers.

It appears the vessel was tried under steam several times at Pembroke, afterward was steamed around to Portsmouth in charge of the admiralty officials, and has subsequently been under steam on several occasions.

On the 14th July it was arranged to make the official trial. The vessel lay at Spithead, fires were lighted at 10 o'clock, steam was slowly raised in all the boilers, and about 1 o'clock the engines were started for a preparatory run before going on the measured-mile full-power trial.

In about eight or ten minutes afterward the foremost starboard boiler in the after stoke-hole exploded. The vessel was towed into Portsmouth Harbor and the stoke-holes were strictly guarded until the official inspection took place. I was present at the first inspection and have been present at all subsequent examinations.

On examining the exploded boiler, it was found that almost the whole of the front plates above the smoke-box doors had been forced away, and lay, together with the smoke-box doors and other *débris*, in a distorted mass on the stoke-hole floors. The plates themselves were torn through the first seam of rivets above the smoke-boxes for almost the whole length of the boiler, rending in a diagonal direction through the solid pla e at the manhole, and again tearing through the seams of rivets connecting them with the crown and other pa ts of the shell. The front of the uptake was bulged, torn, and driven back against the back plates. The plates were drawn over the stays supporting these flat surfaces. The T-iron fastenings to the stays above the uptake plates were sheared through the pinholes. The top of the boiler immediately over the uptake was bulged upward. The sides of the smoke-boxes below the ruptured parts, together with the flat surfaces in the combustion-chambers, and two of the furnace-crowns, were bulged between the stays. Altogether the boiler showed decided symptoms of having been subjected to excessive pressure.

There were no signs of shortness of water.

The workmanship appeared to be of the best description, and from the proof-mark on the front of the boiler, it had been tested in November, 1870, by hydraulic pressure to 60 pounds per square inch.

Both stop valves on this boiler were found to be shut, and the steam-gauge was entirely destroyed.

The safety-valve chest lay on the stoke-hole plates in exactly the position it had fallen. It contained two valves, each 5¼ inches in diameter, one of which could be lifted by means of a screw beneath it, while the other is inclosed and entirely out of the control of the engineer. The cap over the spindle of the hand-lifting valve was carried away, and the top of the spindle broken off.

The stop-valves being closed, it was evident that none of the steam generated in this boiler, from the time the fires were lighted to the time of the explosion, could have gone to the engines. It must therefore have been accumulating pressure in the boiler all that time, or escaping by the safety-valves. The safety-valves were stated to have been loaded to 30 pounds per square inch, and the pressure-gauge was said to have been observed out of order more than an hour before the explosion occurred.

It was considered very important to make a minute examination of the remains of the safety-valves, to see if there was any evidence of their being jammed or otherwise not in working order. The chest was opened in the presence of all the inspectors, and there were no such things as wedges, stops, or other means for jamming the valves found. Both valve-spindles were found broken and bent, the valves were perfectly free in their seats, but the feathers of these valves were battered and distorted in such a manner as to destroy all evidence of their condition at the time of the explosion.

If the safety-valves were not jammed or set fast, the boiler must have exploded owing to structural weakness; and in order to ascertain as far as possible whether it could have arisen from this cause, the whole of the other boilers, with the exception of one small boiler in the forward stoke-hole, were tested by hydraulic pressure to from 60 pounds to 65 pounds per square inch. They were quite tight, and after being carefully gauged before, during, and after the test, they showed no signs of weakness.

Strips of plate were cut from the ruptured parts, and tested by the testing-machine in the dock-yard. The iron was found to be above the average strength of iron used in boiler-making, the mean results of these experiments showing that it would stand a tensile stress of 21 tons per square inch of section.

As a guide to the pressure which this boiler was capable of bearing, a chamber was constructed to represent to some extent the part of the boiler which gave way. From its form it did not, in my opinion, give exact information on that point, seeing that the side plates of the uptake, which very materially add to the strength, were omitted, and consequently the whole strength of the structure was reduced. It was burst at various pressures under four or five different conditions, and although it did not exactly represent the exploded part of the boiler, the results of these experiments show almost with certainty that the pressure within the boiler at the moment of rupture could not have been less than 100 pounds per square inch.

The whole question of the cause of this explosion therefore centers itself upon the condition of the safety-valves at the time of the explosion, and as they were so destroyed by the accident as to make it impossible from their appearance to speak of their condition at that time, it was deemed advisable to make an examination of all the safety-valves of the other boilers. This examination was made, and they were all found to be loaded to about 30 pounds per square inch. They were free to move, and to all appearance were in a workable condition.

The feathers of safety-valves made on this principle require to be an easy fit in their seats; if too slack, the rolling of the vessel shifts them on their faces and causes them to leak. In this case they appeared to be a good working fit, and it was suggested to try them under steam. One of the valve-chests containing two valves was accordingly attached to a land-boiler in the dock-yard. It was found that when steam was raised, one of the valves, although loaded to only 30 pounds per square inch and quite free, when cold, did not lift until the pressure reached 52 pounds per square inch, but after blowing a short time and uniformly heating the cast-iron chest, it rose and fell at about the working-pressure.

This conclusively proved that the expansion of the feathers of the brass valve was sufficient to make them grip or set fast, when fitted so close as this valve was fitted.

Although this single valve did stick under this trial, the whole of the others were tried afterward under similar conditions and were perfectly free; the experiment therefore affords no certain evidence of the exact condition of the safety-valves on the exploded boiler at the time of the explosion. But as there can be little doubt of their inoperativeness, it can be inferred that had they been a little tighter than the valve that was tried and found to stick, or if they had been neglected and scale or dirt allowed to get between the working surfaces, the expansion of the valve would have caused them to jam or set fast, and allow the pressure to accumulate until it overcame the strength of the boiler.

The facts of the case may be summarized as follows:

1. The stop-valves were shut, so that the steam generated in the boiler could not escape to the engines.

2. The steam-gauge was discarded as being out of order, and therefore it did not make known any abnormal pressure in the boiler.

3. The fires were burning briskly, and therefore the pressure must have been rapidly increasing in the boiler.

4. The boiler presents unmistakable symptoms that it burst from excessive pressure, and not from faulty construction.

Under these circumstances, I am of opinion that the safety-valves of the boiler in question at the time of the explosion were set fast in their seats, and that the cause of the explosion was excessive pressure, owing to the stop-valves being shut and there being no outlet for the steam which the fires were generating.

Having, therefore, arrived at this conclusion, I think it my duty to urge the necessity of the greatest care being taken (1) in the designing of safety-valves, and (2) in the absolute necessity of their being frequently examined. There ought to be a certain means of preventing any increase of pressure beyond the stipulated amount.

Since this explosion I have carefully considered this subject, and, with a view to reducing the chances of safety-valves sticking or otherwise being inoperative, I would suggest—

1. That the valve with a central guiding-spindle is less liable to set fast by unequal expansion than a feather-guided valve.

2. That the spindles which support the weights of these valves should be detached from the valves themselves, or that a guide should be fitted between the valve and the weights, so that any inclination of the weights caused by the motion of the vessel should not tend to cant the valves and make them grip in their seats.

3. Easing-gear should be fitted to all the valves, and arranged to lift the valves themselves.

4. That all safety-valve spindles should extend through the covers, and be fitted with sockets and cross-handles, so arranged that it would be impossible to add any

extra weight or to control their action, but at the same time allowing them to be lifted and turned round on their seats, and their efficiency tested by the engineer at any time, whether the steam is up or not.

I would, in conclusion, add that loading safety-valves for marine purposes by means of dead-weights is now in the merchant service almost entirely superseded by springs.

WILLIAM PARKER,
Chief Engineer Surveyor to Lloyd's Register of Shipping.

(Lloyd's Register of British and Foreign Shipping, 2 White Lion Court, Cornhill, London, August 16, 1876.)

Of the four several official reports on this explosion the above is given as the most practical.

BOILERS OF THE MERCANTILE MARINE.

Through the kind attention of the general inspecting engineer of Lloyd's, I had the privilege of observing the practice and inspection applied to the boilers and machinery of the commercial marine of Great Britain. Steamers were visited after long voyages; some vessels in which the machinery and boilers were comparatively new, others having boilers in different stages of deterioration, and still others in which the machinery was undergoing construction. All that relates to the corrosion of boilers will be found already under that head in this report. As to the kind of boilers now employed, no material changes have taken place in the last six years, but the old type of box or wagon shaped boiler, so familiar to the marine engineer of former days, is no longer seen. It is obsolete, reckoned among the inventions of the past, and years hence will probably be shown in the South Kensington Patent Museum near the marine side-lever engine and numerous other obsolete curiosities stored in that interesting institution.

The pressure of steam carried at the present day on marine boilers varies from 60 to 70 pounds per square inch, and the boilers used to generate it have shells either cylindrical or elliptical. More care is exercised than was formerly considered necessary in the details of construction, and better workmanship in fitting and riveting the parts together.

Composition tubes are being used to a considerable extent where iron tubes were formerly employed. Safety-valves, having levers with loaded weights attached, have, in consequence of the higher pressures used than formerly, and of the difficulty with them, arising from the pitching and rolling of the vessel in rough weather, been abandoned, and spring safety-valves substituted in all steamers. The valve adopted generally for boilers, both of commercial and naval vessels, is known as Adams's.

In this connection, it may be interesting to quote from Lloyd's Register a section relating to the designing, building, testing, and fittings of boilers; also their formulæ for the strength of cylindrical shells of boilers and of circular flues or furnaces, and the tests required for boiler plates.

Lloyd's Register of British and Foreign Shipping.

[Extract from the rules of the society.]

MACHINERY AND BOILERS OF STEAMSHIPS.

SECTION 73. With respect to the boilers and machinery, the owners are required to submit them to the inspection of the society's engineer surveyors, who will furnish a report to the committee describing their state and condition. * * * The committee will thereupon grant a certificate, and insert in the registry-book the notification "Lloyd's MC." (*in red*), indicating that the boilers and machinery have been inspected by the engineer-surveyors, and certified to be in good order and safe working condition. * * * * * * *

The society's engineer-surveyors are to examine the plans of the boilers, and approve of the strengths for the intended working pressure.

Any great novelty in the construction of the machinery or boilers to be reported to the committee.

The boilers, together with the machinery, to be inspected at different stages of construction.

The boilers to be tested by hydraulic pressure in the presence of the surveyor to twice the intended working pressure.

Two safety-valves to be fitted to each boiler; if common valves are used, their combined areas to be half a square inch to each square foot of fire-grate surface; if improved valves are used, their efficiency to be tested under steam in the presence of the surveyor, and in all cases set to the working pressure.

A stop-valve to be fitted, so that each boiler can be worked separately.

Each boiler to be fitted with a separate steam-gauge, to accurately indicate the pressure.

With a view to insuring better control over all cocks, valves, and pipes connecting the engines and boilers with the sea, they are to be fixed as follows, viz: All cocks fitted on the plating of steam-vessels to be raised above the level of the stoke-hole plates *on the turn of the bilge*, or attached to Kingston valves of sufficient height to lift them up to the level of the stoke-hole plates, so that they can at all times be seen and attended to. All discharge-pipes to be, if possible, carried above the deep load-line, with discharge-valves fitted on the plating of the vessel. No pipes to be carried through the bunkers without properly-constructed wrought-iron casings around them. The bilge suction-pipes to be fitted to pump from each hold, and suction-pipes and roses to be fitted in the bilges and amidships in the engine-room; these roses to be fitted in a convenient place, where they can at all times be accessible; the cocks and valves connecting these pipes to be fixed above the stoke-hole plates. All blow-off cocks to be constructed so that the spanner or key can only be fixed or taken off when the cock is shut.

The arrangement of bilge-pumps, bilge-injections, suction and delivery pipes, also the non-return valves, to be inspected and approved by the surveyors; any defective arrangement to be reported to the committee.

Lloyd's formulæ for cylindrical shells of boilers.

$$\frac{51520 \times 2\,T \times C}{D \times 6.5 \times 100} = \text{working pressure in pounds.}$$

$$\frac{(P - d) \times 100}{P} = \left\{ \begin{array}{l} \text{percentage of strength of plate at joint} \\ \text{compared with solid plate.} \end{array} \right.$$

$$\frac{(a \times N) \times 100}{P \times T} = \left\{ \begin{array}{l} \text{percentage of strength of rivets compared} \\ \text{with solid plate.} \end{array} \right.$$

51520 = tensile strength of iron in pounds per square inch of section.
T = thickness of plate, in inches.
D = diameter of inside of shell, in inches.
6.5 = factor of safety.
C = less of the two above percentages.
d = diameter of rivets.
P = pitch of rivets.
a = area of section of rivets.
N = number of rows of rivets in shear.

Formula for collapsing of circular flues.

$$\frac{89600 \times T^2}{L \times D} = \text{working pressure in pounds.}$$

89600 = constant.
T = thickness of plate, in inches.
L = length of flue or furnace, in feet.
D = inside diameter of flue or furnace in inches.

The stays in the flat surfaces to take up the whole strain, and not to be subjected to a greater strain than 5,000 pounds per square inch, calculated from the weakest part of the stay or fastening.

TESTS REQUIRED BY LLOYD'S FOR STEEL PLATES (SIEMENS-MARTIN PROCESS) FOR MARINE BOILERS MADE IN 1877.

Tensile and extension tests.

1. Strips cut lengthwise or crosswise of the plates, to have an ultimate tensile strength of not less than 26 and not exceeding 30 tons per square inch of section, with an elongation of 20 per cent. in a length of 8 inches.

2. A specimen of the riveted longitudinal joint is to be tested and shown to have a percentage of strength of, at least, 74 per cent. of the solid plate. Either iron or steel rivets may be used.

Tempering tests.

1. Strips cut lengthwise of the plate, 1½ inches wide, heated uniformly to a low cherry-red and cooled in water of 82° Fahrenheit, must stand bending in a press to a curve of which the inner radius is not greater than one and a half times the thickness of the plates tested. A shearing of *every* plate used in the construction of the furnaces, combustion-chambers, and tube-plates to be subjected to this tempering test with satisfactory results.

2. The strips are all to be cut in a planing-machine, and to have the sharp edges taken off.

Buckling tests.

1. It is to be shown by actual experiment that the flat plates with the proposed reduction of thickness, stayed in the usual manner, are as strong to resist buckling by hydraulic pressure as the ordinary wrought-iron plates.

The material is to withstand, satisfactorily, the above-mentioned tests. The boilers to be constructed under the inspection of the society's engineer surveyors, and when completed to be tested in their presence, by hydraulic pressure, to twice the working pressure, so as to be entitled to the reduction of 25 per cent. in thickness.

These extreme tests have been considered desirable in view of this being the first set of boilers made of this material; no doubt, as confidence is gained in the reliability of this material, these tests may with safety be modified.

On the 26th of January, 1878, the following additional conditions were issued:

3. All the holes to be drilled, or if they be punched the plates to be afterward annealed.

4. All plates, except those that are in compression, that are dished or flanged, or in any way worked in the fire, to be annealed after the operations are completed.

5. The boilers upon completion to be tested in the presence of one of the society's engineer surveyors to not less than twice the intended working pressure.

EXTRACTS FROM RULES BY WHICH THE SURVEYORS OF THE BOARD OF TRADE ARE GUIDED IN THEIR INSPECTION OF BOILERS.

On flat surfaces the pressure allowed should not exceed 5,000 pounds to each effective square inch of sectional area of stay, but if in any case a greater pressure is asked where the flat surfaces are stiffened by T or L irons, the mode of stiffening must be submitted to the board of trade for their consideration and approval before a greater pressure than that mentioned above is allowed for effective sectional area of stay.

The areas of diagonal stays are found in the following way: Find the area of a direct stay needed to support the surface, multiply this area by the length of the diagonal stay, and divide the product by the length of a line drawn at right angles to the surface supported, to the end of the diagonal stay; the quotient will be the area of the diagonal stay required.

When gusset-stays are used their area should be in excess of that found in the above way.

When the tops of combustion-boxes or other parts of a boiler are supported by solid rectangular girders, the following formula, which is used in the board of trade, will be useful for finding the working pressure to be allowed on the girders, assuming that they are not subjected to a greater temperature than the ordinary heat of steam, and in the case of combustion-chambers that the ends are fitted to the edges of the tube-plate and the back plate of the combustion-box.

$$\frac{C \times d^2 \times T}{(W - P)\, D \times L} = \text{working pressure.}$$

W = width of combustion-box in inches.
P = pitch of supporting-bolts in inches.
D = distance between the girders from center to center in inches.
L = length of girder in feet.
d = depth of girder in inches.
T = thickness of girder in inches.
C = 500 when the girder is fitted with one supporting-bolt.
C = 750 when the girder is fitted with two or three supporting-bolts.
C = 850 when the girder is fitted with four supporting-bolts.

The working pressure for the supporting-bolts and for the plate between them shall be determined by the rule for ordinary stays.

The pressure on plates forming flat surfaces will be easily found by the following formula which is used in the board of trade:

$$\frac{C \times (T + 1)^2}{S - 6} = \text{working pressure.}$$

T = thickness of the plate in sixteenths of an inch.
S = surface supported in square inches.
C = constant according to the following circumstances:
C = 100 when the plates are not exposed to the impact of heat or flame, and the stays are fitted with nuts and washers, the latter being at least three times the diameter of the stay, and two-thirds the thickness of the plates they cover.
C = 90 when the plates are not exposed to the impact of heat or flame, and the stays are fitted with nuts only.
C = 60 when the plates are exposed to the impact of heat or flame and steam in contact with the plates, and the stays fitted with nuts and washers, the latter being at least three times the diameter of the stay, and two-thirds the thickness of the plates they cover.
C = 54 when the plates are exposed to the impact of heat or flame and steam in contact with the plate, and stays fitted with nuts only.
C = 80 when the plates are exposed to the impact of heat or flame, with water in contact with the plates, and the stays screwed into the plate and fitted with nuts.
C = 60 when the plates are exposed to the impact of heat or flame, with water in contact with the plate, and the stays screwed into the plate, having the ends riveted over to form a substantial head.
C = 36 when the plates are exposed to the impact of heat or flame, and steam in contact with the plates, with the stays screwed into the plate, and having the ends riveted over to form a substantial head.

When the riveted ends of the screwed stays are much worn, or when the nuts are burned, the constants should be reduced, but the surveyor must act according to the circumstances that present themselves at the time of survey, and it is expected that in cases where the riveted ends of screwed stays in the combustion-boxes and furnaces are found in this state, it will be often necessary to reduce the constant 60 to about 36.

When cylindrical boilers are made of the best material, with all the rivet-holes drilled in place, and all the seams fitted with double butt-straps each of at least ⅝ the thickness of the plates they cover, and all the seams at least double-riveted, with rivets having an allowance of not more than 50 per cent. over the single shear, and provided that the boilers have been open to inspection during the whole period of construction, then 6 may be used as the factor of safety; but the boilers must be tested by hydraulic pressure to twice the working pressure, in the presence and to the satisfaction of the board's surveyors. But when the above conditions are not complied with the additions in the following scale must be added to the factor 6, according to the circumstances of each case:

A	.15	To be added when all the holes are fair and good in the longitudinal seams, but drilled out of place after bending.
B	.3	To be added when all the holes are fair and good in the longitudinal seams, but drilled out of place before bending.
C	.3	To be added when all the holes are fair and good in the longitudinal seams, but punched after bending instead of drilled.
D	.5	To be added when all the holes are fair and good in the longitudinal seams, but punched before bending.
E*	.75	To be added when all the holes are not fair and good in the longitudinal seams.
F	.1	To be added if the holes are all fair and good in the circumferential seams, but drilled out of place after bending.
G	.15	To be added if the holes are fair and good in the circumferential seams, but drilled before bending.

H	.15	To be added if the holes are fair and good in the circumferential seams, but punched after bending.
I	.2	To be added if the holes are fair and good in the circumferential seams, but punched before bending.
J*	.2	To be added if the holes are not fair and good in the circumferential seams.
K	.2	To be added if double butt-straps are not fitted to the longitudinal seams, and the said seams are lap and double-riveted.
L	.1	To be added if double butt-straps are not fitted to the longitudinal seams, and the said seams are lap and treble-riveted.
M	.3	To be added if only single butt-straps are fitted to the longitudinal seams, and the said seams are double-riveted.
N	.15	To be added if only single butt-straps are fitted to the longitudinal seams, and the said seams are treble-riveted.
O	.1	To be added when any description of joint in the longitudinal seams is single-riveted.
P	.1	To be added if the circumferential seams are fitted with single butt-straps and are double-riveted.
Q	.2	To be added if the circumferential seams are fitted with single butt-straps and are single.riveted.
R	.1	To be added if the circumferential seams are fitted with double butt-straps and are single-riveted.
S	.1	To be added if the circumferential seams are lap-joints and are double-riveted.
T	.2	To be added if the circumferential seams are lap-joints and are single-riveted.
U	.25	To be added when the circumferential seams are lap and the streaks or plates are not entirely under or over.
V	.3	To be added when the boiler is of such a length as to fire from both ends, or is of unusual length, such as flue-boilers; and the circumferential seams are fitted as described opposite P, R, and S, but of course when the circumferential seams are as described opposite Q and T, V .3 will become V .4.
W*	.4	To be added if the seams are not properly crossed.
X*	.4	To be added when the iron is in any way doubtful, and the surveyor is not satisfied that it is of the best quality.
Y	1.65	To be added if the boiler is not open to inspection during the whole period of its construction.

Where marked * the allowance may be increased still further if the workmanship or material is very doubtful or very unsatisfactory.

The strength of the joints is found by the following method:

$$\frac{(\text{Pitch} - \text{diameter of rivets}) \times 100}{\text{Pitch}} = \left\{ \begin{array}{l}\text{percentage of strength of plate at joint as com-}\\ \text{pared with the solid plate.}\end{array}\right.$$

$$\frac{(\text{Area of rivets} \times \text{No. of rows of rivets}) \times 100}{\text{Pitch} \times \text{thickness of plate}} = \left\{ \begin{array}{l}\text{percentage of strength of rivets as}\\ \text{compared with the solid plate.†}\end{array}\right.$$

Then take iron as equal to 23 tons, and use the smallest of the two percentages as the strength of the joint, and adopt the factor of safety as found from the preceding scale:

$$\frac{51520 \times \text{percentage of strength of joint}) \times \text{twice the thickness of the plate in inches}}{\text{Inside diameter of the boiler in inches} \times \text{factor of safety } [\times 100]}$$

$=$ pressure to be allowed per square inch on the safety-valves.

Plates that are drilled in place *must* be taken apart and the burr taken off, and the holes slightly countersunk from the outsides.

Butt-straps *must* be cut from plates and *not* from bars, and must be of as good a quality as the shell-plates, and for the longitudinal seams *must* be cut across the fiber. The rivet-holes may be punched or drilled when the plates are punched or drilled OUT of place, but when drilled in place must be taken apart and the burr taken off and slightly countersunk from the outside.

When single butt-straps are used, and the rivet-holes in them punched, they *must* be one-eighth thicker than the plates they cover.

† If the rivets are exposed to double shear, multiply the percentage, as found, by 1.5.

The diameter of the rivets *must* not be less than the thickness of the plates of which the shell is made, but it will be found when the plates are thin, or when lap-joints or single butt-straps are adopted, that the diameter of the rivets should be in excess of the thickness of the plates.

Dished ends that are not truly hemispherical must be stayed; if they are not theoretically equal in strength to the pressure needed they must be stayed as flat surfaces, but if they are theoretically equal in strength to the pressure needed, the stays may have a strain of 10,000 pounds per effective square inch of sectional area.

Surveyors will remember that the strength of a sphere to resist internal pressure is double that of a cylinder of the same diameter and thickness.

All manholes and openings must be stiffened with compensating rings of at least the same effective sectional area as the plates cut out, and in no case should the plate-rings be less in thickness than the plates to which they are attached. The openings in the shells of cylindrical boilers should have their shorter axes placed longitudinally. It is very desirable that the compensating-rings around openings in flat surfaces be made of L or T iron.

Circular furnaces with the longitudinal joints welded or made with a butt-strap:

$$\frac{90000 \times \text{the square of the thickness of the plate in inches}}{(\text{Length in feet} + 1) \times \text{diameter in inches}} = \left\{ \begin{array}{l} \text{working pressure per} \\ \text{square inch.} \end{array} \right.$$

Without the board's special approval of the plans the pressure is in no case to exceed

$$\frac{8000 \times \text{thickness in inches}}{\text{Diameter in inches}}.$$

The length to be measured between the rings, if the furnace is made with rings.

If the longitudinal joints, instead of being butted, are lap-jointed in the ordinary way, then 70,000 is to be used instead of 90,000, excepting only where the lap is beveled and so made as to give the flues the form of a *true* circle, when 80,000 may be used.

When the material or the workmanship is not of the best quality, the constants given above must be reduced, that is to say, the 90,000 will become 80,000; the 80,000 will become 70,000; the 70,000 will become 60,000; and when neither the material nor workmanship are of the best quality, such constants will require to be further reduced, according to circumstances and the judgment of the surveyor, as in the case of old boilers. One of the conditions of best workmanship must be that the joints are either double-riveted with single butt-straps, or single-riveted with double butt-straps, and the holes drilled after the bending is done and when in place, and afterward taken apart, the burr on the holes taken off, and the holes slightly countersunk from the outside.

PART XXIII.

CONCLUSIONS.

CONCLUSIONS.

In considering the foregoing report, it has been seen that the *modus operandi* of naval warfare in all European countries has of recent years undergone important and radical changes.

Ships with auxiliary steam-power, in common with their predecessors of the sailing period and subsequent paddle-wheelers, are *obsolete*, and wooden vessels of all classes are gradually becoming antiquated, and are being relegated to harbor service or abandoned.

Heavily-armored ships, of great fighting power and considerable speed, now constitute the line-of-battle ships of all important naval powers. Low free-board non-sea-going armored vessels are provided for coast defense, and the modern fleets of *cruising* vessels of all navies are now being constructed of either iron or steel, having their bottoms sheathed in wood and covered with copper or zinc, and built with improved structural disposition of materials to utilize the power of the ram and to sustain for a lengthened period the immense engine-power necessary for high speeds.

Again, that ships of a special class, known in England as armed dispatch-vessels, and in France as the rapid type, having a speed of from 16 to 17 knots per hour at sea, are fast gaining favor. Also, that a limited number of small vessels, having large engine-power, are being built exclusively for projecting the Whitehead self-moving torpedo, and armored and unarmored ships are also being fitted with the necessary apparatus for using this locomobile or self-moving weapon.

Finally, that recently-constructed war-vessels of all classes are engined on the compound system.

Captain Simpson, U. S. N., who, during his official tour abroad, made a careful research into the system of artillery employed in European countries, has shown in his very able and exhaustive report, *Mission to Europe*, that cast-iron guns are *obsolete*, and that the ships of all European navies are now armed with rifled guns manufactured wholly of steel or of steel barrels bound with wrought-iron coils. At the date of that report, less than five years ago, the heaviest gun mounted on shipboard in Europe was 35 tons in weight, with a 12-inch caliber. Since that time the weight and power have increased up to 100-ton guns, using charges of 350 pounds of powder and throwing projectiles of from 2,000 to 2,500 pounds; and still heavier guns are likely to be manufactured. With these facts and figures before us, we find that our Navy, although in as efficient a condition as economical Congresses authorize, lags decidedly behind the navies of most other powers, and, excepting perhaps for coast defense, is by no means commensurate with the wealth, extent, and dignity of the country; and should, unfortunately, the necessity for war arise with any considerable power, our real weakness will be exposed, and we will be placed in a condition of humiliation—a condition certainly not to be desired by the people of this great and rapidly-growing republic, having large and increasing commercial interests with all civilized nations.

Heretofore we have been the pioneers, and the main features of very many of the improvements introduced in European naval warfare owe their origin to American genius. The beautiful outlines of American fast-sailing vessels were copied in Europe. The first war-ship propelled by the screw was built in Philadelphia. Shell-fire, and subsequently heavy guns, were first introduced here. The torpedo is an American invention, and the revolving turret for vessels of war originated on this continent. It remained for European naval powers, having large appropriations at command, to develop and expand American inventions. The ideas for the present powerful mastless sea-going armored ships of the English grew out of the visits of our turret-vessel *Miantonomoh* to British ports; and the unarmored fleet of fast ships, of which the *Inconstant* was the first in Europe, owe their development to the building of the *Wampanoag.*

During the time of inactivity and comparative non-progress in American naval construction, the work of reconstructing the navy of Great Britain has been vigorously pushed forward, and in the effort to make it efficient from the modern point of view, there has been no hesitation in weeding out vessels of obsolete types; thus, in the five years ending April 1, 1875, no less than eighty-four vessels of obsolete types had been disposed of by sale, viz: Twelve screw line-of-battle ships; eight screw-frigates; three screw-corvettes; seven screw-sloops; twenty-three gun-vessels; twenty-two sailing-vessels, and nine paddle-wheel steamers; besides which there have been condemned as no longer fit for active service, eight frigates, eight corvettes, nine sloops, and ten gunboats, all wooden screw-vessels.

It has already been seen what kinds of vessels are being added to the navy in lieu of the obsolete wooden types, and it is well to remember that the new fleet given under the head of unarmored ships do not represent the total available force of fast cruisers. The British mercantile marine is possessed of 419 steamers above 1,200 tons and under 5,000 tons register, very many of which ships have high speeds, and some of which can be relied upon for 14 and 15 knots per hour in good weather at sea, for seven or eight days consecutively, and they have sufficient coal-carrying capacity. In the event of war any of these ships are at the command of the government, and in view of utilizing this source of great naval strength, the admiralty have carefully considered the subject of arming with light rifled guns, and more especially with Whitehead torpedoes, such vessels as may be suitable, and in order to act understandingly, they have required the builders or owners to furnish necessary drawings of their vessels.

While lamenting the want of superiority in our individual ships, a careful review of the situation brings home the inevitable conclusion that the interregnum with us will in the end result signally to our advantage. If, after our civil war, we had gone on building armored ships designed for sea-going purposes, each succeeding vessel would have been superseded in offensive and defensive powers by productions on the other side of the Atlantic. The result would have been an antiquated armored fleet, for it has been seen that the power of the gun and the weight and thickness of armor have continued to increase; which, together with additional mechanical appliances, has made each succeeding ship built an improvement over preceding types, and even yet no law of finality seems to be reached either in regard to the weight and power of the gun or the thickness and weight of the armor. With the Sheffield Works promising armor-plates of still greater thickness, and Herr Krupp and Sir William Armstrong proposing guns of 150 tons each, it would be unsafe to regard the limit as having been reached, and unwise to accept even the last productions as the future types. In view of the value given to small fast vessels by the invention of the self-moving torpedo and the risks to be encountered from this terrible weapon, as well as the ram, it is not probable that hereafter any war-vessel will be built larger or as large as those now in process of construction, but it is reasonable to anticipate heavier guns in less numbers mounted on vessels of smaller dimensions Considering these facts and conclusions, the enormous cost and length of time necessary to build an armored sea-going ship in this country, and the further fact that in Europe the advocates for abandoning armor altogether are increasing, any administration may well pause before it sanctions the expenditure of three or four millions of dollars for such a vessel.

Turning now to the policy of maintaining an efficient cruising-fleet for the police of the seas, for the training of men, for the purpose of exhibiting the American flag in foreign ports, and especially in harbors of semi-barbarous countries where they can scarcely realize the existence of a force unless it be visibly present to their gaze, for the repression of piracy and slavery, and for the punishment of offending savage tribes, the ship is the distant representative of our national power, and it should have no superior of its type belonging to any other country. With the view, therefore, of reconstructing and adding to our unarmored fleets, the time is near at hand, if not already present, for reaching judicious conclusions as to the immediate future types of our vessels. We are now at liberty to take advantage of the results of the most expensive and exhaustive experiments made by foreign powers in the construction of ships, of machinery of various kinds for naval purposes, and in the manufacture of weapons.

To the European view it is strange that the nation which first devised and attempted to construct a squadron of powerful, fast, unarmored cruisers, has now nothing to match the British *Raleigh*, *Boadicea*, and *Euryalus*, or fast cruisers of other countries; and stranger still, that the naval force to which American commerce is obliged to look for protection is composed of a small number of corvettes and sloops of very moderate speed and power, mainly of the types and armed with the weapons (torpedoes excepted) of the period before our civil war.

All agree in the necessity for action, but there is a hopeless want of unanimity among our officers as to the types that should be built, and as the prerogative of deciding questions so important, as well as the province of giving shape in detail to designs, belongs to the controlling powers, it may not be out of place to suggest that, in considering the subject, it will be wise, as before said, to go carefully over the ground covered by possible enemies and take advantage of the results of their experience.

Under the head of British unarmored ships, it may be seen that the costly experiment of building the first fast cruisers proved, to say the least, unsatisfactory. They were too costly to build, are too unwieldy to handle, and are too costly to maintain.

Again, it may be seen that the French, in their endeavor to outdo the English in the element of speed, have fallen into the same general faults in building their first two cruisers of the rapid type, the *Duquesne* and *Tourville*.

It may also be seen that the *Rover* is perhaps as small a ship as with our present knowledge can be built, to have the speed of fifteen knots per hour, to have sufficient structural strength to stand the engine-power necessary for this speed, to utilize the power of the ram, to carry sufficient battery, to have sail-power, to be provided with machinery for ejecting torpedoes, and to possess all necessary requirements for keeping the sea.

Lessons can therefore be drawn from these practical illustrations.

Considering now our immediate future wants, two classes described and illustrated under the head of British unarmored ships, viz, the *Rover* and *Garnet*, approach the size and type of cruising-vessels desirable to possess. Besides which, we need vessels of the rapid type in which offensive and defensive capacity must be sacrificed for the express purpose of obtaining what heavily-armed vessels do not possess, extreme speed. This type must have exquisite lines, fine entrance, clear run, the exceptional engine-power of nearly two horses to every ton of displacement, and great structural strength of hull combined with lightness of material.

By trade and commerce we grow to be a rich and powerful people, and by their decay we would grow poor and impotent; and as trade and commerce enrich, so they fortify our country. To the American Congress commercial interests must look for the encouragement afforded by naval protection.

PART XXIV.

APPENDIX.

ROYAL NAVAL COLLEGE AT GREENWICH.

REGULATIONS FOR ADMISSION OF ENGINEER STUDENTS; OBSERVATIONS; RANK, PAY, ETC.

NAVAL MODELS AT GREENWICH.

SCIENTIFIC APPARATUS AT THE SOUTH KENSINGTON MUSEUM; CONSERVATOIRE DES ARTS ET MÉTIERS.

---

24 K

ROYAL NAVAL COLLEGE AT GREENWICH.

By invitation of Vice-Admiral Sir A. Cooper Key, K. C. B., F. R. S.,* the distinguished officer in charge of the college, I visited that institution, and through his personal kind attention I received the privilege of observing the details of instruction in all departments and branches, and had the system of education fully explained. This establishment stands on the south bank of the river Thames, six miles below London Bridge, on the site of the old royal palace in which Henry VIII. was born. The present buildings consist of four masses, forming an architectural group unparalleled in modern England. They were erected under the reigns of different sovereigns, and are known respectively as King Charles's, King William's, Queen Anne's, and Queen Mary's. The first was erected in 1665 and occupied as a palace. Other buildings, additions, and reconstructions followed in 1667, 1669, 1712, 1710, 1750, 1769, 1789, and 1811. Except the originals and the last, they were the work of the celebrated architect Sir Christopher Wren. It was after the battle of La Hogue, in 1691, that the buildings were prepared as a hospital, and subsequently became the asylum for the old and disabled seamen of the royal navy. They were used for this purpose until about six years ago, when the government decided to abolish the system of providing a home for their old seamen, and in lieu thereof to grant them liberal pensions, and to permit every man entitled to its benefits to live with his family or elsewhere, to suit his own convenience.

It was after the buildings were vacated as an asylum, that the Queen, by her order in council dated January 16, 1873, sanctioned the founding of a royal naval college at Greenwich. To meet the new requirements, each building was, when necessary, internally altered; but no external alterations or additions were permitted.

The site with its terrace, 860 feet long, by the river, is attractive, and the location is advantageous in many respects. The buildings are commodious, the several departments admirably arranged and fitted for utility and comfort. The college is provided with chemical and physical laboratories, instruments, and appliances embodying the latest improvements, and on a scale which their lordships state has hitherto not been possible in any naval establishment. In fitting and equipping this important school, cost has not been considered. The expenditures under the direction of the president for the alterations of buildings internally, outfits and appliances of all kinds, during the first year, amounted to the sum of $408,240.

For the management of the college there is nominally a governor, who is first lord of the admiralty and an M. P.; a president, who is a vice-admiral, assisted by a captain in the navy in matters affecting discipline, &c., *unconnected with study;* and there is a director of studies (a civilian), under the president, who superintends the whole system of instruc tion and the various courses of study.

There are two naval officers, instructors in nautical astronomy and navigation; two naval instructors, corresponding to our professors of

* Now in command of the North America and West Indies Station.

mathematics; and four engineer officers, instructors in steam, in marine engineering, and applied mechanics. The remaining members of the staff of instructors, twenty in number, are civilians, and most of them are distinguished scientific professors. Besides the above named, there is one medical officer and one storekeeper, who also acts as cashier. The president, his assistant, and secretary reside in the building; all instructors reside outside the walls.

All officers known in our service as line officers are in the British navy denominated executive officers. They are appointed at an early age and placed on board training-ships as naval cadets. It is not intended to provide at Greenwich for the education of these cadets. The naval college is fitted in every respect for the education of officers of all ranks above that of midshipman, in all branches of theoretical and scientific study bearing upon their profession; and it is stated that while every possible advantage in respect to scientific education will be given, no arrangement will be allowed in any manner prejudicing the all-important practical training in the active duties of the profession. It is after the cadets have been passed to acting sub-lieutenants and acting navigating sub-lieutenants that they enter the college for final instruction and graduation. The college is also open, under limitations, to all executive officers below flag rank, to improve or prepare for promotion. It is at this college that all the officers of the engineering branch of the royal navy are now and are hereafter to be graduated. To this corps of officers my attention was especially invited, and I propose to give particulars relating to this branch only.

The college was opened February 1, 1873, to sub-lieutenants and gunnery lieutenants, and on the 1st of October following it was opened to officers of the engineering branch. The institution is therefore in its infancy. Its foundation is, however, laid, looking to a great future, as the fountain from which are to emanate the officers who are to construct and to control all the ships, all the dock-yards, and all the naval operations of the most powerful navy in the world. It was as early as 1811 that the British Government commenced to encourage the instruction of naval architects in a primary school at the Portsmouth dock-yard; but this school had a brief existence. A second school was by order of the admiralty opened in the same dock-yard in 1848; and as a result of the instruction given here, a number of able scientists were turned out, among them Mr. Isaac Watts and Mr. E. J. Reed, both of whom became chief constructors of the navy.

Subsequently four schools were established for the purpose of preparing students as assistant engineers for the navy. These schools are still in existence at the dock-yards of Chatham, Portsmouth, Sheerness, and Devonport. Finally, the South Kensington School of Naval Architecture was opened to the most meritorious graduates of the students from the dock-yard schools. But it was only when the Royal Naval College at Greenwich was opened, that the admiralty decided systematically to strengthen and improve the engineering branch, by raising the standard of education to a degree previously unknown to that class of officers. It was accordingly determined, January 30, 1873, and announced by regulation, that thereafter all engineer students who successfully passed through the six years' course at the dock-yards would be sent to the Greenwich college for a period of study, and none are now appointed to vessels until passed by the director of studies at Greenwich.

The following regulations will give a clear understanding of the system of entrance and instruction at the dock-yards. The system of instruction and examination at the naval college was submitted to the department by me for our Naval Academy at the beginning of 1876:

ADMIRALTY, *June* 8, 1877.

Regulations for admission of engineer students in Her Majesty's dock-yards.

My lords commissioners of the admiralty are pleased to direct that the following regulations for the admission of engineer students in Her Majesty's dock-yards shall be substituted for those now in force.

2. Vacancies for appointments as engineer students in the dock-yards are open to public competition. The dock-yard at which engineer students are entered each year will be fixed by their lordships.

3. The list of candidates for these appointments will be kept at the admiralty in London. All applications for the forms to be filled up by persons who wish to compete, must be addressed to the secretary of the admiralty before the 1st of March in each year. Such applications should state the place at which the candidate desires to be examined.

4. Candidates must not be less than fourteen nor more than sixteen years of age on the first day of the examination. Proof of age will be required by the production of a certificate of birth, or by declaration before a magistrate. Evidence of respectability and good character must also be produced. All candidates must be children of British subjects.

5. Candidates are to understand clearly that they will be first required to satisfy the admiralty as regards their age, respectability, good character, and physical fitness, before they can be eligible for entry into the dock-yard; and if these conditions are satisfactory, they will then be examined by the civil-service commissioners in educational subjects.

6. Candidates in or near London will be medically examined by the medical director-general of the navy at the admiralty. Those residing near one of Her Majesty's dock-yards, or one of the first-reserve ships, will be examined by the medical officers attached thereto.

7. The examination will commence on the first Tuesday in May in each year, and will be held by the civil-service commissioners in London, Liverpool, Portsmouth, Devonport, Bristol, Leeds, Newcastle-on-Tyne, Edinburgh, Glasgow, Aberdeen, Dublin, Belfast, and Cork.

8. The following will be the subjects of examination, and the maximum number of marks for each subject:

*Arithmetic		300
English:		
* Writing from dictation	100	
* Composition	100	
Grammar	150	
		350
French:		
Translation into English	100	
Grammar	50	
		150
Geography		100
Algebra (up to and including quadratic equations)		300
Geometry (the subjects of the first six books of Euclid's Elements)		300
Total		1,500

Candidates will also be tested as to their ability to read aloud with clearness, distinctness, and accuracy, and without hesitation. Stammering, or any imperfection of utterance, will be regarded as a disqualification.

9. Candidates who fail to pass in the first three subjects (those marked with an asterisk), or in reading aloud, will be disqualified, and their other papers will not be examined. The candidates who display a competent knowledge of all those subjects, and who obtain not less than 750 marks in the aggregate, will be classed in one general list in order of merit, according to the number of marks gained, and will be eligible for appointment as engineer students in one of the dock-yards, according to the number of appointments which it may be decided to make that year.

10. The successful candidates will be entered as engineer students before the 1st of July in each year, and must join with their parents or guardians in a bond for £300 to enter, if required, into Her Majesty's naval service as assistant engineers, if at the expiration of their training they should obtain certificates of good conduct and efficiency for admission in that capacity.

11. The parents or guardians of all engineer students admitted in future, will be required to pay the sum of £25 a year for each student during the first three years of his training.

12. The first payment of £25 is to be made before the student is entered in the

yard, and the second and third payments of £25 each are to be made on or before the 30th day of June in each of the two succeeding years. The payments are to be made to the cashier of the yard to which the student is appointed. In case of failure of payment the student will be discharged.

13. Board and lodging will be provided for engineer students, and they will be required to reside in one of the dock-yards.*

14. The weekly pay of engineer students during their training will be as follows, provided they are well reported on by the officers:

First year	One shilling a week.
Second year	Two shillings a week.
Third year	Three shillings a week.
Fourth year	Five shillings a week.
Fifth year	Eight shillings a week.
Sixth year	Ten shillings a week.

15. Engineer students will be under the supervision of the captain of the steam reserve and a staff of competent officers, and subject to such rules and regulations as their lordships may deem necessary.

16. Special regulations will be made for engineer students in the dock-yards, so as to make a distinction between them and the workmen.

17. Engineer students will remain for six years at one of the dock-yards for practical training in the workshops, and to receive instruction in iron-ship building. They will attend the dock-yard schools for such periods, and to pursue such studies as may from time to time be determined on; they will also pass a portion of their time in the drawing-office. Means will be afforded them of acquiring the groundwork of the knowledge required by a naval engineer respecting the working of marine engines and boilers, including those repairs which can be carried out afloat, the practical use of the various instruments used in the engine-room, including the indicator, and of becoming generally acquainted with the duties of a naval engineer.

18. Engineer students will be examined once a year under the direction of the president of the Royal Naval College. They will be examined by the engineer officers of the admiralty at the end of the fourth, fifth, and sixth years of their service as to their practical acquirements and knowledge of steam-machinery. Two prizes will be given annually at each dock-yard to the engineer students most highly reported on as regards their skill as workmen. Practical engineering will be considered an essential subject at examinations, and in the lists showing the results of the examinations, the numbers obtained in practical subjects will be shown distinct from those obtained in educational subjects. No engineer student will be granted a qualifying certificate for admission at the Royal Naval College unless he obtains at least 50 per cent. of the total number of marks for practical engineering on his final examination.

Admission into the college.

19. The examination of the sixth year students is to be held in time to allow the result to be known by the 1st of July in each year, and it will include tests of their skill as workmen. Those found qualified will, on the completion of their term of service at the dock-yards, proceed to the Royal Naval College, at Greenwich, as acting assistant engineers on probation on the 1st of October succeeding the examination, where they will pass through a course of higher instruction during one term.

20. Those engineer students who fail to pass the examination at the end of their six years' service will be allowed to remain one year longer at the dock-yards, and will then be re-examined, when, if they are unable to pass, they will cease to be eligible for the rank of naval engineer. The pay of a student during such year of probation will be the same as during the sixth year.

21. Engineer students will not be admitted as acting assistant engineers until they have been pronounced fit for Her Majesty's service by the medical officers, and have learned to swim.

22. Acting assistant engineers will be provided with quarters while at Greenwich. During their first term they will be paid 6*s.* a day and 1*s.* 6*d.* a day toward the mess ex-

*As a place of residence at Portsmouth for the students, the old line-of-battle ship *Marlborough* has been fitted up. The ship has undergone a complete transformation, all the machinery and bulkheads having been removed, and gas and water introduced throughout. Dormitories supplied with iron bedsteads and other fixtures, a dining-room, kitchen, class-room, study and library have been provided. The accommodations are sufficient for one hundred students, forty-one having been entered January 1, 1878, and the remaining fifty-nine vacancies will be competed for in May next. Cabins have been built under the poop for a chief engineer and two assistants who will, under the admiral superintendent, be intrusted with the discipline of the students when on board, their professional education being provided for in the dock-yard.

penses. Those selected for further study will receive their full pay and 1*s.* 6*d.* a day toward the mess.

23. The term for study at Greenwich will be from the 1st of October to the 30th of June following. All will be examined under the direction of the president of the Royal Naval College on the completion of their term at Greenwich, and will receive certificates according to their merit, in three classes. Those who obtain first-class certificates will receive commissions dated the same day as their acting appointments. Those who obtain second-class certificates will receive commissions dated six months after the date of their acting appointments, and those who obtain third-class certificates will receive commissions dated the day after their discharge from the Royal Naval College. The additional time given for first-class certificates and second-class certificates will reckon in all respects as time served as assistant engineers.

24. Two assistant engineers will be selected annually from those who take the highest place at the examination on the completion of their term at Greenwich, to pass through a further course of scientific instruction if they desire it. These two will be allowed to remain two more terms at Greenwich, on the completion of which they will be sent to sea as assistant engineers, and after one year's service at sea, they will be considered eligible to fill positions in the dock-yards and at the admiralty.

25. Those passing the second and third terms at Greenwich will be attached during the vacations between the 30th of June and 1st of October, to the dock-yards or steam reserves, where they will attend trials of new and repaired engines, and obtain experience respecting the duties they will have to perform at sea.

26. No assistant engineer who has passed three terms at Greenwich will be allowed to leave Her Majesty's service within seven years of the completion of his term at Greenwich, unless he shall pay the sum of £500 to defray the charges of his education. Such resignation to be subject in each case to their lordships' approval.

The number of students for the mechanical branch entered at the Greenwich Royal College in the first three years was 137, including those—24 in number—transferred from the South Kensington school; this institution having ceased to exist when Greenwich opened.

Of the foreign students permitted to enjoy the benefits of this excellent college, there were in 1876 three Russians, two Italians, two Danes, one Spaniard, one Norwegian, and one Brazilian; all studying engineering and naval architecture, under the rules and laws of the institution.

It will be observed from the foregoing regulations that the system of education provided for the British royal naval engineers is as follows:

1st. They are received into the dock-yards between the ages of fourteen and sixteen, by public competition, after evidence is produced of respectability and good character.

2d. They are required to serve six years in these yards, during which time they are employed for stated periods in the several branches of the engineering factories, on board ships, on the hulls of ships, and in the draughting-rooms. While occupied in this practical training, they are superintended by competent leading men, under the directions of the factory officers. They are also required to attend the dock-yard schools on appointed afternoons and in the evenings for theoretical instruction. They are examined regularly, and those who do not make satisfactory progress are dismissed. At the end of the fourth, fifth, and sixth years they are examined by the engineer officers at the admiralty, as well as by the dock-yard officers and instructors, and all those who pass satisfactory examinations at the end of six years, and are found qualified, are appointed acting assistant engineers, and entered at Greenwich College on the 1st of October succeeding the examination.

3d. All assistant engineers are required to serve one collegiate term at Greenwich, on the completion of which they are examined and receive certificates in three classes. But *two* out of the whole class of each year who take the highest honors may optionally pass through a second and a third collegiate course. It will thus be seen that the greatest number of assistant engineers receive seven years' government instruction before entering on the duties for which they have been educated, and that two are graduated after nine years' instruction. These long-term graduates are required to go to sea for one year only, after which they are eligible for positions at the admiralty and in the dockyards. The training and education are most thorough and complete, and it is believed that from the number graduated some constructing engineers of marked ability will be developed. The college is also open to engineer officers in the service who have not had the advantage of study at Greenwich or South Kensington, but, as with executive officers, the number that can be annually admitted is limited. These officers remain one term, and a small number may elect by permission to remain a second term for study. Their pay while at the college is the same as when at sea, with 1*s.* 6*d.* each toward the mess. If this advantage could be granted the passed and assistant engineers of our Navy, of a term of study at the Naval Academy before examination for promotion, it would result in much good to the service.

All the students reside in the college except the higher grades, who are permitted to reside in the town.

France, Germany,* and other continental countries have also government schools for the education of cadet engineers, but there is no system of instruction for this class of officers so thoroughly practical and scientific as that of the English.

RANK, PAY, ETC., OF ROYAL NAVAL ENGINEERS.

In the autumn of 1875, the lords commissioners of the admiralty appointed a committee or board, consisting of Vice-Admiral Sir A. Cooper Key, K. C. B., F. R. S., Captain Sir John E. Commerell, K. C. B., Captain W. M. Dowell, R. N., C. B., the engineer-in-chief of the navy, and a chief inspector of machinery, royal navy, to investigate the claims of the royal naval engineers for increased rank, pay, &c. This committee was in session in London for several months, carefully taking testimony, all of which was printed in like manner with the proceedings of a military court. The result of their investigations may be found in the report submitted to the admiralty in March, 1876. The first recommendation reads as follows:

"Engineer officers for the future to be *executive officers*, [*i. e.*, line officers], but not to command." Increased rank and pay were recommended; also increased encouragement to retire.

The pay and position of the engineers of the royal navy have recently been attracting considerable attention in England. The well-known author, Hon. E. J. Reed, ex-chief constructor of the British navy, and now member of Parliament, in articles to the press suggests radical changes. He says:

> It is at least a plausible view that we must look to those upon whom the management of the navy as a fighting machine most nearly depends, for the most valuable assistance. A ship of war is now a machine of the greatest intricacy, set to work mainly, if not exclusively, by the power of steam; it is a steam being, and the man who understands it, can work it with safety, can control it efficiently, can use it, preserve it, repair it, renew it, is the engineer.

The London *Times*, in a leader commenting on Mr. Reed's letters, says:

> Every one who wants to be informed of the condition of the navy, and to obtain an opinion on the value of any proposed changes in the design and arrangement of our

* The German school for the education of naval engineers is at Kiel. It is divided into four classes. The instruction comprises machinery, mathematics, mechanics, physics, chemistry, and the English and French languages.

ships, takes counsel of the engineers among the foremost; and if the admiralty and Parliament would get from the service itself the most trustworthy views of naval policy, they must consent to make the fullest recognition of the value and importance of the functions of the engineers.

And adds:

Mr. Reed is altogether right in insisting that it is a matter of prime interest for England that the class [corps] of naval engineers should be raised to a level corresponding to the greatness of their present trust, and to the weight of their enlarged responsibilities.

NAVAL MODELS AT GREENWICH.

The models of the vessels of the British navy for many years were not open to the general public. They were kept in a room at the admiralty, Somerset House, and afterward removed to the South Kensington Museum. On the establishment of the naval college at Greenwich, in 1873, it was determined that they should be sent there for the purpose of being easily accessible to the students. At Greenwich they are exhibited to great advantage and admirably arranged in a building which for generations has been associated with the British navy, and they are within easy reach of the classes most interested in the objects of the exhibition. The subject of study here presented is exceedingly interesting to persons desiring to be informed of the history of naval construction, for here can be seen the models of the vessels of the British navy from its earliest time, step by step, down to the present period. Here are seen the successive development of wooden sailing-vessels from the earliest war period to the date of the introduction of steam; the first paddle-wheel steamer and the subsequent changes in form of this kind of vessel; the first screw-vessel, with the changes that followed the introduction of motive power in that form; the first iron vessel; the first iron-clad or armored ship, and the still later type of fast iron vessel sheathed in wood.

It is believed that the first ship in the royal navy was built in the reign of Henry VII., but the first man-of-war of respectable size, as shown by the models, was the *Great Harry*, or, as she was at first called, the *Henri Grâce à Dieu*, laid down by Henry VII. in 1512, and launched June, 1514. It was under this sovereign that the British navy was established. He not only built the first man-of-war, but also founded the three royal dock-yards at Portsmouth, Woolwich, and Deptford; and at the close of his reign there were fifty-eight ships in the navy, of an aggregate tonnage of 12,000. The tonnage of the *Great Harry* is stated at 1,000, but the dimensions are not precisely known. She had much top-hamper, and history says that, notwithstanding her crankness, she performed good service, and was ultimately burned accidentally at Woolwich in 1553. In the same case with the *Great Harry* is a model of another remarkable ship, *The Sovereign of the Seas*. She was built in 1637. Upon comparing the models of the two ships, one is impressed with the improvement effected in a hundred years. The *Royal William*, of which there are two models, one full-rigged, representing her as originally built in 1670; the other, showing her as altered in 1692, is a better exemplification of the ships of the seventeenth century than is the *Royal Sovereign*.

The *Royal Sovereign* was the first ship in the British navy with three decks. She was afterward cut down, and, it is said, did good service. She was destroyed by accidental fire at Chatham in 1695. The next model is that of the notable *Britannia*, of 100 guns and 1,700 tons, built in 1682. These ships of the seventeenth century furnished the type for line-of-battle ships for nearly a century and a half, and the earliest designs are noticeable as having considerable sheer, the effect of which was further increased by the fact of there being in the after part of the ship two decks more than amidships. There are models of two foreign ships in this room, the *Commerce de Marseilles*, a French vessel of 120 guns, captured at Toulon in 1793; and the *Salvador del Mundo*, captured from the Spaniards in 1797. They are both models of ships superior to those of the English of the same date. The second room contains the models of a few famous ships; among others that of the first *Victory*, built in 1737, and lost seven years after in the English Channel, with her admiral, officers, and crew of 1,000 men. There is also the *Royal George*, famous as having capsized in dock, "with all her crew complete," in 1782.* On the walls there are large numbers of half-models of frigates captured from the French and Spaniards.

Among the later vessels noticed, whose class will soon only be represented by the models in the Royal Museum, are the screw line-of-battle ships *Royal Albert* and *Howe;* the former launched at Woolwich in 1854, and the latter at Pembroke in 1860. The armament of the *Howe* consisted of 121 guns, and her complement of crew was intended to be 1,130 men, but she never made a cruise, and it is not probable she will ever be sent out of the harbor of Devonport, where she now lies.

Going back a little, to the introduction of steam, the first steamer built in the British navy is represented by a model of a paddle-wheeler called the *Gulnare*, a name afterward changed to *Gleaner*. She was launched at Chatham in 1833, was 120 feet long, 23 feet 3 inches broad, and drew 13 feet of water. Two years previous to this, however, a small purchased steamer, the *Black Eagle*, carrying one gun, was employed in the navy. The first screw-vessel was the *Dwarf*, the next was the *Rattler*, both small vessels, built immediately after the successful adaptation of the screw-propeller to the *Princeton* in our own Navy. The first war screw-ship is represented by a model of the *Sans Pareil*, of 80 guns, ordered in 1848 as a sailing-vessel, altered and launched in 1851 as a screw-ship, eight years after our screw-sloop *Princeton* was built, and four or five years after the French admiralty had sent out the *Napoléon.*

The first iron vessels are represented by the *Jackal* and the *Simoom*, both still in commission; the former built in 1844, and the latter, as a war steam-frigate, in 1849. Here may be found reason for comment on the wisdom of naval authorities. It was only about as late as the year 1840 that wise and distinguished old admirals were invited to witness the performance of a little screw-vessel on the Thames, with the view to the introduction of the screw into the navy. They inspected its performance, and reported the screw-propeller unsuitable for war-vessels. Again, the *Simoom*, the first iron war-ship built, was pronounced by the board of admiralty to be unfit for war purposes in consequence of the material of which the hull was composed, and as a result they altered her into a troop-ship. Thirty years after this short-sighted decision the *Simoom* is still sound and doing good service; all the first-rates are built of iron, and all naval vessels are propelled by the same instrument—the screw—pronounced unsuited for war-vessels by the officers of the sailing period. While citing the *Simoom* as proof of the durability of iron vessels, it is not out of place to mention the troop-ship *Himalaya*, of 3,453 tons, B. M. This iron vessel, built about 1850, and purchased by

* It was at Spithead that the *Royal George* capsized.—AN ENGLISH NAVAL ARCHITECT.

the admiralty, in July, 1854, at a cost of $631,800, has seen about twenty-eight years' work, and is still sound in hull and serviceable.

We come next to the collection of models of iron-clad ships. The Russian war brought into existence the use of armor on the seas. Three floating batteries, built by the French, silenced the forts at Kinburn. They were named the *Lave*, *Tonnante*, and the *Dévastation*. There are no models of these in the museum, but there is one of the English batteries subsequently built and used against the Russian forts.

The model of the *Warrior*, the first British armored ship, is to be seen, and the various types that followed her are represented.

In the upper rooms are seen illustrations of the old and new systems of construction of men-of-war, materials and equipments of various kinds, and a full-rigged model of Nelson's *Victory*, the original being still preserved and doing duty as a guard-ship at Portsmouth. In a separate room there is also a large model showing the stations of the ships at the battle of Trafalgar.

The above brief outline of models has been drawn up simply as being of historic interest, and on account of the lessons for students to be derived therefrom.

Several visits were also made to the collection of models of naval vessels, &c., at the Louvre in Paris. There is here seen a succinct history of naval construction in the French service from its earliest period to the present time, but no useful purpose would be served by entering into details of them.

THE EXHIBITION OF SCIENTIFIC APPARATUS AT THE SOUTH KENSINGTON MUSEUM, LONDON, 1876.

England does not possess a patent office furnished with models like ours in Washington. Patentees have only to deposit complete drawings and descriptions of their inventions. These are to be found at the Free Library, Chancery Lane. There is, however, an interesting permanent collection of original machines, instruments, &c., at the South Kensington Museum, where besides, of late years, an annual international exhibition took place, of inventions, productions, manufactures, &c., of more interest to the curious public at large than to mechanics and scientific men. In the beginning of 1875, however, it was determined to step out of the beaten track and to do something of a really scientific nature. Consequently, it was resolved to form a loan collection of scientific apparatus, which was to include not only apparatus of all kinds for investigation, but also such as possessed historic interest on account of the persons by whom, or the researches in which, they had been employed. It was at first intended to devote only a small space in the museum to this collection, which was not expected to be a large one. The idea, however, found such favor with scientific men, both in England and other countries, that it was soon apparent that some building much larger must be provided. Consequently, the galleries of the horticultural gardens of the international buildings were obtained. A general committee was formed in January, 1875, consisting of more than one hundred of the leading men of science in England, and including the presidents of all the learned and scientific bodies. At their first meeting it was decided to divide the whole exhibition into the five following sections, viz: Mechanics, including pure and applied mathematics; physics; chemistry, including metallurgy, geology and mineralogy; geography, and biology. As soon as the programme had been definitely settled, steps were taken to interest foreign countries in the matter, and men of science in nearly all European countries were invited to join the general committee, and also to form special sub-committees to further the due representation of science of their respective countries; all of which was promptly responded to.

The exhibition was opened in May, 1876, and was known as the "Loan Collection of Scientific Apparatus." It was doubtless the most extensive and best collection of the kind ever seen. By mechanics, professional and scientific men, food for study and thought could be found there for all spare time.

Entering at the door nearest the Kensington Museum for a general survey in methodical order, the educational collection is the first reached. Here Germany and Russia divide the palm, England being completely distanced. The most general collection is that contributed by the committee of the Pedagogical Museum of Russia, which indicates most clearly that in no country is the value of scientific instruction more appreciated than in Russia.

On leaving the educational collections, the most notable objects on the right and left respectively are Stephenson's "Rocket" and "Puffing Billy," lent by the commissioners of patents. The next objects on the left and right were then exhibited for the first time in England. They were a steam-cylinder by Papin, bearing the date 1699, sent over from the Royal Museum of Cassel; and a collection of Watt's original models of various parts of steam-engines, contributed by Mr. Hamilton. The steam-cylinder, which was made at Cassel, is almost the only remaining witness of the works of Papin, from which a series of inventions has sprung, which have completely changed in a few decades our modes of life. How exactly Papin knew the importance of his idea of employing steam as a motive power is clearly demonstrated by his writings. They contain the invention of the piston steam-engine and its application to steam-vessels. The cylinder exhibited, undoubtedly the first ever made, was to be employed for a steam-engine of peculiar construction, with which a canal connecting Karlshafen with Cassel, on the summit of Hofgeismar, was to be supplied with water. The model of the pumping-machine was completed, and some parts of the machinery cast at Veckerhagen, when a machine, by means of which he was making experiments on throwing bomb-shells by steam, exploded in Papin's laboratory, and he was obliged to take to flight.

Newcomen's engine and Captain Savery's engine also found places here, as did Bramah's first hydraulic press. Mr. Bennett Woodcroft contributed a large number of models, and the Royal Mining Academy at Berlin and the School of Mines vied with each other in their collections of instruments to aid in teaching. In the series of rooms devoted to naval architecture and marine engineerng, the first object on the left, was a model of the *Faraday*, and, after a long line of the other models, we came, in the next room, to Mr. Froude's apparatus, illustrating his method of ascertaining the resistance of ships by measuring the resistance of their models. This method is now used for British naval vessels, and experiments have shown that the results obtained with the models accord very closely with those obtained from the ship itself; so that the admiralty is no longer obliged to make experiments with the ships themselves. It had never before been exhibited except to those who have been privileged to see Mr. Froude's laboratory at Torquay. The London *Times* thus describes the manner of making the experiments:

We may state that the models, from 6 feet to 16 feet in length, are made of hard paraffine; the experimental apparatus employed in working the model includes appliances for designing, molding, and casting the models, shaping them by automatic machinery, moving them through the water at the required speeds, and automatically recording the leading phenomena of the trial—namely, the speed, the resistance, and the change of level induced by the speed at each end of the model. When the model exactly represents the lines of the ship the form of which has to be studied, it is put into a tank and connected with a dynamometric truck, which runs on a railway about 200 feet in length, suspended over a water-way 36 feet wide and 10 feet deep. The model floating in the water is, as it were, "harnessed" to the truck and travels with it. The towing strain—*i. e.*, the force necessary to make the model accompany the truck in its longitudinal progress—is taken during the experiment by a spiral spring, the extension of which, measuring the towing force, is indicated on a large scale by a pen on a recording cylinder. The recording cylinder is driven by the truck-wheels, and thus its circumferential travel indicates distance run; at the same time another pen, jerked at half-second intervals by a clock, records time. Other pens actuated by strings led over pulleys record the change of level of the ends of the model. Thus the diagrams made furnish an exact measure of the speed and a continuous record of the resistances and of the change of level of the model throughout the experimental run at steady speed. The truck is connected by a wire rope with a winding-drum, driven by a small stationary double-cylinder steam-engine.

The contributions of two other men, famous, the one in mechanical arts, the other in science, have been described as follows:*

Sir Joseph Whitworth has contributed a collection * * of measuring-instruments of the most perfect accuracy of workmanship. Among them is a machine for measuring thicknesses up to one ten-thousandth of an inch; and, wonder greater still, also a machine on the same principle, by means of which a distance of one millionth of an inch can be made perceptible. The latter had never previously been shown anywhere. The collection also includes specimens of the true plane surface, which, as Sir J. Whitworth insists, is the only means of obtaining perfect power of accurate measurement.

To the scientific man, if not to the general public, the most interesting objects in an exhibition of this kind are the actual instruments by which celebrated investigators have discovered the truths with which their name is associated. Of this nature is the apparatus we have here, by which Dr. Joule ascertained the mechanical equivalent of heat, or, in other words, the actual work which a given quantity of heat would accomplish. The two forms of apparatus exhibited are vastly like vertical churns, and the process employed by Joule consisted in churning water till its temperature was thereby raised, and carefully noting the exact rise of temperature and the amount of work performed.

To the eyes of mechanical engineers, the exhibition acquired a new value in the exceedingly interesting collection of original steam-engines, commencing with the production of Papin, the very first of which we have knowledge (and referred to elsewhere in this description), passing through various stages of progression, and culminating in the last type of compound engines introduced into the ships of the British royal navy. This collection of historic machines was placed mainly in the room set apart for applied mechanics. Many of them were moved from their usual dingy resting-places in the patent museum, and are historically, at least, familiar to American engineers who visit South Kensington. They represent the brain-work of the men who laid the very foundation of our profession—quaint old constructions, but which, in semblance or reality, have been the friends of all our generation of engineers since their boyhood.

Papin's cylinder is a really notable memorial of that genius to whom we owe so much (for the piston was also introduced by him), and who died at last penniless and unfortunate in England. Not far from Papin's cylinder stood a model of the engine of his more fortunate successor, Newcomen; it is dated 1705, and is said to have been made by the inventor, and by him presented to King George III. It is now the property of King's College, London, is beautifully made, and is in admirable preservation. Being in all probability the only authenticated

* From the London *Times*.

example of Newcomen's own handiwork in existence, it must be looked upon as possessing much interest. At the back of the case which contained this model hung a print which represents the steam-engine near Dudley Castle, invented by Captain Savery and Newcomen, executed by the latter in 1712. Here was seen the engine of Savery, made in 1718, and used to raise water in the garden of Peter the Great, Emperor of Russia; also the little engine used in the University of Glasgow in 1765, which James Watt was employed to repair, and from which he conceived the idea of the separate condenser that subsequently identified his name with the steam-engine all over the civilized world.

Passing across the street to the patent museum, we find the original old Cornish pumping-engine made by Newcomen and Cawley, which was in operation at Mr. Boulton's Works, Soho, near Birmingham, and to which Watt in 1777 applied for the first time his separate condenser and air-pump; also his first sun and planet engine, erected at Soho in 1788. Returning to the loan exhibition, we found models to illustrate mechanism proposed by Watt; notable among these were a Bell pumping-engine, and an engine with a T-ended beam, having two separate connecting-rods, and intended for driving the shafting of two separate mills. There were a number of modifications of the sun and planet gear, including one in which the connecting-rod carries an elliptical annular wheel. There was also a model representing the improvements carried out by James Watt in the steam-engine; they are as follows:

1st. Making the engine double-acting.
2d. Steam-jacketing the cylinder.
3d. The separate air-pump and condenser.
4th. The parallel motion.
5th. The D slide-valve.
6th. The governor.

A machine which excited very considerable interest was the little engine made by William Symington, for driving the now historic pleasure-boat of Mr. Miller on his lake at Dalswinton. It has two vertical single-acting cylinders (of brass), standing side by side, their piston-rods attached to the two ends of a long chain carried around a central pulley (above and between the cylinders), around two more pulleys which were upon the shafts of the two paddle-wheels, and further, by means of guide-pulleys, along the whole length of the engine, in one piece. The pistons, which are about 4 inches in diameter and 15 inches in stroke, are single-acting, and work alternately, the whole chain being pulled first in one direction and then in the other. By means of a clutch arrangement the paddle-shafts receive motion from it in one direction only. The valves are worked by a plug-rod driven by a chain from the central pulley. This queer contrivance, the "parent of modern steam navigation," was finished in 1788, and drove the little pleasure-boat (25 feet long and 7 feet broad) on which it was placed, at the rate of five miles an hour.

Among the other exhibits of historic interest temporarily in this section may be mentioned two models of Stirling's air-engine, made by the inventor, and belonging to the universities of Edinburgh and Glasgow respectively; the original model of Trevithick's locomotive, lent by Mr. Woodcroft; a very old model of Savery's engine, lent by the council of King's College; a sketch made in 1842, by Mr. William Howe, of the link-motion, together with a little rough wooden model of it made soon after; a set of light models of Brunel's celebrated block-making machinery, lent from the Royal Naval Museum, Greenwich, and which has been at work in the Portsmouth dock-yard for upward of forty years;

also a model of Cugnot's steam-carriage (1769), which succeeded a century ago in carrying four passengers at the rate of 2½ miles an hour, in stages of from twelve to fifteen minutes; this was lent by the *Conservatoire des Arts et Métiers.*

We come next to the engine of the *Comet*, made by Henry Bell, and remarkable as being the first steamboat in Europe advertised for the conveyance of passengers. It was completed in January, 1812, and on the 5th of August of the same year Bell issued a circular announcing that his vessel would run on the Clyde between Glasgow and Greenock. The *Comet* was of 30 tons burden, 42 feet long and 14 feet wide. The engine was estimated at four horse-power. It is compactly arranged, but, as a matter of fact, it is not so well suited to the propulsion of a boat as Symington's engine in the *Charlotte Dundas*, which was built in 1803. Bell's engine has undergone various vicissitudes, having been once at the bottom of the sea; it was eventually used for blowing a smith's fire in Glasgow, where it was discovered by Mr. Napier, who purchased it and presented it to the commissioners of patents. In the department of the exhibition devoted to naval architecture there was a small model of the *Comet*, said to be correct. The original vessel was wrecked off the West Highland coast, and the second *Comet* met a similar fate by collision with the steamer *Ayr*.

Americans will remember that the first application of steam to navigation which showed evidences of success was made on the Delaware River, in 1788, by John Fitch. The boat was 60 feet long, 8 feet wide, and 4 feet deep. The project was not successful, and it was left to Fulton to carry out the idea.*

We next reached historical locomotives, beginning with Trevithick's model. Trevithick's claim to be considered the inventor or originator of the locomotive is based upon his patent taken out in conjunction with Vivian in 1802. The model in question had been contributed for exhibition by Mr. Woodcroft, and had been for some years in the South Kensington Museum. Although this machine was not adapted to run on rails, it is obvious that only the substitution of a pair of wheels for a single one is necessary to enable it to do so. It is a matter of history that Trevithick made a locomotive which, in February, 1804, drew a train of five trucks, carrying ten tons of iron and seventy men, nine miles in five hours. This was at Penydaran, South Wales, over an iron tramway with inclines of 1 in 50; and the exploit indubitably proved that the friction between ordinary smooth rails and similar driving-wheels was sufficient to effect propulsion. This engine continued to work for five months.

---

*The first patents granted by the United States for the propulsion of boats by steam were issued to John Fitch, of Philadelphia, and to James Rumsey, of Pennsylvania, August 26, 1791.

In 1804 and 1805 John Stevens, of Hoboken, N. J., built a boat and steam-engine with a single-screw propeller, and a second with twin screws, and worked them successfully. Subsequently he built the paddle-wheel boat *Phœnix*, and sent her from New York to Philadelphia by sea. She was the first steamer on the ocean, and was for some years employed on the Delaware.

Fulton's first experiments were made on the Seine, at Paris, in 1803, with a paddle-boat 60 feet long. It was unsuccessful. In 1807 he launched the *Clermont* on the East River, New York. This vessel was about 132 feet long and 18 feet wide, fitted with a Boulton & Watt engine, and propelled by paddles. She made regular trips between New York and Albany. He built several other steamers, and in 1811 he constructed at Pittsburgh the first steamer that navigated the Western waters. To him more than to any other one man is due the credit of the introduction of steam navigation.

The first voyage made across the ocean by a steam-vessel was that of the *Savannah*, of 300 tons burden, in 1819. She sailed from New York to Liverpool, thence to St. Petersburg, and returned in safety. This voyage created a great sensation. In 1838 the steamer *Sirius* crossed the Atlantic, and the *Great Western* began to make regular trips.

We came next to the Wylam engine, which had been removed for exhibition from the patent-office museum. It appeared to far greater advantage in its temporary position than it did in its permanent home. It was built at Wylam, about 1813, from the designs of William Hedley, and forms a most interesting link in the history of the locomotive. Under the name of "Puffing Billy," or, as some have it, "Puffing Dilly," it was at work on a colliery-line almost uninterruptedly from the above date until June, 1862. The word "dilly," we may mention, is well known in the north of England in the sense of a carriage, and is said by Hallowell, in his dictionary of archaic words, to be derived from the French *diligence.*

The next is the "Sans Pareil," constructed by Timothy Hackworth of Darlington, in 1829, to compete in the trials on the Liverpool and Manchester Railway. This is followed by Stephenson's first locomotive, the "Rocket," also made in 1829 to compete on the Liverpool and Manchester Railway. This locomotive, unlike all which preceded it, has outside inclined cylinders, the connecting-rods being connected directly to the driving-wheels. The boiler is horizontal. In fact, the system originated by Stephenson is that employed at the present day, though, it is needless to say, vastly improved upon. This original locomotive took the prize and laid the foundation for the fame and fortune of the Stephenson family. Among the many old machines and instruments of peculiar interest in the patent museum, is the first screw-propeller that was applied to a vessel of war in Europe. It is made of gun-metal; is a two-bladed true screw, 10 feet 1 inch in diameter, with 11 feet pitch; has a length on the shaft of 18 inches, and is correctly proportioned. This screw was made for and was used to propel the British war-steamer *Rattler*, from the time that vessel was built, in 1843, until she was broken up. So perfect is the construction of this simple true screw, that in the thousands made since then, in the numerous patents that have been granted for screw-propellers, and in the volumes that have been written on the subject, scarcely any improvement has been effected.*

Leaving these curious and interesting specimens of historic engines and instruments, we came to models in steel and iron of engines of modern date. Here were found, in glass cases, models of a number of the engines of the ships of the British royal navy. They consisted of those in the *Monarch* and *Prince Albert*, by Messrs. Humphrys, Tennant & Co.; of the *Nelson*, *Conqueror*, and *Tamar*, by Messrs. Ravenhill & Co.; a fine collection by Messrs. Maudslay, Sons & Field, consisting of those in the *Agincourt*, *Prince Consort*, *Caledonia*, and *Ocean;* and last, the model, beautifully executed, of the compound engines of the *Boadicea* and *Bacchante*, by Messrs. J. &. E. Rennie. All of these several types of engines are so well known that descriptions are needless.

The next interesting subject for inspection was the collection of models of war and mercantile vessels of various types, among which were noticed not only models of ships built for the British navy and merchant marine, but also of ships built for the navies of Germany, Russia, Spain, Turkey, Holland, Brazil, and other nations.

* Captain Ericsson instituted experiments with his screw-propeller vessel *Francis B. Ogden*, on the Thames, in 1836. His second screw-vessel, the *Robert F. Stockton*, was launched on the Mersey in 1838 and crossed to the United States in 1839. He afterward built the *Enterprise*, and designed the machinery for the first screw-propeller ship of war floated on the ocean. This was the *Princeton*, built in Philadelphia in 1842. She was in continuous commission from 1843 until 1849. In 1847 she crossed the Atlantic, being the first screw war-steamer that made the passage. She cruised for two years in the Mediterranean, and was visited by thousands of persons curious to see the propelling power. She returned to Boston in 1849, the writer being one of the officers on board.

The brief outline of the machines, mechanical appliances, &c., above noted, seemed desirable to be given as a matter of historical interest, principally to the engineering profession.

Professor Edward S. Holden, U. S. N., who was detailed by the Department to inspect and study the astronomical instruments, &c., of the exhibition, has, in his able and instructive report on the subject, printed with the report of the honorable Secretary of the Navy and accompanying documents for 1876, given much valuable information that may be used to the advantage of our government.

CONSERVATOIRE DES ARTS ET MÉTIERS.

To the engineer this is one of the most interesting institutions of Paris, or, indeed, of France. Several visits of observation and study were made to its wonderful collection of models, which have been gathered from all directions, and which represent every department of industry. Among them are many specimens of early engineering; the original looms of Vaucanson and Jacquard are preserved here in the *Salle des Filatures*, and in other departments are almost equally interesting relics.

M. Tresca, the sub-director, has a mechanical laboratory in which are many of the larger objects belonging to the institution, but it is mainly occupied by apparatus for testing the efficiency of machinery and with illustrative models driven by power.

This great school has been liberally aided by the government and its growth has been gradual, but now its collection is unexampled in extent and completeness.

Professor R. H. Thurston, A. M., C. E., late an engineer officer, United States Navy, made several visits to it, and in his able and instructive report, as a member of the scientific commission of the United States to the International Exposition held at Vienna, 1873, gives the following account of its history:

Descartes, the distinguished philosopher, is claimed to have been the earliest to propose public instruction for working-people.* He proposed to build a large lecture-hall for each trade, annexing to each a cabinet containing the apparatus appropriate to that department, and to place each of these lecture-rooms in charge of a professor familiar with the subject there to be taught, who should present to the students the principles of his art in proper form, and who should be capable of answering the questions addressed him by his pupils in relation to all details of practice.

It was a century later, however, that this project of Descartes took shape, and the actual commencement of the work is attributed to the great mechanic, Vaucanson.

This distinguished man, previous to 1775, had gathered together, at *l'Hôtel de Montagne*, the first collection of machinery and apparatus which was ever devoted to public use in the manner proposed by Descartes. At his death, Vaucanson bequeathed this collection to the state, and it thus became the germ of this splendid institution which is now so famous.

M. de Vandermonde, the first director, added five hundred machines to the collection between 1785 and 1792.

In 1793 a "*commission temporaire des arts*" was formed, by decree of the *Convention Nationale*, consisting of MM. Vandermonde, J. P. Leroy, Conte, Beuvelot, Molard, l'Abbé Gregoire, and the celebrated physician, Charles. This commission did a noble work in collecting valuable apparatus and models, and in preserving them from injury during the riots and the turmoil of that sad period in French history.

By a decree of the convention it was soon after ordered that a "*conservatoire des arts et métiers, un dépôt public de machines, modèles, outils, dessins,*" &c., should be established, and that three "*démonstrateurs*" and a designer should be employed. After some delays the new institution was established in the old priory of *Saint-Martin-des-Champs*.

The school has experienced the vicissitudes always to be anticipated in such cases;

* The Marquis of Worcester, the distinguished inventor of one of the earlier forms of steam-engine, two hundred years ago, earnestly urged the establishment of a definitely-arranged system of technical education, which should combine instruction in science and in its useful application in the arts.

but its collections have never ceased growing, and its field has been extended by the addition of new departments and the establishment of new professorships, until it now has a faculty of fifteen members.

Many of the most noted French savants have been members of its councils or of its faculty. Thénard, Charles, Darcet, Dupin, Say, Clement, Berthollet, Chaptal, Gay-Lussac, Arago, Pouillet, Poncelet, Morin, Tresca, Ollivier, Becquerel, Payen, Peligot, Moll, Alcan, and others have all been, or are at present, on the list.

NOTE.—For valuable assistance in preparing the foregoing report, I am indebted to Chief Engineer Frederick G. McKean, United States Navy, attached to the Bureau of Steam-Engineering; for prompt attention at the Government Printing Office, credit is due to the intelligent foreman, Major A. H. S. Davis; and for distinctness and uniformity, the illustrations by Mr. W. F. Merrill, of Malden, Mass., speak for themselves.

○